Computational Intelligence

Computational Intelligence

A Logical Approach

David Poole
Alan Mackworth
Randy Goebel

New York   Oxford
Oxford University Press
1998

Oxford University Press

Oxford   New York
Athens   Auckland   Bangkok   Bogota   Bombay   Buenos Aires
Calcutta   Cape Town   Dar es Salaam   Delhi   Florence   Hong Kong
Istanbul   Karachi   Kuala Lumpur   Madras   Madrid   Melbourne
Mexico City   Nairobi   Paris   Singapore   Taipei   Tokyo   Toronto   Warsaw

and associated companies in
Berlin   Ibadan

Published by Oxford University Press, Inc.,
198 Madison Avenue, New York, New York, 10016
http://www.oup-usa.org

Oxford is a registered trademark of Oxford University Press

Library of Congress Cataloging-in-Publication Data

Poole, David (David Lynton), 1958–
Computational intelligence: a logical approach /
David Poole, Alan Mackworth, Randy Goebel.
p. cm.
Includes bibliographical references and index.
ISBN 0-19-510270-3 (cloth)
1. Artificial Intelligence,
I. Mackworth, Alan K. II. Goebel, Randy G. III, Title.
Q335.P657   1997
006.3—dc21   97-9075   CIP

The authors and publisher of this book have used their best efforts in
preparing this book. These efforts include the development, research, and
testing of the theories and programs to determine their effectiveness. The
authors and publisher make no warranty of any kind, expressed or implied,
with regard to these programs or the documentation contained in this book.
The author and publisher shall not be liable in any event for incidental or
consequential damages in connection with, or arising out of, the
furnishing, performance, or use of these programs.

9 8 7 6 5 4

Printed in the United States of America
on acid-free paper

To our families for their love, support, and patience

Jennifer and Alexandra

Marian and Bryn

Lanie, Kari, Jodi, and Kelly

Contents

List of Figures

Preface

Computational Intelligence: A Logical Approach is a book about a new science—computational intelligence. More commonly referred to as artificial intelligence, computational intelligence is the study of the design of intelligent agents. The book is structured as a textbook, but it is accessible to a wide audience.

We wrote this book because we are excited about the emergence of computational intelligence as an integrated science. As with any science worth its salt, computational intelligence has a coherent, formal theory and a rambunctious experimental wing. We balance theory and experiment, and show how to link them intimately together. We develop the science of CI together with its engineering applications. We believe the adage, "There is nothing so practical as a good theory." The spirit of our approach is captured by the dictum, "Everything should be made as simple as possible, but not simpler."

The book works as an introductory text on artificial intelligence for advanced undergraduate or graduate students in computer science, or related disciplines such as computer engineering, philosophy, cognitive science, or psychology. It will appeal more to the technically-minded; parts are technically challenging, focusing on learning by doing: designing, building, and implementing systems. Any curious scientifically-oriented reader will profit from studying the book. Previous experience with computational systems is desirable, but prior study of logic is not necessary, as we develop the concepts as required.

The serious student will gain valuable skills at several levels ranging from expertise in the specification and design of intelligent agents to skills for implementing, testing, and improving real software systems for several challenging application domains. The thrill of participating in the emergence of a new science of intelligent agents is one of the attractions of this approach. The practical skills of dealing with a world of ubiquitous, intelligent, embedded agents are now in great demand in the marketplace.

The focus is on an intelligent agent acting in an environment. We start with simple agents acting in simple, static environments and gradually increase the power of the agents to cope with more challenging worlds. We make this concrete by repeatedly illustrating the ideas with three different agent tasks: a delivery robot, a diagnostic assistant, and an information slave (the infobot).

Our theory is based on logic. Logic has been developed over the centuries as a formal (that is, precise not obtuse) way of representing assumptions about a world and the process of deriving the consequences of those assumptions. For simple agents in simple worlds we start with a highly restricted simple logic. Then as our agent/environment requires, we increase the logical power of the formalism. Since a computer is simply a symbol-manipulation engine, we can easily map our formal theories into computer programs that can control an agent or be used to reason about an agent. Everything we describe is implemented that way.

Any high-level symbol-manipulation language, such as LISP or Prolog, can be used to support these implementations. We also provide complete implementations in Prolog, a language that is designed around the idea of computations as proof in a simple logic. But this connection is not essential to an understanding or use of the ideas in this book.

Our approach, through the development of the power of the agent's representation language, is both simpler and more powerful than the usual approach of surveying and cataloging various applications of artificial intelligence. However, as a consequence, some applications such as the details of computational vision or the pragmatics of low-level robot control are not covered in this book. We do show how our theory allows us to design situated robots performing in dynamically changing worlds.

We invite you to join us in an intellectual adventure: building a science of intelligent agents.

David Poole
Alan Mackworth
Randy Goebel

Acknowledgments

We are very grateful to Valerie McRae for contributing her proofreading and for-matting skills and for helping in many other ways. We also gratefully acknowledge the helpful comments on earlier drafts of this book received from Carl Alphonce, Brock Atherton, Veronica Becher, Wolfgang Bibel, Craig Boutilier, Jason Burrell, Hugh Clapin, Keith Clark, Jim Delgrande, Koichi Furukawa, Chris Geib, Holger Hoos, Michael Horsch, Ying Li, Greg McClement, Claude Sammut, Julie Schlifske, Daniele Theseider Dupré, Paul van Arragon, Bonnie Webber, Bob Woodham, and Ying Zhang. Thanks to our editors at OUP, Bill Zobrist, Karen Shapiro, and Krysia Bebick.

The four-limbed robot shown on the cover is a *Platonic Beast*, developed by Dinesh Pai, Rod Barman, and Scott Ralph at the University of British Columbia's Laboratory for Computational Intelligence. Based on the spherical symmetry of the tetrahedron, one of the five symmetrical Platonic polyhedra, the Beast can walk, roll over, and climb over obstacles. Aidan Rowe and John Chong provided the technical graphical expertise to complete the cover design.

Chapter 1

Computational Intelligence and Knowledge

1.1 What Is Computational Intelligence?

Computational intelligence is the study of the design of intelligent agents. An **agent** is something that acts in an environment—it does something. Agents include worms, dogs, thermostats, airplanes, humans, organizations, and society. An **intelligent agent** is a system that acts intelligently: What it does is appropriate for its circumstances and its goal, it is flexible to changing environments and changing goals, it learns from experience, and it makes appropriate choices given perceptual limitations and finite computation.

The central scientific goal of computational intelligence is to understand the principles that make intelligent behavior possible, in natural or artificial systems. The main hypothesis is that reasoning is computation. The central engineering goal is to specify methods for the design of useful, intelligent artifacts.

Artificial or Computational Intelligence?

Artificial intelligence (AI) is the established name for the field we have defined as computational intelligence (CI), but the term "artificial intelligence" is a source of much confusion. Is artificial intelligence real intelligence? Perhaps not, just as an artificial pearl is a fake pearl, not a real pearl. "Synthetic intelligence" might be a better name, since, after all, a synthetic pearl may not be a natural pearl but it is a real pearl. However, since we claimed that the central scientific goal is to understand both natural

1

and artificial (or synthetic) systems, we prefer the name "computational intelligence." It also has the advantage of making the computational hypothesis explicit in the name.

The confusion about the field's name can, in part, be attributed to a confounding of the field's purpose with its methodology. The purpose is to understand how intelligent behavior is possible. The methodology is to design, build, and experiment with computational systems that perform tasks commonly viewed as intelligent. Building these artifacts is an essential activity since computational intelligence is, after all, an empirical science; but it shouldn't be confused with the scientific purpose.

Another reason for eschewing the adjective "artificial" is that it connotes simulated intelligence. Contrary to another common misunderstanding, the goal is not to simulate intelligence. The goal is to understand real (natural or synthetic) intelligent systems by synthesizing them. A simulation of an earthquake isn't an earthquake; however, we want to actually create intelligence, as you could imagine creating an earthquake. The misunderstanding comes about because most simulations are now carried out on computers. However, you shall see that the digital computer, the archetype of an interpreted automatic, formal, symbol-manipulation system, is a tool unlike any other: It can produce the real thing.

The obvious intelligent agent is the human being. Many of us feel that dogs are intelligent, but we wouldn't say that worms, insects, or bacteria are intelligent (Exercise 1.1). There is a class of intelligent agents that may be more intelligent than humans, and that is the class of *organizations*. Ant colonies are the prototypical example of organizations. Each individual ant may not be very intelligent, but an ant colony can act more intelligently than any individual ant. The colony can discover food and exploit it very effectively as well as adapt to changing circumstances. Similarly, companies can develop, manufacture, and distribute products where the sum of the skills required is much more than any individual could understand. Modern computers, from the low-level hardware to high-level software, are more complicated than can be understood by any human, yet they are manufactured daily by organizations of humans. Human *society* viewed as an agent is probably the most intelligent agent known. We take inspiration from both biological and organizational examples of intelligence.

Flying Machines and Thinking Machines

It is instructive to consider an analogy between the development of flying machines over the last few centuries and the development of thinking machines over the last few decades.

First note that there are several ways to understand flying. One is to dissect known flying animals and hypothesize their common structural features as necessary fundamental characteristics of any flying agent. With this method an examination of birds, bats, and insects would suggest that flying involves the flapping of wings made of some structure covered with feathers or a membrane. Furthermore, the hypothesis could be verified by strapping feathers to one's arms, flapping, and jumping into the air, as Icarus did. You might even imagine that some enterprising researchers would claim that one need only add enough appropriately layered feather structure to achieve the desired flying competence, or that improved performance required more detailed modeling of birds such as adding a cloaca.

An alternate methodology is to try to understand the principles of flying without restricting ourselves to the natural occurrences of flying. This typically involves the construction of artifacts that embody the hypothesized principles, even if they do not behave like flying animals in any way except flying. This second method has provided both useful tools, airplanes, and a better understanding of the principles underlying flying, namely *aerodynamics*.

It is this difference which distinguishes computational intelligence from other cognitive science disciplines. CI researchers are interested in testing general hypotheses about the nature of intelligence by building machines which are intelligent and which don't simply mimic humans or organizations. This also offers an approach to the question "Can computers really think?" by considering the analogous question "Can airplanes really fly?"

Technological Models of Mind

Throughout human history, people have used technology to model themselves. Consider this Taoist parable taken from the book *Lieh Tzu*, attributed to Lieh Yu-Khou:

> "Who is that man accompanying you?" asked the king. "That, Sir," replied Yen Shih, "is my own handiwork. He can sing and he can act." The king stared at the figure in astonishment. It walked with rapid strides, moving its head up and down, so that anyone would have taken it for a live human being. The artificer touched its chin, and it began singing, perfectly in tune. He touched its hand and it began posturing, keeping perfect time .... The king, looking on with his favorite concubine and other beauties, could hardly persuade himself that it was not real. As the performance was drawing to an end, the robot winked its eye and made advances to the ladies in attendance, whereupon the king became incensed and would have had Yen Shih executed on the spot had not the latter, in mortal fear, instantly taken the robot to pieces to let him see what it really was. And, indeed, it turned out to be only a construction of leather, wood, glue and lacquer, variously colored white, black, red and blue. Examining it closely, the king found all the internal organs complete—liver, gall, heart, lungs, spleen, kidneys, stomach and intestines; and over these again, muscles, bones and limbs with their joints, skin, teeth and

hair, all of them artificial. Not a part but was fashioned with the utmost nicety and skill; and when it was put together again, the figure presented the same appearance as when first brought in. The king tried the effect of taking away the heart, and found that the mouth could no longer speak; he took away the liver and the eyes could no longer see; he took away the kidneys and the legs lost their power of locomotion. The king was delighted.

This story, dating from about the third century B.C., is one of the earliest written accounts of building intelligent agents, but the temples of early Egypt and Greece also bear witness to the universality of this activity. Each new technology has been exploited to build intelligent agents or models of mind. Clockwork, hydraulics, telephone switching systems, holograms, analog computers, and digital computers have all been proposed both as technological metaphors for intelligence and as mechanisms for modeling mind.

Parenthetically, we speculate that one reason for the king's delight was that he realized that functional equivalence doesn't necessarily entail structural equivalence. In order to produce the functionality of intelligent behavior it isn't necessary to reproduce the structural connections of the human body.

This raises the obvious question of whether the digital computer is just another technological metaphor, perhaps a fad soon to be superseded by yet another mechanism. In part, the answer must be empirical. We need to wait to see if we can get substantial results from this approach, but also to pursue alternate models to determine if they are more successful. We have reason to believe the answer to that question is "no." Some reasons are empirical: The results to date are impressive but not, of course, conclusive. There are other reasons. Consider the following two hypotheses. The first is called the **symbol-system hypothesis**:

Reasoning is symbol manipulation.

The second hypothesis is called the **Church–Turing thesis**:

Any symbol manipulation can be carried out on a Turing machine.

A Turing machine is an idealization of a digital computer with an unbounded amount of memory. These hypotheses imply that any symbol manipulation, and so any reasoning, can be carried out on a large enough deterministic computer.

There is no way you can prove these two hypothesis mathematically. All you can do is empirically test then by building reasoning systems. Why should you believe that they are true or even reasonable? The reason is that language, which provides one of the few windows to the mind, is inherently about transmission of symbols. Reasoning in terms of language has symbols as inputs and outputs, and so the function from inputs to outputs can be described symbolically, and presumably can be implemented in terms of symbol manipulation. Also the intelligence that is manifest in an organization or in society is transmitted by language and other signals. Once you have expressed something in a language, reasoning about it is symbol manipulation. These hypotheses don't tell us how to implement arbitrary reasoning on a computer—this is CI's task. What it does tell us is that computation is an appropriate metaphor for reasoning.

This hypothesis doesn't imply that every detail of computation can be interpreted symbolically. Nor does it imply that every machine instruction in a computer or the function of every neuron in a brain can be interpreted symbolically. What it does mean is that there is a level of abstraction in which you can interpret reasoning as symbol manipulation, and that this level can explain an agent's actions in terms of its inputs.

Before you accept this hypothesis, it is important to consider how it may be wrong. An alternative is that action is some continuous function of the inputs to an agent such that the intermediate values don't necessarily correspond to anything meaningful. It is even possible that the functionality can't be interpreted symbolically, without resorting to using meaningless numbers. Alternative approaches are being pursued in both neural networks (page 408) and in building reactive robots (page 443) inspired by artificial insects.

Science and Engineering

As suggested by the flying analogy, there is tension between the science of CI, trying to understand the principles behind reasoning, and the engineering of CI, building programs to solve particular problems. This tension is an essential part of the discipline.

As CI is a science, its literature should manifest the scientific method, especially the creation and testing of refutable theories. Obvious questions are, "What are CI theories about?" and "How would I test one if I had one?" CI theories are about how interesting problems can be represented and solved by machine. Theories are supported empirically by constructing implementations, part of whose quality is judged by traditional computer science principles. You can't accomplish CI without specifying theories and building implementations; they are inextricably connected. Of course, not every researcher needs to do both, but both must be done. An experiment means nothing without a theory against which to evaluate it, and a theory without potentially confirming or refuting evidence is of little use. Ockham's Razor is our guide: Always prefer simple theories and implementations over the more complex.

With these thoughts in mind, you can quickly consider one of the most often considered questions that arises in the context of CI: "Is human behavior algorithmic?" You can dispense with this question and get on with your task by acknowledging that the answer to this question is unknown; it is part of cognitive science and CI's goal to find out.

Relationship to Other Disciplines

CI is a very young discipline. Other disciplines as diverse as philosophy, neurobiology, evolutionary biology, psychology, economics, political science, sociology, anthropology, control engineering, and many more have been studying intelligence much longer. We first discuss the relationship with philosophy, psychology, and other disciplines which study intelligence; then we discuss the relationship with computer science, which studies how to compute.

The science of CI could be described as "synthetic psychology," "experimental philosophy," or "computational epistemology"—**Epistemology** is the study of knowledge. It can be seen as a way to study the old problem of the nature of knowledge and intelligence, but with a more powerful experimental tool than was previously available. Instead of being able to observe only the external behavior of intelligent systems, as philosophy, psychology, economics, and sociology have traditionally been able to do, we are able to experiment with executable models of intelligent behavior. Most importantly, such models are open to inspection, redesign, and experiment in a complete and rigorous way. In other words, you now have a way to construct the models that philosophers could only theorize about. You can experiment with these models, as opposed to just discussing their abstract properties. Our theories can be empirically grounded in implementation.

Just as the goal of aerodynamics isn't to synthesize birds, but to understand the phenomenon of flying by building flying machines, CI's ultimate goal isn't necessarily the full-scale simulation of human intelligence. The notion of psychological validity separates CI work into two categories: that which is concerned with mimicking human intelligence—often called *cognitive modeling*—and that which isn't.

To emphasize the development of CI as a science of intelligence, we are concerned, in this book at least, not with psychological validity but with the more practical desire to create programs that solve real problems. Sometimes it will be important to have the computer to reason through a problem in a human-like fashion. This is especially important when a human requires an explanation of how the computer generated an answer. Some aspects of human cognition you usually do not want to duplicate, such as the human's poor arithmetic skills and propensity for error.

Computational intelligence is intimately linked with the discipline of computer science. While there are many non-computer scientists who are researching CI, much, if not most, CI (or AI) research is done within computer science departments. We believe this is appropriate, as the study of computation is central to CI. It is essential to understand algorithms, data structures, and combinatorial complexity in order to build intelligent machines. It is also surprising how much of computer science started as a spin off from AI, from timesharing to computer algebra systems.

There are other fields whose goal is to build machines that act intelligently. Two of these fields are control engineering and operations research. These start from different points than CI, namely in the use of continuous mathematics. As building real agents involves both continuous control and CI-type reasoning, these disciplines should be seen as symbiotic with CI. A student of either discipline should understand the other. Moreover, the distinction between them is becoming less clear with many new theories combining different areas. Unfortunately there is too much material for this book to cover control engineering and operations research, even though many of the results, such as in search, have been studied in both the operations research and CI areas.

Finally, CI can be seen under the umbrella of *cognitive science*. Cognitive science links various disciplines that study cognition and reasoning, from psychology to linguistics to anthropology to neuroscience. CI distinguishes itself within cognitive science because it provides tools to build intelligence rather than just studying the external behavior of intelligent agents or dissecting the inner workings of intelligent systems.

1.2 Agents in the World

There are many interesting philosophical questions about the nature and substance of CI, but the bottom line is that, in order to understand how intelligent behavior might be algorithmic, you must attempt to program a computer to solve actual problems. It isn't enough to merely speculate that some particularly interesting behavior is algorithmic. You must develop a theory that explains how that behavior can be manifest in a machine, and then you must show the feasibility of that theory by constructing an implementation. We are interested in practical reasoning: reasoning in order to do something. Such a coupling of perception, reasoning, and acting comprises an *agent*. An agent could be, for example, a coupling of a computational engine with physical actuators and sensors, called a *robot*. It could be the coupling of an advice-giving computer—an *expert system*—with a human who provides the perceptual information and who carries out the task. An agent could be a program that acts in a purely computational environment—an *infobot*.

Figure 1.1 shows the inputs and outputs of an agent. At any time the agent has:

- Prior knowledge about the world
- Past experience that it can learn from
- Goals that it must try to achieve or values about what is important
- Observations about the current environment and itself

and it does some action. For each agent considered, we specify the forms of the inputs and the actions. The goal of this book is to consider what is in the black box so that the action is reasonable given the inputs.

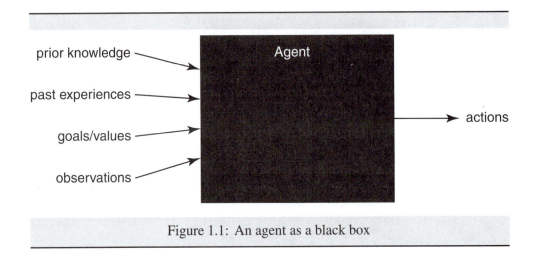

prior knowledge

past experiences

goals/values

observations

Agent

actions

Figure 1.1: An agent as a black box

For our purpose, the world consists of an agent in an environment. The agent's environment may well include other agents. Each agent has some internal state that can encode beliefs about its environment and itself. It may have goals to achieve, ways to act in the environment to achieve those goals, and various means to modify its beliefs by reasoning, perception, and learning. This is an all-encompassing view of intelligent systems varying in complexity from a simple thermostat to a team of mobile robots to a diagnostic advising system whose perceptions and actions are mediated by human beings.

Success in building an intelligent agent naturally depends on the problem that one selects to investigate. Some problems are very well-suited to the use of computers, such as sorting a list of numbers. Others seem not to be, such as changing a baby's diaper or devising a good political strategy. We have chosen some problems that are representative of a range of applications of current CI techniques. We seek to demonstrate, by case study, CI's methodology with the goal that the methodology is transferable to various problems in which you may be interested. We establish a framework that places you, the reader, in a position to evaluate the current CI literature and anticipate the future; and, most importantly, we develop the concepts and tools necessary to allow you to build, test, and modify intelligent agents. Finally we must acknowledge there is still a huge gulf between the dream of computational intelligence and the current technology used in the practice of building what we now call intelligent agents. We believe we have many of the tools necessary to build intelligent agents, but we are certain we don't have all of them. We could, of course, be on the wrong track; it is this fallibility that makes CI science and makes the challenge of CI exciting.

1.3   Representation and Reasoning

Experience shows that the performance of tasks that seem to involve intelligence also seem to require a huge store of knowledge. A major thesis of this book is that CI is the study of knowledge. This raises the question which is part of our subject material, "What is knowledge?" Informally, knowledge is information about some domain or subject area, or about how to do something. Much of our effort will be devoted to formalizing and refining a common-sense notion of knowledge, with the motivation of developing both a theoretical and practical framework for representing and using knowledge.

Humans require and use a lot of knowledge to carry out even the most simple common-sense tasks. Computers are very good at tasks which do not require much knowledge, such as simple arithmetic, symbolic differentiation, or sorting. They aren't, as yet, very good at many knowledge-intensive tasks at which humans excel, such as recognizing faces in a picture, medical diagnosis, understanding natural language, or legal argumentation. At the heart of this book is the design of computational systems that have knowledge about the world and that can act in the world based on that knowledge. The notion of knowledge is central to this book. The systems we want to develop should be able to acquire and use knowledge to solve the problems at hand. The main issues are how to acquire and represent knowledge about some domain and how to use that knowledge to answer questions and solve problems.

You will notice that we make a strong commitment to *logic* approach in this book. Our commitment is really to a precise specification of meaning rather than to any particular syntax. We have no great commitment to any particular syntax. Many different notations are possible. Sometimes we will write sentences, sometimes we will use diagrams. In order to represent anything, you have to commit to some notation, and the simpler the better. We use Prolog's syntax, not because we particularly like Prolog or its syntax, but because it is important for scholars of CI to get experience with using logic to solve problems, and Prolog is probably the most accessible system that allows you to do this.

Representation and Reasoning System

In order to use knowledge and reason with it, you need what we call a *representation and reasoning system* (RRS). A representation and reasoning system is composed of a language to communicate with a computer, a way to assign meaning to the language, and procedures to compute answers given input in the language. Intuitively, an RRS lets you tell the computer something in a language where you have some meaning associated with the sentences in the language, you can ask the computer questions, and the computer will produce answers that you can interpret according to the meaning associated with the language.

At one extreme, the language could be a low-level programming language such as Fortran, C++, or Lisp. In these languages the meaning of the sentences, the programs, is purely in terms of the steps the computer will carry out to execute the program. What computation will be carried out given a program and some input, is straightforward to determine. How to map from an informal statement of a problem to a representation of the problem in these RRSs, programming, is a difficult task.

At the other extreme, the language could be a natural language, such as English, where the sentences can refer to the problem domain. In this case, the mapping from a problem to a representation is not very difficult: You need to describe the problem in English. However, what computation needs to be carried out in the computer in response to the input is much more difficult to determine.

In between these two extremes are the RRSs that we consider in this book. We want RRSs where the distance from a natural specification of the problem to the representation of the problem is not very far. We also want RRSs where the appropriate computation, given some input, can be effectively determined. We consider languages for the specification of problems, the meaning associated with such languages, and what computation is appropriate given input in the languages.

One simple example of a representation and reasoning system between these two extremes is a database system. In a database system, you can tell the computer facts about a domain and then ask queries to retrieve these facts. What makes a database system into a representation and reasoning system is the notion of *semantics*. Semantics allows us to debate the truth of information in a knowledge base and makes such information knowledge rather than just data. In most of the RRSs we are interested in, the form of the information is more flexible and the procedures for answering queries are more sophisticated than in a database. A database typically has table lookup; you can ask about what is in the database, not about what else must be true, or is likely to be true, about the domain.

Chapter 2 gives a more precise definition an RRS and a particular RRS that is both simple and yet very powerful. It is this RRS that we build upon throughout this book, eventually presenting RRSs that can reason about such things as time, typicality, uncertainty, and action.

Ontology and Conceptualization

An important and fundamental prerequisite to using an RRS is to decide how a task domain is to be described. This requires us to decide what kinds of things the domain consists of, and how they are to be related in order to express task domain problems. A major impediment to a general theory of CI is that there is no comprehensive theory of how to appropriately conceive and express task domains. Most of what we know about this is based on experience in developing and refining representations for particular problems.

Despite this fundamental problem, we recognize the need for the following commitments.

- The world can be described in terms of **individuals** (things) and relationships among individuals. An **ontology** is a commitment to what exists in any particular task domain. This notion of relationship is meant to include propositions that are true or false independently of any individuals, properties of single individuals, as well as relationships between pairs or more individuals. This assumption that the world can be described in terms of things is the same that is made in logic and natural language. This isn't a strong assumption, as individuals can be anything nameable, whether concrete or abstract. For example, people, colors, emotions, numbers, and times can all be considered as individuals. What is a "thing" is a property of an observer as much as it is a property of the world. Different observers, or even the same observer with different goals, may divide up the world in different ways.

- For each task or domain, you need to identify specific individuals and relations that can be used to express what is true about the world under consideration. How you do so can profoundly affect your ability to solve problems in that domain.

For most of this book we assume that the human who is representing a domain decides on the ontology and the relationships. To get human-level computational intelligence it must be the agent itself that decides how to divide up the world, and which relationships to reason about. However, it is important for you to understand what knowledge is required for a task before you can expect to build a computer to learn or introspect about how to solve a problem. For this reason we concentrate on what it takes to solve a problem. It should not be thought that the problem of CI is solved. We have only just begun this endeavor.

1.4 Applications

Theories about representation and reasoning are only useful insofar as they provide the tools for the automation of problem solving tasks. CI's applications are diverse, including medical diagnosis, scheduling factory processes, robots for hazardous environments, chess playing, autonomous vehicles, natural language translation systems, and cooperative systems. Rather than treating each application separately, we abstract essential features of such applications to allow us to study principles behind intelligent reasoning and action.

This section outlines three application domains that will be developed in examples throughout the book. Although the particular examples presented are simple—for otherwise they wouldn't fit into the book—the application domains are representative of the sorts of domains in which CI techniques can be, and have been, used.

The three application domains are:

- *An autonomous delivery robot* that can roam around a building delivering packages and coffee to people in the building. This delivery agent needs to be able to, for example, find paths, allocate resources, receive requests from people, make decisions about priorities, and deliver packages without injuring people or itself.

- *A diagnostic assistant* that helps a human troubleshoot problems and suggests repairs or treatments to rectify the problems. One example is an electrician's assistant that can suggest what may be wrong in a house, such as a fuse blown, a light switch broken, or a light burned out given some symptoms of electrical problems. Another example is of a medical diagnostician that finds potential diseases, possible tests, and appropriate treatments based on knowledge of a particular medical domain and a patient's symptoms and history. This assistant needs to be able to explain its reasoning to the person who is carrying out the tests and repairs, as that person is ultimately responsible for what they do. The diagnostic assistant must add substantial value in order to be worth using.

- *An "infobot"* that can search for information on a computer system for naive users such as company managers or people off the street. In order to do this the infobot must find out, using the user's natural language, what information is requested, determine where to find out the information, and access the information from the appropriate sources. It then must report its findings in an appropriate format so that the human can understand the information found, including what they can infer from the lack of information.

These three domains will be used for the motivation for the examples in the book. In the next sections we discuss each application domain in detail.

The Autonomous Delivery Robot

Imagine a robot that has wheels and can pick up objects and put them down. It has sensing capabilities so that it can recognize the objects it needs to manipulate and so it can avoid obstacles. It can be given orders in natural language and obey them, making common sense assumptions about what to do when its goals conflict. Such a robot could be used in an office environment to deliver packages, mail, or coffee. It needs to be useful as well as safe.

In terms of the black box definition of an agent in Figure 1.1, the autonomous delivery robot has as inputs:

- Prior knowledge in terms of knowledge about its capabilities, what objects it may encounter and need to differentiate, and perhaps some prior knowledge about its environment, such as maps.

- Past experience about, for instance, which actions are useful in which situations, what objects are in the world, how its actions affect its position, and experience about previous requests for it to act.

- Goals in terms of what it needs to deliver and when, as well as values that specify tradeoffs such as when it must forgo one goal to pursue another or the tradeoff between acting quickly and acting safely.

- Observations about its environment from such input devices as cameras, sonar, sound, laser range finders, or keyboards for requests.

The robot's output is motor controls that specify where its wheels should turn, where its limbs should move, and what it should do with it grippers.

In order for this robot to be able to function, it has to be able to:

- Determine where individuals' offices are, where to get coffee, how to estimate the length of a trip, and so on. This involves being able to infer information from a database of facts about the domain. How to infer implicit information from a knowledge base is explored in Chapters 2 and 3.

- Find a path between different locations. It may want the shortest, the quickest, or the safest path. This involves searching as developed in Chapter 4.

- Be able to represent knowledge about the domain so that inference can be quick, so that knowledge can be easily acquired, and so that the appropriate knowledge is represented. Such issues are discussed in Chapter 5.

- Plan how to carry out multiple goals, even when they use the same resources, for example, when the robot's carrying capacity is limited. Planning is discussed in Chapter 8.

- Make default assumptions—for example, about where people will be or where coffee can be found. See Chapter 9.

- Make tradeoffs about plans even though there may be uncertainty about what is in the world and about the outcome of its actions. Such reasoning under uncertainty is discussed in Chapter 10.

- Learn about features of its domain, as well as learn about how its actions affect its position and its rewards. See Chapter 11.

- Sense the world, know where it is, steer around the corridors (avoiding people and other objects), and pick up and put down objects. See Chapter 12.

Figure 1.2 depicts a typical laboratory environment for a delivery robot. This environment consists of four laboratories and many offices arranged in a grid. We assume that the robot can only push doors, and the directions of the doors in the

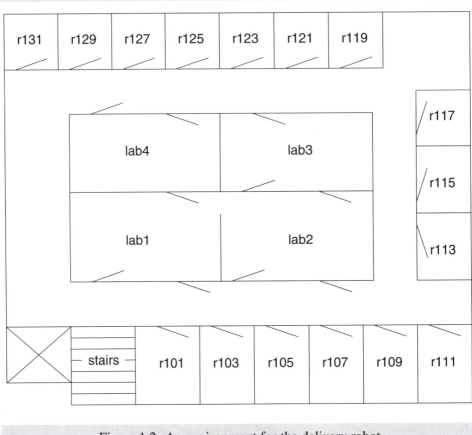

Figure 1.2: An environment for the delivery robot

diagram reflect the directions where the robot can travel. We also assume that rooms need keys, and that keys can be obtained from various sources. The robot needs to deliver parcels and letters from room to room. The environment also contains a stairway that can be hazardous to the robot.

The Diagnostic Assistant

A diagnostic assistant is intended to advise a human about some particular artifact, such as a medical patient, the electrical system in a house, or an automobile, when symptoms are manifest. It should advise about potential underlying faults or diseases, what tests to carry out, and what treatment to prescribe. In order to give such advice the assistant needs to have some model of the system, knowledge of potential causes, available tests, available treatments, and observations about a particular artifact. Assisting a human involves making sure that the system provides added value, is easy for a human to use, and isn't more trouble than it is worth. It must be able to justify why the suggested diagnoses or actions are appropriate. Humans are, and should be, suspicious of computer systems that are impenetrable. When humans are responsible for what they do, even if it is based on a computer system's advice, they need to have reasonable justifications for the suggested actions.

In terms of the black box definition of an agent in Figure 1.1, the diagnostic assistant has as inputs:

- Prior knowledge such as what's normal and what's abnormal about how switches and lights work, how diseases or malfunctions manifest themselves, what information tests provide, and the side effects of repairs or treatments.

- Past experience in terms of data of previous cases that include the effects of repairs or treatments, the prevalence of faults or diseases, the prevalence of symptoms for these faults or diseases, and the accuracy of tests.

- Goals of fixing the device and tradeoffs such as between fixing or replacing different components, or whether a patient prefers to live longer if it means they will be less coherent.

- Observations of symptoms of a device or patient.

The output of the diagnostic assistant is in terms of recommendations of treatments and tests, along with rationales for its recommendations.

In order for the diagnostic assistant to be useful it must be able to:

- Derive the effects of faults and interventions (Chapter 3).

- Search through the space of possible faults or disease complexes (Chapter 4).

- Explain its reasoning to the human who is using it (Chapter 6).

- Derive possible causes for symptoms; rule out other causes based on the symptoms (Chapter 7).

- Plan courses of tests and treatments to address the problems (Chapter 8).

- Hypothesize problems and use default knowledge that may not always be true (Chapter 9).

- Reason about the uncertainties about the artifact given only partial information about the state of the artifact, the uncertainty about the effects of the treatments, and the tradeoffs between the alternate courses of action (Chapter 10).

- Learn about what symptoms are associated with the faults or diseases, the effects of treatments, and the accuracy of tests (Chapter 11).

Figure 1.3 shows a depiction of the electrical distribution in a house. In this house, power comes into the house through circuit breakers, and then it goes to power outlets or to lights through light switches. For example, light l_1 is on if there is power coming

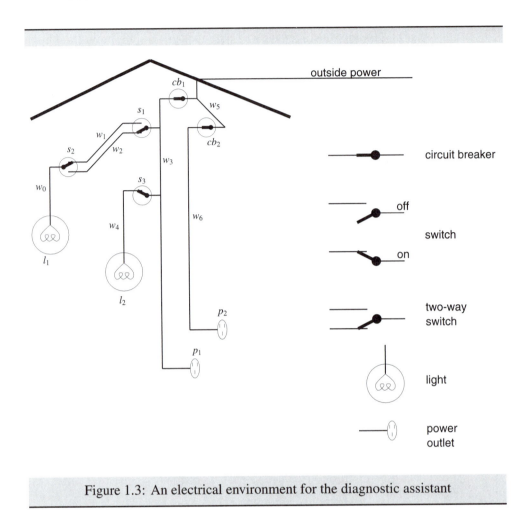

outside power

circuit breaker

off

switch

on

two-way
switch

light

power
outlet

Figure 1.3: An electrical environment for the diagnostic assistant

into the house, if circuit breaker cb_1 is *on*, and if switches s_1 and s_2 are either both up or both down. This is the sort of model that a normal householder may have of the electrical power in the house which they could use to determine what is wrong given evidence about the position of the switches and which lights are on and which are off. The diagnostic assistant is there to help the householder or an electrician to troubleshoot electrical problems.

The Infobot

An **infobot** is like a robot, but instead of interacting with a physical environment, it interacts with an information environment. Its task is to extract information from a network of diverse information sources such as the Internet or a multimedia encyclopedia. The infobot must determine what information is needed from a query in a formal language, from a sophisticated user, or from a natural language query from a casual user such as a manager or person off the street. It must determine where the information may be obtained, retrieve the information, and present it in a meaningful way to the user.

In terms of the black box definition of an agent in Figure 1.1, the infobot has as inputs:

- Prior knowledge about the meaning of words, the types of information sources, and how to access information.
- Past experience about where information can be obtained, the relative speed of various servers, and information about the preferences of the user.
- Goals in terms of what information it needs to find out and tradeoffs about how much expense should be involved to get the information and the tradeoff between the volume and quality of information.
- Observations about what information is at the current sites, what links are available, and the load on various connections.

The output of the infobot is information presented so that the user can understand what is there and the significance of missing information.

The infobot needs to be able to:

- Derive information that is only implicit in the knowledge base(s), as well as interact in natural language (Chapter 3).
- Search through a variety of knowledge bases looking for relevant information (Chapter 4).
- Find good representations of knowledge so that answers can be computed efficiently (Chapter 5).
- Explain how an answer was derived or why some information was unavailable (Chapter 6).
- Make conclusions about lack of knowledge, determine conflicting knowledge, and be able to conclude disjunctive knowledge (Chapter 7).
- Use default reasoning about where to obtain different information (Chapter 9).
- Make tradeoffs between cheap but unreliable information sources and more expensive but more comprehensive information sources (Chapter 10).
- Learn about what knowledge is available where, and what information the user is interested in (Chapter 11).

We consider two different infobots: the "unibot" and the "webbot." The **unibot** interacts with a database of information about courses, scheduling, degree requirements, and grades. The **webbot** interacts with the World Wide Web, finding information that may be of use to a user. One of the most interesting aspects of an infobot is that it ought to be able to volunteer information that users don't know exists, and so can't be expected to ask for even though they may be interested.

Common Features

These three examples have common features. At one level of abstraction, they each have four tasks:

Modeling the environment The robot needs to be able to model the physical environment, its own capabilities, and the mechanics of delivering parcels. The diagnostic assistant needs to be able to model the general course of diseases or faults, know how normal artifacts work, know how treatments work, and know what information tests provide. The infobot needs to be able to model how information can be obtained, what are legal answers to questions, and what information is actually needed, based on a request.

Evidential reasoning or perception This is what control theorists call "system identification" and what doctors call "diagnosis." Given some observations about the world, the task is to determine what is really in the world. This is most evident in the diagnostic assistant, where the system is given symptoms (observations) and has to determine the underlying faults or diseases. The delivery robot must try to determine where it is and what else is in its environment based on limited sensing information such as touch, sonar, or vision. The infobot has to determine where information is available, given only partial information about the contents of information sources.

Action Given a model of the world and a goal, the task is to determine what should be done. For the delivery robot, this means that it must actually do something, such as rove around the corridors and deliver things. For the diagnostic assistant, the actions are treatments and tests. It isn't enough to theorize about what may be wrong, but a diagnostician must make tests and has to consider what it will do based on the outcome of tests. It isn't necessary to test if the same treatment will be carried out no matter what the test's outcome, such as replacing a board on a computer or giving a patient an antibiotic. The actions of the infobot are computational, such as, consulting a knowledge base in order to extract some information.

Learning from past experience This includes learning what the particular environment is like, the building the delivery robot is in, the particular patient being diagnosed, or the communication bottlenecks of a computer network; learning general information, how the robot sensors actually work, how well particular diseases respond to particular treatments, or how fast different types of computer connections are; and learning how to solve problems more efficiently.

These tasks cut across all application domains. It is essentially the study of these four tasks that we consider in this book. These four tasks interact. It is difficult to study one without the others. We have decided that the most sensible organization is to build the tools needed from the bottom up and to show how the tools can be used for each task and, through these tasks, demonstrate the limitations of the tools. We believe this organization will help in understanding the commonalities across different domains and in understanding the interaction among the different tasks.

1.5 Overview

Our quest for a unified view of CI is based on the fundamental nature of the concepts of representation and reasoning. We seek to present these techniques as an evolution of ideas used to solve progressively more difficult problems.

Chapter 2 starts with a simple representation and reasoning system, where we assume that the agents have complete knowledge about the world and that the world never changes. Subsequent chapters discuss the removal of such constraints in terms of their effect on representation and reasoning. In Chapter 3, we give specific examples of using a definite knowledge encoding for various useful applications. In Chapter 4, we show how many reasoning problems can be understood as search problems. We review some standard approaches to search-based problem solving, including various kinds of informed and uninformed search. Chapter 5 discusses knowledge representation issues and explains how they are manifest in the ideas developed in the first three chapters. Chapter 6 provides further details about knowledge-based systems and presents an overview of a system architecture of a typical expert system, including tools to enable an expert system to explain its reasoning. Chapter 7 removes the assumptions about definite knowledge, by allowing disjunctive and negative knowledge, culminating in full first-order predicate calculus and some aspects of modal logic. Chapter 8 removes the assumption that the world is static. To represent a dynamic world requires some notion of time or state change, which, in turn, introduces us to the planning problem. Chapter 9 discusses hypothetical reasoning and its application to default reasoning, diagnostic reasoning, and recognition. Chapter 10 introduces reasoning under uncertainty, representations for uncertain knowledge, and decision making under uncertainty. Chapter 11, which discusses learning, shows how previous experience can be used by an agent. Chapter 12 shows how the reasoning capabilities can be put together to build agents that perceive and interact in an environment.

Three appendices provide supplementary material including a glossary of terms used in CI, a Prolog tutorial, and implementations of various system components presented in the main text.

1.6   References and Further Reading

The ideas in this chapter have been derived from many sources. Here we will try to acknowledge those that are explicitly attributable to other authors. Most of the other ideas are part of AI folklore. To try to attribute them to anyone would be impossible.

Minsky (1986) presents a theory of intelligence as emergent from a "society" of unintelligent agents. Haugeland (1997) contains a good collection of articles on the philosophy behind computational intelligence.

Turing (1950) proposes an objective method for answering the question "Can machines think?" in terms of what is now known as the *Turing test*.

The symbol-system hypothesis is due to Newell & Simon (1976). See also Simon (1996) who discusses the role of symbol systems in a multi-disciplinary context. The distinctions between real, synthetic and artificial intelligence are discussed by Haugeland (1985), who also provides useful introductory material on interpreted, automatic formal symbol systems, and the Church–Turing thesis. For a critique of the symbol-system hypothesis see Winograd (1990). Wegner (1997) argues that computers that interact with the world may be more powerful than Turing machines, thus that the Church–Turing thesis is in fact false.

The Taoist story is from Needham's classic study of science and technology in China (Ronan, 1986).

For discussions on the foundations of AI and the breadth of research in AI see Kirsh (1991a), Bobrow (1993), and the papers in the corresponding volumes, as well as Schank (1990) and Simon (1995). The importance of knowledge in AI is discussed in Lenat & Feigenbaum (1991) and Smith (1991).

For overviews of cognitive science and the role that AI and other disciplines play in that field see Gardner (1985), Posner (1989), and Stillings, Feinstein, Garfield, Rissland, Rosenbaum, Weisler & Baker-Ward (1987).

A number of AI texts are valuable as reference books complementary to this book, providing a different perspective on AI. See classic books by Nilsson (1971; 1980), Genesereth & Nilsson (1987), Charniak & McDermott (1985) and more recent books including Ginsberg (1993), Russell & Norvig (1995) and Dean, Allen & Aloimonos (1995).

The Encyclopedia of Artificial Intelligence (Shapiro, 1992) is an encyclopedic reference on AI written by leaders in the field. There are a number of collections of classic research papers. The general collections of most interest to readers of this book include Webber & Nilsson (1981) and Brachman & Levesque (1985). More specific collections are given in the appropriate chapters.

There are many journals that provide in-depth research contributions and conferences where the most up-to-date research can be found. These include the journals

Artificial Intelligence, Journal of Artificial Intelligence Research, IEEE Transactions on Pattern Analysis and Machine Intelligence, Computational Intelligence, International Journal of Intelligent Systems, and *New Generation Computing*, as well as more specialized journals such as *Neural Computation, Computational Linguistics, Machine Learning, Journal of Automated Reasoning, Journal of Approximate Reasoning, IEEE Transactions on Robotics and Automation*, and the *Logic Programming Journal. AI Magazine*, published by the *American Association for Artificial Intelligence (AAAI)*, often has excellent overview articles and descriptions of particular applications. There are many conferences on Artificial Intelligence. Those of most interest to a general audience are the biennial *International Joint Conference on Artificial Intelligence (IJCAI)*, the *European Conference on AI (ECAI)*, the *Pacific Rim International Conference on AI (PRICAI)*, and various national conferences, especially the *American Association for Artificial Intelligence National Conference on AI*, and innumerable specialized conferences and workshops.

1.7   Exercises

Exercise 1.1   For each of the following, give five reasons why:

(a) A dog is more intelligent than a worm.

(b) A human is more intelligent that a dog.

(c) An organization is more intelligent than an individual human.

Based on these, give a definition of what "more intelligent" may mean.

Exercise 1.2   Give as many disciplines as you can whose aim is to study intelligent behavior of some sort. For each discipline find out where the behavior is manifest and what tools are used to study it. Be as liberal as you can as to what defines intelligent behavior.

Exercise 1.3   Choose a particular world, for example, what is on some part of your desk at the current time.

i) Get someone to list all of the things that exist in this world (or try it yourself as a thought experiment).

ii) Try to think of twenty things that they missed. Make these as different from each other as possible. For example, the ball at the tip of the right-most ball-point pen on the desk, or the spring in the stapler, or the third word on page 21 of a particular book on the desk.

iii) Try to find a thing that can't be described using natural language.

iv) Choose a particular task, such as making the desk tidy, and try to write down all of the things in the world at a level of description that is relevant to this task.

Based on this exercise, discuss the following statements:

(a) What exists in a world is a property of the observer.

(b) You need general constructors to describe individuals, rather than expecting each individual to have a separate name.

(c) What individuals exist is a property of the task as well as the world.

(d) To describe the individuals in a domain, you need what is essentially a dictionary of a huge number of words and ways to combine them to describe individuals, and this should be able to be done independently of any particular domain.

Chapter 2

A Representation and Reasoning System

2.1 Introduction

For a computer to be intelligent it has to be programmed appropriately. Ideally you would like to tell it only as much as it needs to know in a high-level language, and you would like for it to be able to reason and learn how to act intelligently. The language you use to communicate with the computer, the meaning associated with that language, together with a specification of what the computer does in response makes up a representation and reasoning system (RRS) (page 9).

This chapter develops the general idea of an RRS. We present the syntax, semantics, and two different proof procedures for one simple representation and reasoning system, namely the definite clause language. This chapter should be read in conjunction with Chapter 3, which discusses how this particular RRS can be used to represent a variety of problems.

2.2 Representation and Reasoning Systems

Before providing a more detailed description of a particular representation and reasoning system, we investigate the features we want in an RRS and how to use an RRS when you have one.

- First, begin with a task domain that you want to characterize. This is a domain in which you want the computer to be able to answer questions, solve problems, or interact with directly. For example, the domain may be an office layout, a house's electrical wiring system, or a university's course structure.

- Second, distinguish things you want to talk about in the domain, namely the ontology (page 10). The different wires and switches may be individuals, but you may not care about the decorative plates for the switches or each individual nail holding the walls together.

- Third, use symbols in the computer to represent the objects and relations in the domain. We say the symbols *denote* the objects in the world. The symbol s_2 may refer to a particular light switch, or the symbol *david* may denote a particular person.

- Fourth, tell the computer knowledge about that domain. This may include knowledge about what is true in that domain as well as knowledge about how to solve problems in that domain. You may tell the system that switch s_2 is connected to wire w_0 that is in turn connected to light l_1.

- Fifth, ask the RRS a question which prompts the RRS to reason with its knowledge to solve problems, produce answers, or sanction an agent's actions. You may want to know whether light l_1 is lit based on the knowledge given to the system.

In order to tell the computer anything, you have to have a language of communication, you need a way to give meaning to the sentences, and you need a way for the system to compute answers. Together these form a a representation and reasoning system:

A **Representation and Reasoning System (RRS)** is made up of a formal language, a semantics, and a reasoning theory, where:

- A **formal language** specifies the legal sentences you can use to express knowledge about a task domain. The language defines the legal symbols and how they can be put together. A language is specified by a **grammar**. A **knowledge base** is a set of sentences in the language.

- A **semantics** specifies the meaning of sentences in the language. It specifies a commitment to how symbols in the language relate to the task domain. It permits you to explain the meaning of sentences in the language in terms of the task domain, as well as to understand the meaning of the computer's answers. With this semantic commitment, you can discuss correctness, or truth, of knowledge independently of how it is used. You can define the meaning of a computed answer, independently of how it was computed. It is important that the semantics be natural and intuitive, so that the result of a computation can be easily interpreted.

- A **reasoning theory**, or **proof procedure**, is a possibly nondeterministic specification of how an answer can be derived from the knowledge base. This is typically a set of inference rules that sanction consequences left implicit, or is a nondeterministic specification of how an answer is computed. A reasoning theory may be inadequate for the semantics you've defined. A reasoning theory is said to be *sound* if it only generates correct answers with respect to the semantics, and it is said to be *complete* if it generates all correct answers or guarantees to produce one if one exists. (See page 46 for more details on soundness and completeness.)

An **implementation of an RRS** consists of a language parser and a reasoning procedure:

- A **language parser** is a program that distinguishes legal sentences of the formal language and, given legal sentences, produces some internal form to be stored as data structures in the computer.

- A **reasoning procedure** is an implementation of a reasoning theory together with a search strategy. You can't implement a reasoning theory without committing to some way to resolve the nondeterminism in the reasoning theory; this is the role of a search strategy. A reasoning procedure manipulates the data structures produced by the parser, to produce answers. There can be many different implementations of the same reasoning theory such as where the reasoning theory is implemented in different languages, using different encoding techniques, or using different search strategies. A reasoning procedure may be incomplete either because there is some conclusion that the semantics specifies should exists but which the reasoning theory cannot derive, or because the search strategy can't find an answer sanctioned by the reasoning theory.

The semantics isn't reflected in the definition of the implementation above, but there are corresponding elements to the formal language and the reasoning theory. The semantics isn't something that is actually implemented; it gives a meaning to the symbols for an external viewer. The semantics also specifies what is a correct conclusion. Thus you can study the soundness and completeness of a reasoning procedure with respect to the semantics.

Figure 2.1 illustrates the relationship between the components. On the left is a computer implementing the reasoning procedure. Sentences in the language are fed into it and answers come out. The computer doesn't know what the symbols mean. The person who supplied the information has an associated meaning with the symbols. They can check the truth of the inputs according to the semantics and can interpret the output according to that semantics.

Associated with an RRS is a **representational methodology**. This is a specification of how the RRS can be used. It could be provided as a user manual for the RRS. The methodology is derived through design and through experience. The value of a

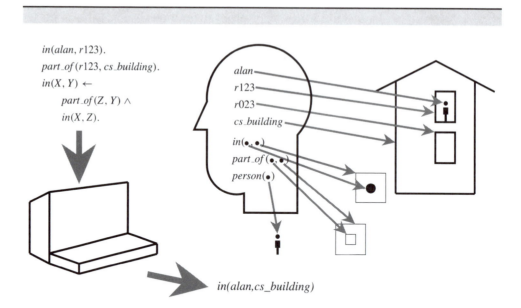

$in(alan, r123).$
$part\_of(r123, cs\_building).$
$in(X, Y) \leftarrow$
$\quad part\_of(Z, Y) \wedge$
$\quad in(X, Z).$

alan
r123
r023
cs_building
$in(\bullet, \bullet)$
$part\_of(\bullet, \bullet)$
$person(\bullet)$

$in(alan, cs\_building)$

The meaning of the symbols are in the user's head. The computer takes in symbols and outputs symbols. The output can be interpreted by the user according to the meaning they place on the symbols.

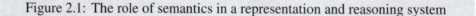

Figure 2.1: The role of semantics in a representation and reasoning system

methodology is best understood if you consider that many RRSs are equivalent to a Turing machine in that they can compute any computable function; any differences between these RRSs arise in what they can do naturally, simply, and efficiently. Often the form and the assumptions behind an RRS impose constraints on what can be represented, and thus may make the RRS well-suited for some domains and inappropriate for others. For example, some RRSs allow the representation of uncertainty in a domain, whereas others can represent only certain knowledge.

Just as the reasoning procedure was divided into a proof procedure and a search strategy, there is a corresponding distinction in the programming methodology between specification of the **logic** of the domain and the specification of the **control** to find a solution to the problem. Different semantics may impose different views of the world and, therefore, different ways to conceptualize the problem. Similarly, different search procedures may provide different ways to specify control information to improve reasoning efficiency. For example, some RRSs allow the order of the sentences to contain heuristic control information without having any corresponding semantic information. For any RRS, you need programming methodologies that guide both the logical specification and the implicit and explicit control knowledge.

While these aspects of an RRS might appear in different guises, all are important. You can't record anything in the computer without syntax, though it may take various forms, such as text, graphics, or even sound. A semantics without a proof procedure is useless since you actually want to compute. A reasoning procedure without a semantics is also useless, since it fails to provide any way to relate computed consequences to the task domain you are trying to characterize. A reasoning procedure which isn't sound can't be relied on to maintain fidelity with the task domain. A reasoning procedure which is sound but not complete may produce useful consequences, but no meaning can be assigned to its failure to establish the truth of an answer; it may be that an answer exists, but the reasoning procedure can't find it.

This notion of an RRS should be compared with more familiar programming language definitions, which typically consist of a formal language, a parser, and a reasoning procedure but no semantics. Often there is a reasoning theory in terms of a procedural semantics, but usually there is a less formal specification of program execution. There is usually no way to interpret the result of a program independently of how it is computed.

The above definitions of semantics and reasoning theory correspond to the notions of a Tarskian semantics and proof in mathematical logic. With logic, you can define knowledge independently of how it is used. You can verify the correctness of knowledge since you know its meaning. Under a particular domain and interpretation, you can debate the truth of sentences in the language. You can use the same semantics to establish the correctness of an implementation.

The separation of a program's meaning from how it's computed lets you exploit performance ideas such as parallelism or domain-specific heuristics with the confidence of correct results. A program's correctness is defined by the semantics, not by a particular control strategy. As long as you are faithful to the semantics, the control can be optimized for the machine used.

Logic provides the specification of meaning in terms of the relationship between the symbols in a computer and the task domain. We make no assumption that the reasoning procedures are just doing deduction. It is the commitment to a semantics that is inherited from logic, not any particular class of proof procedures.

2.3 Simplifying Assumptions of the Initial RRS

Scientific progress on difficult problems often results from solving simpler, more abstract but related problems. For example, classical physics developed under the unrealistic but extremely useful assumption of a friction-less world.

Similarly, to begin the logical analysis of computational intelligence we shall make various explicit assumptions. Given the view of an intelligent agent in an environment, there are three kinds of assumptions you can make. First, you could assume certain properties are true of the agent; these are type A assumptions. Second, you could assume certain properties are true of the environment; these are type E assumptions. Third, you could make assumptions, of type AE, about the relationship between an agent and its environment.

First we restate the ontological assumption (page 10):

Assumption IR   **(Individuals and Relations)** An agent's knowledge can be usefully described in terms of *individuals* and *relations* among individuals.

The notion of relations includes nullary relations that are true or false, unary relations that are properties of single individuals such as their color or height, binary relations between pairs of individuals, ternary relations among triples of individuals, and relations amongst more individuals.

Assumption IR underlies the approach described in this book. This is an AE assumption. It says essentially that linguistic or symbolic knowledge is all that an agent needs to act. Of course, to build a real robot, you need some nonlinguistic knowledge such as how to interpret sounds as speech or how to interpret two-dimensional stimuli on the retina or a camera into features. This assumption implies that results of these perceptive tasks should be symbolic knowledge. We don't focus on these perceptual tasks because they are well covered elsewhere. This assumption isn't saying that the world is made up of things and relations among things, but rather that the world can be described in these terms.

To the assumption IR we now add two quite restrictive assumptions, both of which are waived or relaxed as we develop our approach.

Assumption DK   **(Definite Knowledge)** An agent's knowledge base consists of *definite* and *positive* statements.

We assume that the sentences in the knowledge base aren't vague or negative. Knowing that either "Craig is in room r113" or "Craig is in the lab," but not which, is vague. Knowing that "Craig isn't in room r117" is negative. This chapter defines *definite knowledge* precisely. Since the assumption applies only to the form of the agent's knowledge base, it is of type A. This assumption is relaxed in Chapter 7.

One assumption that we're *not* making at this point is that the agent's knowledge is complete. We don't assume the agent knows everything. Section 7.4 (page 248) considers agents with complete knowledge that conclude that a sentence is false because they have no knowledge to the contrary.

Assumption SE   **(Static Environment)** The environment is *static*.

This assumption allows you to ignore change. We assume that no actions occur which change the environment while the agent is constructing its representation or reasoning about it. The knowledge could be about a world at one instant in time, and not account for how things have changed from the past or how things might change in the future. This excludes statements such as "Craig was in room 117" or "After the robot has picked up a box, it is carrying the box." Assumption SE allows you to ignore time. This assumption is lifted in Chapter 8.

Our motivation is simplicity. We want to begin with a simple framework which allows you to describe a task domain and then reason to conclusions entailed by the description. Once conceptualized, you want to be able to describe relationships between individuals in the world and describe how these relations depend on each other. The initial language allows knowledge of the form, "if one set of relations in the world is true, then another relation is true."

Having committed to some initial assumptions about task domains, we can define the first RRS. We begin with a simple yet powerful language in which to state knowledge about the world. The system we choose is sufficient for use under the assumptions. When you understand its deficiencies, you can address them, together with a review of the assumptions, in order to progress to more complex domains and the attendant extension of the RRS.

The first RRS is the logical language of *definite clauses*. One particular reasoning procedure built upon definite clauses is the reasoning procedure *pure Prolog*, which makes a commitment to the definite clause language and to a particular proof procedure and search strategy. There are other reasoning procedures for definite clauses—for example, constraint logic programming languages, parallel logic programming systems, and deductive database systems. We don't want to exclude these from consideration.

We introduce the RRS in two phases. In the first phase, we assume:

Assumption FD   **(Finite Domain)** There are only a finite number of individuals of interest in the domain. Each individual can be given a unique name.

The RRS that makes these assumptions is called Datalog. The last assumption is lifted when we consider function symbols (page 58).

2.4 Datalog

This section gives the language's syntax. The syntax is based on normal mathematical notation for predicate symbols but follows Prolog's convention for variables. The **syntax** of **Datalog** is given by the following:

- A logical **variable** is a word consisting of a sequence of letters, digits, or the underscore "_" starting with an uppercase letter or the underscore.
  For example *X*, *Room*, *B4*, *Raths*, *The_big_guy* are all variables.

- A **constant** is a word starting with a lowercase letter. Constants can also be words consisting only of digits; these are called numerals.

- A **predicate symbol** is a word starting with a lower case letter.

     For example, *alan, r123, f, grandfather*, and *borogroves* can be constants or predicate symbols, depending on the context. 725 is a constant.

- A **term** is either a variable or a constant.

     For example *X, alan, cs422, mome*, or *Raths* can be terms.

- An **atomic symbol**, or simply an **atom**, is of the form p or $p(t_1, \ldots, t_n)$, where p is a predicate symbol and each t_i is a term.

     For example, *teaches(sue, cs422), in(alan, r123), happy(C), father(bill, Y), sunny*, and *outgrabe(mome, Raths)* can all be atoms.

- A **body** is of the form $a_1 \wedge \ldots \wedge a_m$, where each a_i is an atom. When $m = 1$, there are no $\wedge$s. Defined recursively, a body is either an atom or of the form $a \wedge b$, where a is an atom and b is a body. The symbol "$\wedge$" is read as "and." The body is said to be a **conjunction** of atoms.

- A **definite clause** is either an atom, a, called a **fact**, or of the form $a \leftarrow b$, called a **rule**, where a, the **head**, is an atom and b is a body. The symbol "$\leftarrow$" should be read as "if."

- A **query** is of the form $?b$, where b is a body. The symbol "?" should be read as "prove."

- An **expression** is either a term, an atom, a definite clause, or a query.

- A **knowledge base** is a set of definite clauses.

The syntax is constructed from names, variables, and a limited collection of special constants: periods, commas, parentheses, and those mentioned above, namely "$\leftarrow$," "$\wedge$," and "?".

Anything following a "%" to the end of the line is a **comment** that is not part of the formal language and is only there to convey information to a reader.

As far as the syntax is concerned, constants and predicate symbols can only be distinguished by their context.

Example 2.1

Suppose you are given the knowledge base:

$in(alan, R) \leftarrow teaches(alan, cs422) \wedge in(cs422, R).$

$grandfather(william, X) \leftarrow father(william, Y) \wedge parent(Y, X).$

$slithy(toves) \leftarrow mimsy \wedge borogroves \wedge outgrabe(mome, Raths).$

From context, *alan, cs422, william, toves*, and *mome* are constants; *in, teaches, grandfather, father, parent, slithy, mimsy, borogroves*, and *outgrabe* are predicate symbols; and *X, Y*, and *Raths* are variables.

The first two clauses about Alan and William may make some intuitive sense, even though we have not explicitly provided any formal specification for the meaning of sentences of the definite clause language. However, regardless of the mnemonic names' suggestiveness, as far as the computer is concerned, the first two clauses have no more meaning than the third. Meaning is provided only by virtue of a semantics.

An expression is **ground** if it doesn't contain any variables.
The next section defines the semantics. We first consider only ground expressions and then extend the semantics to include variables.

2.5   Semantics

In this section we outline the general idea behind semantics. We develop what is essentially a standard Tarskian semantics. This is followed by a formal definition of semantics. We then give the user's view of semantics which explains how semantics can be used to state knowledge, followed by the computer's view of semantics that shows how a computer can use semantics even though it may not know the meaning of the symbols.

The description of an RRS (page 23) included a specification of how sentences in the language could be interpreted in a semantics. The semantics concerns two things: first, a set of individuals in the world, the domain, and relations between them, and second, a correspondence between the language's symbols and those individuals and relations.

A semantics tells you how to put symbols of the language into a correspondence with the task domain, and then it tells you how to use that commitment to understand sentences of the language. The semantics depends both on the world you are trying to characterize and on how the constants and predicate symbols in the syntax correspond to individuals and properties in the world.

The correspondences between symbols of the language and objects in a world define an *interpretation* for the language which the syntax describes.

Example
2.2

Consider the domain of Figure 2.1 (page 26). This is intended to depict a domain where there is a person who is in a room in a building. In order to be able to represent this domain, you name, using a constant, each individual you are interested in. Suppose you decide to use the constant *alan* to denote the person, the constant *r*123 to denote the room, and *cs_building* to denote the building. Intuitively you are naming anything that you may want to refer to. You have to be careful to distinguish the constant *alan* from the person with that name. You can have the symbol *alan* in the computer, but you can't put the person into the computer.

With this same correspondence in mind, you might pick the unary predicate symbol *person* to correspond to the property of being a person. Alan is a person, so *person(alan)* would be true in this world and under this correspondence. *r123* isn't a person, so *person(r123)* would be false under this interpretation.

Similarly, you may choose the binary predicate symbol *in* to denote the containment relationship between an object and a container that it is in. In this correspondence, *in(alan, r123)* is true and *in(alan, cs_building)* is true because the person Alan is in both the room *r123* and the building *cs_building*. The atom *in(cs_building, r123)* is false under this interpretation, because the CS building is not in the room *r123*.

It is important to remember that you interpret the symbols with a particular intended interpretation in mind. This interpretation includes the correspondence between symbols and individuals and relations. The intended interpretation is in the designer's mind, and not in the computer. You don't put people or buildings in the computer, but only symbols. These symbols may denote people, buildings or anything else.

An interpretation that makes a clause or set of clauses true is known as a *model* of those clauses.

These intuitions are formalized in the next section, and then expounded upon in the subsequent sections. We first describe a formal semantics, and then show how it can be used to give meaning.

Formal Semantics

In this section we give a formal semantics for the language. In the following two sections we show how this formal semantics can be viewed from the user's viewpoint and from the computer's viewpoint.

An **interpretation** is a triple $I = \langle D, \phi, \pi \rangle$, where

- D is a nonempty set. D is the *domain*. Elements of D are *individuals*.

- ϕ is a mapping that assigns to each constant an element of D.

- π is a mapping that assigns to each n-ary predicate symbol a function from D^n into {*TRUE*, *FALSE*}.

ϕ is a function from names into individuals. The constant c is said to **denote** the individual $\phi(c)$. c is a symbol, but $\phi(c)$ can be a real physical object such as a person or a course. It is often difficult to write down the definition of ϕ without pointing at the object being denoted.

What Does "Formal" Mean?

We have used the term "formal" to describe the mathematical definition of semantics, so what does "formal" mean in this context?

The use of the term "formal" as in "formal definition," "formal theory," or "formal semantics" is often intimidating, with the perception that "formal" means "mathematically obscure and unintuitive." *Formal* should really mean the opposite, namely that the definition is precise and the concept is well-defined. There should be no undefined terms. A proof should be possible just using the form of the definitions, without having to appeal to intuition about the subject matter. A formal theory needn't be unintuitive. You should always be suspicious that an *unintuitive* formal theory does not formalize the intuitions you want to formalize. The advantage of a formal theory is that it should clarify what is or isn't part of the theory.

The essence of a formal definition or a formal semantics is that whether something fits the definition should not be a matter of debate or conjecture. The definition should be such that what is defined is unambiguous. Formal definitions should be familiar to those who program computers. You can't leave anything undefined, as computers only use the form of the program. You can't define a procedure in an informal or hand-waving way.

The opposite of formal is informal. An informal definition is one that uses terms that are left undefined or not precisely defined. This may be adequate for cases where the definition is clear, but when the definition needs to be used for the boundary cases, informal definitions are usually not adequate.

Often formal is equated with mathematical. This should not be surprising because mathematics is the study of common structures and is built up formally from simpler components. Lakatos (1976) shows convincingly how even mathematics isn't as formal as it might seem.

The mathematical concepts we build upon include:

sets Sets have elements (members). You write $s \in S$ if s is an element of set S. The elements in a set define the set, so that two sets are equal if they have the same elements.

tuples Tuples are ordered grouping of elements. Two tuples are equal if they have the same members in the corresponding positions. If S is a set, S^n is the set of n-tuples where each element is a member of S. $S_1 \times S_2$ is the set of ordered pairs with the first element in S_1 and the second element in S_2.

relations A relation is a set of tuples. An n-ary relation is a subset of S^n. The tuples in the relation are said to be *true* of the relation.

functions A function, or mapping f from set D, the domain of f, into set R, the range of f, written $f : D \rightarrow R$, is a subset of $D \times R$ such that for every $d \in D$ there is a unique $r \in R$ such that $\langle d, r \rangle \in f$. You usually write $f(d) = r$ if $\langle d, r \rangle \in f$.

While these may seem like obscure definitions for common-sense concepts, and these would hardly be called formal definitions, you can now use the common-sense concepts comfortable with the knowledge that if you are unsure about something, you can check the definitions.

$\pi(p)$ specifies whether the relation denoted by the n-ary predicate symbol p is true or false for each n-tuple of individuals. If predicate symbol p has no arguments, then $\pi(p)$ is either *TRUE* or *FALSE*.

Example 2.3

Let's formalize the interpretation of Figure 2.1 (page 26).

D is the set with four elements: the person Alan, room 123, room 023, and the CS building. This isn't a set of four symbols, but is the set containing the actual person, the actual rooms, and the actual building. It is difficult to write down this set, and fortunately you never really have to. Ideally you want to point to the physical objects, as is done in Figure 2.1.

The constants are *alan*, *r123*, *r023*, and *cs_building*. The mapping ϕ is defined by: $\phi(alan)$ is the person called Alan, $\phi(r123)$ is room 123, $\phi(r023)$ is the room directly below room 123, and $\phi(cs\_building)$ is the building. ϕ is a mapping from symbols onto real physical objects.

The predicate symbols are *person*, *in*, and *part_of*. The extension of the predicate symbol *person* is given by:

$$\pi(person)(alan') = TRUE$$
$$\pi(person)(r123') = FALSE$$
$$\pi(person)(r023') = FALSE$$
$$\pi(person)(cs\_building') = FALSE$$

where *alan'* is the person denoted by the symbol *alan*; that is, *alan'* is $\phi(alan)$. Similarly for the other individuals. The extension of the predicate symbol *in* is given by:

$$\pi(in)(alan', r123') = TRUE$$
$$\pi(in)(alan', cs\_building') = TRUE$$

and $\pi(in)$ applied to any other pair of individuals has value *FALSE*. This means that the person called Alan is in the room *r123* and is also in the CS building, and these are the only instances of the *in* relation that are true. Similarly,

$$\pi(part\_of)(r123', cs\_building') = TRUE$$
$$\pi(part\_of)(r023', cs\_building') = TRUE$$

and $\pi(part\_of)$ applied to any other pair of individuals has value *FALSE*. This means the two rooms are part of the CS building, and there are no other *part_of* relationships that are true in this interpretation.

It is important to emphasize that the elements of D are the real physical objects, and not their names. The name *alan* isn't in the name *r123*, but rather the person denoted by *alan* is in the room denoted by *r123*.

p	q	$p \wedge q$	$p \leftarrow q$
true	true	true	true
true	false	false	true
false	true	false	false
false	false	false	true

Figure 2.2: Truth table defining $\wedge$ and $\leftarrow$

Each ground term denotes an individual in an interpretation. A constant c denotes in I the individual $\phi(c)$.

Each ground atom is either true or false in an interpretation. Atom $p(t_1, \ldots, t_n)$ is **true** in I if $\pi(p)(t'_1, \ldots, t'_n) = TRUE$, where t'_i is the individual denoted by term t_i, and is **false** in I otherwise.

Example 2.4

The atom $in(alan, r123)$ is true in the interpretation of Example 2.3, as the person denoted by *alan* is indeed in the room denoted by $r123$. Similarly, *person(alan)* is true, as is *part_of*$(r123, cs\_building)$. The atoms $in(cs\_building, r123)$ and *person(r123)* are false in this interpretation.

You can now use the definition of truth and falsity of atoms in an interpretation to determine the truth or falsity of ground clauses in an interpretation. If P and Q are ground formulae, the truth of a composite formula can be computed using the truth table of Figure 2.2.

Thus, a ground clause $h \leftarrow b_1 \wedge \ldots \wedge b_m$ is false in interpretation I if h is false in I and each b_i is true in I, and is true in interpretation I otherwise.

In order to define semantics of variables, define a **variable assignment**, ρ, to be a function from the set of variables into the domain. Thus a variable assignment assigns an element of the domain to each variable. Given ϕ and a variable assignment ρ, each term denotes an individual in the domain. If the term is a constant, the individual denoted is given by ϕ. If the term is a variable, the individual denoted is given by ρ. Given an interpretation and a variable assignment, each atom is either true or false using the same definition as above; all we have done is expanded the notion of denotation of terms. Similarly, given an interpretation and a variable assignment, each clause is either true or false.

A clause is true in an interpretation if it is true for all variable assignments. This is called a **universal quantification**. The variables are said to be universally quantified in the scope of the clause. Thus a clause is false in an interpretation means there is a variable assignment under which the clause is false. Notice how the scope of the variable is the whole clause; you must use the same variable assignment for the head and the body of the clause.

Example
2.5

The clause

$$part\_of(X, Y) \leftarrow in(X, Y)$$

is false in the interpretation of Example 2.3, as under the variable assignment with X denoting the individual $alan'$ and Y denoting the individual $r123'$, the clause's body is true, as $\pi(in)(alan', r123') = TRUE$, and the clause's head is false, as $\pi(part\_of)(alan', r123') = FALSE$.

The clause

$$in(X, Y) \leftarrow part\_of(Z, Y) \wedge in(X, Z)$$

is true, as in all variable assignments where the body is true, the head is also.

A set of clauses is true in an interpretation if and only if every element of the set is true in that interpretation. The clauses are thus implicitly conjoined.

A **model** of a set of clauses is an interpretation in which all the clauses are true.

If *KB* is a set of clauses and *g* is an atom or conjunction of atoms, then *g* is a **logical consequence** of *KB*, or *g* **logically follows** from *KB*, or *KB* **entails** *g*, written

$$KB \models g$$

if *g* is true in every model of *KB*. That is, there is no interpretation in which *KB* is true and *g* is false.

The symbol "$\models$" is a semantic relation that holds between a set of clauses and the formulae they entail. Both *KB* and *g* are symbolic, and so can be represented in the computer, whereas the semantic relation is defined in terms of models, which aren't necessarily syntactic. The $\models$ relation isn't about computation, proof, derivation, or rules of inference; it provides the specification of what follows from some truths.

Example
2.6

Let the knowledge base *KB* be:

$$in(alan, r123).$$
$$part\_of(r123, cs\_building).$$
$$in(X, Y) \leftarrow$$
$$\qquad part\_of(Z, Y) \wedge$$
$$\qquad in(X, Z).$$

The interpretation defined in Example 2.3 is a model of *KB*, as each clause is true in that interpretation.

$KB \models in(alan, r123)$, as this is stated explicitly as a fact. If every clause of *KB* is true in an interpretation, then $in(alan, r123)$ must be true in that interpretation.

$KB \not\models in(alan, r023)$. The interpretation defined in Example 2.3 is a model of *KB*, in which $in(alan, r023)$ is false.

$KB \not\models part\_of(r023, cs\_building)$. Although $part\_of(r023, cs\_building)$ is true in the interpretation of Example 2.3, there is another model of KB in which $part\_of(r023, cs\_building)$ is false. In particular, the interpretation which is like the interpretation of Example 2.3, but where
$$\pi(part\_of)(\phi(r023), \phi(cs\_building)) = FALSE,$$
is a model of KB in which $part\_of(r023, cs\_building)$ is false.

$KB \models in(alan, cs\_building)$. If the clauses in KB are true in interpretation I, it must be the case that $in(alan, cs\_building)$ is true in I, otherwise there is an instance of the third clause of KB that is false in I, a contradiction to I being a model of KB.

User's View of Semantics

The formal description of semantics doesn't tell you why semantics is interesting or how it can be used as a basis to build intelligent systems. The basic idea behind the use of logic is that when you have a particular world you want to characterize, you can choose that world as an *intended interpretation*, choose denotations for the symbols with respect to that interpretation, and write, as clauses, what is true in that world. When the system computes a logical consequence of a knowledge base, you can interpret this answer with respect to the intended interpretation.

Informally, the methodology for designing a representation of the world and how it fits in with the formal semantics can be expressed as:

Step 1 Choose the task domain or world you want to represent. You must have an intended interpretation to capture a world you want to describe and ask questions about. This could be some aspect of the real world (for example, the structure of courses and students at a university, or a laboratory environment at a particular point in time), some imaginary world (for example, the world of Alice in Wonderland, or the state of the electrical environment if a switch breaks), or an abstract world (for example, the world of numbers and sets). Within this world, let the domain D, be the set of all individuals or things that you want to be able to refer to and reason about. Also choose a set of relations among these individuals.

Step 2 Associate constants in the language with individuals in the world that you want to name. For each element of D you want to refer to by name, assign a constant in the language. For example, you may choose the name "*alan*" to denote a particular professor, the name "*cs322*" for the introductory AI course, the name "*two*" for the number which is the successor of the number one, and the name "*red*" for the color of stop lights. Each of these names denotes the corresponding object in the world.

Step 3   For each relation that you may want to represent, associate a predicate symbol in the language. Each *n*-ary predicate symbol denotes a function from D^n into {*TRUE*, *FALSE*}, which specifies the subset of D^n for which the relation is true. For example, the predicate symbol "*teaches*" of two arguments (a teacher and a course) may correspond to the binary relation which is true when the individual denoted by the first argument teaches the course denoted by the second argument. These relations need not be binary. They could have any number (zero or more) of arguments. For example, "*is_red*" may be a predicate that has one argument.

Step 4   You can now write, as clauses, statements that are true in the intended interpretation. This is often called **axiomatizing** the domain, where the given clauses are the **axioms** of the domain. If the person who is denoted by the symbol *alan* actually teaches the course denoted by the symbol *cs*502, then you can assert the clause *teaches*(*alan*, *cs*502) as being true in the intended interpretation.

Step 5   You can now ask questions about the intended interpretation, and interpret the answers using the meaning that you have assigned to the symbols.

Within this methodology, the user doesn't actually tell the computer anything until step four. The first three steps are carried out in the head of the user. Of course, the user should document the denotations in order to make their programs understandable to other people, so that they remember each symbol's denotation, and so that they can check the truth of the clauses. This isn't something that the computer has access to.

The world itself doesn't prescribe what the individuals are.

Example
2.7

In one conceptualization of a domain, *pink* may be a predicate symbol of one argument that is true when the individual denoted by that argument is pink. In another conceptualization, *pink* may be an individual that is the color pink, and it can be used as the second argument to a binary predicate *color* which says that the object denoted by the first argument has the color denoted by the second argument. Alternatively, you may want to describe the world at a level of detail where you don't distinguish between various shades of red—and so not include the color pink—or you may describe the world in more detail where pink is too general a term.

The problem of choosing between several possible conceptualizations is non-trivial, and it may be affected by the intended interpretation or the problem to be solved within that interpretation. (See Chapter 5 for a more in-depth discussion of these issues.)

It is important to realize that the denotations are only in the head of the user. The denotations aren't written down, except maybe in natural language to convey the meaning to other people. Often the individuals in the domain are real physical objects, so it is usually difficult to give the denotation without pointing at the object. When the individual is an abstract object—for example, a university course or the concept of love—it is virtually impossible to write the denotation. This doesn't stop you from representing and reasoning about such concepts. With a robot that perceives and interacts with its environment, there can be a tighter relationship between the symbols and their denotation.

The Computer's View of Semantics

The one who provides information to the system has, in her head, an intended interpretation, and interprets symbols according to that intended interpretation. She states knowledge, in terms of clauses, about what is true in the intended interpretation. The computer does not have access to the intended interpretation, but only to the clauses. As you'll see below, the computer is able to tell whether some statement is a logical consequence of the axioms. Under the assumption that the intended interpretation is a model of the clauses—the user has written true clauses according to the meaning assigned to the symbols—if a statement is a logical consequence of the clauses it is true in the intended interpretation as it is true in all models of the axioms. If the user tells lies—the axioms aren't all true in the intended interpretation—the computer's answers may be false in the intended interpretation.

The concept of logical consequence seems like exactly the right tool to derive implicit information from an axiomatization of a world. Suppose *KB* represents the knowledge about the intended interpretation. If $KB \models g$, then *g* must be true in the intended interpretation. If $KB \not\models g$, that is, *g* isn't a logical consequence of *KB*, there is a model of *KB* in which *g* is false. As far as the computer is concerned, the intended interpretation may be the model of *KB* in which *g* is false, and so doesn't know whether *g* is true in the intended interpretation.

Given a set of axioms, the models of the axioms correspond to all of the ways that the world could be given the axioms are true. Axioms restrict the set of worlds that are possible.

It is very important to understand that, until you consider computers with perception, the computer doesn't know the meaning of the symbols. It is the human that gives the symbols meaning. All the computer knows about the world is what it is told about the world. All it can be expected to do is provide logical consequences of the knowledge base. All logical consequences of the knowledge base are true in the world under consideration if the world under consideration is a model of the knowledge base. (Chapters 9 and 10 show how the computer can jump to plausible conclusions and also reason about its uncertainty.)

Example 2.8

Suppose the system is provided with the knowledge base:

$$two\_doors\_east(r103, r107) \leftarrow$$
$$imm\_east(r103, r105) \land$$
$$imm\_east(r105, r107).$$
$$imm\_east(r103, r105).$$

imm_east(*r*105, *r*107).

two_doors_east(*r*103, *r*107) is a logical consequence of the knowledge base. If the user had a particular denotation of the symbols and these clauses were indeed true in the intended interpretation, then *two_doors_east*(*r*103, *r*107) must be true in that interpretation. The computer doesn't know any more about the intended interpretation than is given in the knowledge base. *two_doors_east*(*r*203, *r*207) isn't a logical consequence of the knowledge base and so may or may not be true of the intended interpretation; the system has no way of telling.

2.6   Questions and Answers

One reason to build a formal description of a world is to be able to determine what follows from that description. Having stated some facts about a particular domain, you might like to ask questions about whether certain unstated relationships are implied by those facts. The RRS we have in mind is one whose reasoning theory is specified as a question answering procedure which accepts queries and attempts to show whether instances of the query follow from what has been asserted. The meaning of an answer depends on the concept of truth just defined.

Queries

A **query** is a way to ask whether something is a logical consequence of a knowledge base. Once the system has been provided with a set of clauses claimed to be true, the knowledge base, you use a query to ask whether a formula is a logical consequence of the knowledge base. Queries have the form

?*body*.

A ground query is a question that has the **answer** "*yes*" if the body is a consequence of the knowledge base, or the answer "*no*" if the body is not a consequence of the knowledge base. The latter doesn't mean that *body* is false in the intended interpretation, but rather that you don't know whether it is true or false based on the knowledge provided.

For each query of the above form, it is as though you carried out a query transformation by adding the clause

yes ← *body*

where *yes* is a predicate symbol that doesn't appear in the knowledge base, to the knowledge base and then trying to show whether the introduced atom *yes* is a logical consequence of the knowledge base. The answer *"yes"* means that *yes*, and so also *body*, is a logical consequence of the knowledge base with the above clause added. The answer *"no"* means that *yes*, and so also *body*, isn't a logical consequence of the knowledge base.

The method of replacing a general body with a clause whose head is *yes* is one way of providing a syntactic uniformity in answers. Although this uniformity isn't necessary to define the meaning of questions and answers, this uniformity makes it easier to understand more complex answers.

Interpreting Variables

The formal semantics (page 32) provides the formal meaning of variables. In this section, we show how the RRS can express and use generalized knowledge by using variables.

When a variable appears in a clause, the clause is true in an interpretation only if the clause is true for all possible values of that variable. The variable is said to be **universally quantified** within the scope of the clause. If a variable X appears in a clause C, then claiming that C is true in an interpretation means that C is true no matter which individual from the domain is denoted by X.

Example 2.9

To write in a knowledge base the clauses

$$two\_doors\_east(E, W) \leftarrow$$
$$imm\_east(E, M) \land$$
$$imm\_east(M, W).$$

is to claim that that the clause is true no matter which individual in the world is denoted by E, no matter which individual is denoted by W, and no matter which individual is denoted by M, as long as the same individuals are used consistently for E, W, and M, respectively. So the clause must be true for *all* choices of individuals E, W, and M.

If in the above example the predicate symbol *two_doors_east* denotes the relation "is two doors east of," and the predicate symbol *imm_east* denotes the relation "is immediately east of," then the above clause means "for all individuals E, W and C, E is two doors east of W if E is immediately east of C and C is immediately east of W." The claim is that there can't be individuals such that the right-hand side is true and the left-hand side is false in the intended interpretation.

Notice how the semantics treats variables that appear in a clause's body but not in its head:

Example
2.10

The variable M in the clause defining *two_doors_east*, is universally quantified at the level of the clause, thus the clause is true for all variable assignments. Consider particular values r_1 for E and r_2 for W, then the clause

$$two\_doors\_east(r_1, r_2) \leftarrow$$
$$imm\_east(r_1, M) \wedge$$
$$imm\_east(M, r_2).$$

is true for all variable assignments to M. Thus if there exists a variable assignment c_1 for M such that $imm\_east(r_1, c_1) \wedge imm\_east(c_1, r_2)$ is true in an interpretation given that variable assignment, then $two\_doors\_east(r_1, r_2)$ must be true in that interpretation. Thus you can read the clause of Example 2.9 as "for all E and for all W, $two\_doors\_east(E, W)$ is true if there exists an M such that $imm\_east(E, M) \wedge imm\_east(M, W)$ is true."

A clause with free variables is true in an interpretation if it is true no matter which individuals the variables denote. In particular, if you uniformly replace the free variables with ground terms in a true clause, the resulting clause must also be true.

It may seem as though there is something peculiar about talking about the clause being true for cases where it doesn't make sense.

Example
2.11

Consider the clause

$$two\_doors\_east(cs422, love) \leftarrow$$
$$imm\_east(cs422, sky) \wedge$$
$$imm\_east(sky, love).$$

where *cs422* denotes a course, *love* denotes an abstract relationship, and *sky* denotes the sky. Here the clause is vacuously true in the intended interpretation, according to the truth table for $\leftarrow$, as the clause's right-hand side is false in the intended interpretation.

As long as whenever the head is nonsensical, the body is also, you'll never use the rule to prove anything nonsensical. When checking for the truth of a clause, you need only be concerned with those cases in which the clause's body is true. Using the convention that a clause is true whenever the body is false, even if it doesn't make sense, makes the semantics simpler and doesn't really cause any problems.

When there are variables in a query, you typically don't want to know whether there is an instance of the query that is a logical consequence of the knowledge base. You want to know which instances of the query are a logical consequence of the knowledge base. An **instance** of a query is obtained by substituting terms for the variables in the query. Different occurrences of the same variable must be replaced by the same term. In this case, an **answer** is either an instance of the query that is a logical consequence of the knowledge base, or is "*no*" meaning that there are no instances of the query that logically follow from the knowledge base. Instances of the query are specified by providing values for the variables in the query.

Variables in queries are translated according to the standard translation (page 40), but *yes* is parameterized with the free variables appearing in the query.

Example 2.12

It may not be enough to say whether room $r107$ has a room two doors east, but you want to know which room is two doors east of room r107. To answer such questions, you translate the query

$$?two\_doors\_east(R, r107)$$

into the answer clause

$$yes(R) \leftarrow two\_doors\_east(R, r107)$$

meaning "the answer is R if R is two doors east of r107," and then try to prove $yes(R)$.

In general, if the query is B, with free variables $V_1, \ldots, V_k$, then the corresponding answer clause is

$$yes(V_1, \ldots, V_k) \leftarrow B.$$

The aim now isn't merely to determine whether *yes* is a consequence of the knowledge base, but to determine which instances of $yes(V_1, \ldots, V_k)$ are consequences of the knowledge base.

Clauses, Questions, and Answers

A representation system of the type described lets a user provide clauses about a world and permits queries as to what else must be true of that world, given the clauses. Queries are used to ask whether some instance of a body is a logical consequence of the clauses.

Not only do you want to know whether some instance of a query is a logical consequence of the knowledge base, but you typically want to know which instances of the query are a logical consequence of the knowledge base. Determining which instances of a query follow from a knowledge base is known as **answer extraction**.

Example 2.13

Consider the clauses of Figure 2.3. The person who wrote these clauses presumably has some meaning associated with the symbols, and has written the clauses because they are true in some, perhaps imaginary, world. The computer knows nothing about rooms or directions. All it knows is the clauses it is given; and it can compute their logical consequences.

% *imm_west*(W, E) is true if room W is immediately west of room E.

>   *imm_west*(r101, r103).
>   *imm_west*(r103, r105).
>   *imm_west*(r105, r107).
>   *imm_west*(r107, r109).
>   *imm_west*(r109, r111).
>   *imm_west*(r131, r129).
>   *imm_west*(r129, r127).
>   *imm_west*(r127, r125).

% *imm_east*(E, W) is true if room E is immediately east of room W.

>   *imm_east*(E, W) ←
>   *imm_west*(W, E).

% *next_door*(R1, R2) is true if room R1 is next door to room R2.

>   *next_door*(E, W) ←
>   *imm_east*(E, W).
>   *next_door*(W, E) ←
>   *imm_west*(W, E).

% *two_doors_east*(E, W) is true if room E is two doors east of room W.

>   *two_doors_east*(E, W) ←
>   *imm_east*(E, M) ∧
>   *imm_east*(M, W).

% *west*(W, E) is true if room W is west of room E.

>   *west*(W, E) ←
>   *imm_west*(W, E).
>   *west*(W, E) ←
>   *imm_west*(W, M) ∧
>   *west*(M, E).

Figure 2.3: Clauses provided by the user

The user can ask as a query

$$?imm\_west(r105, r107)$$

and the answer is *yes*. The user can ask the query

$$?imm\_east(r107, r105)$$

and the answer is *yes*. The user can ask the query

$$?imm\_west(r205, r207)$$

and the answer is *no*. This means that there isn't enough information in the database to determine whether $r205$ is immediately West of $r207$ or not.

Given the query

$$?next\_door(R, r105)$$

there are two answers. One answer for $R = r107$ that means $next\_door(r107, r105)$ is a logical consequence of the clauses. The other answer is for $R = r103$. For the query

$$?west(R, r105)$$

there are two answers: one for $R = r103$, and one for $R = r101$. For the query

$$?west(r105, R)$$

there are three answers: one for $R = r107$, one for $R = r109$, and one for $R = r111$. For the query

$$?next\_door(X, Y)$$

there are 16 answers, including

$$X = r103, Y = r101$$
$$X = r105, Y = r103$$
$$X = r107, Y = r105$$
$$X = r109, Y = r107$$
$$X = r129, Y = r131$$
$$X = r101, Y = r103$$
$$\dots$$

2.7   Proofs

So far we have specified what an answer is, but not how it can be computed. The definition of $\models$ tells us what formulae should be logical consequences, but not how to compute them.

This section describes how to compute logical consequences using the notion of a **proof**. A proof is a mechanically derivable demonstration that a formula logically follows from a knowledge base. A **theorem** is a formula that can be proved. A proof procedure is a possibly nondeterministic algorithm for deriving consequences of a knowledge base.

First, we describe a proof procedure for the ground case with no variables. This is expanded to the case with variables. These proof procedures require choices to be made. (Chapter 4 discusses how to find appropriate choices through search.)

We want some mechanical means by which the logical consequence of a set of clauses can be derived solely on the basis of their form, without needing to consider the intended interpretation. In terms of an RRS, we specify a reasoning theory that serves as a specification for constructing proofs, in the same way that semantics provides a definition of truth. You need a procedure that manipulates symbols and can be carried out by a computer, but where the answer can be interpreted semantically.

A proof procedure's quality can be judged by whether it computes what it is meant to compute. To be able to interpret a derived answer, you must be sure that what is proved is justified by the semantics.

Given a proof procedure, $KB \vdash g$ means g can be computed or **derived** from knowledge base KB.

You need to be able to verify that each computed answer is justified by the semantics:

A proof procedure is **sound** with respect to a semantics if everything that can be derived from a knowledge base is is a logical consequence of the knowledge base. That is, if $KB \vdash g$, then $KB \models g$.

In addition to its accuracy, you would also like to verify that a proof procedure is general enough to be able to construct a proof for any atom that is a logical consequence of the knowledge base:

A proof procedure is **complete** with respect to a semantics if there is a proof of each logical consequence of the knowledge base. That is, if $KB \models g$, then $KB \vdash g$.

We present two different ways to construct proofs in the initial RRS. We first present a bottom-up procedure and then a top-down procedure. Each is developed first for the ground case and then for the case with variables.

```
C := {};
repeat
      select r ∈ KB such that
             r is "h ← b₁ ∧ ... ∧ bₘ"
             bᵢ ∈ C for all i, and
             h ∉ C;
      C := C ∪ {h}
until no more clauses can be selected
```

Figure 2.4: Bottom-up proof procedure for computing consequences of *KB*

Bottom-Up Ground Proof Procedure

The first computational method we consider is a bottom-up procedure to derive logical consequences. It is called bottom-up with an analogy to building a house, where you build on top of the structure you already have. In the bottom-up proof procedure you are building on facts that have already been established. It should be contrasted with a top-down approach that starts from a query and tries to find clauses that support the query (page 49.)

The general idea is based on one **rule of derivation**, a generalized form of the rule of inference called *modus ponens*:

> If "$h \leftarrow b_1 \wedge \ldots \wedge b_m$" is a clause in the knowledge base, and each b_i has been derived, then h can be derived.

Sometimes you say that you are **forward chaining** on this clause. This is in the sense of going forward from what is known, rather than going backwards from the query.

This rule also covers the case when $m = 0$. You can always immediately derive the atom at the head of a clause with no body.

Figure 2.4 gives a procedure for computing the **consequence set** C of a set KB of clauses. You write $KB \vdash g$ if $g \in C$ at the end of the above procedure.

Example 2.14

Suppose you are given the knowledge base:

$$a \leftarrow b \wedge c.$$
$$b \leftarrow d \wedge e.$$
$$b \leftarrow g \wedge e.$$
$$c \leftarrow e.$$
$$d.$$
$$e.$$

$f \leftarrow a \wedge g.$

The bottom-up procedure can have the following sequence of elements of C, each time through the loop:

{}

$\{d\}$

$\{e, d\}$

$\{c, e, d\}$

$\{b, c, e, d\}$

$\{a, b, c, e, d\}$

The last rule is never used. You can't derive f. This is as it should be as there is a model of the rules in which f and g are both false.

There are a number of properties that can be established for the proof procedure of Figure 2.4:

Soundness Every atom in C is a logical consequence of KB. That is, if $KB \vdash g$ then $KB \models g$. To show this, assume that there is an atom in C that isn't a logical consequence of KB. If there is such an element, there must be a first element added to C that isn't in every model of KB. Suppose this is h, and suppose it isn't true in model I of KB. h must be the first element generated that is false in I. As h has been generated, there must be some clause in KB of the form $h \leftarrow b_1 \wedge \ldots \wedge b_m$ such that $b_1 \ldots b_m$ are all in C. The b_i are generated before h and so are all true in I. This clause's head is false in I, and its body is true in I, so by the definition of truth of clauses, this clause is false in I. This is a contradiction to the fact that I is a model of KB. Thus every element of C is a logical consequence of KB.

Complexity The algorithm of Figure 2.4 halts, and the number of times the loop is repeated is bounded by the number of clauses in KB. This can be easily seen, as each clause can only be used once. Thus the complexity of the above algorithm is linear in the size of the knowledge base, if you can index the clauses so that the inside loop can be carried out in constant time.

Fixed Point The final C generated in the algorithm of Figure 2.4 is called a **fixed point** because any further application of the rule of derivation won't change C. Let I be the interpretation in which every atom in the fixed point is true, and every atom not in the fixed point is false. I is a model of KB. To prove this, suppose "$h \leftarrow b_1 \wedge \ldots \wedge b_m$" $\in KB$ is false in I. The only way this could occur is if $b_1, \ldots, b_m$ are in C, and h isn't in C. By construction, h can be added to C the next time through the loop, and so the algorithm won't halt, a contradiction to C being the fixed point. Thus there can be no clause in KB that is false in interpretation I.

Completeness Suppose $KB \models g$. Then g is true in every model of KB, so it is true in the model defined by the fixed point, and so it is in C, and so $KB \vdash g$.

The model I defined by the fixed point above is a **minimal model** in the sense that it has the least number of true propositions. Every other model must also assign the atoms in the fixed point to be true.

Top-Down Ground Proof Procedure

An alternative proof method is to search *backwards* or *top-down* from a query to determine if it is a logical consequence of the given clauses.

We first give the definition of **definite clause resolution** for the ground case (without variables) and later give the general case with variables (page 56).

An **answer clause** is of the form

$$yes \leftarrow a_1 \wedge a_2 \wedge \ldots \wedge a_m$$

Given an answer clause, you select (page 50) an element of the answer clause's body. The **SLD resolution**, linear resolution with selector functions for definite clauses, of the above answer clause on the selection a_i with the clause

$$a_i \leftarrow b_1 \wedge \ldots \wedge b_p$$

is the answer clause

$$yes \leftarrow a_1 \wedge \ldots \wedge a_{i-1} \wedge b_1 \wedge \ldots \wedge b_p \wedge a_{i+1} \wedge \ldots \wedge a_m.$$

That is, a_i in the answer clause is replaced by the resolved clause's body.

An **answer** is an answer clause with $m = 0$. That is, it is the answer clause $yes \leftarrow$.

A **derivation** of a query "$?q_1 \wedge \ldots \wedge q_k$" from knowledge base KB is a sequence of answer clauses $\gamma_0, \gamma_1, \ldots, \gamma_n$ such that

- γ_0 is the answer clause corresponding to the original query, namely the answer clause $yes \leftarrow q_1 \wedge \ldots \wedge q_k$,

- γ_i is obtained by resolving γ_{i-1} with a clause in KB, and

- γ_n is an answer.

The procedure in Figure 2.5 specifies an interpreter for solving a query. It follows the definition of a derivation. It is nondeterministic in that there is a point in the algorithm where you have to *choose* a clause to resolve against. There may be many, or perhaps no, choices available. You can find all derivations by systematically considering all the different choices.

Nondeterministic Choice

In many AI programs, you want to separate the definition of what a solution is from how it is computed. Usually the algorithms are *nondeterministic*. This means that there are choices in the program that are left unspecified. There are two sorts of nondeterminism:

- The first, sometimes called "don't-care nondeterminism," is exemplified by the "select" in Figure 2.4. In this form of nondeterminism, if one selection doesn't lead to a solution there is no point in trying any other alternatives. Don't-care nondeterminism is exemplified by the problem of resource allocation, where you have a number of requests for a single resource and have to select who gets the resource at each time. The correctness isn't affected by the selection, but efficiency and termination may be. When you have an infinite sequence of selections, you have to be careful to make fair selections. A **fair selection** is one where a request isn't kept waiting indefinitely. If a request is repeatedly available to be selected it will eventually be selected. The problem of an element that is repeatedly not selected is called *starvation*.

- The second, sometimes called "don't-know nondeterminism," is exemplified by the "choose" in Figure 2.5. Just because one choice did not lead to a solution doesn't mean that other choices will be as futile. An implementation needs to search all alternatives. Often you speak of an *oracle* that can specify, at each point, which choice will lead to a solution. As you don't have such an oracle, you have to search through the space of all alternatives. Chapter 4 presents many algorithms that can be used to search the space of different choices.

  Don't-know nondeterminism plays a large role in computational complexity theory. The class of P problems contains those problems that can be solved in polynomial time; there is an algorithm that solves the problem with time complexity polynomial in the size of the problem. The class of problems known as NP problems are those that could be solved in polynomial time if you had an oracle that chooses the correct value at each time, or, equivalently, if a solution can be verified in polynomial time. It is widely conjectured that $P \neq NP$, which means that no such oracle can exist. One great result of complexity theory is that the hardest problems in the NP class of problems are all equally complex; if one can be solved in polynomial time, they all can. This is known as the class of NP-complete problems. A problem is NP-hard if it is at least as hard as an NP-complete problem. A decision problem is co-NP-complete if the inverse problem—the one that answers "yes" whenever the original answers "no"—is NP-complete. If $P \neq NP$, then search is inevitable. See Garey & Johnson (1979) for a detailed discussion of computational complexity.

In this book we consistently use the term "**select**" for don't-care nondeterminism and "**choose**" for don't-know nondeterminism.

$solve(q_1 \wedge \ldots \wedge q_k)$:

    $ac :=$ "$yes \leftarrow q_1 \wedge \ldots \wedge q_k$"

    repeat

        select a conjunct a_i from the body of ac;

        choose clause C from KB with a_i as head;

        replace a_i in the body of ac by the body of C

    until ac is an answer.

Figure 2.5: Top-down definite clause interpreter, without variables

Example 2.15

Suppose you are given

$$a \leftarrow b \wedge c.$$
$$b \leftarrow d \wedge e.$$
$$b \leftarrow g \wedge e.$$
$$c \leftarrow e.$$
$$d.$$
$$e.$$
$$f \leftarrow a \wedge g.$$
$$?a.$$

The following shows a sequence of assignments to ac in the repeat loop of Figure 2.5 for $solve(a)$:

$$yes \leftarrow a.$$
$$yes \leftarrow b \wedge c.$$
$$yes \leftarrow d \wedge e \wedge c.$$
$$yes \leftarrow e \wedge c.$$
$$yes \leftarrow c.$$
$$yes \leftarrow e.$$
$$yes \leftarrow .$$

The following shows a sequence of choices, where the second clause for b was chosen, that fails to provide a proof:

$$yes \leftarrow a.$$
$$yes \leftarrow b \wedge c.$$
$$yes \leftarrow g \wedge e \wedge c.$$

If g is selected, there there are no rules that can be chosen. This proof attempt is said to *fail*.

When you have derived the answer, you can read a bottom-up proof in the opposite direction. Every top-down derivation corresponds to a bottom-up proof, and every bottom-up proof has a corresponding top-down derivation. This equivalence can be used to show the soundness and completeness of the derivation procedure.

Substitutions

So far the notion of a proof has ignored the issue of variables. We have only presented the reasoning procedure for ground clauses.

An **instance** of a clause is obtained by uniformly substituting terms for variables in the clause. All instances of a particular variable are replaced by the same term.

By the formal semantics of a clause with variables (page 32), if a clause is true in an interpretation, any instance of it is also true in that interpretation, as a clause with universal variables is true no matter which individuals are denoted by the variables.

The proof procedure extended for variables must account for the fact that a free variable in a clause means that any instance of the clause is true. Intuitively, a proof can use any instance of a clause. Moreover, a proof may need to use different instances of the same clause. We need to develop some way of representing and manipulating the many possible instances of clauses, in order to extend the proof procedure for variables.

An instance of a clause is represented as the original clause, together with a substitution.

A **substitution** is a finite set of the form $\{V_1/t_1, \ldots, V_n/t_n\}$, where each V_i is a distinct variable and each t_i is a term. The element V_i/t_i is a **binding** for variable V_i. A substitution is in **normal form** if no V_i appears in any t_j. We assume that all substitutions are in normal form.

Example 2.16

For example, $\{X/Y, Z/a\}$ is a substitution in normal form that binds X to Y and binds Z to a. The substitution $\{X/Y, Z/X\}$ isn't in normal form, as the variable X occurs both on the left and on the right of a binding.

The **application** of a substitution $\sigma = \{V_1/t_1, \ldots, V_n/t_n\}$ to expression e, written $e\sigma$, is the expression which is like the original expression e but with every occurrence of V_i in e replaced by the corresponding t_i. $e\sigma$ is called an **instance** of e. If $e\sigma$ is ground it is called a ground instance of e.

Example 2.17

Some applications of substitutions are

$$p(a, X)\{X/c\} = p(a, c).$$

$$p(Y, c)\{Y/a\} = p(a, c).$$
$$p(a, X)\{Y/a, Z/X\} = p(a, X).$$
$$p(X, X, Y, Y, Z))\{X/Z, Y/t\} = p(Z, Z, t, t, Z).$$

Substitutions can apply to clauses, atoms, and terms. For example, the result of applying the substitution $\{X/Y, Z/a\}$ to the clause

$$p(X, Y) \leftarrow q(a, Z, X, Y, Z)$$

is the clause

$$p(Y, Y) \leftarrow q(a, a, Y, Y, a).$$

The concept of a substitution allows you to specify when two expressions can be made the same:

A substitution σ is a **unifier** of expressions e_1 and e_2 if the substitution applied to both results in the same expression. That is, $e_1\sigma$ is identical to $e_2\sigma$.

Example 2.18

$\{X/a, Y/b\}$ is a unifier of $t(a, Y, c)$ and $t(X, b, c)$ as

$$t(a, Y, c)\{X/a, Y/b\} = t(X, b, c)\{X/a, Y/b\} = t(a, b, c).$$

In general, expressions can have many unifiers:

Example 2.19

Atoms $p(X, Y)$ and $p(Z, Z)$, have, as unifiers, $\{X/b, Y/b, Z/b\}$, $\{X/c, Y/c, Y/c\}$, and $\{X/Z, Y/Z\}$. The third unifier is more general than the first two, as both have X the same as Z and Y the same as Z, but make more commitments as to what these values are.

Substitution σ is a **most general unifier** (mgu) of expressions e_1 and e_2 if σ is a unifier of the two expressions; and if there is another substitution σ' that is also a unifier of e_1 and e_2, then $e\sigma'$ is an instance of $e\sigma$ for all expressions e.

Expression e_1 is a **renaming** of e_2 if they differ only in the names of variables. In this case they are both instances of each other.

If two terms have a unifier, they have at least one most general unifier. The expressions resulting from applying the most general unifiers to the terms are all renamings of each other.

Example 2.20

$\{X/Z, Y/Z\}$ and $\{Z/X, Y/X\}$ are both most general unifiers of $p(X, Y)$ and $p(Z, Z)$. The resulting applications

$$p(X, Y)\{X/Z, Y/Z\} = p(Z, Z)$$
$$p(X, Y)\{Z/X, Y/X\} = p(X, X)$$

are renamings of each other.

Bottom-Up Procedure with Variables

You can do a bottom-up proof procedure for the case with variables by carrying out the bottom-up procedure (page 47) on the set of all ground instances of the clauses. A **ground instance** of a clause is obtained by uniformly substituting constants appearing in the knowledge base or in any query for the variables in the clause. If there are no constants, one needs to be invented.

Example 2.21

Suppose the knowledge base is

$q(a)$.
$q(b)$.
$r(a)$.
$s(W) \leftarrow r(W)$.
$p(X, Y) \leftarrow q(X) \wedge s(Y)$.

The set of all ground instances is

$q(a)$.
$q(b)$.
$r(a)$.
$s(a) \leftarrow r(a)$.
$s(b) \leftarrow r(b)$.
$p(a, a) \leftarrow q(a) \wedge s(a)$.
$p(a, b) \leftarrow q(a) \wedge s(b)$.
$p(b, a) \leftarrow q(b) \wedge s(a)$.
$p(b, b) \leftarrow q(b) \wedge s(b)$.

The bottom-up proof procedure will derive $q(a)$, $q(b)$, $r(a)$, $s(a)$, $p(a, a)$, and $p(b, a)$.

Example 2.22

Suppose the knowledge base is

$p(X, Y)$.
$g \leftarrow p(W, W)$.

You need to invent a new constant symbol, say c. The set of all ground instances is then

$p(c, c)$.
$g \leftarrow p(c, c)$.

The bottom-up proof procedure will derive $p(c, c)$ and g. Note that if a query contained constants, these constants appear in the knowledge base when you add the answer clause, and so the set of ground instances would change to reflect these constants.

It is easy to see that the procedure is sound, as each instance of each rule is true in every model. This procedure is essentially the same as the variable-free case, but it uses the set of ground instances of the clauses, all of which are true by definition.

This procedure, if fair, is also complete for ground atoms. That is, if the procedure is fair and a ground clause is a consequence of the clauses, it will be eventually derived. To prove this, as in the ground case (page 49), you construct a particular generic model. To construct a model, you have to determine what the constants denote. We use a **Herbrand interpretation** where the domain is symbolic and consists of all constants of the language. You have to invent an individual if there are no constants. In a Herbrand interpretation each constant denotes itself. Consider the Herbrand interpretation where the ground instances of the relations that are eventually derived by the bottom-up procedure with a fair selection rule are true. It is easy to see that this is a model of the rules given. As in the variable-free case (page 47), it is a **minimal model** in that it has the fewest true atoms of any model. If $KB \models g$ for ground atom g, then g is true in the minimal model and thus is eventually derived.

Example 2.23	Consider the clauses of Figure 2.3 (page 44). You can immediately derive each instance of *imm_west* given as a fact. Then you can add the *imm_east* clauses:

> $imm\_east(r103, r101)$
>
> $imm\_east(r105, r103)$
>
> $imm\_east(r107, r105)$
>
> $imm\_east(r109, r107)$
>
> $imm\_east(r111, r109)$
>
> $imm\_east(r129, r131)$
>
> $imm\_east(r127, r129)$
>
> $imm\_east(r125, r127)$

Next, the *next_door* relations that follow can be added, including:

> $next\_door(r101, r103)$
>
> $next\_door(r103, r101)$

The *two_door_east* relations can be added, including:

> $two\_door\_east(r105, r101)$
>
> $two\_door\_east(r107, r103)$

Finally, the *west* relations that follow can be added to C.

To solve query B with variables $V_1, \ldots, V_k$:

Set ac to generalized answer clause $yes(V_1, \ldots, V_k) \leftarrow B$;
While ac is not an answer do
      Suppose ac is $yes(t_1, \ldots, t_k) \leftarrow a_1 \wedge a_2 \wedge \ldots \wedge a_m$
      Select atom a_i in the body of ac;
      Choose clause $a \leftarrow b_1 \wedge \ldots \wedge b_p$ from KB;
      Rename all variables in $a \leftarrow b_1 \wedge \ldots \wedge b_p$ to have new names;
      Let θ be the most general unifier of a_i and a. Fail if they don't unify;
      Set ac to $(yes(t_1, \ldots, t_k) \leftarrow a_1 \wedge \ldots \wedge a_{i-1} \wedge b_1 \wedge \ldots \wedge b_p \wedge$
                        $a_{i+1} \wedge \ldots \wedge a_m)\theta$
end while.

Figure 2.6: Top-down definite clause interpreter, with variables

Definite Resolution with Variables

The top-down derivation procedure can be extended to the case with variables by allowing instances of rules to be used in the derivation.

A **generalized answer clause** is of the form

$$yes(t_1, \ldots, t_k) \leftarrow a_1 \wedge a_2 \wedge \ldots \wedge a_m,$$

where $t_1, \ldots, t_k$ are terms and $a_1, \ldots, a_m$ are atoms. Suppose you select the atom a_i. The **SLD resolution** of the above generalized answer clause on a_i with the chosen clause

$$a \leftarrow b_1 \wedge \ldots \wedge b_p,$$

where a_i and a have most general unifier θ, is the answer clause

$$(yes(t_1, \ldots, t_k) \leftarrow a_1 \wedge \ldots \wedge a_{i-1} \wedge b_1 \wedge \ldots \wedge b_p \wedge a_{i+1} \wedge \ldots \wedge a_m)\theta.$$

An **SLD derivation** is a sequence of generalized answer clauses $\gamma_0, \gamma_1, \ldots, \gamma_n$ such that

- γ_0 is the answer clause corresponding to the original query. If the query is B, with free variables $V_1, \ldots, V_k$, then the initial generalized answer clause γ_0 is

  $$yes(V_1, \ldots, V_k) \leftarrow B.$$

- γ_i is obtained by selecting an atom in the body of γ_{i-1}, choosing a clause in the knowledge base, and resolving γ_{i-1} with a *copy* of the chosen clause, renaming the variables in the clause such that the introduced variables don't appear anywhere else. The difference between this and the previous definition of a derivation is that for the variable case you must take copies of clauses from the knowledge base. This is to remove name clashes between variables. It is important because a single proof may use many different instances of a clause.

- γ_n is an answer. That is, it is of the form

$$yes(t_1, \ldots, t_k) \leftarrow .$$

When this occurs you have the answer $V_1 = t_1, \ldots, V_k = t_k$.

This shows how **answer extraction for definite clauses** can be carried out as a side effect of proving the query. The arguments to *yes* keep track of the answers to the initial query.

Example 2.24

Consider the database of Figure 2.3 (page 44), and the query

$$?two\_doors\_east(R, r107).$$

The following is a derivation:

$$yes(R) \leftarrow two\_doors\_east(R, r107)$$
$$\text{resolve with } two\_doors\_east(E_1, W_1) \leftarrow$$
$$imm\_east(E_1, M_1) \wedge imm\_east(M_1, W_1).$$
$$\text{substitution: } \{E_1/R, W_1/r107\}$$
$$yes(R) \leftarrow imm\_east(R, M_1) \wedge imm\_east(M_1, r107)$$
$$\text{select leftmost conjunct}$$
$$\text{resolve with } imm\_east(E_2, W_2) \leftarrow imm\_west(W_2, E_2)$$
$$\text{substitution: } \{E_2/R, W_2/M_1\}$$
$$yes(R) \leftarrow imm\_west(M_1, R) \wedge imm\_east(M_1, r107)$$
$$\text{select leftmost conjunct}$$
$$\text{resolve with } imm\_west(r109, r111)$$
$$\text{substitution: } \{M_1/r109, R/r111\}$$
$$yes(r111) \leftarrow imm\_east(r109, r107)$$
$$\text{resolve with } imm\_east(E_3, W_3) \leftarrow imm\_west(W_3, E_3)$$
$$\text{substitution: } \{E_3/r109, W_3/r107\}$$
$$yes(r111) \leftarrow imm\_west(r107, r109)$$
$$\text{resolve with } imm\_west(r107, r109)$$
$$\text{substitution: } \{\}$$
$$yes(r111) \leftarrow$$

An answer is thus $R = r111$.

Note that this derivation used two different instances of the rule

$$imm\_east(E, W) \leftarrow imm\_west(W, E).$$

One eventually substituted $r111$ for E, and one substituted $r109$ for E.

Some choices of which clauses to resolve against would have resulted in a partial derivation which could not be completed.

2.8   Extending the Language with Function Symbols

Datalog requires you to name, using a constant, every individual you may want to reason about. Often you want to be able to identify an individual in terms of components, rather than needing a separate constant for each individual.

Example 2.25

In many domains you want to be able to refer to a time as an individual. You may want to say that some course is held at 11:30 a.m. You don't want to have to choose a constant for each different time. It is better to define times in terms of, say, the number of hours past midnight and the number of minutes past the hour. Similarly, you may want to reason with facts which mention particular dates. You don't want to give a constant for each date. It is easier to define a date in terms of the year, the month, and the day.

Using a constant to name each individual also means that you can only name a finite number of individuals, and the number of individuals is fixed when designing the knowledge base. There are, however, many cases where you want to reason about a potentially infinite set of individuals.

Example 2.26

Suppose you want to build an infobot that takes questions in English and that answers them by consulting an online database. In this case, each sentence can be considered to be an individual. You don't want to have to give each sentence its own name, as there are too many English sentences to expect to name them all. It may be better to name the words and then to specify a sentence in terms of the sequence of words in the sentence. This may be more practical as there are far fewer words to name than sentences, and each word has it own natural name.

Example 2.27

You may want to reason about lists of students. For example, the infobot may need to derive the average mark of a class of students. A class list of students is an individual that has properties, such as its length, and its seventh element. While it is possible to name each list if you restrict yourself to lists of bounded length made up of a finite number of elements, it is very inconvenient to do so. It is much better to have a way to describe particular lists in terms of their elements.

Function symbols allow you to describe individuals indirectly. Rather than using a constant to describe an individual, it can be described in terms of its properties.

Syntactically a **function symbol** is a word starting with a lowercase letter. We extend the definition of a term (page 30) so that a **term** is either a variable, a constant, or of the form $f(t_1, \ldots, t_n)$, where f is a function symbol and each t_i is a term. Apart from extending the definition of terms, the language stays the same.

Note that terms only appear within predicate symbols. You do not write clauses that imply terms. You may, however, write clauses that imply relations on function symbols.

The semantics must be changed to reflect the new syntax. The only thing we change is the definition of ϕ (page 32). We extend ϕ so that ϕ is a mapping that assigns to each constant an element of D and to each n-ary function symbol a function from D^n into D. Thus ϕ specifies which mapping each function symbol denotes. In particular, ϕ specifies which individual each ground term denotes.

This slight increase in the language has a major impact. With just one function symbol there are infinitely may different terms, and infinitely many different atoms. The infinite number of terms can be used to describe an infinite number of individuals.

Example 2.28

Suppose you want to define times during the day as in Example 2.25. You can use the function symbol *am* so that $am(H, M)$ denotes the time $H{:}M$ am, when H is an integer between 1 and 12 and M is an integer between 0 and 59. For example, $am(10, 38)$ denotes the time 10:38 am. *am* denotes a function from integers into times. Similarly you can define the symbol *pm* to denote the times after noon.

The only way to use the function symbol is to write clauses that define relations using the function symbol. There is no notion of *defining* the *am* function. One reason is that times are not in the computer more than people are.

To write clauses using function symbols, you can write rules that are quantified over the arguments of the function symbol. For example, the following defines the $before(T_1, T_2)$ relation that is true if time T_1 is before time T_2 in a day.

$$before(am(H1, M1), pm(H2, M2)).$$
$$before(am(12, M1), am(H2, M2)) \leftarrow$$
$$\quad H2 < 12.$$
$$before(am(H1, M1), am(H2, M2)) \leftarrow$$
$$\quad H1 < H2 \wedge$$
$$\quad H2 < 12.$$
$$before(am(H, M1), am(H, M2)) \leftarrow$$
$$\quad M1 < M2.$$

$$before(pm(12, M1), pm(H2, M2)) \leftarrow$$
$$H2 < 12.$$
$$before(pm(H1, M1), pm(H2, M2)) \leftarrow$$
$$H1 < H2 \wedge$$
$$H2 < 12.$$
$$before(pm(H, M1), pm(H, M2)) \leftarrow$$
$$M1 < M2.$$

Function symbols can be used to build data structures.

Example 2.29

A tree is a useful data structure. You could use a tree to build a syntactic representation of a sentence for our natural language processing system. We could decide that a labeled tree is either of the form $node(N, LT, RT)$ or of the form $leaf(L)$. Thus *node* is a function from a name, a left tree, and a right tree into a tree. The function symbol *leaf* denotes a function from node into a tree.

The relation $at\_leaf(L, T)$ is true if label L is the label of a leaf in tree T. It can be defined by

$$at\_leaf(L, leaf(L)).$$
$$at\_leaf(L, node(N, LT, RT)) \leftarrow$$
$$at\_leaf(L, LT).$$
$$at\_leaf(L, node(N, LT, RT)) \leftarrow$$
$$at\_leaf(L, RT).$$

This is an example of a structural recursive program. The rules cover all of the cases for each of the structures representing trees.

The relation $in\_tree(L, T)$ is true if label L is the label of an interior node of tree T. It can be defined by

$$in\_tree(L, node(L, LT, RT)).$$
$$in\_tree(L, node(N, LT, RT)) \leftarrow$$
$$in\_tree(L, LT).$$
$$in\_tree(L, node(N, LT, RT)) \leftarrow$$
$$in\_tree(L, RT).$$

More examples of using data structures are given in Section 3.5.

Proof Procedures with Function Symbols

The proof procedures with variables carry over for the case with function symbols. The main difference is that the class of terms is expanded with function symbols.

With function symbols there are infinitely many terms. This means that, when forward chaining on the clauses we have to ensure that the selection criteria for selecting clauses is fair (page 50).

Example 2.30
> To see why fairness is important, consider the following clauses as part of a larger program:
>
> $$num(s(N)) \leftarrow num(N).$$
> $$num(0).$$
>
> An unfair strategy could always choose the first of these clauses to forward chain on. The first clause can always be used to derive a new consequence. It would never select the other clauses and thus never derive the consequences of these other clauses.

This problem of ignoring some clauses forever is known as **starvation**. A **fair** selection criteria is one such that any clause that is available to be chosen will eventually be chosen. The bottom-up proof procedure is complete only if it is fair.

The top-down proof procedure uses the most general unifier as does the case with variables. You have to be careful of one thing. A variable X does not unify with a term t in which X occurs and is not X itself. If these are allowed to unify, the proof procedure becomes unsound, as the following example shows:

Example 2.31
> Suppose you want to unify X and $s(X)$. You can't ensure that the intended interpretation is not the domain of natural numbers where the function symbol s denotes the successor function, where $s(n)$ is the number after n. In this interpretation there is no number X which is equal to its successor. Without the occurs check, X and $s(X)$ unify. This makes the proof procedure unsound:
> Consider the knowledge base with only one fact:
>
> $$lt(X, s(X)).$$
>
> Suppose the intended interpretation is the domain of integers with lt meaning "less than" and $s(X)$ denoting the integer after X. The query $?lt(X, X)$ should fail as it is false in our intended interpretation that there is no number less than itself. However, if X and $s(X)$ could unify, this query would succeed. In this case the proof procedure would be unsound as something could be derived that is false in a model of the axioms.

The following example shows details of SLD resolution with function symbols.

Example
2.32

Consider the clauses

$$append(c(A, X), Y, c(A, Z)) \leftarrow$$
$$append(X, Y, Z).$$
$$append(nil, Z, Z).$$

For now, ignore what this may mean. Like the computer, treat this as a problem of symbol manipulation. Consider the query:

$$?append(F, c(L, nil), c(l, c(i, c(s, c(t, nil)))))$$

The following is a derivation:

$yes(F, L) \leftarrow append(F, c(L, nil), c(l, c(i, c(s, c(t, nil)))))$
  resolve with $append(c(A_1, X_1), Y_1, c(A_1, Z_1)) \leftarrow append(X_1, Y_1, Z_1)$
  substitution: $\{F/c(l, X_1), Y_1/c(L, nil), A_1/l, Z_1/c(i, c(s, c(t, nil)))\}$
$yes(c(l, X_1), L) \leftarrow append(X_1, c(L, nil), c(i, c(s, c(t, nil))))$
  resolve with $append(c(A_2, X_2), Y_2, c(A_2, Z_2)) \leftarrow append(X_2, Y_2, Z_2)$
  substitution: $\{X_1/c(i, X_2), Y_2/c(L, nil), A_2/i, Z_2/c(s, c(t, nil))\}$
$yes(c(l, c(i, X_2)), L) \leftarrow append(X_2, c(L, nil), c(s, c(t, nil)))$
  resolve with $append(c(A_3, X_3), Y_3, c(A_3, Z_3)) \leftarrow append(X_3, Y_3, Z_3)$
  substitution: $\{X_2/c(s, X_3), Y_3/c(L, nil), A_3/s, Z_3/c(t, nil)\}$
$yes(c(l, c(i, c(s, X_3))), L) \leftarrow append(X_3, c(L, nil), c(t, nil))$

At this stage both clauses are applicable. Choosing the first clause gives

  resolve with $append(c(A_4, X_4), Y_4, c(A_4, Z_4)) \leftarrow append(X_4, Y_4, Z_4)$
  substitution: $\{X_3/c(t, X_4), Y_4/c(L, nil), A_4/t, Z_4/nil\}$
$yes(c(l, c(i, c(s, X_3))), L) \leftarrow append(X_4, c(L, nil), nil)$

At this point there are no clauses whose head unifies with the generalized answer clause's body. The proof fails.
  Choosing the second clause instead of the first gives:

  resolve with $append(nil, Z_5, Z_5)$.
  substitution: $\{Z_5/c(t, nil), X_3/nil, L/t\}$
$yes(c(l, c(i, c(s, nil))), t) \leftarrow$

At this point the proof succeeds, with answer $F = c(l, c(i, c(s, nil)))$, $L = t$.

2.9   References and Further Reading

The semantics presented here is due to Tarski in the early 1930s (Tarski, 1956). For introductions to logic see Copi (1982) for an informal overview, Enderton (1972) and Mendelson (1987) for more formal approaches and Bell & Machover (1977) for advanced topics. For in-depth discussion of the use of logic in AI see the multivolume *Handbook of Logic in Artificial Intelligence and Logic Programming* (Gabbay, Hogger & Robinson, 1993).

The notion of resolution and unification is due to Robinson (1965). SLD resolution, and the idea of logic as a programming language is due to Kowalski (1974) and Colmerauer, Kanoui, Roussel & Pasero (1973), building on previous AI work by Green (1969), Hayes (1973), and Hewitt (1969). The fixed point semantics of logic programs is due to van Emden & Kowalski (1976). For more detail on the semantics and formal properties of logic programs see Lloyd (1987).

The distinction between logic and control comes from Kowalski (1979b), which is also the distinction between epistemic and heuristic knowledge of McCarthy & Hayes (1969).

2.10   Exercises

Exercise 2.1

Given the knowledge base:

$$a \leftarrow b \wedge c.$$
$$a \leftarrow e \wedge f.$$
$$b \leftarrow d.$$
$$b \leftarrow f \wedge h.$$
$$c \leftarrow e.$$
$$d \leftarrow h.$$
$$e.$$
$$f \leftarrow g.$$
$$g \leftarrow c.$$

(a) Give a model of the model of the knowledge base.

(b) Give an interpretation that is not a model of the knowledge base.

(c) Give two atoms that are logical consequences of the knowledge base.

(d) Give two atoms that are not logical consequences of the knowledge base.

Exercise 2.2 Consider the language that contains the constant symbols a, b, and c; the predicate symbols p and q; and no function symbols. We might also have the following knowledge bases built from this language:

$$KB_1 = \{\ p(a)\ \}.$$
$$KB_2 = \{\ p(X) \leftarrow q(X)\ \}.$$
$$KB_3 = \{\ p(X) \leftarrow q(X),$$
$$p(a),$$
$$q(b)\ \}.$$

Now consider possible interpretations for this language of the form $I = \langle D, \pi, \phi \rangle$, where D consists of exactly four domain elements, w, x, y, and z.

(a) How many interpretations with the four domain elements exist for our simple language? Give a brief justification for your answer. Hint: Consider how many possible assignments ϕ exist for the constant symbols, and consider how many extensions predicates p and q can have to determine how many assignments π exist. Don't try to enumerate all possible interpretations.

(b) Of the interpretations outlined above, how many are models of KB_1? Give a brief justification for your answer.

(c) Of the interpretations outlined above, how many are models of KB_2? Give a brief justification for your answer.

(d) Of the interpretations outlined above, how many are of KB_3? Give a brief justification for your answer.

Exercise 2.3 Given the knowledge base KB containing the clauses:

$$a \leftarrow b \wedge c.$$
$$b \leftarrow d.$$
$$b \leftarrow e.$$
$$c.$$
$$d \leftarrow h.$$
$$e.$$
$$f \leftarrow g \wedge b.$$
$$g \leftarrow c \wedge k.$$
$$j \leftarrow a \wedge b.$$

(a) Show how the bottom-up proof procedure works for this example. Give all logical consequences of KB.

(b) f isn't a logical consequence of KB. Give a model of KB in which f is false.

(c) a is a logical consequence of KB. Give a top-down derivation for the query $?a$.

Exercise 2.4 Consider the following knowledge base:

$$r(a).$$

$r(e)$.

$p(c)$.

$q(b)$.

$s(a, b)$.

$s(d, b)$.

$s(e, d)$.

$p(X) \leftarrow q(X) \wedge r(X)$.

$q(X) \leftarrow s(X, Y) \wedge q(Y)$.

Show the set of ground atomic consequences derivable from this KB. Assume that a bottom-up proof procedure is used and that at each iteration the first applicable clause is selected in the order shown. Furthermore, applicable constant substitutions are chosen in "alphabetic order" if more than one applies to a given clause; for example, if X/a and X/b are both applicable for a clause at some iteration, derive $q(a)$ first. In what order are consequences derived?

Exercise 2.5   In Example 2.24 (page 57) we fortuitously chose $imm\_west(r109, r111)$ as the clause to resolve against. What would have happened if we had chosen another clause? Show the sequence of resolutions that arise, and either show a different answer or give a generalized answer clause that can't resolve with any clause in the knowledge base.

Exercise 2.6   In Example 2.24, we always selected the leftmost conjunct to resolve on. Is there a selection rule (a selection of which conjunct in the query to resolve against) that would have resulted in only one choice for this example? Give a general rule that, for this example at least, results in fewer failing branches being made. Give an example where this rule fails.

Exercise 2.7   In a manner similar to Example 2.24, show derivations of the queries:

(a) $?two\_doors\_east(r107, R)$.

(b) $?next\_door(R, r107)$.

(c) $?west(R, r107)$.

(d) $?west(r107, R)$.

Give all answers for each query.

Exercise 2.8   Consider the following knowledge base:

$has\_access(X, library) \leftarrow student(X)$.

$has\_access(X, library) \leftarrow faculty(X)$.

$has\_access(X, library) \leftarrow has\_access(Y, library) \wedge parent(Y, X)$.

$has\_access(X, office) \leftarrow has\_keys(X)$.

$faculty(diane)$.

$faculty(ming)$.

> *student(william).*
>
> *student(mary).*
>
> *parent(diane, karen).*
>
> *parent(diane, robyn).*
>
> *parent(susan, sarah).*
>
> *parent(sarah, ariel).*
>
> *parent(karen, mary).*
>
> *parent(karen, todd).*

(a) Provide a top-down SLD derivation of the query ?*has_access(todd, library)*.

(b) The query ?*has_access(mary, library)* has two straightforward, but quite distinct SLD derivations. Give both of them.

(c) Does there exist an SLD derivation for ?*has_access(ariel, library)*? Briefly, why or why not?

(d) Argue that the set of answers to the query ?*has_access(X, office)* is empty.
    If the clause

$$has\_keys(X) \leftarrow faculty(X).$$

is added to KB, what is the set of answers to this query?

Exercise 2.9   What is the result of the following application of substitutions:

(a) $f(A, X, Y, X, Y)\{A/X, Z/b, Y/c\}$.

(b) $yes(F, L) \leftarrow append(F, c(L, nil), c(l, c(i, c(s, c(t, nil)))))$
    $\{F/c(l, X_1), Y_1/c(L, nil), A_1/l, Z_1/c(i, c(s, c(t, nil)))\}$.

(c) $append(c(A_1, X_1), Y_1, c(A_1, Z_1)) \leftarrow append(X_1, Y_1, Z_1)$
    $\{F/c(l, X_1), Y_1/c(L, nil), A_1/l, Z_1/c(i, c(s, c(t, nil)))\}$.

Exercise 2.10   Give a most general unifier of the following pairs of expressions:

(a) $p(f(X), g(g(b)))$ and $p(Z, g(Y))$

(b) $g(f(X), r(X), t)$ and $g(W, r(Q), Q)$

(c) $bar(val(X, bb), Z)$ and $bar(P, P)$

Exercise 2.11   List all of the ground atomic logical consequence of the following KB:

$$q(Y) \leftarrow s(Y, Z) \wedge r(Z).$$
$$p(X) \leftarrow q(f(X)).$$
$$s(f(a), b).$$
$$s(f(b), b).$$
$$s(c, b).$$
$$r(b).$$

Exercise 2.12   Consider the following logic program:

$f(empty, X, X).$

$f(cons(X, Y), W, Z) \leftarrow$
$\quad f(Y, W, cons(X, Z)).$

Give each top-down derivation, showing substitutions (as in Example 2.32) for the query:

$?f(cons(a, cons(b, cons(c, empty))), L, empty).$

What are all of the answers?

Exercise 2.13   Consider the following logic program:

$rd(cons(H, cons(H, T)), T).$

$rd(cons(H, T), cons(H, R)) \leftarrow$
$\quad rd(T, R).$

Give a top-down derivation, showing all substitutions for the query

$?rd(cons(a, cons(cons(a, X), cons(B, cons(c, Z)))), W).$

What is the answer corresponding to this derivation?

Is there a second answer? If there is one, show the derivation; if not, explain why.

Chapter 3

Using Definite Knowledge

3.1 Introduction

Chapter 2 presented the theory of reasoning with definite clauses. This chapter presents the representational methodology for that RRS. It shows how that theory can be used to represent a domain, answer questions, and solve problems in the domain.

An RRS (page 23) was defined as consisting of a syntax, a semantics, a reasoning theory, and a programming methodology. This final component is largely pragmatic, but important. A methodology for using an RRS means that programming experience can be passed along so that new users don't need to reinvent a methodology from scratch. It makes the difference between an RRS that is useful, as opposed to merely theoretically interesting. This chapter provides some guidelines for using the simple definite clause RRS. Essentially we will follow the steps of the users' view of semantics (page 37) for many different applications.

The definite clause language can be viewed in at least four ways, as:

- A database language
- A question-answering system
- A programming language
- A representation and reasoning system

This chapter describes each of these and gives extended examples in knowledge-based systems and natural language understanding.

This chapter presents a number of detailed examples of the RRS in use; for knowledge-based reasoning tasks, representing abstract concepts, and natural language processing. What is important about such applications is that you have an intended interpretation for each symbol and can interpret the clauses as statements about the world you are reasoning about. The types of questions you have to consider in order to build such applications include:

- What are the concepts and individuals you want to represent?
- At what level of detail do you want to represent the world (e.g., when representing knowledge about universities, do you want to represent at the level of individual courses or at the level of detail of individual lectures)?
- Is each clause true in the intended interpretation? If this is the case and the proof procedure is sound, all answers will be true in the intended interpretation.
- Do the rules for the predicates cover all of the cases?

More issues relevant to representing knowledge will be presented in Chapter 5. This chapter concentrates on how to use the above questions to guide the development of a knowledge base.

All of the examples here can be executed by Prolog, which is an implementation of the definite clause RRS that embodies one possible strategy for finding proofs. (Pure) Prolog (see Appendix B) is an implementation of the top-down proof procedure that always selects the leftmost conjunct and does a depth-first search (see Chapter 4). There is, however, nothing in this chapter that relies on this top-down, left-to-right, backtracking strategy.

3.2 Case Study: House Wiring

Our first example will be to axiomatize the electrical environment of Figure 1.3 (page 16), following the methodology for the user's view of semantics (page 37). This axiomatization will allow us to simulate the electrical system, and it will be expanded in later chapters to let us diagnose faults based on observed symptoms. First, however, we will use it to demonstrate various features of the RRS and to demonstrate the methodology for using the RRS.

Assume that the goal of the modeling is to be able to determine whether lights are on or off, based on switch positions and the status of circuit breakers, and, eventually, to be able to diagnose what's wrong with wires, switches, and circuit breakers if something unexpected is observed. Assume you aren't concerned here with the color of the wires, the design of the switches, the length or weight of the wire, the date of manufacture of the lights and the wires, or any of the other myriad of detail one could imagine about the domain. These can be represented in a manner similar to what's presented here.

You need to choose a level of abstraction. You should model the domain at the most general level that will enable you to solve the problems you are interested in.

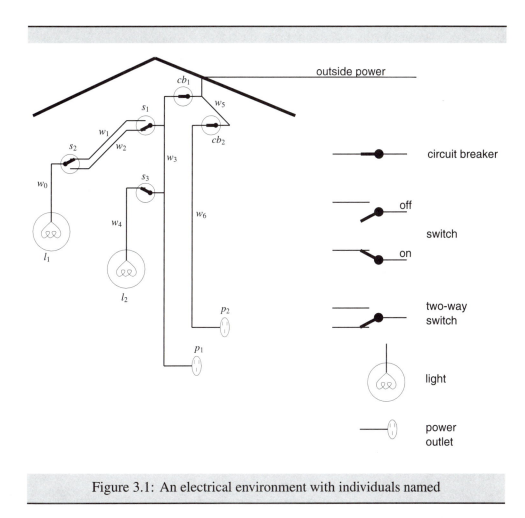

Figure 3.1: An electrical environment with individuals named

You also want to model the domain at a level that you have information about. For example, you could model the domain in terms of the actual voltages and currents, but this would be inappropriate as you don't know them. High-level reasoning can be done without this much detail. Exactly the same reasoning would be done if this were a 12-volt DC system or a 120-volt AC system; the voltages and frequencies are irrelevant. There may be situations for which you actually need the voltages and currents—for example, if you want to know how much the lights will dim if you plug in the overhead projector. What is important is that you choose the level of description that is appropriate. These issues are discussed in more detail in Chapter 5.

Let's model this domain at a common-sense level that non-electricians may use to describe the domain in terms of wires being live and currents flowing from the outside through wires to the lights, and that circuit breakers and light switches connect and disconnect wires. At this level of abstraction, we ignore Kirchhoff's laws and just model the live wires, ignoring the return wires.

The first thing you must do is decide what are the *individuals* in the domain. For the electrical environment you can treat each wire as an individual, as well as each switch, each light, and each power outlet. Then you need to give names to each of these individuals. This is done in Figure 3.1. For example, the individual w_0 is the wire between light l_1 and switch s_2.

Next choose which relationships you want to represent. Assume the following predicates with their associated intended interpretations:

- *light*(L) is true if the individual denoted by L is a light.
- *lit*(L) is true if the light L is lit, and emitting light.
- *live*(W) is true if there is power coming into W—W is live.
- *up*(S) is true if switch S is up.
- *down*(S) is true if switch S is down.
- *ok*(E) is true if E isn't blown; E can either be a circuit breaker or a light.
- *connected_to*(X, Y) is true if component X is connected to Y such that current would flow from Y to X.

At this stage you haven't told the computer anything. It doesn't know what the predicates are, let alone what they mean! Nor does it know what the individuals are or their names.

Once you have decided which symbols you want to use and what they denote, you can write axioms about what's true in the world, in the form of clauses. The simplest forms of clauses are those without bodies, or facts, which form a database such as

$light(l_1)$.
$light(l_2)$.
$down(s_1)$.
$up(s_2)$.
$up(s_3)$.
$ok(l_1)$.
$ok(l_2)$.
$ok(cb_1)$.
$ok(cb_2)$.

Once the computer has been told this, it can answer queries such as

$?light(l_1)$.

for which the answer is *yes*. The query

$?light(l_6)$.

has no answer. The computer doesn't have enough information to know whether l_6 is a light or not.

You can also ask the computer other queries such as "what's up?":

$?up(X).$

which has two answers, $X = s_2$ and $X = s_3$. There are two things that the computer knows are up, namely s_2 and s_3.

The next simplest rules are those where what's true depends on other truths. For example, which wires are connected depends upon the status of switches and circuit breakers:

$connected\_to(l_1, w_0).$

$connected\_to(w_0, w_1) \leftarrow up(s_2).$

$connected\_to(w_0, w_2) \leftarrow down(s_2).$

$connected\_to(w_1, w_3) \leftarrow up(s_1).$

$connected\_to(w_2, w_3) \leftarrow down(s_1).$

$connected\_to(l_2, w_4).$

$connected\_to(w_4, w_3) \leftarrow up(s_3).$

$connected\_to(p_1, w_3).$

$connected\_to(w_3, w_5) \leftarrow ok(cb_1).$

$connected\_to(p_2, w_6).$

$connected\_to(w_6, w_5) \leftarrow ok(cb_2).$

$connected\_to(w_5, outside).$

Given such a conditional database, you can ask queries such as "what's l_1 connected to?":

$?connected\_to(l_1, W).$

which has one answer, $W = w_0$.

You can ask, "what's w_0 connected to?":

$?connected\_to(w_0, W).$

which has one answer, $W = w_1$. w_0 is connected to w_1 because switch s_2 is up.

You can ask, "what's w_1 connected to?":

$?connected\_to(w_1, W).$

which has no answer. The computer doesn't know anything about what w_1 is connected to.

You can ask, "what connects to w_3?":

$?connected\_to(Y, w_3).$

which has three answers, $Y = w_2$, $Y = w_4$, and $Y = p_1$.

You can also ask more general questions, such as what's connected to what:

   $?connected\_to(X, W)$.

which has 10 answers, including $X = l_1$, $W = w_0$, which means $connected\_to(l_1, w_0)$ is a logical consequence of the knowledge base and $X = w_0$, $W = w_1$ which means $connected\_to(w_0, w_1)$ is a logical consequence of the knowledge base.

You can use variables to express general rules such as a light is lit when there is power coming into it and it's OK:

   $lit(L) \leftarrow$

      $light(L) \wedge$

      $ok(L) \wedge$

      $live(L)$.

$light(L)$ is in the body to make sure that the rule is only applied to lights—you don't want to be able to derive that power outlet p_1 is lit. You also need the light to be ok to be able to conclude it's lit.

Just as other programming languages have loops, you can write recursive definitions (page 78) such as defining $live$:

   $live(Y) \leftarrow$

      $connected\_to(Y, Z) \wedge$

      $live(Z)$.

   $live(outside)$.

Under this definition, you can now prove $live(w_5)$, $live(w_3)$, $live(w_4)$, $live(l_2)$, and so also $lit(l_2)$. You can prove $live(w_2)$ but can't prove either $live(w_0)$ or $lit(l_1)$.

The query

   $?live(W)$.

has 9 answers, namely $W = w_2$, $W = l_2$, $W = w_4$, $W = p_1$, $W = w_3$, $W = p_2$, $W = w_6$, $W = w_5$, and $W = outside$.

One advantage of the above representation is that the knowledge is modular. If you were to change the knowledge base so that s_1 were up instead of down, you would be able to prove that l_1 would be lit. Similarly, to extend the example to include the overhead projector, you need only write the local rules about how the overhead projector works, and hopefully don't need to change any of the other clauses (see Exercise 3.3).

The axiomatization is used here for simulation, namely determining what the output would be from the inputs. In later chapters the same axiomatization is used for diagnosis, which is determining the state of components based on given inputs and observed outputs.

3.3   Databases and Recursion

The language presented so far is closely related to other computer science and mathematical concepts. This section shows how the definite clause language is related to database concepts such as projection, union, join, and so on. We also explain how loops can be written using recursion, and we explain the relationship to mathematical induction.

It is important to understand the relationships to these concepts: Those who are familiar with these concepts can relate the language and methodology with what they have learned before. Those who aren't already familiar with the concepts should still be able to grasp the ideas.

Database Operations

Much modern database theory is defined in terms of the relational algebra and the relational calculus. The language presented in Section 2.4 on page 29 can be seen as a basis for the relational algebra, which also includes recursion.

A **relational database**, or just a **database**, is a knowledge base of ground facts.

| Example 3.1 | An example database is given in Figure 3.2. Giving the intended interpretation for the symbols is a good way to document the program. You should also describe the denotation for the constants and the function symbols if there were any. Good documentation in terms of what the symbols denote means that a user can verify the truth of the program by, for example, checking whether the course denoted by 312 is actually offered in the department denoted by *comp_science*. |

You can use datalog rules to define the relational algebra database operations:

Selection   The use of a constant or variables repeated in a query or in a body corresponds to *selection* in the relational calculus. You can do more sophisticated selection by conjoining arbitrary conditions on the variables to the query. For example, in

$$cs\_course(C) \leftarrow department(C, comp\_science).$$
$$math\_course(C) \leftarrow department(C, math).$$

the first clause *selects* those instances of the department relation whose second argument is *comp_science*, and the *math_course(C)* relation is the *selection* of the courses whose department is Mathematics.

% *course*(*C*) is true if *C* is a university course.

> *course*(312).
> *course*(322).
> *course*(315).
> *course*(371).

% *department*(*C*, *D*) is true if course *C* is offered in department *D*.

> *department*(312, *comp_science*).
> *department*(322, *comp_science*).
> *department*(315, *math*).
> *department*(371, *physics*).

% *student*(*S*) is true if *S* is a student.

> *student*(*mary*).
> *student*(*jane*).
> *student*(*john*).
> *student*(*harold*).

% *female*(*P*) is true if person *P* is female.

> *female*(*mary*).
> *female*(*jane*).

% *enrolled*(*S*, *C*) is true if student *S* is enrolled in course *C*.

> *enrolled*(*mary*, 322).
> *enrolled*(*mary*, 312).
> *enrolled*(*john*, 322).
> *enrolled*(*john*, 315).
> *enrolled*(*harold*, 322).
> *enrolled*(*mary*, 315).
> *enrolled*(*jane*, 312).
> *enrolled*(*jane*, 322).

Figure 3.2: Sample database

Union The union of relations is defined by using multiple rules with the same head. Each body defines a relation, and the relation defined by the head is the union of all of the rules. For example, in

$$cs\_or\_math\_course(C) \leftarrow cs\_course(C).$$
$$cs\_or\_math\_course(C) \leftarrow math\_course(C).$$

cs_or_math_course is the *union* of the relations *cs_course* and *math_course*.

Join The conjunction of two relations in the body of a clause defines a database *join*. The join is on the shared variables; that is you must find two instances of the relations such that the values assigned to the same variables unify. For example, the query

$$?enrolled(S, C) \wedge department(C, D).$$

is the join of the *enrolled* and *department* relations on the course attribute.

Projection When there are variables in the body of a clause that don't appear in the head, then you say that the relation is projected onto those variables in the head. For example, in

$$in\_dept(S, D) \leftarrow enrolled(S, C) \wedge department(C, D).$$

the relation *in_dept* is the *projection* onto the student and department attributes of the join of the *enrolled* and *department* relations.

Example 3.2

With the database of Figure 3.2, and the predicates defined in this section, the query

$$?cs\_course(C)$$

has two answers: $C = 312$ and $C = 322$. The query

$$?cs\_or\_math\_course(C)$$

has three answers: $C = 312$, $C = 322$, and $C = 315$.

The join of the *enrolled* and *department* relations given above has eight answers. The query:

$$?in\_dept(S, D)$$

has six different answers. Note that for some answers such as $S = mary$, $D = comp\_science$, there are two distinct proofs (one involving 322 and one involving 312). SLD resolution (page 56) finds answers corresponding to both of these proofs.

Example 3.3

In Figure 2.3 (page 44), the relation *next_door* is the union of the relations *imm_east* and *imm_west*. In the body of the definition of *two_doors_east*, the relation *imm_east* is joined to itself, where the second argument to the first *imm_east* must be the same as the first argument to the second *imm_east*. In the definition of *two_doors_east* in Figure 2.3, the body is a relation of three variables E, M, and W, and you project this relation onto E and W to define the *two_doors_east* relation.

Recursion and Mathematical Induction

The idea behind writing rules is to define a predicate in terms of something simpler: To solve a hard problem, break it down into smaller problems. The idea of **recursion** is to define a predicate in terms of simpler instances of itself, where *simpler* means easier to prove.

Example 3.4

In Figure 2.3 (page 44), *west* is defined recursively. The base case—where the recursion stops—is when W is immediately west of E. Other rules define *west* in terms of a simple instances of itself, where simpler means one step closer to the base case.

In the definition of *live* (page 74), a simpler instance of the *live* predicate is one step closer to outside. Thus you prove $live(Y)$ by finding a Z that is also live, but is closer to the outside. At each stage through the recursion you get one step closer to the outside. The base case is that the outside wire is live.

The general idea of recursion is to have what's called a *well-founded ordering* of instances of relations such that each relation is defined in terms of elements lower in the ordering, and that each decreasing chain eventually reaches an element that is simplest in the ordering—defined by a clause with no body.

Recursion is a way to view mathematical induction top-down. Mathematical induction is a proof procedure which is used to prove universal conclusions about countable structures (e.g., the integers). To prove a theorem about all natural numbers (non-negative integers), you can first prove it's true for 0—the base case—and then prove that if it's true for all integers less than n, then it's true for n—the inductive step. Each recursive definition provides a way to prove the instances of the query by mathematical induction. You need to provide a base case, along with rules that specify how one instance of a relation is implied by simpler instances of relations—those that are lower in the well-founded ordering.

Example 3.5

Continuing Example 3.4, in the definition of *west* in Figure 2.3, *imm_west* provides the base case, and the second rule for *west* gives the inductive step. In terms of the integers, you are doing induction on the number n of steps west the two rooms are. The base case for $n = 1$ is when the rooms are immediately west of each other. The inductive rule specifies how if *west* can be proved for two rooms $n - 1$ steps west of each other and the appropriate *imm_west* instance can be proved, then *west* can be inferred for rooms n steps apart.

The definition of *live* (page 74), is a specification of the mathematical induction proof of $live(W)$ by induction on the number of steps away from the outside. The base case that that the outside is live. If you have proved $live(Z)$ at n steps from the outside, the rule specifies how to derive $live(Y)$ where Y is $n + 1$ steps from the outside.

3.4   Verification and Limitations

You can see the power of the use of semantics when it comes to verifying—proving correct—logic programs. We can use the basic notions of semantics (see page 37) as a basis for verifying logic programs. Related ideas are discussed later when we consider debugging logic programs (page 221). The same ideas of semantics demonstrate profound limitations of the definite clause logic as well.

Verification of Logic Programs

Part of the power provided by representing knowledge with a language with a logical semantics is evident when you attempt to argue the correctness of a set of clauses viewed as a logic program. To establish correctness, you argue that each clause is true in the intended interpretation. Because of the properties established for the reasoning procedure, you know that answers constructed by a sound proof procedure will also be true in the intended interpretation.

Example
3.6

To check the correctness of the clauses in Section 3.2, you need to know what the symbols mean, and you need to be able to access the world to determine if claimed truths are indeed true. For example, if you know the meaning of the predicate *light* and know the denotation of l_1, you can check whether $light(l_1)$ is true by determining in the world whether the individual denoted by l_1 is indeed a light. You can check the clause

$$connected\_to(w_0, w_1) \leftarrow up(s_2)$$

by checking whether the wire denoted by w_0 is indeed connected to the wire denoted by w_1 if the switch denoted by s_2 is up.

This verifiability of logic programs is the prime motivation behind using semantics (page 31). Note that this verifiability means that you have checked that any answer produced is correct. If an incorrect answer—one that is false in the intended interpretation— is returned and the proof procedure is sound, then one of the input clauses must be wrong—false in the intended interpretation. If you can guarantee that the input clauses are true in the intended interpretation and that the proof procedure is sound, you can guarantee that answers will be true in the intended interpretation. This idea can be developed to build algorithmic debuggers for the definite clause RRS (page 221).

This local checking of each clause doesn't guarantee that any answer that could be produced is produced. In order to guarantee this, you have to make sure that the clauses for each predicate cover all of the cases—if an instance of the predicate is true, then one of the clauses is applicable to prove that predicate. You also need to prove that the program halts—that each recursion is well founded (page 78). The semidecidability of the logic means that you may not be able to automatically check whether a given program will halt, but this doesn't mean that you shouldn't take steps to ensure that your programs don't loop forever.

Limitations

In the language and semantics specified so far there is no notion of complete knowledge. (Complete knowledge is introduced as Assumption CK (page 248).) This section outlines some of the consequences of its absence.

Example
3.7

Consider the database of Figure 3.2 (page 76), and the question of whether there is anyone enrolled in course 371. The answer to this question should be "I don't know." There isn't enough information in the database to answer this question. It's quite possible that all of the facts in the database of Figure 3.2 are true and yet there at 57 students enrolled in course 371.

This observation can lead to an argument that there are some predicates that can't be defined from particular knowledge bases.

Example
3.8

Suppose you have a database defining the relation *enrolled*(*S*, *C*), with the intended interpretation that student *S* is enrolled in course *C*. Given such a database, there is no way to define a relation *empty_course*(*C*) with the intended interpretation that the course *C* has no students enrolled in it. Our definite clause logic isn't powerful enough to define such a relation. This is because, given a set of *enrolled* relations, it can never be necessarily true that some course is empty; it may be there is someone enrolled in it, but you don't know who.

Consider course 371 in Figure 3.2. Even though there is no one listed as enrolled in it, it isn't a logical consequence of the database that it's an empty course, as outlined in Example 3.7.

As a programming language, the definite clauses are as powerful as a Turing machine. This means that a sound and complete implementation can compute any computable function. Hence, any computable problem can be encoded so that a solution can be found.

Example
3.9

You could represent the enrolled knowledge in the form *class_list*(*C*, *Ss*), meaning that *Ss* is the list (page 81) of all students enrolled in course *C*. Note that you have explicitly said that all of the students are included in the list. In this representation, *empty_course* is easy to define:

$$empty\_course(C) \leftarrow class\_list(C, nil).$$

where *nil* denotes the empty list. This would mean not only that you can query whether a course is empty, but also that the person axiomatizing the knowledge base must also state that a course has no students enrolled in it—for example, by writing *class_list*(371, *nil*).

For a discussion of a complete knowledge assumption that allows us to infer something from a lack of knowledge, see Section 7.4 (page 248).

3.5 Case Study: Representing Abstract Concepts

This section shows how abstract mathematical constructs can be defined in the definite clause RRS. In some sense, mathematical concepts are simpler than other relations as their meaning is clear cut, and you need not debate whether the clauses are true and cover the cases necessary to define a concept; proving correctness of clauses is more straightforward.

This section defines the concept of a **list**. A list is an ordered sequence of elements. As an abstract data type, you need to be able to access the head of a nonempty list, access the tail of a nonempty list, which is also a list, and construct a list.

Lists are useful for many applications, for example, when representing a university domain, you often want to represent a class list (the list of students enrolled in a class) or the list of courses taken by a student.

To use lists, you have to decide on a representation of lists. You could represent lists by giving each list you may use a name, and then using a relation to specify the elements of the list. This representation of lists is often appropriate and is often the simplest representation available.

Example 3.10

You could define lists in terms of the relations

- $elt(L, I, E)$ that is true if E is the Ith element of list L.
- $size(L, N)$ that is true if list L has N elements.

A class list for course *cs*502 could then be denoted by the constant *cs502_classlist*, and axiomatized using:

$$elt(cs502\_classlist, 0, fred).$$
$$elt(cs502\_classlist, 1, mary).$$

$elt(cs502\_classlist, 2, susan)$.

$elt(cs502\_classlist, 3, harold)$.

$size(cs502\_classlist, 4)$.

Given this representation it's easy to access the members and the size of the list.

Many other representations are possible. One is developed in the rest of this section and another is explored in Exercise 3.7.

One option is to represent the empty list as a constant and then to say that a list is made up of the first element of a list and the rest of the list. Suppose you use

- the constant *nil* to represent the empty list
- the function symbol *cons* so that the term $cons(H, T)$ represents the list whose first element, the **head** of the list, is represented by H and where the rest of the list, the **tail**, is represented by T.

The symbols "*nil*" and "*cons*" are not predefined. They are arbitrary names that we chose to denote the empty list and the list constructor. As far as the computer is concerned, you might as well have chosen the words "*foo*" and "*bar*."

Example 3.11

Given that *nil* denotes the empty list, $cons(harold, nil)$ denotes the list with one element denoted by *harold*. The list $cons(susan, cons(harold, nil))$ denotes a list with two elements. The first element is *susan*, and the tail of the list is the list containing *harold*. The list

$$cons(fred, cons(mary, cons(susan, cons(harold, nil))))$$

denotes a list with four elements; the first element is *fred*, the second element is *mary*, the third element is *susan*, and the fourth element is *harold*.

Given this representation of lists, you can write a predicate $list(L)$ that is true just when L is a list:

$list(nil)$.

$list(cons(H, T)) \leftarrow$

    $list(T)$.

Like most relations on lists, the rules for *list* cover the cases of what a list could be.

Let's now define the ternary relation *append*, such that $append(X, Y, Z)$ is true when X, Y, and Z are lists and Z contains the elements of X, in order, followed by the elements of Y, in order. This can be defined recursively as

$append(nil, Z, Z)$.

$append(cons(A, X), Y, cons(A, Z)) \leftarrow$

    $append(X, Y, Z)$.

Again, we have axiomatized what is true for each case of the first argument to append. *X* can either be *nil* or of the form *cons(A, X)*. The first clause covers the case when *X* is *nil*, and the second clause specifies what happens when *X* is of the form *cons(A, X)*.

This can also be viewed as an inductive definition, based on the length of the first argument of *append*. The first clause defines the case where the first argument is a list of length zero. The second clause defines the case where the length is $n + 1$ in terms of the case where the length is *n*. A top-down reasoning procedure uses the inductive definition as a recursive program. The first clause gives the base case, and the second specifies how an *append* problem can be reduced to a simpler *append* problem: one with a shorter first list. In general, you prove the correctness of such programs by mathematical induction and compute them by recursion.

The definition of *append* doesn't imply any particular style of use:

Example 3.12

The query

$$?append(cons(a, cons(b, nil)), cons(c, cons(d, nil)), L)$$

has the answer $L = cons(a, cons(b, cons(c, cons(d, nil))))$.
The query:

$$?append(cons(a, cons(b, nil)), L, cons(a, cons(b, cons(c, cons(d, nil))))).$$

has answer $L = cons(c, cons(d, nil))$.
The query:

$$?append(A, B, cons(a, cons(b, cons(c, cons(d, nil)))))$$

has five answers:

$A = nil, B = cons(a, cons(b, cons(c, cons(d, nil))))$;
$A = cons(a, nil), B = cons(b, cons(c, cons(d, nil)))$;
$A = cons(a, cons(b, nil)), B = cons(c, cons(d, nil))$;
$A = cons(a, cons(b, cons(c, nil))), B = cons(d, nil)$;
$A = cons(a, cons(b, cons(c, cons(d, nil)))), B = nil$.

Because lists are so useful and because the notation used here is so clumsy, for the rest of this book we will make a special syntactic form for lists. Note that this is purely syntactic; you are just writing *cons* and *nil* in a special way. The empty list, *nil*, is written as a pair of two square brackets, "[ ]." The list with head *H* and tail *T*, *cons(H, T)*, is written [*H*|*T*]. There are two syntactic simplifications:

- $[\alpha | [\beta]]$ is written as $[\alpha, \beta]$.
  Using this rule,
  $[a | [d | nil]]$ is written as $[a, d | nil]$.
  $[a, b, c | [d, e, f]]$ is $[a, b, c, d, e, f]$.
  $[a, b, c | [d, e | F]]$ is $[a, b, c, d, e | F]$.

- [α|nil] is written as [α].

    Using this rule, [a, b, c|nil] is written as [a, b, c].
    Using both rules, [a|[d|nil]] is written as [a, d].

append can be rewritten with this syntax as

> *append*([ ], Z, Z).
> *append*([A|X], Y, [A|Z]) ←
> > *append*(X, Y, Z).

Once you have lists, it's useful to be able to write a procedure to determine whether some element is a member of a list or is not. These can be defined as follows:

member(X, L) is true if X is an element of list L.

> *member*(X, [X|L]).
> *member*(X, [H|R]) ←
> > *member*(X, R).

The member relation can be used in many ways.

Example 3.13

As well as asking whether something is an element of a list, you can ask such queries as

> ?*member*(*size*(*robin*, S), [*size*(*alan*, *big*), *color*(*grass*, *green*),
> > *size*(*robin*, *medium*), *sound*(*dog*, *woof*)]).

This query has one answer: S = *medium*. This is the only instance of the query which follows from the axioms.

You can also write the inverse of member. *notin*(X, L) is true if X isn't an element of list L. This is if X is different from every member of L. Note the relation *different* isn't some special predefined relation; you must write down clauses for its definition (see Exercise 3.11).

> *notin*(A, [ ]).
> *notin*(A, [H|T]) ←
> > *different*(A, H)∧
> > *notin*(A, T).

It can be the case that there are many different axiomatizations of the same relation. Consider the relation *rev*(L, R) that is true if list R contains the same elements as list L, but in reverse order. You can define *rev*, recursively in terms if the structure of the list L, using *append* as follows:

> *rev*([ ], [ ]).

$$rev([H|T], R) \leftarrow$$
$$rev(T, RT) \wedge$$
$$append(RT, [H], R).$$

Consider the length of the proof. A proof for *append* takes time proportional to the length of its first argument. *rev* has the number of iterations that is proportional to the length of the list. This means that the length of the proof is proportional to the square of the length of the list being reversed. This implementation is often referred to as **naive reverse**.

As in most other programming languages, you should be able to reverse a list in time proportional to the length of the list. The following code does this, where *reverse(L, R)* is true if R contains the elements of L, but in reverse order, and *rev3(L, A, R)* is true if R contains the elements of L in reverse order followed by the elements of A.

$$reverse(L, R) \leftarrow$$
$$rev3(L, [\,], R).$$
$$rev3([\,], R, R).$$
$$rev3([H|T], A, R) \leftarrow$$
$$rev3(T, [H|A], R).$$

A proof using *reverse* is linear in the size of the list being reversed. The A in this case is called an **accumulator**. Intuitively, as elements are stripped off the first list, they are added to the accumulator—the list A accumulates the elements that have been taken off the list being reversed.

Another way to look at *rev3(L, A, R)* is that $R - A$ is a difference list.

A **difference list** is a term of the form $L - E$, where L is a list and E is an ending of list L. The elements of L before E are the members of the difference list $L - E$. For example, $[a, b, c, d] - [c, d]$ contains a and b. The difference list $[a, b, c, d] - [a, b, c, d]$ contains no elements. The difference list $[a, b, c, d] - [\,]$ contains a, b, c, and d. Here "$-$" is an infix function symbol.

When E is a free variable, you can treat E as a pointer to the end of the difference list. For example, the difference list $[a, b, c|E] - E$ contains the elements a, b, and c. Similarly, the difference list $[d, e|F] - F$ contains the elements d and e. By treating E as a pointer to the end of the difference list we can add the elements d and e to the end of the list by unifying E with $[d, e|F]$. Then the difference list $[a, b, c, d, e|F] - F$ contains a, b, c, d, and e. We have created this difference list in constant time with one unification, rather than needing to traverse the first list as does *append*. In general we can append difference lists in constant time using the definition:

$$append\_dl(A - B, B - C, A - C).$$

Example
3.14

Consider appending the difference list containing a, b, and c to the difference list containing d and e. This can be done with the query

$$?append\_dl([a, b, c|E] - E, [d, e|F] - F, R).$$

The resulting answer is $R = [a, b, c, d, e|F] - F$. This call also unifies E with $[d, e|F]$, which shows one limitation of difference lists: You can't subsequently append the difference list $[a, b, c|E] - E$ to another difference list that doesn't start with d and e.

In *rev3*, $R - A$ is a difference list that contains the elements of L in reverse order. That is, A is an ending of list R, and the elements of R before A contains the elements of L in reverse order. When A is the empty list, R is the reverse of list L; this is the case in the body of *reverse*. *rev3* is defined inductively on the length of the first list. When L is empty, then $R = A$. When L is of the form $[H|T]$, if reversing T results in $R - [H|A]$, then reversing $[H|T]$ results in $R - A$. In this case, H should be the last element of R before A.

It is worthwhile understanding how difference lists work, as you will see them again when you look at using definite clauses to encode grammars (page 93).

3.6 Case Study: Representing Regulatory Knowledge

The following example shows how to combine databases and sophisticated rules in order to represent nontrivial regulations of a university. It is intended to show a knowledge-based application of the definite clause language which may be appropriate for an infobot.

This example shows the kind of knowledge that can be used to represent regulations about completion of university degrees. The actual courses and regulations are fictional.

Assume you have databases about individual students and about what grades they achieved in different courses. *grade*(*St*, *Course*, *Mark*) is true if student *St* achieved *Mark* in *Course*. An example database for one student is shown in Figure 3.3.

Also assume you have a database containing relations about the structure of the university and about different courses. Such a database is shown in Figure 3.4.

You can use rules to represent degree requirements. Here we give the subset of the degree requirements that are applicable for computer science students. These will be used to illustrate the power of the representation.

satisfied_degree_requirements(*St*, *Dept*) is true if student *St* has completed a four-year degree in department *Dept*. In order to have satisfied these requirements, the

% grade(*St*, *Course*, *Mark*) is true if student *St* achieved *Mark* in *Course*.

    grade(*robin*, *engl*101, 87).
    grade(*robin*, *phys*101, 89).
    grade(*robin*, *chem*101, 67).
    grade(*robin*, *math*101, 77).
    grade(*robin*, *cs*126, 84).
    grade(*robin*, *cs*202, 88).
    grade(*robin*, *geol*101, 74).
    grade(*robin*, *phys*202, 92).
    grade(*robin*, *math*202, 81).
    grade(*robin*, *cs*204, 87).
    grade(*robin*, *hist*101, 66).
    grade(*robin*, *cs*333, 77).
    grade(*robin*, *cs*312, 74).
    grade(*robin*, *math*302, 41).
    grade(*robin*, *math*302, 87).
    grade(*robin*, *cs*304, 79).
    grade(*robin*, *cs*804, 80).
    grade(*robin*, *psyc*303, 85).
    grade(*robin*, *stats*324, 91).
    grade(*robin*, *cs*405, 77).

Figure 3.3: Example database of student record information

% dept(*Dept*, *Fac*) is true if department *Dept* is in faculty *Fac*.

> *dept*(*history*, *arts*).
> *dept*(*english*, *arts*).
> *dept*(*psychology*, *science*).
> *dept*(*statistics*, *science*).
> *dept*(*mathematics*, *science*).
> *dept*(*cs*, *science*).

% course(*Course*, *Dept*, *Level*) is true if *Course* is a course in department *Dept*, at
% level *Level*.

> *course*(*hist*101, *history*, *first*).
> *course*(*engl*101, *english*, *first*).
> *course*(*psyc*303, *psychology*, *third*).
> *course*(*stats*324, *statistics*, *third*).
> *course*(*math*302, *mathematics*, *third*).
> *course*(*phys*101, *physics*, *first*).

Figure 3.4: Example database relations about the university

student must have covered the core courses of the department, satisfied the department's
faculty requirements, fulfilled the electives requirement, and completed enough units.

> *satisfied_degree_requirements*(*St*, *Dept*) ←
> > *covers_core_courses*(*St*, *Dept*) ∧
> > *dept*(*Dept*, *Fac*) ∧
> > *satisfies_faculty_req*(*St*, *Fac*) ∧
> > *fulfilled_electives*(*St*, *Dept*) ∧
> > *enough_units*(*St*, *Dept*).

covers_core_courses(*St*, *Dept*) is true if student *St* has covered the core courses for
department *Dept*.

> *covers_core_courses*(*St*, *Dept*) ←
> > *core_courses*(*Dept*, *CC*, *MinPass*) ∧
> > *passed_each*(*CC*, *St*, *MinPass*).

core_courses(*Dept*, *CC*, *MinPass*) is true if *CC* is the list of core courses for department *Dept*, and *MinPass* is the passing grade for these core courses.

For example, the core courses for a CS degree are *cs202*, *math202*, *cs204*, *cs304*, and *math302*. The student must pass each with at least 65:

$$core\_courses(cs, [cs202, math202, cs204, cs304, math302], 65).$$

passed_each(CL, St, MinPass) is true if student *St* has a grade of at least *MinPass* in each course in list *CL* of courses. This is defined recursively on the length of the list of courses:

> *passed_each*([ ], *S*, *M*).
> *passed_each*([*C*|*R*], *St*, *MinPass*) ←
>   *passed*(*St*, *C*, *MinPass*) ∧
>   *passed_each*(*R*, *St*, *MinPass*).

passed(St, C, MinPass) is true if student *St* has a grade of at least *MinPass* in course *C*.

> *passed*(*St*, *C*, *MinPass*) ←
>   *grade*(*St*, *C*, *Gr*) ∧
>   *Gr* ≥ *MinPass*.

where $X \geq Y$ is true if X is a number that is bigger than the number Y. Assume that someone has axiomatized arithmetic and has provided this predicate (see Exercise 3.10).

The Science faculty requirements are that a student must have passed (have greater than 50% in) *engl101*, *phys101*, *chem101*, *math101*, either *cs101* or *stats101*, and either *biol101*, *geol101*, or *astr101*.

> *satisfies_faculty_req*(*St*, *science*) ←
>   *passed_each*([*engl101*, *phys101*, *chem101*, *math101*], *St*, 50) ∧
>   *passed_one*([*cs126*, *stats101*], *St*, 50) ∧
>   *passed_one*([*biol101*, *geol101*, *astr101*], *St*, 50).

passed_one(CL, St, MinPass) is true if student *St* has a grade of at least *MinPass* in one of the courses in list *CL* of courses.

> *passed_one*(*CL*, *St*, *MinPass*) ←
>   *member*(*C*, *CL*) ∧
>   *passed*(*St*, *C*, *MinPass*).

fulfilled_electives(St, Dept) is true if student *St* has fulfilled the elective requirements for department *Dept*. A student in a science department needs an arts elective and two science electives.

>$fulfilled\_electives(St, SciDept) \leftarrow$
>>$dept(SciDept, science) \land$
>>$has\_arts\_elective(St) \land$
>>$has\_sci\_elective(St, SciDept, Sci1) \land$
>>$has\_sci\_elective(St, SciDept, Sci2) \land$
>>$different(Sci1, Sci2).$

$different(C_1, C_2)$ is true if C_1 is a different course to course C_2. Assume that this is axiomatized as part of the database. You could imagine assuming that different names refer to different individuals (page 238), but this is sometimes not a valid assumption. In general $different$ can be nontrivial to axiomatize when, for example, a course has two different numbers (for example, when the same course is taught in two different departments, each with their own number for the course).

$has\_arts\_elective(St)$ is true if student St has passed an arts course (other than engl101).

>$has\_arts\_elective(St) \leftarrow$
>>$passed(St, ArtsEl, 50) \land$
>>$course(ArtsEl, ArtsDept, L) \land$
>>$dept(ArtsDept, arts) \land$
>>$different(ArtsEl, engl101).$

$has\_sci\_elective(St, Major, SciC)$ is true if $SciC$ is a third-year or fourth-year course passed by student St. $SciC$ must be in a science department other than department $Major$.

>$has\_sci\_elective(St, Major, SciC) \leftarrow$
>>$passed(St, SciC, 50) \land$
>>$course(SciC, Dept, Level) \land$
>>$member(Level, [third, fourth]) \land$
>>$different(Dept, Major) \land$
>>$dept(Dept, science).$

$enough\_units(St, Dept)$ is true if student St has enough units for a degree in department $Dept$. Departments in the faculty of science require 18 courses.

>$enough\_units(St, SciDept) \leftarrow$
>>$dept(SciDept, science) \land$
>>$has\_courses(St, CL, 18, 50).$

has_courses(*St*, *L*, *N*, *MinPass*) is true if *L* is a list of *N* courses that student *St* has passed with a grade of at least*MinPass*.

has_courses(*S*, [ ], 0, *M*).
has_courses(*St*, [*C*|*R*], *N*1, *MinPass*) ←
    plus(*N*, 1, *N*1) ∧
    has_courses(*St*, *R*, *N*, *MinPass*) ∧
    passed(*St*, *C*, *MinPass*) ∧
    notin(*C*, *R*).

where *plus*(*X*, *Y*, *Z*) is true if *X*, *Y* and *Z* are numbers such that $X + Y = Z$ (see Exercise 3.10).

Note that you need to collect the courses into a list in order to make sure you don't count the same course twice.

Given this knowledge base, *satisfied_degree_requirements*(*robin*) can be proved.

3.7 Applications in Natural Language Processing

Natural language processing is an interesting and difficult domain in which to develop and evaluate representation and reasoning theories. All of the problems of CI arise in this domain; solving "the natural language problem" is as difficult as solving "the CI problem" because any domain can be expressed in natural language. The field of **computational linguistics** has a wealth of techniques and knowledge. We can only scratch the surface of these here and, hopefully, give a flavor of how natural language understanding can be carried out by a computer.

There are at least three reasons for studying natural language processing:

- You want a computer to communicate with users in their terms; you would rather not force users to learn a new language. This is particularly important for casual users and those who have neither the time nor inclination to learn new interaction skills, such as managers and children. Users of the infobot, for example, would like to interact in *their* terms.

- There is a vast store of information recorded in natural language that could be accessible via computers. Information is constantly generated in the form of books, news, business and government reports and scientific papers, many of which are online. An infobot needs to be able to process natural language to retrieve much of the information available on computers.

- Many of the problems of CI arise in a very clear and explicit form in natural language processing, and thus it's a very good domain in which to experiment with general theories.

The development of natural language processing provides the possibility of natural language interfaces to knowledge bases and natural language translation. Current applications of natural language have severe restrictions; for example, for interfaces, you may assume that the domain of discourse will be limited to the vocabulary of that domain. This makes it possible to exploit incomplete theories of natural language understanding for real applications.

There are three major aspects of any natural language understanding theory:

Syntax The syntax describes the form of the language. It is usually specified by a grammar. Natural language is much more complicated than the formal languages used for RRSs. The RRS of this chapter is used to define the syntax of a restricted class of natural language.

Semantics The semantics provides the meaning of the utterances or sentences of the language. Defining such a semantics is difficult. A semantics is typically specified as a function from utterances into sentences of an RRS. One design goal for building an RRS is that it be expressive enough to represent the meaning of natural language sentences. The RRS presented in this chapter can represent the meaning of only a very limited set of sentences.

Pragmatics The pragmatic component explains how the utterances relate to the world. Usually you have to consider more than the sentence; you have to take into account the context, the goals of the speaker and the listener, special conventions, and the like.

To understand the difference between these aspects, consider the following sentences which might appear at the start of a CI textbook:

- *This book is about computational intelligence.*
- *Green frogs sleep soundly.*
- *Colorless green ideas sleep furiously.*
- *Sleep ideas green furiously colorless.*

The first sentence would be quite appropriate at the start of such a book; it's syntactically, semantically, and pragmatically well-formed. The second sentence is syntactically and semantically well-formed, but would appear very strange at the start of a CI book; it's thus not pragmatically well-formed for that context. The third sentence—due to the linguist Noam Chomsky—is syntactically well-formed, but it's semantically ill-formed (ideas can't be colorless and green at the same time and be furious and sleep at the same time). The fourth sentence is syntactically ill-formed; it doesn't make any sense, syntactically, semantically, or pragmatically.

We aren't attempting to give an introduction to computational linguistics. To do so would involve introducing many linguistic issues as well as many computational issues. We introduce only enough linguistics to enable us to demonstrate some issues relevant to representation and reasoning.

Using Definite Clauses for Context-Free Grammars

This section shows how to use definite clauses to represent aspects of the syntax and semantics of natural language.

Languages are defined by their legal sentences. Sentences are sequences of symbols. The legal sentences are specified by a grammar. You often say that the grammar *generates* the language.

Our first approximation of a solution to the problem of parsing natural language is to use a context-free grammar. A **context-free grammar** is a set of **rewrite rules**, with **nonterminal** symbols transforming into a sequence of terminal and nonterminal symbols. A sentence of the language is a sequence of **terminal** symbols that can be generated by such rewriting rules. For example, the grammar rule

$$sentence \longmapsto noun\_phrase, verb\_phrase$$

means that a nonterminal symbol *sentence* can be a *noun_phrase* followed by a *verb_phrase*. The symbol "$\longmapsto$" means "can be rewritten as." If you represent sentences of natural language as lists, this rule means that some list is a sentence if it's a noun phrase followed by a verb phrase:

$$sentence(S) \leftarrow noun\_phrase(N), verb\_phrase(V), append(N, V, S).$$

To say that the word "computer" is a noun, you would write

$$noun([computer]).$$

There is an alternative, simpler representation of context-free grammar rules using definite clauses that doesn't need an explicit *append*, known as the **definite clause grammar (DCG)**. You make each nonterminal symbol a predicate and add two extra arguments, which together form a **difference list** (page 85) of words that make the class given by the nonterminal.

Under this representation, $noun\_phrase(T_1, T_2)$ is true if list T_2 is an ending of the list T_1 such that all of the words in T_1 before T_2 form a noun phrase. T_2 is the rest of the sentence. You can think of T_2 as representing a position in a list that is after position T_1. The difference list represents the words between these positions.

Example 3.15

The atomic formula

$$noun\_phrase([the, student, passed, the, course, with, a, computer],$$
$$[passed, the, course, with, a, computer])$$

is true in the normal interpretation as "the student" forms a noun phrase.

The grammar rule

$$sentence \longmapsto noun\_phrase, verb\_phrase$$

means that there is a sentence between some T_0 and T_2 if there is a noun phrase between T_0 and T_1 and a verb phrase between T_1 and T_2:

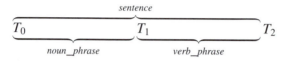

This grammar rule can be specified as the clause:

$$sentence(T_0, T_2) \leftarrow$$
$$\qquad noun\_phrase(T_0, T_1) \wedge$$
$$\qquad verb\_phrase(T_1, T_2).$$

In general, the rule

$$h \longmapsto b_1, b_2, \ldots, b_n$$

says that h is composed of a b_1 followed by a b_2, ..., followed by a b_n, and is written as the definite clause

$$h(T_0, T_n) \leftarrow$$
$$\qquad b_1(T_0, T_1) \wedge$$
$$\qquad b_2(T_1, T_2) \wedge$$
$$\qquad \vdots$$
$$\qquad b_n(T_{n-1}, T_n)$$

using the interpretation

where the T_i are new variables.

To say that nonterminal h gets mapped to the terminal symbols, t_1, ..., t_n, you write

$$h([t_1, \cdots, t_n | T], T)$$

using the interpretation

Thus $h(T_1, T_2)$ is true if $T_1 = [t_1, ..., t_n | T_2]$.

Example 3.16

The rule that specifies that the nonterminal h can be rewritten to the nonterminal a followed by the nonterminal b followed by the terminal symbols c and d, followed by the nonterminal symbol e followed by the terminal symbol f and the nonterminal symbol g, can be written as

$$h \longmapsto a, b, [c, d], e, [f], g$$

and can be represented as

$$h(T_0, T_6) \leftarrow$$
$$\quad a(T_0, T_1) \wedge$$
$$\quad b(T_1, [c, d | T_3]) \wedge$$
$$\quad e(T_3, [f | T_5]) \wedge$$
$$\quad g(T_5, T_6).$$

Note that the translations $T_2 = [c, d | T_3]$ and $T_4 = [f | T_5]$ were done manually.

Figure 3.5 axiomatizes a simple grammar of English. Figure 3.6 gives a simple dictionary of words and their part of speech.

Example 3.17

For the grammar of Figure 3.5 the dictionary of Figure 3.6, the query

$$?noun\_phrase([the, student, passed, the, course, with, a, computer], R).$$

will return

$$R = [passed, the, course, with, a, computer]$$

The sentence "The student passed the course with a computer." has two different parses, one using the clause instance:

$$verb\_phrase([passed, the, course, with, a, computer], [\,]) \leftarrow$$
$$\quad verb([passed, the, course, with, a, computer],$$
$$\qquad [the, course, with, a, computer]) \wedge$$
$$\quad noun\_phrase([the, course, with, a, computer], [\,]) \wedge$$
$$\quad pp([\,], [\,])$$

and one using the instance

$$verb\_phrase([passed, the, course, with, a, computer], [\,]) \leftarrow$$
$$\quad verb([passed, the, course, with, a, computer],$$
$$\qquad [the, course, with, a, computer]) \wedge$$
$$\quad noun\_phrase([the, course, with, a, computer], [with, a, computer]) \wedge$$
$$\quad pp([with, a, computer], [\,]).$$

In the first of these, the prepositional phrase modifies the noun phrase (i.e., the course is with a computer); and in the second, the prepositional phrase modifies the verb phrase (i.e., the course was passed with a computer).

% A sentence is a noun phrase followed by a verb phrase.

$$sentence(T_0, T_2) \leftarrow$$
$$noun\_phrase(T_0, T_1) \wedge$$
$$verb\_phrase(T_1, T_2).$$

% A noun phrase is a determiner followed by modifiers followed by a noun followed
% by an optional prepositional phrase.

$$noun\_phrase(T_0, T_4) \leftarrow$$
$$det(T_0, T_1) \wedge$$
$$modifiers(T_1, T_2) \wedge$$
$$noun(T_2, T_3) \wedge$$
$$pp(T_3, T_4).$$

% Modifiers consist of a sequence of adjectives.

$$modifiers(T, T).$$
$$modifiers(T_0, T_2) \leftarrow$$
$$adjective(T_0, T_1) \wedge$$
$$modifiers(T_1, T_2).$$

% An optional prepositional phrase is either nothing or a preposition followed by a
% noun phrase.

$$pp(T, T).$$
$$pp(T_0, T_2) \leftarrow$$
$$preposition(T_0, T_1) \wedge$$
$$noun\_phrase(T_1, T_2).$$

% A verb phrase is a verb followed by a noun phrase and an optional prepositional
% phrase.

$$verb\_phrase(T_0, T_3) \leftarrow$$
$$verb(T_0, T_1) \wedge$$
$$noun\_phrase(T_1, T_2) \wedge$$
$$pp(T_2, T_3).$$

Figure 3.5: A context-free grammar for a very restricted subset of English

det(T, T).
det([*a*|T], T).
det([*the*|T], T).
noun([*student*|T], T).
noun([*course*|T], T).
noun([*computer*|T], T).
adjective([*practical*|T], T).
verb([*passed*|T], T).
preposition([*with*|T], T).

Figure 3.6: A simple dictionary

Augmenting the Grammar

The assumption that the grammar is context-free doesn't allow us to adequately express the complexity of actual English grammar. Two mechanisms can be added to this grammar to make it more expressive:

- Extra arguments to the nonterminal symbols
- Arbitrary conditions on the rules

The extra arguments will enable us to do several things: to construct a parse tree, to represent the semantic structure of a sentence, to incrementally build a query which represents a question to a database, and to accumulate information about phrase agreement (such as number, tense, gender, and person).

Building Structures for Nonterminals

You can add an extra argument to the predicates to represent a parse tree, forming a rule such as

sentence($T_0, T_2, s(NP, VP)$) ←
    noun_phrase(T_0, T_1, NP) ∧
    verb_phrase(T_1, T_2, VP).

which is to mean that the parse tree for a sentence is of the form $s(NP, VP)$ where NP is the parse tree for the noun phrase and VP is the parse tree for the verb phrase.

This is important if you want some result from the syntactic analysis, and not just to know whether the sentence is syntactically valid. The notion of a parse tree is a simplistic form of what's required as it doesn't adequately represent the meaning or "deep structure" of a sentence. For example, you would really like to recognize that "Alex taught the AI course" and "the AI course was taught by Alex" have the same meaning, only differing in the active or passive voice.

Canned Text Output

There is nothing in the definition of the grammar that requires English input and the parse tree as output. Just as in Section 3.5 where different arguments can be made bound and unbound in *append*, you can use a grammar rule with the meaning of the sentence bound and a free variable representing the sentence. This will produce a sentence that matches the meaning.

One such use of grammar rules is to provide canned text output from logic terms: The output is a sentence in English that matches the logic term. This is useful for producing English versions of atoms, rules, and questions that a user—who may not know the intended interpretation of the symbols, or even the syntax of the formal language—can easily understand.

Example 3.18

Figure 3.7 shows a grammar for producing canned text on schedule information. For example, the query

$$?trans(scheduled(w92, cs422, clock(15, 30), above(csci333)), T, [\,]).$$

produces the answer $T = [the, winter, 1992, session, of, the, advanced, artificial, intelligence, course, is, scheduled, at, 3, :, 30, pm, in, the, room, above, the, computer, science, department, office].$

This grammar would probably not be useful for understanding natural language, as it requires a very stylized form of English; the user would have to use the exact translation of a term to get a legal parse.

Enforcing Constraints

Natural language imposes constraints which, for example, disallow sentences such as "a students eat." Words in a sentence must satisfy some agreement criteria. "A students eat" fails to satisfy the criterion of number agreement, which specifies whether the nouns and verbs are singular or plural.

Number agreement can be enforced in the grammar by parameterizing the non-terminals by the number and making sure that the numbers of the different parts of speech agree. You need only add an extra argument to the relevant nonterminals.

% *trans*(*Term*, *T*0, *T*1) is true if *Term* translates into the words contained in the differ-
% ence list $T0 - T1$.

> $trans(scheduled(S, C, L, R), T_1, T_8) \leftarrow$
> > $trans(session(S), T_1, [of|T_3]) \wedge$
> > $trans(course(C), T_3, [is, scheduled, at|T_5]) \wedge$
> > $trans(time(L), T_5, [in|T_7]) \wedge$
> > $trans(room(R), T_7, T_8).$
>
> $trans(session(w92), [the, winter, 1992, session|T], T).$
> $trans(course(cs422), [the, advanced, artificial, intelligence, course|T], T).$
> $trans(time(clock(0, M)), [12, :, M, am|T], T).$
> $trans(time(clock(H, M)), [H, :, M, am|T], T) \leftarrow$
> > $H > 0 \wedge H < 12.$
>
> $trans(time(clock(12, M)), [12, :, M, pm|T], T).$
> $trans(time(clock(H, M)), [H1, :, M, pm|T], T) \leftarrow$
> > $H > 12 \wedge$
> > $H1$ is $H - 12.$
>
> $trans(room(above(R)), [the, room, above|T_1], T_2) \leftarrow$
> > $trans(room(R), T_1, T_2).$
>
> $trans(room(csci333), [the, computer, science, department, office|T], T).$

Figure 3.7: Grammar for output of canned English

Example 3.19

The grammar of Figure 3.8 doesn't allow "a students," "the student eat," or "the students eats," since all have number disagreement, but allows "a green student eats," "the students," or "the student," since "the" can be either singular or plural.

To parse the sentence "the student eats," you issue the query

> $?sentence([the, student, eats], [\ ], Num, T)$

and the answer returned is

> $Num = singular,$
> $T = s(np(definite, [\ ], student, nopp), vp(eat, nonp, nopp)).$

To parse the sentence "the students eat," you issue the query

> $?sentence([the, students, eat], [\ ], Num, T)$

% A sentence is a noun phrase followed by a verb phrase.

$sentence(T_0, T_2, Num, s(NP, VP)) \leftarrow$
$\quad noun\_phrase(T_0, T_1, Num, NP) \land$
$\quad verb\_phrase(T_1, T_2, Num, VP).$

% A noun phrase is empty or a determiner followed by modifiers followed by a noun
% followed by an optional prepositional phrase.

$noun\_phrase(T, T, Num, nonp).$
$noun\_phrase(T_0, T_4, Num, np(Det, Mods, Noun, PP)) \leftarrow$
$\quad det(T_0, T_1, Num, Det) \land$
$\quad modifiers(T_1, T_2, Mods) \land$
$\quad noun(T_2, T_3, Num, Noun) \land$
$\quad pp(T_3, T_4, PP).$

% A verb phrase is a verb, followed by a noun phrase, followed by an optional prepo-
% sitional phrase.

$verb\_phrase(T_0, T_3, Num, vp(V, NP, PP)) \leftarrow$
$\quad verb(T_0, T_1, Num, V) \land$
$\quad noun\_phrase(T_1, T_2, N2, NP) \land$
$\quad pp(T_2, T_3, PP).$

% An optional prepositional phrase is either nothing or a preposition followed by a
% noun phrase. Only the null case is given here.

$pp(T, T, nopp).$

% Modifiers is a sequence of adjectives. Only the null case is given.

$modifiers(T, T, [\,]).$

% The dictionary.

$det([a|T], T, singular, indefinite).$
$det([the|T], T, Num, definite).$
$noun([student|T], T, singular, student).$
$noun([students|T], T, plural, student).$
$verb([eats|T], T, singular, eat).$
$verb([eat|T], T, plural, eat).$

Figure 3.8: A grammar that enforces number agreement and builds a parse tree

and the answer returned is

$Num = plural$,
$T = s(np(definite, [\,], student, nopp), vp(eat, nonp, nopp))$.

To parse the sentence "a student eats," you issue the query

$?sentence([a, student, eats], [\,], Num, T)$

and the answer returned is

$Num = singular$,
$T = s(np(indefinite, [\,], student, nopp), vp(eat, nonp, nopp))$.

Note that the only difference between the answers is whether the subject is singular and whether the determiner is definite.

Building a Natural Language Interface to a Database

You can augment the preceding grammar to implement a simple natural language interface to a database. The idea is that, instead of transforming sub-phrases into parse trees, you transform them into a form that can be queried on a database. For example, a noun phrase becomes an object with a set of predicates defining it.

Example 3.20

The phrase "a female student enrolled in a computer science course," could be translated into:

$answer(X) \leftarrow$
$\quad female(X) \wedge student(X) \wedge enrolled\_in(X, Y) \wedge course(Y)$
$\quad \wedge department(Y, comp\_science)$.

Let's ignore the problems of quantification, such as how the words "all," "a," and "the" get translated into quantifiers. You can construct a query by allowing noun phrases to return an individual and a list of constraints imposed by the noun phrase on the individual. Appropriate grammar rules are specified in Figure 3.9, and they are used with the dictionary of Figure 3.10.

In this grammar,

$noun\_phrase(T_0, T_1, O, C_0, C_1)$

means that list T_1 is an ending of list T_0, and the words in T_0 before T_1 form a noun phrase. This noun phrase refers to the object O. C_0 is an ending of C_1, and the formulae in C_1, but not in C_0, are the constraints on object O imposed by the noun phrase.

% A noun phrase is a determiner followed by modifiers followed by a noun followed
% by an optional prepositional phrase.

$noun\_phrase(T_0, T_4, Obj, C_0, C_4) \leftarrow$
$\quad det(T_0, T_1, Obj, C_0, C_1) \wedge$
$\quad modifiers(T_1, T_2, Obj, C_1, C_2) \wedge$
$\quad noun(T_2, T_3, Obj, C_2, C_3) \wedge$
$\quad pp(T_3, T_4, Obj, C_3, C_4).$

% Modifiers consist of a sequence of adjectives.

$modifiers(T, T, Obj, C, C).$
$modifiers(T_0, T_2, Obj, C_0, C_2) \leftarrow$
$\quad adjective(T_0, T_1, Obj, C_0, C_1) \wedge$
$\quad modifiers(T_1, T_2, Obj, C_1, C_2).$

% An optional prepositional phrase is either nothing or a preposition followed by a
% noun phrase.

$pp(T, T, Obj, C, C).$
$pp(T_0, T_2, O_1, C_0, C_2) \leftarrow$
$\quad preposition(T_0, T_1, O_1, O_2, C_0, C_1) \wedge$
$\quad noun\_phrase(T_1, T_2, O_2, C_1, C_2).$

Figure 3.9: A grammar that constructs a query

Example 3.21

The query

$\quad ?noun\_phrase([a, computer, science, course], [\,], Obj, [\,], C).$

will return

$\quad C = [course(Obj), dept(Obj, comp\_science)].$

The query

$\quad ?noun\_phrase([a, female, student, enrolled, in, a, computer,$
$\quad\quad science, course], [\,], P, [\,], C).$

returns

$\quad C = [course(X), dept(X, comp\_science), enrolled(P, X), student(P),$
$\quad\quad female(P)].$

$det(T, T, O, C, C)$.

$det([a|T], T, O, C, C)$.

$det([the|T], T, O, C, C)$.

$noun([course|T], T, O, C, [course(O)|C])$.

$noun([student|T], T, O, C, [student(O)|C])$.

$noun([john|T], T, john, C, C)$.

$noun([cs312|T], T, 312, C, C)$.

$adjective([computer, science|T], T, O, C, [dept(O, comp\_science)|C])$.

$adjective([female|T], T, O, C, [female(O)|C])$.

$preposition([enrolled, in|T], T, O_1, O_2, C, [enrolled(O_1, O_2)|C])$.

Figure 3.10: A dictionary for constructing a query

If the elements of the list C are queried against the database in Figure 3.2, you find precisely the female students enrolled in a computer science course, namely Mary and Jane.

Limitations

So far we have assumed a very simple form of natural language. Our aim was to show what could be easily accomplished with simple tools, rather than a comprehensive study of natural language. Useful front-ends to databases can be built with the tools outlined in this section by, for example, constraining the domain sufficiently and asking the user which of two competing interpretations they intend.

The discussion of natural language processing assumes that natural language is compositional. That is, the meaning of the whole can be derived from the meaning of the parts. This is, in general, a false assumption. You usually have to know the context in the discourse and the situation in the world to discern what was meant by an utterance. There are many types of ambiguity that can only be resolved by understanding the context of the words.

For example, you can't always determine the correct reference of a description, without knowledge of the context and the situation. A description doesn't always refer to a uniquely determined individual.

Example 3.22

Consider the following paragraph:

The student took many courses. Two computer science courses and one mathematics course were particularly difficult. *The mathematics course…*

The referent is defined by the context and not just the description "*The mathematics course.*" There could be more mathematics courses, but we know from context that the phrase is referring to the particularly difficult one that the student took.

Many problems of reference arise in database applications if you allow the use of "*the*" or "*it,*" or permit words that have more than one meaning. The use of context to understand what's being referred to is very natural. Consider:

> *Who is the head of the mathematics department?*
> *Who is her secretary?*

Later chapters consider more problems that arise in understanding natural language.

3.8   References and Further Reading

The are many books on Prolog programming that provide further details on using the RRS presented here (e.g., Sterling & Shapiro, 1994; Bratko, 1990; O'Keefe, 1990). Unfortunately, most of the Prolog books concentrate on the peculiarities of Prolog, rather than on logic programming in general. The Prolog-specific features usually don't translate into richer RRSs.

For general issues on natural language processing, see Allen (1994) and Winograd (1983). Grosz, Jones & Webber (1986) contains a collection of classical papers on natural language processing. For a review of interfaces to databases see Perrault & Grosz (1986). For more on the use of logic programming for natural language processing see Pereira & Shieber (1987), Abramson & Dahl (1989), and Dahl (1994).

3.9   Exercises

Exercise 3.1   Consider the following facts:

> *Dogs are animals. All animals get pleasure from feeding. People like the animals they own. People do things that bring pleasure to things they like. Fido is Mary's dog.*

(a) Write the above as a set of (definite) clauses.

(b) Show how to pose the query which asks "What does Mary do?"

(c) Show an SLD derivation (page 56) for the above query, showing variable bindings.

Exercise 3.2   Suppose that a university registration system is implemented as a large knowledge base filled with definite clauses. In it are occurrences of: constant symbols, such as 73485796 denoting a student number and *cs322* denoting a course; predicate symbols, such as *enrolled_in*, *major*, and *owes_money*; and maybe even some function symbols. The clauses in the knowledge base are organized into facts, such as

> *enrolled_in*(73485796, *cs322a*).
>
> *major*(73485796, *comp_sci*).
>
> *owes_money*(23456789, 1000).
>
> *dept*(*cs322a*, *comp_sci*).

together with *rules*, such as

> *elective*(*S*, *C*) ←
>
>   *enrolled_in*(*S*, *C*) ∧ *dept*(*C*, *D*) ∧ *major*(*S*, *M*) ∧ *distinct*(*M*, *D*).
>
> *in_good_standing*(*S*) ←
>
>   *passed_all_courses*(*S*) ∧ *major*(*S*, *M*) ∧
>
>   *enough_courses*(*S*, *M*) ∧ *paid_up*(*S*).
>
> *space_available*(*C*) ←
>
>   *limit*(*C*, *L*) ∧ *curr_enrlmnt*(*C*, *N*) ∧ *less*(*N*, *L*).

Clearly, the knowledge base changes each year, each term, even each day. We will consider how things change from one term to the next.

(a) While there are many possible formal interpretations of this KB (of which many are models), there is a particular *intended* interpretation of this KB at any particular time. This would be the "real world" of containing, for example, students, courses, and money. With respect to this intended interpretation, does the *denotation* of the constant and predicate symbols typically change from one term to the next? Why or why not?

(b) Do the *facts* typically change from one term to the next? Why or why not?

(c) Do the *rules* typically change from one term to the next? Why or why not?

Exercise 3.3   Write axioms to show how the overhead projector can be added to the electrical domain (page 70). The overhead projector needs to be plugged in, the power outlet needs power, and the projector needs to be turned on and not be broken for it to be lit. The mirror needs to be in place and the projector focused in order to see a transparency that is on the projector.

Exercise 3.4   Change the example of Section 3.2 so that wires are connected to switches and so that the connection of the input and output ports of the switches depends on the status of the switch. This would mean that the places where wires are connected to switches—the ports—are individuals that can be referred to.

Exercise 3.5 This exercise will let us explore the standard list representation (page 81).

(a) Given the definition of *append* (page 84) that uses the square bracket notation, what is the result of the following queries:

$?append(A, [a|B], [b, a, d, a, c, a])$.

$?append(X, [A, B|Y], [a, b, c, d, e])$.

(b) Suppose that someone had decided that lists were a good representation for sets and wrote the following code for $subset(L_1, L_2)$ that is true if every element of list L_1 is in list L_2:

$subset([\,], L)$.

$subset([H|T], L) \leftarrow$

$member(H, L) \wedge$

$subset(T, L)$.

What are the answers to the following queries:

$?subset([a, b], [b, a, d])$.

$?subset([X, Y], [b, a, d])$.

$?subset(S, [b, a, d])$.

Comment on the use of this representation for sets. Can you think of a better representation?

(c) The predicates *rev* and *reverse* (page 85) were described in terms of the second argument being the reverse of the first. Of course, being the reverse of a list is a symmetric relation. For the following queries, give the SLD derivation (page 56), and discuss what answers will be produced if you try to find all answers:

$?rev([a, b, c, d], R)$.

$?rev(R, [a, b, c, d])$.

$?reverse([a, b, c, d], R)$.

$?reverse(R, [a, b, c, d])$.

Exercise 3.6 This exercise uses the list representation (page 81) to define some predicates.

(a) Write a relation $shorter(L1, L2)$ that is true if list $L1$ contains the same number or fewer elements than list $L2$.

What are all of the answers to the following queries:

$?shorter([a, g], [b, a, d, a])$.

$?shorter([b, a, d, a], [a, g])$.

$?shorter([a, a, a, a], [a, b])$.

$?shorter(L, [b, a, d, a])$.

$?shorter([b, a, d, a], L)$.

(b) Write a relation $remove(E, L, R)$ that is true if R is the resulting list of removing one instance of E from list L. The relation is false if E isn't a member of L.

What are all of the answers to the following queries:

$?remove(a, [b, a, d, a], R)$.

$?remove(E, [b, a, d, a], R)$.

$?remove(E, L, [b, a, d])$.

$?remove(p(X), [a, p(a), p(p(a)), p(p(p(a)))], R)$.

(c) Write a relation $subsequence(L1, L2)$ that is true if list $L1$ contains a subset of the elements of $L2$ in the same order.

What are all of the answers to the following queries:

$?subsequence([a, d], [b, a, d, a])$.

$?subsequence([b, a], [b, a, d, a])$.

$?subsequence([X, Y], [b, a, d, a])$.

$?subsequence(S, [b, a, d, a])$.

Exercise 3.7   Suppose you decide to have a different representation of nonempty lists, and you use $e(M)$ to be the single element list containing element M, and use $concat(L1, L2)$ to be the list containing the elements of $L1$ followed by the elements of $L2$.

Note that in this representation there may be a number of different ways to represent the same list. For example,

$concat(concat(e(a), e(b)), concat(e(c), e(d)))$ and

$concat(e(a), concat(concat(e(b), e(c)), e(d)))$

both represent the same list.

For such a representation:

(a) Define the relation $member(E, L)$ that is true if E is a member of list L.

(b) Define $append(A, B, C)$ that has the same meaning as the *append* defined (page 84). That is, C contains the elements of A followed by the elements of B.

(c) Define the relation $subset(L1, L2)$ that is true if list $L1$ is a subset of list $L2$ (i.e., every member of $L1$ is a member of $L2$).

(d) Define the relation $same\_members(L1, L2)$ that is true if list $L1$ contains the same elements as list $L2$.

(e) Define the relation $same\_list(L1, L2)$ that is true if list $L1$ contains the same elements, in the same order, as list $L2$.

(f) Discuss the advantages or disadvantages of both representations of lists.

Exercise 3.8   For the example of Section 3.6:

(a) Fix up the knowledge base so that a student can't count core courses as electives.

(b) Explain why *different* is needed in the definition of *fulfilled_electives* (page 89). What happens if it isn't there?

(c) Explain why *enough_units* needs to construct a list of the courses taken. Can you implement this without constructing a list of courses taken? Explain.

Exercise 3.9 Some representations are more efficient for some tasks than others. This exercise explores two representations of student record information of Figure 3.3 (page 87). Consider the definition of *has_courses* (page 91).

(a) How many different answers are there to the query

   ?*has_courses*(*robin*, *L*, 19, 50).

   Don't try to count them by running the program. Think about it!

(b) Should the query

   ?*has_courses*(*robin*, *L*, 20, 50).

   have an answer or not? Explain why an implementation takes so long to execute. *Hint:* Refer to Exercise 3.9(a).

(c) Suppose you decide to represent the student record information database of Figure 3.3 in terms of the relation *taken*(*St*, *Cs*) where *St* is a student and *Cs* is a list of terms of the form *gr*(*Course*, *Mark*), which represents the fact that student *St* received *Mark* in course *Course*. *Cs* represents a list of all courses taken.

   i) Rewrite the database of Figure 3.3 using the new representation.

   ii) Define *enough_units* using the new representation.

   iii) Define *grade*(*St*, *Course*, *Mark*) in terms of *taken*.

   iv) Can you define *taken* in terms of *grade*? Explain.

   How many different proofs (and answers) are there for the new versions of *enough_units*(*St*, *cs*)? How much more efficient is the new representation than the old representation? Is there an advantage of the old representation over the new representation?

Exercise 3.10 This exercise is to axiomatize an abstract domain, namely arithmetic on natural numbers (positive integers).

   Use a binary representation of natural numbers, where a number is either *one*, *b*(*N*, *zero*), or *b*(*N*, *one*), where

- *one* denotes the number 1,
- *b*(*N*, *zero*) denotes the number $2 \times N'$ where N' is the number denoted by N,
- *b*(*N*, *one*) denotes the number $2 \times N' + 1$ where N' is the number denoted by N.

For example, *b*(*one*, *one*) denotes binary 11 which is three, *b*(*b*(*one*, *one*), *zero*) denotes binary 110 or six, and *b*(*b*(*b*(*one*, *one*), *zero*), *one*) denotes binary 1101 or thirteen.

   Define the following predicates, such that all of the operations are linear in the size of the representation, and not linear in the size of the numbers:

(a) *number*(N) is true if N is the form of a number.

(b) *succ*(N, M) is true if M is the number after number N.

(c) *plus*(X, Y, Z) is true if the number denoted by Z is the sum of the numbers denoted by X and Y (i.e., $Z = X + Y$).

(d) *times*(X, Y, Z) is true if Z is the product of numbers X and Y (i.e., $Z = X \times Y$).

(e) *is*(V, E), which is a relation between a number V and an arithmetic expression E, is true if V is the value of expression E. An expression can be a number or of the form $X + Y$, $X * Y$, or $X - Y$ where X and Y are expressions. This is often written as V *is* E.

(f) Define the relation *lt*(N, M) that is true if number N is less than number M. This is often written as $N < M$.

(g) Define the relation *neq*(N, M) that is true if N and M are different numbers. This is often written as $N \neq M$. *Hint:* Use the answer from Exercise 3.10(f).

(h) Suppose someone suggested extending this representation to all non-negative integers simply by adding *zero* as a number. Why might this cause problems? *Hint:* Consider *b*(*zero*, *one*).

Exercise 3.11 Suppose you wanted to axiomatize the relation *different*(C_1, C_2) that is true if course C_1 is a different course to course C_2.

(a) Suppose there are 100 courses (all of them different), how many facts does it take to define *different* as a database?

(b) If there are *n* different courses how many facts does it take to define the database?

(c) Suggest a way so that *different* can be defined using a knowledge base that is linear in the number of courses. *Hint:* Use the solution to Exercise 3.10(g).

Exercise 3.12 Consider a database consisting of the following relations:

- *scheduled*(C, S, T, R,) is true when section S for course C is scheduled at time T in room R. For example, *scheduled*(cs486, 1, pm(2), mc1052) means that section 1 of course cs486 is scheduled at 2 p.m. in room mc1052. (*pm* is a function from numbers to times).

- *enrolled*(P, C, S) is true when student P is enrolled in section S of course C.

- *dept*(C, D) is true when C is a course in department D.

You can also assume the existence of the relations $X \neq Y$ (true whenever X is different from Y) and *plus*(X, Y, Z) (true whenever $X + Y = Z$).

(a) Build such a database (using fabricated data) and pose the following questions to it:

   i) Is Robin enrolled in cs486?

   ii) Is Fred enrolled in a computer science course?

   iii) Is anyone enrolled in 2 sections of the same course?

   iv) Is there any room with three lectures on consecutive hours?

(b) Add the knowledge that cs686 is always scheduled at the same time and place as cs486, in corresponding sections.

(c) Define a relation *clash(S)* which means that student S has two courses she is supposed to take at the same time.

Exercise 3.13 Using only the relations $X \neq Y$ (true whenever X is different from Y), and *plus*(X, Y, Z) (true whenever $X + Y = Z$), answer the following:

(a) Given the database for Exercise 3.12, define a relation *more_students*(C, S, N) which means there are more than N students enrolled in section S of course C. *Hint:* This can be proved by induction on N.

(b) One of the following two predicates can't be expressed in the language of definite clauses (as defined here) and one can. Explain why, and implement the other. Note that they are the inverse of each other.

   i) *commonality(List_of_students)* is true if two of the students in the list have a subject in common.

   ii) *disjoint_subjects(List_of_students)* is true if the courses taken by the students in the list of students are disjoint.

Exercise 3.14 Write a symbolic differentiation program. Algebraic expressions will be arbitrary formulae, made from algebraic variables (which will be represented as lower case constants) and numbers, with standard algebraic connectives (addition, subtraction, multiplication, and division) and trigonometric functions (at least sine and cosine).

For this question, you may assume the relations

- *atom(X)* that is true if X is a constant
- *number(X)* that is true if X is a number
- V *is* E that is true if expression E evaluates to number V
- $E_1 \neq E_2$ that is true if expressions E_1 and E_2 evaluate to different numbers

For *is* and $\neq$ the expressions can only involve numbers and the standard algebraic connectives.

Define the predicates

   diff (X, V, Y)

which means that Y is the derivative of X with respect to V, and

   simplify(X, Y),

which means that Y is a simplified expression equivalent to X. Only basic simplification needs to be done (e.g., addition of zero, multiplication by zero or one, arithmetic expressions involving only integers). For example, the query

   ?*diff* $((x + y) * x, x, A)$.

produces the answer (with no simplification)

   $A = (1 + 0) * x + (x + y) * 1,$

which can be simplified to

$$A = x + x + y.$$

Exercise 3.15 Write a definite clause program to solve the following problem: Given a set of colored building blocks, each a unit cube, construct a tower of a given height such that each block is a different color from the blocks immediately above and immediately below (if they exist). Use the relations $block(X)$, which is true if X is a block, and $color(B, C)$, which is true if block B is of color X. The only predefined relations you may assume are $<$, is, and $\neq$.

Exercise 3.16 Assume that you have a database containing the following relations:

- $done(S, C, M)$ meaning student S has completed course C with mark M.
- $prereqs(C, L)$ meaning L is the list of courses which are prerequisites for course C.

(a) Define a relation $can\_enroll(P, C)$ meaning "P has passed (i.e., grade greater than 49) all of the prerequisites of course C."

(b) Show an SLD derivation (page 56) for the query:

$?can\_enroll(randy, cs489)$

given the facts

$prereqs(cs486, [cs350, cs443])$.

$done(randy, cs340, 93)$.

$done(randy, cs443, 23)$.

$done(randy, cs443, 83)$.

(c) How would you express the fact "the prereqs of cs385 are cs322 and one of cs321, cs327, cs328"?

(d) Assume you are given a database of the relation $prereq(C_1, C_2)$ which means "C_2 is a prerequisite of C_1," as well as the relation $done$.
    Either define can_enroll as in part (a) or explain why it can't be defined.

Exercise 3.17 Think of a problem or domain where definite clauses are inadequate. What else is needed, and why?

Exercise 3.18 How could you change the grammar for building queries so that it takes quantification into account (e.g., the use of the words *all* or *some*).

Discuss the limitations of the restricted formulation of natural language processing presented in this chapter. What forms can't be easily handled? Are the assumptions reasonable for various applications?

Exercise 3.19 The following sentences describe the same transaction:

- John sold the book to Mary.
- Mary bought the book from John.

How could you build a definite clause grammar which recognizes these as equivalent?

Exercise   Build a pseudonatural language interface to a database. To do this, try to work out
3.20   the forms of the sentences that are anticipated, as well as the forms of the queries that
              are allowed. Determine the vocabulary suitable for the domain of the database. Build
              a simple prototype, and test it to determine its quality. What sort of restrictions are
              placed on such an interface? What is hard to build? Is this a suitable interface to a
              database?

Chapter 4

Searching

4.1 Why Search?

Search underlies much of computational intelligence. When you are given a problem, you are usually not given an algorithm to solve it; you have to search for a solution. For example, Chapter 2 shows how the semantic problem of determining logical consequence can be reduced to the syntactic problem of searching for a proof. The best you can hope for is a nondeterministic specification of what is a solution. The existence of NP-complete problems (page 50) with efficient means to test answers but no efficient methods for finding them indicates that searching is, in general, a necessary part of solving problems.

Search is an enumeration of a set of potential partial solutions to a problem so that they can be checked to see if they truly are solutions, or could lead to solutions. In order to carry out a search, you need:

- A definition of a potential solution
- A method of generating the potential solutions, hopefully in a clever way
- A way to check whether a potential solution is a solution

In this chapter, we abstract the general mechanism of searching and present it in terms of searching for paths in directed graphs. In order to solve a problem, you explicate the underlying search space and apply a search algorithm to that search space.

It is often believed that humans are able to use intuition to jump to solutions to difficult problems. However, humans don't tend to solve general problems, but instead solve specific instances about which they may know much more than the underlying search space. Problems where there is little underlying structure or where that structure cannot be related to the physical world are very difficult for humans to solve. The existence of public key encryption codes, where the search space is clear and the test for a solution is given and for which humans have no hope of solving and computers can't solve in a realistic time frame, demonstrates the difficulty of search.

This suggests that computer agents need to exploit knowledge about special cases to guide them to a solution. This extra knowledge beyond the search space is **heuristic knowledge**. This chapter only considers one instance of heuristic knowledge in the form of an estimate of the distance from nodes in a graph to a solution.

We also consider a special case where the multidimensional aspect of the search space is important and can be exploited for computational gain. In such constraint satisfaction problems you assign a value to each of a number of variables so that the resulting assignment either satisfies some hard constraints or optimizes the value of some soft constraints (page 147).

4.2   Graph Searching

Many problem-solving tasks can be transformed to the problem of finding a path in a graph. Searching in graphs forms an appropriate level of abstraction within which to study problem solving independent of a particular domain.

Informally, a graph consists of a set of nodes, along with a set of arcs between nodes. The idea is to find a path along arcs from a start node to a goal node.

Example 4.1

In the robot delivery domain, for the task of finding a path from one location to another, the search space consists of all paths that the delivery robot can take (see Figure 4.1). Within the space of all paths are some that satisfy the goal of getting from outside room $r103$, position $o103$, to room $r123$. It may be the case that any path from $o103$ to $r123$ is acceptable. At the other extreme, you may want the shortest path from $r103$ to $o123$ where shortest could be in terms of time or distance. In between these extremes, you may want a path that's no more than 10% longer than the shortest path. The selection of a method to search such a space to find an appropriate path is the subject of this chapter. Different methods provide different guarantees about the paths found.

Example 4.2

In Section 2.7 (page 46), the semantic problem of finding a logical consequence of a knowledge base was reduced to the search problem of finding a proof for a query. SLD resolution (page 56) was presented as a search problem where you have to search through the space of possible derivations until you find a proof.

There are many ways to represent a problem as a graph, but two deserve special attention:

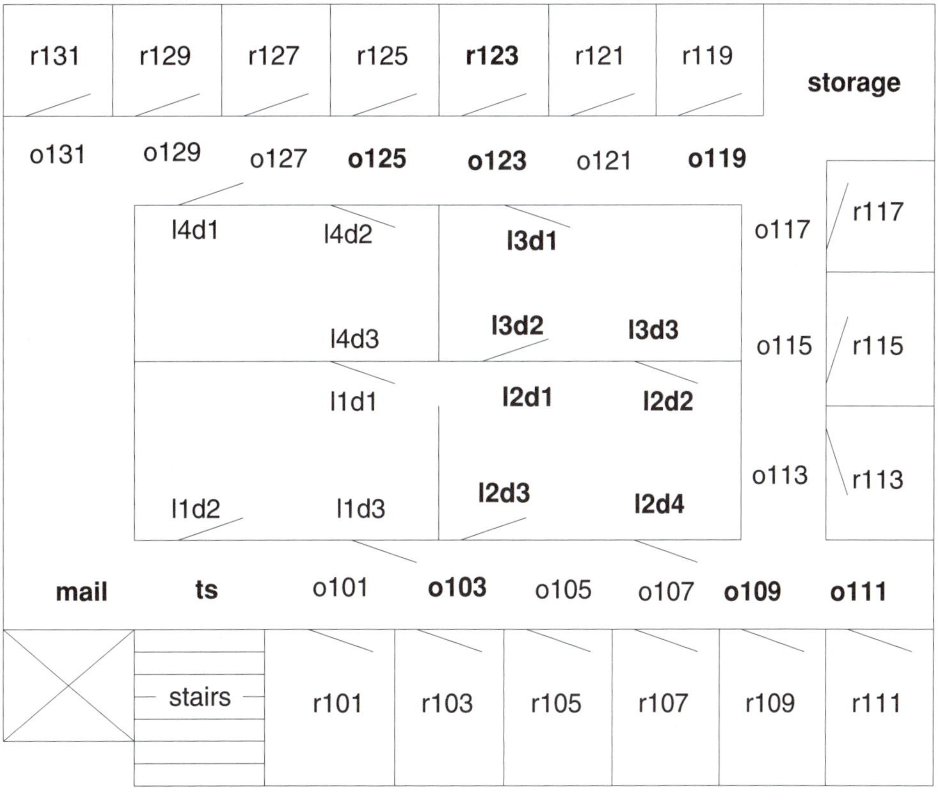

Figure 4.1: The delivery robot domain with interesting locations labeled

- *State-space graph* in which a node represents a state of the world, and an arc represents changing from one state of the world to another. The path finding problem for the delivery robot is of this form, where the state is the position of the robot.

- *Problem-space graph* where a node represents a problem to be solved, and an arc represents alternate decompositions of the problem. The problem of finding an SLD derivation is of this form.

Whether the nodes represent states or problems, essentially the same algorithms can be used to search the space. One exception to this is the class of constraint satisfaction problems where there is no start node and the aim is to find an assignment of values to variables. The assignment should either minimize some value function, forming optimization problems, or satisfy some criteria in satisfiability or constraint satisfaction problems. The multidimensional nature of the problem can be exploited. These problems are explored in Section 4.7 (page 147).

Formalizing Graph Searching

A (directed) **graph** consists of a set N of **nodes** and a set A of (ordered) pairs of nodes called (directed) **arcs**. A node n_2 is a **neighbor** of n_1 if there is an arc from n_1 to n_2. This is, $\langle n_1, n_2 \rangle \in A$. Arcs and nodes may be **labeled**.

A **path** from node s to node g is a sequence of nodes $n_0, n_1, \ldots, n_k$ such that $s = n_0$, $g = n_k$ and there is an arc from n_{i-1} to n_i—that is, if $\langle n_{i-1}, n_i \rangle \in A$. Sometimes it's useful to view a path as the sequence of arcs $\langle n_o, n_1 \rangle, \langle n_1, n_2 \rangle, \ldots, \langle n_{k-1}, n_k \rangle$. The second definition is more appropriate when the arcs are labeled.

A **cycle** is a nonempty path such that the end node is the same as the start node—that is, if $n_0 = n_k$ and $k \neq 0$. A directed graph without any cycles is called a **directed acyclic graph** or **DAG**. This should probably be an "acyclic directed graph," but DAG sounds better than ADG!

In order to encode problems as graphs, one set of nodes is referred to as **start nodes** and another set is called **goal nodes**. A **solution** is a path from a start node to a goal node.

Sometimes there is a **cost**, a positive number, associated with arcs. In this case the cost of a path is the sum of the costs of the arcs in the path.

Given a set of start nodes S and a set of goal nodes G, and a node n such that there is a path found from $s \in S$ to n, define $g(n)$ to be the cost of the path found from s to n.

Example 4.3

Consider the problem of the delivery robot finding a path from location $o103$ to location $r123$ in the domain depicted in Figure 4.1. In this figure the interesting locations are named. For simplicity, we only consider the locations written in bold and initially limit the directions that the robot can travel. Figure 4.2 shows the resulting graph where the nodes represent locations, and the arcs represent possible single steps between locations.

In this graph the nodes are $N = \{main, ts, o103, l2d3, o109. \ldots\}$ and the arcs are $A = \{\langle ts, mail \rangle, \langle o103, ts \rangle, \langle o103, l2d3 \rangle, \langle o103, o109 \rangle, \ldots\}$. The node $o125$ has no neighbors. The node ts has one neighbor, namely $mail$, and the node $o103$ has three neighbors, namely ts, $l2d3$, and $o109$.

One path from $o103$ to $r123$ is the sequence of nodes: $o103, o109, o119, o123, r123$.

If $o103$ were a start node and $r123$ were a goal node, the above path would be a solution to the graph-searching problem.

Example 4.4

Finding a proof using SLD resolution is a graph-searching problem. The nondeterministic SLD resolution algorithm of Figure 2.5 (page 51) together with a selection strategy implies a search graph. Each node represents an answer clause. The neighbors of a node $yes \leftarrow a_1 \wedge \ldots \wedge a_m$, with a_i as the selected atom, represent all of the possible answer clauses obtained by resolving on a_i. There is a neighbor

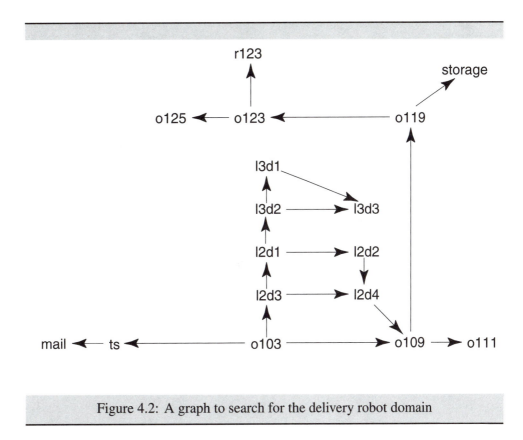

Figure 4.2: A graph to search for the delivery robot domain

for each clause whose head unifies with a_i. The goal node is of the form *yes* ← .
Figure 4.3 shows the search graph for a particular query for a particular knowledge
base, always selecting the leftmost atom in an answer clause.

You would not expect that the search graphs would be given statically, as
this would entail anticipating every possible query. When there are variables and
function symbols there are infinitely many possible queries. It is more realistic
that the search graph is dynamically constructed as needed. All that is needed is
a way to generate the neighbors of a node. Determining the neighbors is a simple
procedure, namely, resolution on the selected atom.

The **forward branching factor** of a node is the number of arcs going out of the
node, and the **backward branching factor** of a node is the number of arcs going into
the node. These provide measures of the complexity of graphs. When we discuss
the time and space complexity of the search algorithms we assume that the branching
factors are bounded.

Example
4.5

In the graph of Figure 4.2, the forward branching factor of node $o103$ is three;
there are three arcs coming out of node $o103$. The backward branching factor of

Given the knowledge base

$$a \leftarrow b \land c. \quad a \leftarrow g. \quad a \leftarrow h.$$
$$b \leftarrow j. \quad\quad b \leftarrow k. \quad d \leftarrow m.$$
$$d \leftarrow p. \quad\quad f \leftarrow m. \quad f \leftarrow p.$$
$$g \leftarrow m. \quad\quad g \leftarrow f. \quad k \leftarrow m.$$
$$h \leftarrow m. \quad\quad p.$$

and the query

$$?a \land d.$$

The search graph for an SDL derivation assuming the leftmost atom is selected in each answer clause is

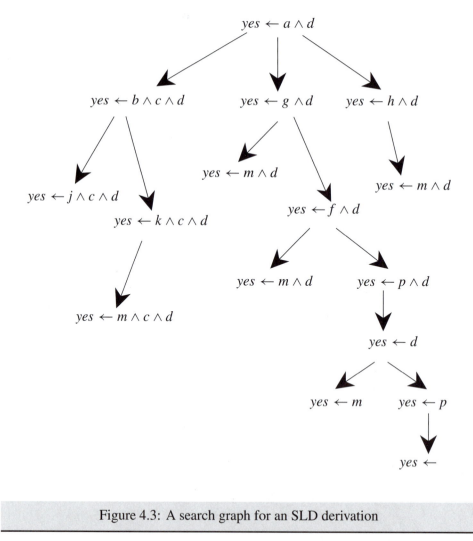

Figure 4.3: A search graph for an SLD derivation

node $o103$ is zero; there are no arcs coming into node $o103$. The forward branching factor of *mail* is zero and the backward branching factor of *mail* is one. The forward branching factor of node $l2d3$ is two and the backward branching factor of $l2d3$ is one.

The branching factor is important because it's a key component in the size of the graph. If the forward branching factor for each node is b, and the graph is a tree, there are b^n nodes that are n steps away from a node.

4.3 A Generic Searching Algorithm

This section describes a general algorithm to search for a solution path in a graph. The algorithm is independent of any particular search strategy and any particular graph. Different search strategies are obtained by providing appropriate definitions for two of the undefined predicates. The generic algorithm lets you understand the idea behind different searching algorithms before investigating any particular search strategy.

The intuitive idea behind the generic search algorithm is, given a graph, a set of start nodes, and a set of goal nodes, to incrementally explore paths from the start nodes. This is done by maintaining a **frontier** or **fringe** of paths from the start node that have been explored. The frontier contains all of the paths that could form initial segments of paths from a start node to a goal node. As search proceeds, the frontier expands into the unexplored nodes until a goal node is encountered. The way in which the frontier is expanded defines the search strategy. Sometimes it's useful to think about a frontier being the set of nodes reached (see Figure 4.4), and sometimes it's more useful to think of a frontier as a set of paths from the start node.

At each step, you advance the frontier by removing a node from the frontier, extending the path to that node by all of its neighbors, and adding these new paths to the frontier. Different search strategies are defined by which element of the frontier is selected at each stage. If a path on the frontier doesn't end at a goal node and the end node of the path has no neighbors, extending the node by its neighbors means removing the node. This is reasonable as this node couldn't be part of path from the start node to a goal node.

Associated with each element n of the frontier is the cost $g(n)$ of the corresponding path to that node.

The first axiomatization of the general graph search algorithm treats the frontier as a set of nodes and ignores the paths found to the nodes on the frontier:

$$search(F_0) \leftarrow$$
$$select(Node, F_0, F_1) \wedge$$
$$is\_goal(Node).$$

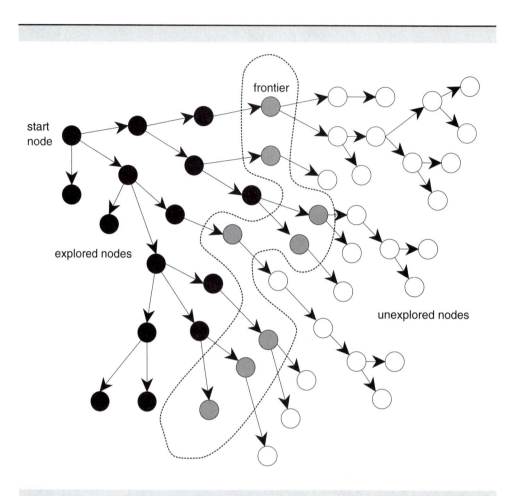

Figure 4.4: Problem solving by graph searching

$search(F_0) \leftarrow$
    $select(Node, F_0, F_1) \wedge$
    $neighbors(Node, NN) \wedge$
    $add\_to\_frontier(NN, F_1, F_2) \wedge$
    $search(F_2).$

with the following intended interpretation of the predicates:

- $search(Frontier)$ is true if there is a path from one element of the *Frontier* to a goal node.

- $is\_goal(N)$ is true if N is a goal node.

- $neighbors(N, NN)$ is true if NN is the list of neighbors of node N.

- *select*(N, F_0, F_1) means N is some element selected from frontier F_0, and F_1 is the set of nodes remaining when N is removed. This fails if F_0 is empty (in which case the search has failed).

- *add_to_frontier*(NN, F_1, F_2) means that F_2 is the frontier made by adding the list of nodes NN to frontier F_1. In particular, something is an element of F_2 if it's an element of NN or an element of F_1.

The relations *select* and *add_to_frontier* together define the search strategy. They tell us which node will be expanded at each step. The relation *neighbors* defines the graph, and the relation *is_goal* defines what is a solution.

The searching process is initiated by a query of the form

?*search*(*S*)

where *S* is a list of starting nodes. A proof of the query means that there exists some path from a node in *S* to a goal node.

Example
4.6

The graph of Figure 4.2 (page 117) can be represented as:

neighbors($o103, [ts, l2d3, o109]$).

neighbors($ts, [mail]$).

neighbors($mail, [\,]$).

neighbors($o109, [o111, o119]$).

neighbors($o111, [\,]$).

neighbors($o119, [storage, o123]$).

neighbors($storage, [\,]$).

neighbors($o123, [r123, o125]$).

neighbors($o125, [\,]$).

neighbors($l2d1, [l3d2, l2d2]$).

neighbors($l2d2, [l2d4]$).

neighbors($l2d3, [l2d1, l2d4]$).

neighbors($l2d4, [o109]$).

neighbors($l3d2, [l3d3, l3d1]$).

neighbors($l3d1, [l3d3]$).

neighbors($l3d3, [\,]$).

neighbors($r123, [\,]$).

The goal of getting to room $r123$ is specified as the fact

is_goal($r123$).

To use the preceding algorithm to search starting at $o103$, you can pose the query

$?search([o103])$.

The correctness of the search algorithm can be argued by showing that both clauses are true in the intended interpretation. The first clause can be seen to be correct by noticing that if *Node* is an element of F_0 (which *select* guarantees) and *Node* is a goal, then there is an element of F_0 which is a goal node, and so there is an (empty) path to a goal. Thus the first clause is true in the intended interpretation.

To show the second clause is true, suppose the right-hand side of the clause is true; then there is a path from an element n of F_2 to a goal. This element of F_2 is either an element of *NN* or an element of F_1, by definition of *add_to_frontier*. If n is an element of F_1, it's also an element of F_0 and so the left-hand side of the clause is true. Alternatively, if n is an element of *NN*, there is an arc from *Node* to n, by definition of *neighbors*, and a path from n to a goal, so there is a path from *Node* to a goal, hence the left-hand side of the clause is true.

To argue completeness of this algorithm, suppose there is a path from the frontier to a goal. If you select an element of that path, you will be one step closer to the goal. If the correct element of the frontier is chosen at any stage in the future, you will be one step closer to the goal. If the selection criteria is fair (page 50), each element of the frontier will eventually be chosen, and you are guaranteed to find a path to the goal. Note that not all search strategies are fair.

One subtle detail of the above algorithm should be noted: A node is only checked to determine if it's a goal *after* it has been selected from the frontier. An alternative would be to check a node to determine if it's a goal *before* it's added to the frontier. There are two main reasons for checking whether a node is a goal after it is selected:

- Sometimes there is a very costly arc from a node on the frontier to a goal node. You don't always want to return this as the path found, as there may be a much shorter path that needs a few steps. This is crucial when you require the least cost path to be found.

- You often need some computation to determine the label of a node. However, you don't actually need to examine a node to add it to the frontier. You can add a promise to generate the appropriate node. For example, in the search space for a top-down derivation (page 49), generating a node requires unification and building a new goal clause. All you need for the search is to be able to generate the details of a node when needed. You only need the details of a node when you check whether it's a goal node or when you find its neighbors. Generating the details of the node early may waste time if a solution can be found without a node being generated, and it may waste space in storing information before it's needed.

Costs

Some search algorithms use path lengths and thus need arc costs. Assume a relation

$$cost(N_1, N_2, C)$$

that's true if C is the length of the arc $\langle N_1, N_2 \rangle$.

Example 4.7

Assume that the arc costs for the graph of Example 4.6 are given by the relation below:

$cost(o103, ts, 8)$. $cost(o103, l2d3, 4)$.
$cost(o103, o109, 12)$. $cost(ts, mail, 6)$.
$cost(o109, o111, 4)$. $cost(o109, o119, 16)$.
$cost(o119, storage, 7)$. $cost(o119, o123, 9)$.
$cost(o123, r123, 4)$. $cost(o123, o125, 4)$.
$cost(l2d1, l3d2, 3)$. $cost(l2d1, l2d2, 6)$.
$cost(l2d2, l2d4, 3)$. $cost(l2d3, l2d1, 4)$.
$cost(l2d3, l2d4, 7)$. $cost(l2d4, o109, 7)$.
$cost(l3d2, l3d3, 6)$. $cost(l3d2, l3d1, 4)$.
$cost(l3d1, l3d3, 8)$.

Note that each arc is represented in two different ways here. One is is in terms of the *neighbors* relation and the other is in the *cost* relation. There are other representations that don't have such redundancy. Our representation was chosen to keep the relations simple.

Finding Paths

One problem with the above search algorithm is that although paths are implicitly discovered, they aren't accessible. Instead of having the frontier consist of nodes, it is better if the elements on the frontier include the paths found from the starting node. This allows a solution path to be returned and allows access to the paths on the frontier.

To implement this you need more structure in the elements of the frontier. You can treat the frontier as a list of terms of the form

$$node(Node, Path, PathCost)$$

where *Node* is a node in the graph, *Path* is the list of the nodes in the path up to, but not including, *Node*, and *PathCost* is the total cost of the path to *Node*. A path is represented by the list of its nodes in reverse order. Adding a node to the end of a path involves adding an element to the front of the list.

The following code implements this, where $psearch(F_0, Path)$ is true if F_0 is a frontier and *Path* is a path from a start node to a goal node that passes through one of the nodes on the frontier:

$psearch(F_0, [N|P]) \leftarrow$
  $select(node(N, P, PC), F_0, F_1) \wedge$
  $is\_goal(N).$
$psearch(F_0, Path) \leftarrow$
  $select(node(Node, P, PC), F_0, F_1) \wedge$
  $neighbors(Node, Neighbors) \wedge$
  $add\_paths(Neighbors, node(Node, P, PC), NF) \wedge$
  $add\_to\_frontier(NF, F_1, F_2) \wedge$
  $psearch(F_2, Path).$

$add\_paths(Ns, node(N, P, PC), NF)$ is true if NF is the list of new frontier elements to replace frontier element $node(N, P, PC)$ for each node in Ns, where Ns is a list of neighbors of node N.

$add\_paths([\,], FE, [\,]).$
$add\_paths([M|R], node(N, P, PC), [node(M, [N|P], NPC)|FR]) \leftarrow$
  $cost(N, M, C) \wedge$
  NPC is $PC + C \wedge$
  $add\_paths(R, node(N, P, PC), FR).$

To search from starting nodes $S = \{s_1, \ldots, s_m\}$, and produce *Path* as a path found to a goal, issue the query:

$?psearch([node(s_1, [\,], 0), \ldots, node(s_m, [\,], 0)], Path).$

Example 4.8

To use the path-finding search algorithm to find a path starting at $o103$ in the graph of Example 4.6, pose the query

$?psearch([node(o103, [\,], 0)], P).$

The *Path* returned is a path from $o103$ to $r123$ represented as the list of nodes on the path in reverse order. The three answers are:

$P = [r123, o123, o119, o109, o103]$
$P = [r123, o123, o119, o109, l2d4, l2d3, o103]$
$P = [r123, o123, o119, o109, l2d4, l2d2, l2d1, l2d3, o103]$

These answers will be returned for any search strategy. Only the order will change.

4.4   Blind Search Strategies

Notice that the problem determines the graph and the goal, and thus what the neighbors of a node are, but *not* how to select a node or how to add a node to the frontier. This is the job of a search strategy. A search strategy specifies which elements are selected from the frontier. Different strategies are obtained by modifying *select* and *add_to_frontier*.

This section presents three **blind search strategies** that don't take into account where the goal is. They can't see where they are going. When they happen to stumble on a goal, they report success.

Depth-First Search

The first strategy is **depth-first search**. In depth-first search, paths are pursued in a depth-first manner, searching one path to its completion before trying an alternative path. The new path is developed from the last choice point, which is the last node that has neighbors which haven't been investigated. Some paths might be infinite, in which case a depth-first search might never succeed.

This method involves *backtracking*: The algorithm selects the first alternative at each node, and it backtracks to the next alternative when it has pursued all of the paths from the first choice.

Depth-first search is implemented by treating the frontier as a stack; the element that's removed at any time is the last one that was added:

$$select(Node, [Node|Frontier], Frontier).$$
$$add\_to\_frontier(Neighbors, Frontier_1, Frontier_2) \leftarrow$$
$$append(Neighbors, Frontier_1, Frontier_2).$$

where *append* is defined before (page 84).

Depth-first search selects the first node in the frontier at each choice point and adds that node's neighbors to the front of the frontier.

Example 4.9   Consider depth-first search on the graph axiomatized in Example 4.6 (page 121). Initially the frontier is [o103]. At the next stage the frontier is

$$[ts, l2d3, o109].$$

Next *ts* is removed from the frontier, and replaced by its neighbors, resulting in the frontier

$$[mail, l2d3, o109].$$

Next *mail* is removed from the frontier and is replaced by its neighbors (of which there are none), resulting in the frontier

[*l2d3*, *o*109].

At this stage, *l2d3* is the top of the stack. Notice what has happened: You pursued all paths from *ts*, and when all of these were exhausted (there was only one), you backtracked to the next element of the stack, namely *l2d3*. Next, *l2d3* will be selected and paths from it pursued. To do this, *l2d3* is replaced in the frontier by its neighbors, resulting in the frontier

[*l2d1*, *l2d4*, *o*109].

Then *l2d1* is selected from the frontier and is replaced by its neighbors resulting in the frontier:

[*l3d2*, *l2d2*, *l2d4*, *o*109].

Now *l3d2* is selected from the frontier, and is replaced by its neighbors, resulting in the frontier:

[*l3d3*, *l3d1*, *l2d2*, *l2d4*, *o*109].

l3d3 has no neighbors, and thus the search "backtracks" to the last alternative that hasn't been pursued, namely from the node *l3d1*.

Example 4.10

Consider the trace of the path-finding search algorithm *psearch* on the same problem. At the stage where *l3d3* is about to be selected, the frontier is

[*node*(*l3d3*, [*l3d2*, *l2d1*, *l2d3*, *o*103], 17),

    node(*l3d1*, [*l3d2*, *l2d1*, *l2d3*, *o*103], 15),

    node(*l2d2*, [*l2d1*, *l2d3*, *o*103], 14),

    node(*l2d4*, [*l2d3*, *o*103], 11),

    node(*o*109, [*o*103], 12)].

Notice how each element of the frontier represents a path from the start with the property that the path up to the last node is an initial segment of the path [*l3d2*, *l2d1*, *l2d3*, *o*103] at the front of the frontier. This always holds for depth-first search.

To understand the complexity of depth-first search, consider an analogy with family trees where the neighbors of a node correspond to the children. Suppose the elements of the path at the head of the frontier correspond to the father, the grandfather, the great grandfather, and so forth. The final nodes of each path of the frontier correspond to the brothers, uncles, great uncles, and so on. If there is a branching factor of b and the first element of the list has length n, there can be at most $n \times (b - 1)$ other elements

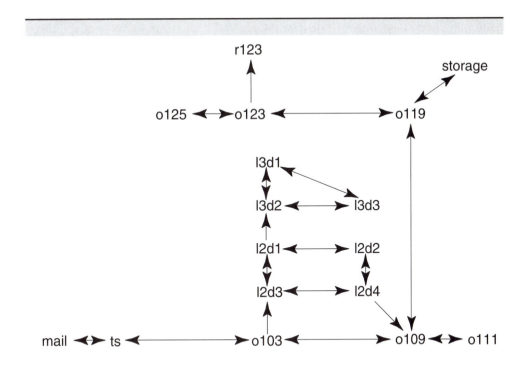

Links of the form $X \longleftrightarrow Y$ means there is an arc from X to Y and an arc from Y to X. That is, $\langle X, Y \rangle \in A$ and $\langle Y, X \rangle \in A$.

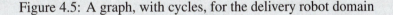

Figure 4.5: A graph, with cycles, for the delivery robot domain

of the frontier. These elements correspond to the $b - 1$ alternative paths that could be pursued by changing one value on the path. Thus, for depth-first search the space used is linear in the depth of the path length from the start to a node.

If there is a solution on the first branch searched, then the time complexity is linear in the length of the path length—it only considers those elements on the path, along with their siblings. The worst case complexity is infinite. Depth-first may get trapped on infinite branches and never find a solution, even if one exists, for infinite graphs or for graphs with loops.

Example 4.11   Consider the modification of the delivery graph presented in Figure 4.5. There is an infinite path leading from *ts* to *mail*, back to *ts*, back to *mail*, and so forth. As presented, depth-first search follows this path forever, never considering alternative paths from *l2d3* or *o109*. The frontier for the first five iterations of the path-finding search algorithm using depth-first search is

$$[node(o103, [], 0)]$$

$[node(ts, [o103], 8), node(l2d3, [o103], 4), node(o109, [o103], 12)]$

$[node(mail, [ts, o103], 14), node(o103, [ts, o103], 16),$
    $node(l2d3, [o103], 4), node(o109, [o103], 12)]$

$[node(ts, [mail, ts, o103], 20), node(o103, [ts, o103], 16),$
    $node(l2d3, [o103], 4), node(o109, [o103], 12)]$

$[node(mail, [ts, mail, ts, o103], 26), node(o103, [ts, mail, ts, o103], 28),$
    $node(o103, [ts, o103], 16), node(l2d3, [o103], 4),$
    $node(o109, [o103], 12)]$

Depth-first search can be improved by not considering paths with cycles (page 138). Depth-first search is appropriate when

- Space is restricted.
- There are many solutions, perhaps with long path lengths.
- The order of nodes in the list of neighbors of a node can allow you to tune the system so that solutions are found on the first try.
- You can easily determine when you are on the wrong path.

It's a poor method when

- It's possible to get caught in infinite paths.
- There are cycles in the graph.
- There are solutions at shallow depth.
- There is heuristic knowledge indicating when one path is likely to be better than another.

Breadth-First Search

The second search method considered is breadth-first search where the paths from the start node are generated in order of the number of arcs in the path. This is accomplished by always selecting one of the paths with the fewest arcs. This strategy is implemented by treating the frontier as a queue: The element that is removed at any time is the element of the queue that was added earliest.

To implement breadth-first search you can represent the queue as a list such that the first element of the list is the earliest element of the list. The neighbors are added to the end of the list instead of the beginning as in depth-first search:

$select(Node, [Node|Frontier], Frontier)$.

$add\_to\_frontier(Neighbors, Frontier_1, Frontier_2) \leftarrow$
    $append(Frontier_1, Neighbors, Frontier_2)$.

Notice how we have merely changed the order of arguments to *append* to transform depth-first search into breadth-first search.

Example
4.12

Consider breadth-first search on the graph of Example 4.6 (page 121). Initially the frontier is [*o*103]. This is replaced by its neighbors, making the frontier [*ts*, *l*2*d*3, *o*109]. These are the nodes one step away from *o*013. The next three elements of the frontier chosen are *ts*, *l*2*d*3 and *o*109, at which stage the frontier contains the neighbors of *ts* followed by the neighbors of *l*2*d*3 followed by the neighbors of *o*109, namely

[*mail*, *l*2*d*1, *l*2*d*4, *o*111, *o*119].

These are the nodes that are two steps away from *o*103. These five nodes are the next elements of the frontier chosen, at which stage the frontier contains the elements three steps away from *o*103, namely

[*l*3*d*2, *l*2*d*2, *o*109, *storage*, *o*123].

The same frontier for the path-finding search algorithm is

[*node*(*l*3*d*2, [*l*2*d*1, *l*2*d*3, *o*103], 11),

   node(*l*2*d*2, [*l*2*d*1, *l*2*d*3, *o*103], 14),

   node(*o*109, [*l*2*d*4, *l*2*d*3, *o*103], 18),

   node(*storage*, [*o*119, *o*109, *o*103], 35),

   node(*o*123, [*o*119, *o*109, *o*103], 37)]

Note how each of the paths on the frontier has the same number of steps. For breadth-first search, the number of steps in the paths on the frontier always differ by at most one.

Suppose the branching factor of the search is b. If the first path of the frontier contains n arcs, there are at least b^n elements of the frontier. All of these paths contain n or $n + 1$ arcs. Thus, both the space and the time complexities are exponential in the length of the shortest path to a goal, where the length is measured in the number of arcs. This method is, however, guaranteed to find a solution and will find a solution with the fewest arcs.

Breadth-first search is useful when

- Space is not a problem.
- You want to find the solution containing the fewest arcs.
- There may be few solutions, but with a short path length.
- There may be infinite paths.
- There is no extra (heuristic) knowledge to guide the search.

It is a poor method when all solutions have a long path length or there is some heuristic knowledge available. It isn't used very often because of its space complexity.

Comparing Algorithms

It seems obvious that we want to compare different search algorithms. This raises the issues of what criteria should be used to compare algorithms. You can compare algorithms on at least

- The time taken
- The space used
- The accuracy of the results

The time and the space used by an algorithm is a function of the inputs to the algorithm. In computer science, you often talk about the **asymptotic complexity** of algorithms. This specifies how the time or space grows with the input size of the algorithm. We say an algorithm has time, or space, complexity $O(f(n))$—read "big-oh of $f(n)$"—for input size n, where $f(n)$ is some function of n, if there exist constants n_0 and k such that the time, or space, of the algorithm is less than $k \times f(n)$ for all $n > n_0$. The most common types of functions are exponential functions such as 2^n, 3^n, or 1.015^n; polynomial functions such as n^5, n^2, n, or $n^{1/2}$; and logarithmic functions, $\log n$. In general exponential algorithms get worse quicker than polynomial algorithms which are, in turn worse than logarithmic algorithms.

An algorithm has time, or space, complexity $\Omega(f(n))$ for input size n, if there exists constants n_0 and k such that the time, or space, of the algorithm is greater than $k \times f(n)$ for all $n > n_0$. An algorithm has time or space complexity $o(n)$ if it has complexity $O(n)$ and $\Omega(n)$. Typically you cannot give an $o(f(n))$ complexity on an algorithm, as most algorithms take different times for different inputs. Thus, when comparing algorithms, you have to specify the class of problems that will be considered. When comparing two algorithms A and B, you can say A is better than B, where better is a measure of time, space, or accuracy, if

- The worst case of A is better than the worst case of B
- A works better in practice, or the average case of A is better than the average case of B, where you average over typical problems
- You characterize the class of problems for which A is better than B, so that which algorithm is better depends on the problem
- For every problem, A is better than B

The worst-case asymptotic complexity is often the easiest to show, but it is the least useful of these comparisons. Characterizing the class of problems for which one algorithm is better than another is usually the most useful, particularly if it is easy to determine which class a given problem is in. Unfortunately, this characterization is usually very difficult.

Characterizing when one algorithm is better than the other can be done either theoretically using mathematics, or empirically by building implementations. Theorems are only as valid as the assumptions on which they are based. Similarly, empirical investigations are only as good as the suite of test cases. It is easy to disprove a conjecture that one algorithm is better for some class of problems than another by showing a counterexample, but it is much more difficult to prove such a conjecture.

Lowest-Cost-First Search

When a non-unit cost is associated with arcs, you often want to find the solution that minimizes the total cost of the path. For example, costs may be distances and you may want to minimize the total distance; costs may be resources required by a robot to carry out the step represented by the arc, and you may want to minimize the resources used to solve the goal.

The search algorithms considered so far aren't guaranteed to find the minimum cost paths; they haven't used the arc cost information at all. Breadth-first search will find the solution with the fewest arcs first, but the distribution of arc costs may be such that the path of fewest arcs isn't the one of minimal cost.

The simplest search method guaranteed to find a minimum cost path is similar to breadth-first search, but instead of expanding the path with the minimum number of steps, you select the path with the minimum cost. This is implemented by treating the frontier as a priority queue, ordered by the function $g(n)$. Recall that $g(n)$ is a function that gives the cost of the path generated from the start node to node n. The following logic program assumes the priority queue is represented as a list sorted by $g(n)$. (This isn't the most efficient implementation; see Exercise 4.9.)

$$select(Node, [Node|Frontier], Frontier).$$
$$add\_to\_frontier(Neighbors, Frontier_1, Frontier_3) \leftarrow$$
$$append(Frontier_1, Neighbors, Frontier_2) \land$$
$$sort\_by\_g(Frontier_2, Frontier_3).$$

Example 4.13

Consider lowest-cost-first search with the graph axiomatized in Example 4.6 (page 121).

Initially the frontier is $[o103]$. At the next stage it is $[l2d3_4, ts_8, o109_{12}]$ (the subscripts are the path cost to the node). $l2d3$ is selected, with the resulting queue

$$[l2d1_8, ts_8, l2d4_{11}, o109_{12}].$$

$l2d1$ is selected resulting in frontier

$$[ts_8, l3d2_{11}, l2d4_{11}, o109_{12}, l2d2_{14}].$$

Then ts is selected, and the resulting priority queue is

$$[l3d2_{11}, l2d4_{11}, o109_{12}, mail_{14}, l2d2_{14}].$$

Then $l3d2$ is selected, and so forth. Note how lowest-cost-first search is growing many paths incrementally, always expanding the path with lowest cost.

If the costs of the arcs are bounded below by a positive constant and the branching factor is finite, lowest-cost-first search is guaranteed to find an **optimal solution**—a solution with lowest path cost—if one exists. Moreover, the first path to a goal that's found is a path with least cost. Such a solution is optimal, as you are generating the paths from the start in order of path cost. If there were a better path than the first solution found, it would have been selected from the frontier earlier.

Like breadth-first search, lowest-cost-first is typically exponential in both space and time.

4.5   Heuristic Search

All of the search methods in the preceding section are blind in that they didn't take into account the goal. They don't use any information about where they are trying to go until they happen to stumble on a goal. This section presents methods that use heuristic information about which nodes are the most promising to pursue at any stage. This information is in the form of a heuristic function $h(n)$, from node n into a non-negative real number. It is an estimate of the path length from node n to a goal node. $h(n)$ is an *underestimate* if $h(n)$ is less than or equal to the actual cost of the shortest-cost path from node n to a goal.

There is nothing magical about a heuristic function. It must only use information that can be quickly obtained about a node. There is a tradeoff between the amount of work it takes to derive a heuristic value for a node and how accurately the heuristic value of a node mirrors the actual path length from the node to a goal. For all of the examples, we assume that the heuristic value can be simply derived from the properties of a node.

Example
4.14

For the graph of Figure 4.2 (page 117) you can use the straight-line distance between each node and the goal position to build a heuristic function.

The examples below assume the following heuristic function:

$h(mail, 35)$. $h(ts, 29)$. $h(o103, 21)$.
$h(o109, 29)$. $h(o111, 33)$. $h(o119, 13)$.
$h(o123, 4)$. $h(o125, 8)$. $h(r123, 0)$.
$h(l2d1, 13)$. $h(l2d2, 19)$. $h(l2d3, 17)$.
$h(l2d4, 22)$. $h(l3d1, 8)$. $h(l3d2, 10)$.
$h(l3d3, 16)$. $h(storage, 12)$.

This h-function is an underestimate as the h value is less than or equal to the exact distance. It's the exact distance for node $o123$, but it's very much an underestimate for node $l3d1$.

Example 4.15 | For an SLD search graph, such as in Figure 4.3, one possible heuristic function is the number of atoms in the answer clause associated with the node. This is an underestimate of the path length, as resolving each atom away will need a path length of at least this number.

It is possible to carry out a more detailed analysis. For each predicate you can derive a lower bound on how many steps it takes to prove any instance of that predicate. The value of a node would then be the sum of the values of the predicates in the answer clause. This would take more work to derive the heuristic value, but it would result in a more accurate heuristic function than summing the number of atoms. It is not really sensible for the ground case, because in this case you may as well derive the atoms that follow from a knowledge base. The extra work taken to derive the heuristic function may pay off for the non-ground case.

The heuristic function is a way to quickly inform the search about the direction to a goal. It provides an informed way to guess which neighbor of a node will lead to a goal.

Best-First Search

The simplest way to use the heuristic function is to always choose the element of the frontier that appears to be closest to the goal. In other words, choose the node n with the lowest value of $h(n)$. This is called **best-first search**. This strategy is implemented by treating the frontier as a priority queue, ordered by the function $h(n)$. Thus this is like lowest-cost-first search, but choosing the element with minimal h-value rather than the element of the queue with minimal g-value.

A naive way to implement best-first search is to append the neighbors to the frontier and then sort this new frontier by h:

$select(Node, [Node|Frontier], Frontier)$.

$add\_to\_frontier(Neighbors, Frontier_1, Frontier_3) \leftarrow$
$\quad append(Frontier_1, Neighbors, Frontier_2) \wedge$
$\quad sort\_by\_h(Frontier_2, Frontier_3)$.

A better way to do this is to implement the frontier as a heap (see exercise 4.9), in which case you can insert and remove nodes in $\log(n)$ time, where n is the size of the frontier.

Example 4.16 | Consider first the case of best-first search with the graph axiomatized in Example 4.6 (page 121).

Initially the frontier is $[o103]$. At the next stage the frontier is $[l2d3_{17}, o109_{29}, ts_{29}]$, where the subscripts are the heuristic value of the node. $l2d3$ is selected, with the resulting queue $[l2d1_{13}, l2d4_{22}, o109_{29}, ts_{29}]$. Then $l2d1$ is selected, and the resulting priority queue is $[l3d2_{10}, l2d2_{19}, l2d4_{22}, o109_{29}, ts_{29}]$. Then $l3d2$ is selected. Best-first search grows many paths incrementally, always expanding the node which appears closest to the goal.

This search is focused in a limited way. Best-first search pursues a number of paths, expanding the node that seems locally the best at the time. Best-first search has the disadvantage, like breadth-first search, of using space that's exponential in the path length. However, unlike breadth-first search, it isn't guaranteed to find a solution even if one exists. It doesn't necessarily find the shortest path first.

Heuristic Depth-First Search

Heuristic depth-first is a way to use heuristic knowledge in depth-first search. It retains the space advantages of depth-first search, while using problem-specific information to guide the search.

The idea is to make the locally best choice according to the heuristic function by ordering the neighbors before adding them to the front of the frontier.

Heuristic depth-first is implemented by using the heuristic function $h(n)$ to sort the neighbors of the node before adding them to the frontier, which still acts as a stack:

$select(Node, [Node|Frontier], Frontier)$.

$add\_to\_frontier(Neighbors, Frontier_1, Frontier_2) \leftarrow$
$\quad sort\_by\_h(Neighbors, Sorted\_neighbors) \wedge$
$\quad append(Sorted\_neighbors, Frontier_1, Frontier_2)$.

Heuristic depth-first locally chooses which subtree to develop. It chooses a neighbor of the current node to pursue, and it only considers other neighbors if all paths from the chosen node end without finding a solution. Heuristic depth-first search is a variation of hill climbing (page 156).

Example 4.17

Consider heuristic depth-first search on Example 4.6 (page 121). The frontier is initially $[o103]$. The neighbors of $o103$ are sorted according to the heuristic function resulting in the queue $[l2d3, o109, ts]$. Then it does a depth-first search from $l2d3$. If the paths from $l2d3$ fail, it searches from $o109$. If all of these paths fail, it searches from ts. If these searches fail, the search from $o103$ fails.

Heuristic depth-first travels locally towards the goal, but once it has chosen a direction, it pursues all paths from that choice before trying another path. Thus, it can be "led up the garden path." If there is an infinite path from the first neighbor chosen, it would never choose the second neighbor. This may preclude finding a solution.

Example
4.18

> Once it has chosen $l2d3$ as the best neighbor of $o103$, it pursues all paths from $l2d3$ before choosing another neighbor of $o103$. Even if the paths from $l2d3$ get arbitrarily bad, it still doesn't consider the paths from $o109$ until all of the paths from $l2d3$ have been exhausted. If there were an infinite path from $l2d3$, as in Figure 4.5, this would be followed, and so it would never try going from $o103$ to $o109$.

A^* Search

A^* is an enhancement of lowest-cost-first and best-first searches that considers both the current path cost as well as heuristic information in its selection of current best path. For each path on the frontier, A^* uses an estimate of the total path length from a start node to a goal node constrained to start along that path. It uses $g(n)$, the length of the path found from a start node to frontier node n, as well as the heuristic function $h(n)$, the estimated path length from node n to the goal.

For any node n on the frontier, define $f(n) = g(n) + h(n)$ as the estimate of the total path length from the start node to a goal node, constrained to start via the path found to node n.

$$\underbrace{start \xrightarrow{\text{actual}} n}_{g(n)} \underbrace{\xrightarrow{\text{estimate}} goal}_{h(n)}$$
$$\underbrace{\phantom{start \xrightarrow{\text{actual}} n \xrightarrow{\text{estimate}} goal}}_{f(n)}$$

A^* is implemented by treating the frontier as a priority queue ordered by $f(n)$:

$select(Node, [Node|Frontier], Frontier)$.

$add\_to\_frontier(Neighbors, Frontier_1, Frontier_3) \leftarrow$
$\quad append(Frontier_1, Neighbors, Frontier_2) \wedge$
$\quad sort\_by\_f(Frontier_2, Frontier_3)$.

Example
4.19

> Let's see what happens using A^* search on Example 4.6 (page 121). The frontier is initially $[o103]$, with $o103$ having an f-value of 21. It is replaced by its neighbors, forming the frontier
>
> $$[l2d3_{21}, ts_{37}, o109_{41}]$$
>
> (with the paths of the frontier indicated by their final node, subscripted by the f-values of the path). Note that $f(l2d3) = g(l2d3) + h(l2d3) = 4 + 17 = 21$. Next $l2d3$ is selected and replaced by its neighbors, forming the frontier
>
> $$[l2d1_{21}, l2d4_{33}, ts_{37}, o109_{41}].$$

Then $l2d1$ is selected and replaced by its neighbors forming the frontier

$$[l3d2_{21}, l2d2_{33}, l2d4_{33}, ts_{37}, o109_{41}].$$

Then $l3d2$ is selected and replaced by its neighbors, forming

$$[l3d1_{23}, l3d3_{33}, l2d2_{33}, l2d4_{33}, ts_{37}, o109_{41}].$$

Up to this stage you have been continually exploring what seems to be the direct path to the goal. Next $l3d1$ is selected and replaced by its neighbors, forming the frontier

$$[l3d3_{33}, l2d2_{33}, l2d4_{33}, ts_{37}, l3d3_{39}, o109_{41}].$$

At this stage there are two different paths to the node $l3d3$ on the queue: The one that doesn't go through $l3d1$ has a lower f-value than the one that does. Later (page 139) we consider when you can prune one of these paths. The shortest path to $l3d3$ is selected next and replaced by its neighbors, forming

$$[l2d2_{33}, l2d4_{33}, ts_{37}, l3d3_{39}, o109_{41}].$$

Then $l2d2$ is selected and replaced by its neighbors, forming

$$[l2d4_{33}, ts_{37}, l2d4_{39}, l3d3_{39}, o109_{41}].$$

Note that here you are pursuing many different paths from the start.

The lowest cost path is eventually found. The algorithm was forced to try many different paths, since several of them temporarily seemed to be the shortest. It still does better than either lowest-cost-first search or best-first search.

The property that A^* always finds an optimal path, if one exists, and the first path found to a goal is optimal is called the **admissibility** of A^*. Admissibility means that, even in the case where the search space is infinite, if solutions exist, an optimal solution—a shortest path from a start node to a goal node—will be found.

Proposition 4.1   (A^* admissibility): If there is a solution, A^* always finds a solution, and the first solution found is optimal, if

- the branching factor is finite (each node has only a finite number of neighbors)

- arc costs are bounded above zero (there is some $\epsilon > 0$ such that all of the arc costs are greater than ϵ), and

- $h(n)$ is a lower bound on the actual minimum cost of the shortest path from n to a goal node.

Proof: If the arc costs are bounded above $\epsilon > 0$, eventually $g(n)$ will exceed any finite number, and thus will exceed a solution length if there is one (at depth in the search tree no greater than m/ϵ, where m is the solution length). Therefore cycles in the search graph will increase the value of $g(n)$ for nodes that are repeatedly inserted in the frontier, so that picking the frontier node with minimum value of $f(n) = g(n) + h(n)$ will eventually pick a frontier node on the solution path, if a solution exists.

Let's argue that the first path to a goal selected is an optimal path. The f-value for any node on an optimal solution path is less than or equal to the f-value of an optimal solution. This is because h is an underestimate of the actual cost from a node to a goal. Thus the f-value of a node on an optimal solution path is less than the f-value for any non-optimal solution. Thus a non-optimal solution can never be chosen while there is a node on the frontier that leads to an optimal solution (as an element with minimum f-value is chosen at each step). So before we can select a non-optimal solution, you will have to pick all of the nodes on an optimal path, including each of the optimal solutions.

Q.E.D.

It should be noted that the admissibility of A^* doesn't ensure that every intermediate node selected from the frontier is on the optimal path from the start node to a goal node (see multiple-path pruning on page 139).

Admissibility relieves you from worrying about cycles and ensures that the first solution found will be optimal. It doesn't ensure that the algorithm won't change its mind about which partial path is the best while it's searching.

Summary of Search Strategies

The table in Figure 4.6 gives a summary of the searching strategies presented.

The depth-first methods are linear in space with respect to the path length found, but aren't guaranteed to find a solution if one exists. Breadth-first, lowest-cost-first and A^* may be exponential in both space and time, but are guaranteed to find a solution if one exists, even if the graph is infinite (as long as there is a finite branching factor). With perfect heuristics, heuristic depth-first, best-first, and A^* all lead directly to solutions, but with no heuristics they degenerate to depth-first search, random search (any node on the frontier could be chosen), and lowest-cost-first search, respectively.

Strategy	Selection from Frontier	Halts?	Space
Depth-first	Last node added	No	Linear
Breadth-first	First node added	Yes	Exponential
Heuristic depth-first	Locally minimal $h(n)$	No	Linear
Best-first	Globally minimal $h(n)$	No	Exponential
Lowest-cost-first	Minimal $g(n)$	Yes	Exponential
A^*	Minimal $f(n) = h(n) + g(n)$	Yes	Exponential

"Halts?" means "Is the method guaranteed to halt if there is a path to a goal on a (possibly infinite) graph with a finite number of neighbors for each node and with arc lengths bounded above zero?" When an algorithm halts, it has worst case exponential complexity in the size of the path length, and those algorithms that are not guaranteed to halt have infinite worst-case complexity.

Space is either *linear* in path length or is *exponential* in path length.

Figure 4.6: Summary of search strategies

4.6   Refinements to Search Strategies

This section describes some refinements that can be made to the above strategies. We don't extend the implementation to cover these refinements but the extensions are relatively straightforward.

There are two methods that are applicable when there are cycles in the graph: One checks explicitly for cycles while the other, more general method checks for multiple paths to a node. We then consider a general method to have the systematic search of breadth-first search or A^* search, but with the space advantages of depth-first search. Finally we consider methods to break down a bigger search problem into a number of smaller search problems, each of which may be much easier to solve.

Cycle Checking

It's possible for a graph representing a search space to include cycles. For example, in the robot delivery domain of Figure 4.5, the robot can go back and forth between $o103$ and $o109$. Some of the search methods above can get trapped in cycles, continuously repeating the cycle and never finding an answer even in finite graphs.

The simplest method of pruning the search tree and guaranteeing that you will find a solution in a finite graph is to ensure that you don't consider neighbors that are already on the path from the start. This is known as a *cycle check*. You can implement a cycle check by pruning a node that already appears on the path from the start node to that node, either when it's added to the frontier or when it's selected.

The computational complexity of a cycle check depends on the search method being used. For depth-first methods where the graph is explicitly stored, the overhead can be as low as a constant factor; you can simply add a bit to each node in the graph, set it when the node is expanded, and reset it on backtracking. You avoid cycles by never expanding a node with a set bit. This can be done because depth-first search maintains a single current path. The elements on the frontier are alternative branches from this path.

For the search strategies that maintain multiple paths, namely all of those with exponential space in Figure 4.6, a cycle check takes time linear in the length of the path being searched. You essentially have to search up the partial path being considered, checking to ensure you don't add a node that already appears.

Multiple-Path Pruning

There is another way in which the search space can be pruned. There is often more than one path to a node. If you don't need to find more than one path, then you can prune from the frontier any node to which you have already found a path. If you want to find the shortest path to a goal, then you have to ensure that the first path found to any node is the shortest path; or if you find a shorter path to a node, then you have to either remove all paths which used the longer path to the node (as these can't be optimal) or incorporate a new initial section on these paths.

In breadth-first and lowest-cost-first search, the first path found to a node (when the node is selected from the frontier) is the shortest path to the node; pruning subsequent paths can't remove a shorter path. This can be implemented by keeping a *closed* list of nodes that have been selected, and checking whether a node is on the closed list before expanding the node.

As described (page 135), with A^* it's possible that when a node is selected for the first time, the associated path isn't the shortest path to that node. To see when this arises, suppose you have selected a path to node n for expansion, but there is a shorter path to node n, which you haven't found yet. Then there is another node n' on the frontier, through which that shorter path passes. You must have $f(n) \leq f(n')$ because n was selected before n'. This means

$$g(n) + h(n) \leq g(n') + h(n').$$

If the path to n via n' is shorter than the path found to n, then you have

$$g(n') + d(n', n) < g(n),$$

where $d(n', n)$ is the actual distance from node n' to n. From these two equations, you can derive

$$d(n', n) < g(n) - g(n') \leq h(n') - h(n).$$

You can thus ensure that the first path found to any node is the shortest path if you put the restriction on h that $|h(n') - h(n)| \leq d(n', n)$ for any two nodes n and n'. This is called the *monotone restriction*. It is applicable to, for example, the heuristic function of Euclidean distance (the straight line distance in an n-dimensional Euclidean space) between two points when the cost function is distance.

In essence the monotone restriction ensures that there is no special undiscovered shortcut from n' to n. With the monotone restriction, the f-values on the frontier are monotonically nondecreasing. When the frontier is expanded, the f-values don't get smaller.

Multiple-path pruning subsumes a cycle check, since a cycle is another path to a node and is therefore pruned. This can be done in constant time, if you store all of the nodes that you have looked at. Multiple-path pruning is preferred over cycle checking for breadth-first methods, where you have to store virtually all of the nodes considered anyway. For depth-first search strategies, however, you don't want to store all of the nodes already considered (as this then makes the method exponential in space), and so cycle checking is preferred over multiple-path checking for those methods.

Iterative Deepening

So far none of the methods have been ideal; the only ones that guarantee that a path will be found need exponential space. (See Figure 4.6.) One way to combine the space efficiency of depth-first search with the optimality of breadth-first methods is to use **iterative deepening**. The idea is to recompute the elements of the frontier rather than storing them. Each recomputation can be a depth-first search which will thus use less space.

Consider making a breadth-first search into an iterative deepening search. This is carried out by having a depth-first searcher which searches only to a limited depth. You can first do a depth-first search to depth one by building paths of length one in a depth-first manner. Then you build paths to depth two, and so on. You throw away all of the previous computation each time and start again. Eventually you will find a solution, if one exists, and as you are enumerating paths in order, the shortest (in terms of number of arcs) will always be found first.

When implementing an iterative deepening search, you have to distinguish between

- failure to find a solution because the depth bound was reached
- failure of the depth-bound search that didn't involve reaching the depth bound

In the first case you need to retry the path at a higher depth. In the second case it's a waste of time, as you know there is no path no matter what the depth.

An implementation of iterative-deepening search is presented in Figure 4.7. This is a modification of the generic search algorithm that uses depth-first search and a richer representation of the nodes on the frontier. The iterative-deepening searcher does a depth-first search, not pursuing paths that exceed the depth bound. This program fails whenever breadth-first search would fail and only returns a path once, even though it may be rediscovered in subsequent iterations:

- The depth bound is increased if the depth bound search was truncated by reaching the depth-bound. In this case the search failed *unnaturally*. There are no more answers if the search didn't prune any paths due to the depth bound. In this case the search failed *naturally*. This ensures that when the depth bound was not reached, the program will stop and report no more paths.

- It only reports a solution path if that path would not have been reported in the previous iteration. Thus it only reports paths whose length equals the depth bound.

The obvious problem with iterative deepening is the wasted computation which occurs at each step. This, however, may not be as bad as may be at first thought, particularly if the branching factor is high. Consider the running time of the algorithm. Assume you have a constant branching factor of $b > 1$. Consider the search where the bound is k. At path length k, you have b^k nodes; each of these has been generated once. The nodes at depth $k - 1$ have been generated twice, those at depth $k - 2$ have been generated three times, etc., and the nodes at depth 1 have been generated k times. The total number of nodes generated is thus

$$
\begin{aligned}
& b^k + 2b^{k-1} + 3b^{k-2} + \cdots + kb \\
= \; & b^k(1 + 2b^{-1} + 3b^{-2} + \cdots + kb^{1-k}) \\
\leq \; & b^k \left(\sum_{i=1}^{\infty} ib^{(1-i)} \right) \\
= \; & b^k \left(\frac{b}{b-1} \right)^2.
\end{aligned}
$$

So there is a constant overhead $(b/b - 1)^2$ times the cost of just generating the nodes at depth n. When $b = 2$ there is an overhead factor of 4, and when $b = 3$ there is an overhead of 2.25 over just generating the frontier. In fact, this algorithm is $O(b^k)$, and there can't be an asymptotically better uninformed search strategy.

If the branching factor is close to one, increasing the depth bound by one each time may not be a sensible solution. In particular, with a branching factor of 1 at depth k, iterative deepening will have searched $k + (k - 1) + (k - 2) + \cdots + 1 = k(k + 1)/2$ nodes, whereas the other search methods will have searched k nodes. One way to

% *dbsearch*(F, DB, Q, *How*1, P) is true if a depth-bound search from frontier F can
% find a path P of length $\geq DB$. Q is the initial frontier to (re)start from. *How* specifies
% whether the explored branches for the current depth bound all failed *naturally*,
% without the depth bound being reached, or some failed *unnaturally* and were pruned
% due to the depth bound. The frontier is a list of *node*(*Node*, *Path*, *PathLength*).
% *Node* is a node. *Path* is the list of nodes on the path found to Node, in reverse order.
% *PathLength* is the length of Path.

> *dbsearch*([*node*(N, P, DB)|_], DB, _, _, [$N|P$]) $\leftarrow$
>> *is_goal*(N).
>
> *dbsearch*([*node*(N, P, PL)|F_1], DB, Q, H, S) $\leftarrow$
>> $PL < DB \wedge$
>>
>> *neighbors*(N, NNs) $\wedge$
>>
>> $PL1$ is $PL + 1 \wedge$
>>
>> *add_paths_db*(NNs, [$N|P$], $PL1$, F_1, F_2) $\wedge$
>>
>> *dbsearch*(F_2, DB, Q, H, S).
>
> *dbsearch*([*node*(_, _, PL)|F_1], DB, Q, _, S) $\leftarrow$
>> $PL \geq DB \wedge$
>>
>> *dbsearch*(F_1, DB, Q, *unnaturally*, S).
>
> *dbsearch*([ ], DB, Q, *unnaturally*, S) $\leftarrow$
>> $DB1$ is $DB + 1 \wedge$
>>
>> *dbsearch*(Q, $DB1$, Q, *naturally*, S).

% *add_paths_db*(NNs, *Path*, *PL*, F_0, F_1) adds the list *Ns* of nodes to the front of fron-
% tier F_0 forming frontier F_1. The nodes need to be converted to the form of elements
% of the frontier. *Path* is the path found, and *PL* is the path's length.

> *add_paths_db*([ ], _, _, F, F).
>
> *add_paths_db*([$NN|R$], *Path*, *PL*, F_0, [*node*(NN, *Path*, *PL*)|F_1]) $\leftarrow$
>> *add_paths_db*(R, *Path*, *PL*, F_0, F_1).

% *idsearch*(Ns, P) is true if P is a path found from starting nodes Ns using iterative
% deepening search.

> *idsearch*(Ns, P) $\leftarrow$
>> *add_paths_db*(Ns, [ ], 0, [ ], Fr) $\wedge$
>>
>> *dbsearch*(Fr, 0, Fr, *natural*, P).

Figure 4.7: Iterative deepening searcher

reduce the overhead, particularly if you aren't concerned with finding a shortest path solution, is to double the depth bound at each step, instead of increasing the bound by one at each step. The resulting overhead is then $O(k \log k)$.

Iterative deepening can also be applied to A^* search. **Iterative deepening A^*** (IDA*) does repeated depth-bounded depth-first searches. Instead of the bound being on the number of arcs in the path, it is a bound on the value of $f(n)$. The threshold starts at the value of $f(s)$, where s is the starting node with minimal h-value. IDA* then carries out a depth-first depth-bounded search, but never expands a node with a higher f-value than the current bound. If the depth-bounded search fails *unnaturally*, the next bound is the minimum of the f-values that exceeded the previous bound. IDA* thus checks the same nodes as A^*, but recomputes them with a depth-first search instead of storing them.

Direction of Search

The general searching problem is exponential in the path length. Anything that can be done to reduce the exponent or to reduce the factor that you are taking to the exponent will give great saving. This section investigates methods to reduce the combinatorial explosion inherent in search.

The abstract definition of the graph-searching method of problem solving is symmetric in the sense that it doesn't matter if you begin with a start node and search forward for a goal node, or begin with a goal node and search backward for a start node. However, in many applications the goal is determined implicitly by a predicate and not explicitly as a set of nodes, so backward search may be impossible.

For those cases where the goal nodes are explicit, it may be more efficient to search in one direction than the other. The complexity of the search space is exponential in the branching factor. It's typically the case that forward and backward search have different branching factors. A general principle is to search forward or backward depending on which has the smaller branching factor.

There are many cases where you can do better then this. This section considers three ways in which search efficiency can be improved for many search spaces.

Bidirectional Search

The idea of a bidirectional search is to reduce the search by searching forward from the start and backward from the goal simultaneously. When the two search frontiers intersect, you reconstruct a single path which extends from the start state through the frontier intersection to the goal.

A new problem arises: ensuring that the two search frontiers actually meet. For example, a depth-first search in both directions is likely to be disastrous since its small search frontiers are likely to pass each other by. Breadth-first search in both directions would be guaranteed to meet.

A combination of depth-first search in one direction and breadth-first search in the other would guarantee the required intersection of the search frontiers, but the choice of which to apply in which direction may be difficult. The decision depends on the cost of saving the breadth-first frontier and searching it to check when the depth-first method will intersect one of its elements.

There are situations where bidirectional search can result in substantial savings. For example, if the forward and backward branching factors of the search space are both b, and the goal is at depth k, then breadth-first search will take time proportional to b^k, whereas a symmetric bidirectional search will take time proportional to $2b^{k/2}$, an exponential savings in time; however, the time complexity is still exponential. Note that this complexity analysis assumes that finding the intersection of frontiers is free; this may not be a valid assumption for many applications.

Island-Driven Search

One of the ways that search may be made more efficient is to identify a limited number of places where the forward search and backward search can meet. For example, in searching for a path from two rooms on different floors, it may be appropriate to constrain the search to first go to the elevator on one level, then to the elevator on the goal level. Intuitively, these designated positions are **islands** in the search graph, which are constrained to be on a solution path from the start node to a goal node.

When islands are specified, you can decompose the search problem into several search problems—for example, one from the initial room to the elevator, one from the elevator on one level to the elevator on the other level, and one from the elevator to the destination room. This reduces the search space by having three simpler problems to solve. Having smaller problems helps to reduce the combinatorial explosion of large searches and is an example of how extra knowledge about a problem can be used to improve efficiency of search.

To find a path between s and g using islands:

1. Identify a set of islands $i_0, ..., i_k$;
2. Find paths from s to i_0, from i_{j-1} to i_j for each j from 1 to k, and from i_k to g.

Each of these searching problems should be correspondingly simpler than the general problem and therefore easier to solve.

The identification of islands is extra knowledge beyond that which is in the graph. The use of inappropriate islands may make the problem more difficult (or even impossible to solve). Typically you don't want to have to search through a large space of possible islands. It may also be possible to identify an alternate decomposition of the problem by choosing a different set of islands. For example, you could choose the bottom and top of the stairs if you had a robot that could climb stairs.

Searching in a Hierarchy of Abstractions

The notion of islands can be used to define a form of problem abstraction, where you consider problem-solving strategies that work at multiple levels of detail or multiple levels of abstraction.

The idea of searching in a hierarchy of abstractions is to first abstract the problem, leaving out as many details as possible. A partial solution to a problem may be found, one that requires further details to be worked out. For example, the problem of getting from one room to another requires the use of many instances of turning, but you would like to reason about the problem at a level of abstraction where the details of the actual steering are omitted. It's expected that an appropriate abstraction solves the problem in broad strokes, leaving only minor problems to be solved. The route planning problem for the delivery robot is too difficult to solve by searching without leaving out details until you need to consider them.

Searching in a hierarchy of abstractions depends very heavily on *how* one decomposes and *abstracts* the problem to be solved. Once the problems are abstracted and decomposed, any of the search methods can be used to solve them. It isn't easy, however, to recognize useful abstractions and problem decompositions.

Dynamic Programming

One other method deserves attention because it's important in many optimization problems, particularly those involving decision making under uncertainty (see Chapter 10). One intuition behind **dynamic programming** is to construct the perfect heuristic function so that heuristic depth-first is guaranteed to find a solution without ever backtracking. The heuristic function constructed isn't really a heuristic function but represents the exact costs of a minimal cost path from each node to the goal. From this value function, you can construct a specification of which arc to take at every step, called a **policy**.

Let $dist(n)$ be the actual distance of the shortest path from node n to a goal. $dist(n)$ can be defined as

$$dist(n) = \begin{cases} 0 & \text{if } is\_goal(n), \\ \min_{\langle n,m\rangle \in A}(|\langle n, m\rangle| + dist(m)) & \text{otherwise.} \end{cases}$$

The general idea is to start at the goal and build a table of $dist(n)$ for each node. This can be done by carrying out an algorithm similar to lowest-cost-first search, with multiple-path pruning, from the goal nodes in the *inverse graph*, which is the graph with all arcs reversed. Rather than having a goal to search for, the dynamic programming algorithm records the *dist* values for each node found. It uses the inverse graph, because you want the distances from each node to the goal, and not the distances from the goal to each node. In essence, dynamic programming works backwards from the goal, trying to build the shortest path to the goal from each node in the graph.

For a particular goal, once the *dist*-value for each node has been recorded, you can use the *dist*-value to go from each node n to its neighbor m which minimizes $|\langle n, m \rangle| + dist(m)$. This policy will tell us how to get from any node to the goal in the shortest path. Each step in the policy can be executed in constant time with respect to the size of the graph, assuming a bounded number of neighbors for each node. It takes time and space linear in the size of the graph to build the *dist* table.

Dynamic programming is useful when:

- You want the shortest path
- The graph is finite and small enough to be able to store the *dist*-value for each node
- The goal does not change very often
- The policy is used a number of times for each goal, so that the cost of generating the *dist*-values can be amortized over many instances of the problem

The main problems with dynamic programming are:

- It only works when the graph is finite
- You need to recompute a policy for each different goal
- The time and space required is linear in the size of the graph; the graph size for finite graphs is typically exponential in the path length

| Example 4.20 | For the graph axiomatized in Example 4.6 (page 121), the distance from $r123$ to the goal is 0, thus you have trivially |

$$dist(r123) = 0.$$

Continuing with a shortest first search from $r123$, you can derive:

$$dist(o123) = 4$$
$$dist(o119) = 13$$
$$dist(o109) = 29$$
$$dist(l2d4) = 36$$
$$dist(l2d2) = 39$$
$$dist(o103) = 41$$
$$dist(l2d3) = 43$$
$$dist(l2d1) = 45$$

At this stage the backward search halts. There are two things that can be noticed here. First is that if a node doesn't have a *dist* value, then there is no path to the goal from that node. Second, you can quickly determine the next arc on the shortest path to the goal for any node. For example, to determine the shortest path from $o103$ to $r123$, you need to compare $4 + 43$ (the cost of going via $l2d3$) with $12 + 29$

> (the cost of going straight to $o109$) and can quickly determine that the robot should go directly to $o109$ from $o103$.

When building the *dist* function, you have implicitly determined which neighbor leads to the goal, and so you probably should store the cost of the node as well as which next step leads to the shortest distance to the goal.

4.7 Constraint Satisfaction Problems

In **constraint satisfaction problems** (CSPs) or what could be called multidimensional selection problems, you are given a set of variables, a domain for each variable, and a set of constraints or an evaluation function. These problems involve choosing a value for each variable so that the total assignment satisfies the constraints or optimizes the evaluation function. The multidimensional aspect of these problems, where each variable can be sees as a separate dimension, make them difficult but also provide structure that can be exploited. While such problems can be regarded as graph-searching problems, they are interesting as a special case as they arise in many applications and have some associated special algorithms.

CSPs can be divided into two main classes:

- *Satisfiability problems,* where the goal is to find an assignment of values to variables that satisfies some constraints. We assume that a constraint depends only on the assignment of values to variables; an assignment of values to the variables either satisfies the constraints or not. Often the constraints only depend on the values of a subset of the variables; this fact can be exploited for efficiency.

- *Optimization problems,* where each assignment of a value to each variable has a cost or an objective value associated with it. The goal is to find an assignment with the least cost or with the highest objective value. This is called the **optimal** assignment.

The constraints of satisfiability problems, which have to be met, are often called **hard constraints**. The costs in optimization problems, which specify preferences rather than what has to be met, are often referred to as **soft constraints**. Many problems are a mix of these two, for example, the robot deliverer may have some duties that it has to do, some other preferences on what it can do, and some things it can't do.

Sometimes the values are from a discrete set with no structure, but often the values come from a more structured set such as the the integers or the reals where it makes sense to talk about local values in each dimension.

CSPs can be considered as graph-searching problems in at least two different ways:

- In the first, a node corresponds to an assignment of a value to all of the variables, and the neighbors of a node correspond to changing one variable value to a local value. These problems differ from the graph-searching problems presented so far in that you aren't interested in the path, there is no starting node, and you can easily generate an arbitrary node (by choosing an assignment of values to variables), so that any node can be used as a starting point.

- In the second case, you totally order the variables as $\langle X_1, \ldots, X_n \rangle$. A node corresponds to an assignment of values to the first k variables, as in $X_1 = v_1 \wedge \cdots \wedge X_k = v_k$ for some values $v_1, \ldots, v_k$. The neighbors of a node that represents $X_1 = v_1 \wedge \cdots \wedge X_k = v_k$ are all of the nodes that assign the same variables to the first k variables and assign different values to variable X_{k+1}. That is, the neighbors of node $X_1 = v_1 \wedge \cdots \wedge X_k = v_k$ are the nodes $X_1 = v_1 \wedge \cdots \wedge X_k = v_k \wedge X_{k+1} = v_{k+1}$ for each value v_{k+1} in the domain of X_{k+1}. The start node is where $k = 0$; it doesn't assign a value to any node, and a goal node is one where all of the variables have been assigned a value and the variable assignment either satisfies the constraints or has a value that is optimal.

When the variables all are discrete with a finite number of values, these problems can be solved by considering all of the assignments of a value to each variable, but often there are too many such assignments (there are exponentially many assignments in the number of variables). Often, however, the variables don't have a finite number of possible values (e.g., they may take real values), and so each value can't be tested in turn.

For these classes of problems we will consider two quite different classes of solution methods. The first exploits the multidimensional aspect of these problems and concise encoding of the constraints, and it still searches the space of possibilities systematically but uses smart techniques to prune off large chunks of the search space. This is exemplified by the arc consistency algorithm (page 153). The second gives up on systematically searching the space; instead it does iterative improvement of a current assignment of values to variables, random assignments of values to variables, or a mixture of both.

Posing a Constraint Satisfaction Problem

A CSP is characterized by a set of variables $V_1, V_2, \ldots, V_n$. Each variable V_i has an associated domain $\mathbf{D}_{V_i}$ of possible values. For satisfiability problems, there are constraint relations on various subsets of the variables which give legal combinations of values for these variables. These constraints can be specified by subsets of the Cartesian products of the domains of the variables involved. A solution to the CSP is an n-tuple of values for the variables that satisfies all the constraint relations. For optimization problems there is a function that gives a cost for each assignment of a value to each variable. A solution to an optimization problem is an n-tuple of values for the variables that optimizes the cost function.

Consider, for example, solving problems that require temporal reasoning. Suppose you wish to schedule a set of activities. The activities could be meetings of people, assembly actions for a complex object such as a water pump or the manufacturing processes, such as casting, milling, and drilling, involved in machining a part like a connecting rod. For each activity you are given the times at which it may start and you are told each activity will take one unit of time to complete. You are also told to satisfy various constraints arising from prerequisite requirements and resource use limitations. For an activity x, constraints may be placed on the variable X that represents the starting time for x. For a pair of activities x and y, you may be told that x must occur before y ($X < Y$), that x and y mustn't occur at the same time ($X \neq Y$), or that x and y must occur at the same time ($X = Y$).

Example 4.21

Suppose that the delivery robot needs to schedule delivery activities a, b, c, d, and e, and that each activity happens at either time 1, 2, 3, or 4. Let A be the variable representing the time that activity a will occur, and similarly for the other activities. Suppose it's given these initial variable domains, which represent possible times for each of the deliveries:

$$\mathbf{D}_A = \{1, 2, 3, 4\}, \mathbf{D}_B = \{1, 2, 3, 4\}, \mathbf{D}_C = \{1, 2, 3, 4\},$$
$$\mathbf{D}_D = \{1, 2, 3, 4\}, \mathbf{D}_E = \{1, 2, 3, 4\}$$

and the following constraints to satisfy:

$$(B \neq 3) \wedge (C \neq 2) \wedge (A \neq B) \wedge (B \neq C) \wedge (C < D) \wedge (A = D) \wedge$$
$$(E < A) \wedge (E < B) \wedge (E < C) \wedge (E < D) \wedge (B \neq D).$$

Various restrictions on the general CSP definition are possible. In a finite CSP (FCSP) the domains are required to have a finite number of discrete values. Each k-ary relation can then be specified extensionally as the set of all k-tuples that satisfy it. An FCSP can be specified as a set of clauses and a query in Datalog (page 29) without rules, such that in a *clause* each *term* is a constant, while in the *query* each *term* is a *variable*. In a *binary* CSP the relations are unary and binary, constraining only individual variables or pairs of variables. The scheduling problem posed above is a binary FCSP. In terms of the assumptions about an agent let's strengthen DK on definite knowledge to:

Assumption PF

(Positive Facts) The agent's knowledge of the world consists only of positive function-free ground facts.

A representation for a binary FCSP query is a query consisting of two parts: the domain declarations for each variable of the form $d(V)$, where d is a predicate defined by a ground database, and a conjunction of constraints each over some subset of

the variables. The constraints can be arbitrary predicates that refer only to variables
defined by domain declarations.

Example
4.22

For Example 4.21, the following defines the ground database:

> $time(1)$.
>
> $time(2)$.
>
> $time(3)$.
>
> $time(4)$.
>
> $1 < 2$. $1 < 3$. $1 < 4$. $2 < 3$. $2 < 4$. $3 < 4$.
>
> $1 \neq 2$. $1 \neq 3$. $1 \neq 4$. $2 \neq 3$. $2 \neq 4$. $3 \neq 4$.
>
> $2 \neq 1$. $3 \neq 1$. $4 \neq 1$. $3 \neq 2$. $4 \neq 2$. $4 \neq 3$.
>
> $1 = 1$. $2 = 2$. $3 = 3$. $4 = 4$.

The query takes the following form:

$$?time(A) \wedge time(B) \wedge time(C) \wedge time(D) \wedge time(E) \wedge$$
$$B \neq 3 \wedge C \neq 2 \wedge$$
$$(A \neq B) \wedge (B \neq C) \wedge (C < D) \wedge (A = D) \wedge (E < A) \wedge$$
$$(E < B) \wedge (E < C) \wedge (E < D) \wedge (B \neq D)$$

Given the ubiquity of CSPs in various guises, it's worth trying to find relatively
efficient ways to solve them. Since the general FCSP problem is NP-hard, there are
really only four strategies:

- Try to find algorithms that work well on typical cases even though the worst
  case may be exponential
- Try to find special cases that have efficient algorithms
- Try to find efficient approximation algorithms
- Develop parallel and distributed algorithms

The next sections present solution methods based on generate-and-test, hill climbing,
randomized strategies, backtracking and consistency techniques. These methods are
appropriate for all four strategies, but we focus on the first and third strategies.

Generate-and-Test Algorithms

Any FCSP can be solved by an exhaustive generate-and-test algorithm. The search
space is the *assignment space* **D**, which is the set of n-tuples formed by taking the
Cartesian product of the variable domains:

$$\mathbf{D} = \mathbf{D}_{V_1} \times \mathbf{D}_{V_2} \times \ldots \times \mathbf{D}_{V_n}.$$

Example
4.23

> In the scheduling example we have
>
> $$\begin{aligned}
> \mathbf{D} &= \mathbf{D}_A \times \mathbf{D}_B \times \mathbf{D}_C \times \mathbf{D}_D \times \mathbf{D}_E \\
> &= \{1, 2, 3, 4\} \times \{1, 2, 3, 4\} \times \{1, 2, 3, 4\} \\
> &\quad \times \{1, 2, 3, 4\} \times \{1, 2, 3, 4\} \\
> &= \{\langle 1, 1, 1, 1, 1\rangle, \langle 1, 1, 1, 1, 2\rangle, ..., \langle 4, 4, 4, 4, 4\rangle\}.
> \end{aligned}$$
>
> On each element of $\mathbf{D}$ the body of the query can be tested. In this case there are $|\mathbf{D}| = 4^5 = 1024$ different assignments to be tested. In the crossword example of Exercise 4.10 there are $40^6 = 4,096,000,000$ possible assignments.

To implement the generate-and-test algorithm you can carry out the top-down procedure of Figure 2.5 (page 51), always selecting the domain declarations before the constraints (e.g., as does the leftmost selection rule where the domain constraints are written before the constraints as in Example 4.22).

If each of the n variable domains has size d, then $|D| = d^n$, and if there are e relations in the query, then the total number of relations tested is $O(ed^n)$. As n becomes large, this very quickly becomes intractable, and so you need to find alternative solution methods.

Backtracking Algorithms

Generate-and-test involves assigning values to all variables before checking the constraints. Typically, constraints can be tested before all of the variables have been assigned values. Many constraints only involve a few variables.

You can systematically explore $\mathbf{D}$ by instantiating the variables in some order and evaluating each constraint predicate as soon as all its variables are bound. If any constraint evaluates to false, the current partial assignment can't be part of any total valid assignment. You can then try another partial assignment of values. The efficiency gain is that a single predicate failure eliminates a potentially huge subspace of $\mathbf{D}$, namely, the product space of the currently uninstantiated variables.

Example
4.24

> In the scheduling problem the assignments $A = 1$ and $B = 1$ are inconsistent with the constraint $A \neq B$ regardless of the values of the other variables. If the variables are assigned values in the order $\langle A, B, C, D, E\rangle$, this inconsistency can be discovered and that pair of values rejected before any values are assigned to C, D, or E, thus saving a large amount of work.

Under this view, a CSP can be viewed as a graph-searching algorithm. First totally order the variables, as $V_1, \ldots, V_n$. A node in the graph consists of a substitution that assigns values to the first j variables for some j, such as $\{V_1/v_1, \ldots, V_j/v_j\}$ where $v_i \in \mathbf{D}_{V_i}$. The neighbors of node $\{V_1/v_1, \ldots, V_j/v_j\}$ are the nodes $\{V_1/v_1, \ldots, V_j/v_j,$

$V_{j+1}/v_{j+1}\}$ for each $v_{j+1} \in \mathbf{D}_{V_{j+1}}$. The start node is the empty substitution, and a goal node is substitution that grounds every variable, and satisfies the constraints. You prune any node in the graph which fails the constraints, as any of its descendants also fail the constraints. Any of the graph-searching algorithms may be used, although depth-first search is typically used.

Backtracking corresponds to solving a CSP using the depth-first search with an SLD derivation (page 56), where you select a constraint if all of the variables it refers to are bound, otherwise you select the domain constraint for the next variable that is unbound. This is equivalent to using the leftmost selection rule on the query where the constraints are considered immediately after all of the variables in the query are bound.

Example 4.25

For the scheduling Example 4.22, backtracking would be equivalent to using the leftmost selection rule (always selecting the leftmost conjunct of a query) on the query

$$?time(A) \wedge time(B) \wedge A \neq B \wedge$$
$$time(C) \wedge B \neq C \wedge$$
$$time(D) \wedge B \neq D \wedge C < D \wedge A = D \wedge$$
$$time(E) \wedge E < A \wedge E < B \wedge E < C \wedge E < D.$$

The efficiency of backtracking depends critically on the ordering of the variables. One ordering may be much more efficient than others.

Consistency Algorithms

Although backtracking is usually a substantial improvement over generate-and-test, it still has various inefficiencies which can be overcome.

Example 4.26

Consider, in the scheduling example, variables C and D. The assignment $C = 4$ is inconsistent with each of the possible assignments to D since $\mathbf{D}_D = \{1, 2, 3, 4\}$ and $C < D$. In the course of the backtrack search this fact can be rediscovered for very many different assignments to A, B and, possibly, E. This inefficiency can be avoided by the simple expedient of deleting 4 from $\mathbf{D}_C$, once and for all. This idea is the basis for the consistency algorithms.

The consistency algorithms are best thought of as operating over the network of constraints formed by the CSP. Each variable corresponds to a vertex of the graph with its attached domain. Each constraint relation $P(X, Y)$ corresponds to the two arcs $\langle X, Y \rangle$ and $\langle Y, X \rangle$ in the graph. Such a network is called a **constraint network**.

A node in a constraint network is **domain consistent** if no value in the domain of the node is ruled impossible by any of the constraints.

Input:
> a set of variables
> a domain D_X for each variable X
> relations P_X on variable X that must be satisfied
> relations P_{XY} on variables X and Y that must be satisfied

Output: arc consistent domains for each variable

Algorithm:
> for each variable X
>> $D_X := \{x \in D_X | P_X(x)\}$
>>> % make domain consistent

> $TDA = \{\langle X, Y \rangle \,|P_{XY}$ is a binary constraint$\}$
> $\cup \{\langle Y, X \rangle \,|P_{XY}$ is a binary constraint$\}$
>>> % TDA is the set of to-do arcs

> repeat
>> Select any arc $\langle X, Y \rangle \in TDA$;
>> $TDA := TDA - \{\langle X, Y \rangle\}$;
>> $ND_X := D_X - \{x|x \in D_X$ and there is no $y \in D_Y$ such that $P_{XY}(x, y)\}$;
>>> % ND_X is the pruned version of D_X

>> if $ND_X \neq D_X$ then
>>> $TDA := TDA \cup \{\langle Z, X \rangle \,|Z \neq Y\}$;

> until TDA is empty.

Figure 4.8: Arc consistency algorithm AC-3

Example 4.27

$\mathbf{D}_B = \{1, 2, 3, 4\}$ isn't domain consistent as $B = 3$ violates the constraint that specifies $B \neq 3$.

Domain consistency doesn't take into account any of the values for any of the other variables.

An arc $\langle X, Y \rangle$ is **arc consistent** if for each value of X in $\mathbf{D}_X$ there is some value for Y in $\mathbf{D}_Y$ such that $P(X, Y)$ is satisfied. A network is arc consistent if all its arcs are arc consistent.

If an arc $\langle X, Y \rangle$ is *not* arc consistent, all values of X in $\mathbf{D}_X$ for which there is no corresponding value in $\mathbf{D}_Y$ may be deleted from $\mathbf{D}_X$ to make the arc $\langle X, Y \rangle$ consistent.

In the research literature a series of algorithms for achieving network arc consistency, known as AC-i, $i = 1, 2, \ldots$, have been proposed.

The arc consistency algorithm AC-3 is given in Figure 4.8. It makes the entire network arc consistent by considering a queue of potentially inconsistent arcs, the *to-do* arcs. These initially consist of all the arcs in the graph. Note that any relation

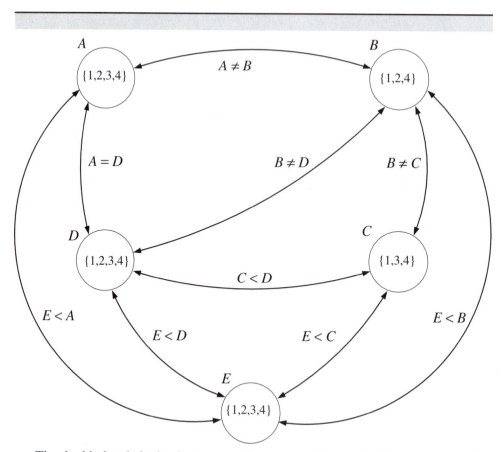

The double-headed edge between nodes X and Y stands for the two arcs $\langle X, Y \rangle$ and $\langle Y, X \rangle$. The label on the edge is the relation that must be satisfied.

Figure 4.9: Domain-consistent constraint network for the scheduling problem

P_{XY} between variables X and Y makes two arcs, $\langle X, Y \rangle$ and $\langle Y, X \rangle$. Until the queue is empty, an arc is removed from the queue and considered. If it isn't consistent, it's made consistent and all consistent arcs that could, as a result, have become inconsistent are placed back on the queue.

Example 4.28

Consider applying AC-3 to the scheduling example as shown in network form in Figure 4.9. The network as shown has already been made domain consistent. Suppose arc $\langle D, C \rangle$ is considered first. The relation to be tested is $C < D$. The arc is inconsistent because $D = 1$ isn't consistent with some value in $\mathbf{D}_c$, so 1 must be deleted from $\mathbf{D}_D$. $\mathbf{D}_D$ becomes $\{2, 3, 4\}$ and arcs $\langle A, D \rangle$, $\langle B, D \rangle$, and $\langle E, D \rangle$ could be added to TDA, but they are on it already.

If arc $\langle C, E \rangle$ is considered next, then $\mathbf{D}_C$ is reduced to $\{3, 4\}$ and arc $\langle D, C \rangle$ goes back on the queue to be reconsidered. If $\langle D, C \rangle$ is next, then $\mathbf{D}_D$ is further reduced to the singleton $\{4\}$. Processing arc $\langle C, D \rangle$ prunes $\mathbf{D}_C$ to $\{3\}$. Making arc $\langle A, D \rangle$ consistent reduces $\mathbf{D}_A$ to $\{4\}$. Processing $\langle B, D \rangle$ reduces $\mathbf{D}_B$ to $\{1, 2\}$. Then arc $\langle B, E \rangle$ reduces $\mathbf{D}_B$ to $\{2\}$. Finally, arc $\langle E, B \rangle$ reduces $\mathbf{D}_E$ to $\{1\}$. All arcs remaining on the queue are consistent, and so the algorithm terminates with an empty queue.

Regardless of the order in which the arcs are considered, the algorithm will terminate with the same result, namely, an arc consistent network and the same set of reduced domains. There are three possible cases depending on the state of the network on termination:

- In the first case, each domain is empty, indicating there is no solution to the CSP. In this case, as soon as any one domain becomes empty, all the domains will become empty before the algorithm terminates.

- In the second case, each domain has a singleton value, indicating that there is a unique solution to the CSP, as in Example 4.28.

- In the third case, every domain is nonempty and at least one has multiple values left in it. In this case, any nonsingleton domain may be split in half and the algorithm applied recursively to the two CSPs that result. Splitting the smallest nonsingleton domain is usually most effective. Notice that all the arcs for these CSPs are known to be consistent initially, except those pointing directly at the vertex whose domain has been split.

These ideas are all implemented in the binary FCSP solver, *csp*, defined as a Prolog program in Appendix C (page 508). It is a complete solver that uses AC-3 and domain splitting to solve the problem. A simple CSP and the scheduling example are solved there.

If each variable domain is of size d and there are e relations to be tested then *AC-3* is $O(ed^3)$. For some CSPs, for example, if the constraint graph is a tree, *AC-3*, alone solves the CSP, and does it in time linear in the number of variables.

Various extensions to the arc consistency technique are also possible. Extending it from binary to k-ary relations is straightforward. The arc consistency approach need only be slightly modified. The domains needn't be finite: They may be specified using descriptions, not just lists of their values. Higher-order consistency techniques, such as **path consistency**, that consider k-tuples of variables at a time, not just pairs, are also available but often are less efficient for solving a problem than arc consistency alone.

Hill Climbing

The previous sections have considered algorithms that systematically search the space: If the space is finite, you will either find a solution or finish and report that there is no solution. Unfortunately, many search spaces are too big for systematic search, possibly even infinite. In any reasonable time, systematic search will have failed to consider enough of the search space to give any meaningful results. The remainder of this chapter considers methods that are intended to work in these very large spaces. The methods don't systematically search the search space, but are intended to find solutions or optimize a value fast. They can't guarantee that they will find a solution even if one exists, or that the value found is the optimal, but they are very useful in practice.

Hill climbing is a solution method that works by choosing a value for each variable and iteratively improving its assignment. It requires a heuristic value for each total assignment. You can model hill climbing as a graph-searching procedure where a node in a graph corresponds to an assignment of a value to each variable as a node. The neighbors of a node correspond to the assignments that are close to the assignment represented by the node (see below for a more precise characterization). Initially you select a single node to start; and maintaining a single node at each stage, you select the neighbor of the node with the highest heuristic value and use that as the next node to search from. You stop when no neighbor has a higher value than the current node.

Hill climbing can be implemented as follows, where $hill\_climb(N, S)$ is true if hill climbing from node N results in local maxima S:

$$hill\_climb(N, N) \leftarrow$$
$$\qquad neighbors(N, NN) \land$$
$$\qquad best(N, NN, N).$$
$$hill\_climb(N, S) \leftarrow$$
$$\qquad neighbors(N, NN) \land$$
$$\qquad best(N, NN, M) \land$$
$$\qquad M >_h N \land$$
$$\qquad hill\_climb(M, S).$$

where $M >_h N$ is true if the heuristic value of M is greater than the heuristic value of N, and similarly for other comparisons. $best(N, L, M)$ is true if M is the maximal h-value node that's either the node N or an element of list L:

$$best(N, [\,], N).$$
$$best(N, [M|R], B) \leftarrow$$
$$\qquad N \geq_h M \land$$
$$\qquad best(N, R, B).$$

$$best(N, [M|R], B) \leftarrow$$
$$N <_h M \wedge$$
$$best(M, R, B).$$

This is called hill climbing, since the problem in two dimensions is similar to the problem of finding the highest point in a terrain, and the algorithm works by taking a single step in whichever of the four compass directions goes up the most. The algorithm halts when none of the steps leads up.

Hill climbing is usually carried out in domains where there are many dimensions and any systematic search would take too long. There are a number of variants that depend on the structure of the values of the variables:

- When the domains are unordered, the neighbors of a node correspond to choosing another value for one of the variables. Thus, if there are n variables and each has m values, there are $n \times (m - 1)$ neighbors for each node. The number of neighbors is still typically much less than the m^n nodes in the graph.

- When the domains are ordered or when there is other structure in the domains, the neighbors of a node can be the adjacent values for one of the dimensions. Thus, if there are n variables and each has m values, there are typically $2 \times n$ neighbors for each node (the exceptions occur when a variable has its maximum or minimum value).

- If the domains are continuous, then you have to be more careful about what are the neighbors of a node. In this case you can use what is known as **gradient descent** when trying to find a minimum value, or **gradient ascent** when trying to find a maximum value. Gradient descent is like tracking the path of an object rolling down a hill under the force of gravity, without momentum—although momentum can be added and often improves the performance of gradient descent. The general idea is that the neighbor of a node is chosen so that it changes each dimension in proportion to the slope of h, the function you are trying to maximize (or minimize for the case of gradient descent).

  If $\langle X_1, \ldots, X_n \rangle$ are the variables that have to be assigned values, then a node corresponds to a tuple of values $\langle v_1, \ldots, v_n \rangle$. The neighbors of the node $\langle v_1, \ldots, v_n \rangle$ are obtained by moving in each direction in proportion to the slope of h in that direction. The new value for X_i is

$$v_i - \eta \frac{\partial h}{\partial X_i}$$

for gradient descent. The "−" becomes "+" for gradient ascent. Here η is the constant of proportionality that determines how fast you approach the maximum. If η is too large, you can overshoot the maximum; if η is too small, progress becomes very slow. This also assumes that h is differentiable; if the formula for the derivative of h isn't given, the partial derivative in each direction can be estimated by evaluating h at a close point for each dimension.

Gradient descent is used for parameter learning, where there can be thousands of real-valued parameters to be optimized (page 408).

Hill climbing has some important difficulties. Consider the direct analog of climbing a mountain; you can find the following problems:

Foothills The search has found a local maxima, but hasn't found a global maximum. Every direction you look in may be downhill, but this doesn't mean that you have found the maximum value.

Plateaus The search has reached a point where no neighbor is better, but some are the same as the current node. Maybe stepping in one of those directions will then lead to a path up. No single step can improve the situation.

Ridges A ridge is a special form of a foothill where the search has reached a point where all of the directions point down, but by making a combination of steps you may be able to do better. Consider the problem of a steep ridge running up to the northeast, and you can move in the directions north, south, east and west. There is a path that rises to the northeast, but every neighbor is downhill (see Figure 4.10). Ridge problems can often be addressed by having more than one step look ahead. Programs that do only one step look ahead are often called **myopic** or **greedy**. With two step look ahead, a northeast ridge can be detected.

Ignorance of the peak You may be at a local maxima, a foothill or a plateau, and not know whether you are at a global maxima. Unless you know some particular feature of the peak, you may not know you are at a global maxima when you reach it.

One of the ways to overcome these problems is to systematically search the space when a goal isn't reached, as does heuristic best search. Heuristic best search can be seen as a variant of hill climbing that searches the space by backtracking. This is often not a good idea, as typically the spaces are much too big for an exhaustive search. You need to find better methods that, while they may not always find the optimal solutions, work well in practice.

Randomized Algorithms

Consider two quite different ways of finding a maximum value:

- Hill climbing, starting from some position, keep moving uphill, and report the maximum value found
- Keep picking values at random and report the maximum value found

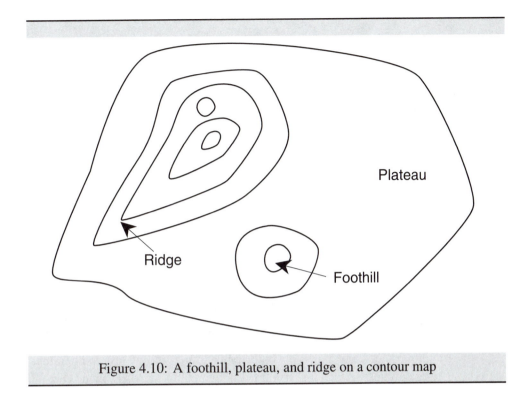

Figure 4.10: A foothill, plateau, and ridge on a contour map

Which would you expect to do better? If you think of the climbing a hill analogy, hill climbing would at least find a local maximum, but it may not find the mountain. The random searching would tend to not find a local maximum, but at least it would locate a point on the mountain.

It would seem that a combination of these two techniques may be much better at finding global optima. There are a number of ways these can be combined including:

- random-restart hill climbing, in which values are chosen at random, and from each of these values, you do hill climbing to find the corresponding local maxima. This is equivalent to picking a random value, hill climbing from that value, and, when the hill climbing finds a local maxima, restarting at another random position. The problem is that much time is spent on the hill climbing for values that won't lead to a global maxima.

- two-phase search in which you first select values at random, and then do hill climbing from the maximum value found. It is hoped that the random phase will find the mountain, and then the hill climbing will find the peak of the mountain. Unfortunately, the second phase of hill climbing still tends to find local maxima that are smaller peaks on the mountain.

Let n be a random assignment of values to all variables;
Let T be a (high) temperature;
repeat
    Select neighbor n' of n at random;
    if $h(n') \geq h(n)$
        then $n := n'$
        else $n := n'$ with probability $e^{(h(n')-h(n))/T}$;
    reduce T
until some stopping criteria is reached.

Figure 4.11: Simulated annealing algorithm

For optimization problems, **simulated annealing** is a method that combines random selection and hill climbing to find global maxima. In particular, it does a random walk, choosing neighbors at random and deciding at random whether to visit that neighbor. Simulated annealing is typically described in terms of thermodynamics. The random movement corresponds to high temperature, and at low temperature there is little randomness. Annealing is a process in metallurgy where metals are slowly cooled to make them reach a state of low energy where they are very hard. Simulated annealing is a process where the temperature is reduced slowly, starting from a random search at high temperature and doing pure hill climbing at zero temperature. The randomness will tend to find the mountains and to jump out of local maxima, and the hill climbing will lead us to local maxima. The question is how to find a mix of the two that's parameterized by the temperature and that smoothly ranges from the random search to the pure hill climbing. The solution proposed by simulated annealing is like hill climbing where you iteratively improve a single value. You choose neighbors at random, If the neighbor has a better heuristic value, you continue from the neighbor. If the neighbor has a worse heuristic value, it still could be chosen with a probability depending on the temperature and the difference in the heuristic values. This is shown in Figure 4.11.

How the temperature is reduced and when to stop is left undefined in this procedure. When $T = \infty$, this chooses neighbors at random; in the limit as T approaches zero, this chooses only neighbors that improve the value. If the temperature is reduced slowly enough, this guarantees to find the optimal result. Unfortunately, reducing it slowly enough to guarantee optimal results may need to be as slow as generate-and-test, but often a quicker cooling results in good performance.

One method which has proved quite successful for satisfiability problems is **GSAT**, standing for "greedy satisfiability," in which a node is an assignment of a value to each variable. The heuristic function is the number of constraints that are unsatisfied by the variable assignment; a variable assignment is a solution when this is zero. The neighbors of a node are obtained by changing the value of one of the variables in the assignment. A variable can't, however, have its value changed twice in a row. GSAT does hill climbing, trying to minimize the heuristic function. It selects a neighbor with minimal heuristic value, even if it's worse than the current node. Of the neighbors with minimal heuristic value, one is selected at random as the next node to search from. From a random starting position the search is carried out for a selected number of iterations. This search is repeated a limited number of times, each from a random starting position. If it finds a solution, it can report the solution, but no information can be obtained from failure to find a solution.

Example 4.29

Consider the scheduling Example 4.21 (page 149). Suppose you start with the node $A = 2, B = 2, C = 3, D = 2, E = 1$. Its heuristic value is 3, as it doesn't satisfy $A \neq B$, $B \neq D$, and $C < D$. Its neighbor with the minimal heuristic value changes B to 4, and the resulting node has a heuristic value of 1, as only $C < D$ is unsatisfied. This is now a local minima. You can now change D to 4, which has a heuristic value of 2. You can change A to 4, with a heuristic value of 2. You can then change B to 2 with a heuristic value of zero, and a solution is found.

Beam Search and Genetic Algorithms

The preceding hill climbing and randomized algorithms maintain a single current node. This section considers algorithms that maintain multiple nodes. The first method, beam search, maintains the best k nodes, but the nodes don't interact. We then consider a stochastic beam search, leading into genetic algorithms where, inspired by evolution, the k nodes forming a population interact in various ways to produce the new population.

In hill climbing and the stochastic algorithms considered here, you always maintain a single current node. **Beam search** is a method that's like hill climbing, but where you maintain up to k nodes instead of just one. You report success if you have found a goal node in the set of current nodes. At each stage of the algorithm, you find the set of all of the neighbors of all of the current nodes, select the k best of these (or all of them if there are less than k), and repeat with this new set of current nodes.

Beam search lets us consider multiple paths at the same time. When $k = 1$, it's hill climbing. When $k = \infty$, it's breadth-first search. Beam search is useful for memory-bounded cases, where you can choose k so that you don't run out of memory.

An alternative to beam search is **stochastic beam search**, where, instead of choosing the best k nodes, you choose k nodes at random, with the nodes with a higher heuristic value being more likely to be chosen. Stochastic beam search tends to allow more diversity in the k nodes than does plain beam search. If you think in terms of evolution, the heuristic value reflects the fitness of the node; the fitter the individual, the more likely it is to pass onto the next generation. Stochastic beam search is like asexual reproduction; each node leads to its local mutations and then you proceed with survival of the fittest. Note that under stochastic beam search, it's possible that a node is selected multiple times at random.

Genetic algorithms pursue the analogy with evolution further. Genetic algorithms are like stochastic beam search, but where the new elements of the population can be combinations of pairs of nodes. In particular, you choose pairs of nodes, and then create new offspring by taking some of the values for the offspring's variables from one of the parents and some from the other parent, loosely analogous to how DNA is spliced in sexual reproduction.

The new operation that occurs in genetic algorithms is called **cross-over**. In uniform cross-over you select two nodes (the parents), and you create two new nodes (the children). For each variable, you randomly assign the values from the parents to the children. That is, for each variable, you randomly choose which parent's value goes to which child.

Assume that you have a population of k nodes. These are often called *individuals*, but we won't use that term as it's very different to our use of the term. Evolution proceeds by maintaining k nodes as a generation and then using these nodes to generate a new generation by following the steps:

1. Randomly choose pairs of nodes, where the fitter individuals are more likely to be chosen.

2. For each pair, perform a cross-over.

3. Randomly mutate some (very few) values by choosing other values for some randomly chosen variables.

You do this until you have created k nodes, and then the operation proceeds to the next generation.

Example 4.30

Consider Example 4.21 (page 149). Suppose you use the same heuristic function as in Example 4.29. The node $A = 2, B = 2, C = 3, D = 1, E = 1$ has a heuristic value of four. It has such a low value, mainly because $E = 1$. Its offspring that preserve this property will tend to have a lower heuristic value than those that don't, and thus will be more likely to survive. Other nodes may have low values for different reasons, for example the node $A = 4, B = 2, C = 3, D = 4, E = 4$ also has a heuristic value of four, mainly because of the assignment of values to the first four variables. Again offspring that preserve this property will be fitter and more likely to survive than those that don't. If these two were to mate, some of the offspring would inherit the bad properties of both and would die off. Some, by chance, would inherit the good properties of both. These would then have a better

> chance of survival. Maintaining multiple nodes lets genetic algorithms maintain multiple reasons why some individuals survive.

The main problems with genetic algorithms are that they are extremely slow. This should not be surprising because the process of natural evolution is very slow. The other problem is in finding a heuristic function that doesn't have local minima or plateaus. Also, convergence is very sensitive to exactly which variables are used to describe the problem. Slight variations in the problem specification can alter the convergence.

4.8 References and Further Reading

State space and problem space searching are discussed in Nilsson (1971). For a detailed analysis of heuristic search see Pearl (1984).

Depth-first iterative deepening is described in Korf (1985).

Constraint satisfaction techniques are described in Mackworth (1992). See Van Hentenryck (1989) and Jaffar & Maher (1994) for ways to use constraint satisfaction techniques in logic programming.

Simulated annealing is due to Kirkpatrick, Gelatt & Vecchi (1983). GSAT is due to Selman, Levesque & Mitchell (1992). See Gent & Walsh (1993) for an analysis of GSAT, and see Selman, Kautz & Cohen (1994) for related methods.

Genetic algorithms were pioneered by Holland (1975). For more recent developments see Goldberg (1989) and Koza (1992).

4.9 Exercises

Exercise 4.1 Comment on the following quote: "One of the main goals of AI should be to build general heuristics which can be used for any graph-searching problem."

Exercise 4.2 Which of the path-finding search procedures are fair in the sense that any element on the frontier will eventually be chosen? Consider this for finite graphs without loops, finite graphs with loops and infinite graphs (with finite branching factors).

Exercise 4.3 Assume you are designing a compiler for a machine with one register and the following instructions:

ONE	Set the value of the register to 1
DOUBLE	Double the contents of the register
ADD	Add one to the contents of the register
SUB	Subtract one from the contents of the register
DIVIDE	Divide the value of the register by 3

The first four instructions can only be executed when the register's contents aren't divisible by 3. When the register's contents are divisible by 3, the only instruction available is *DIVIDE*.

The problem to consider is how to put a constant into the register. (You may assume that the register begins with a value of 1.)

(a) Convert this problem into a graph-searching problem. What's a node? What's an arc? What are the neighbors of an arbitrary node?

(b) Given the goal of having 27 in the register and the heuristic function $h(n) = |27 - n|$, draw the portion of the search tree generated in the heuristic depth-first search (with cycle check) forward search. What's the solution found?

(c) Suppose that the costs of the operators on the register containing the number n are:

ONE	1
DOUBLE	n
ADD	1
SUB	1
DIVIDE	$2n/3$

(that is, each operator has a cost equivalent to the difference they make to the register's value) and using the same heuristic function as in (b), draw the A* search tree (with multiple-path pruning). What's the path actually found?

Exercise 4.4 Consider the problem of finding a path in the grid shown in Figure 4.12 from the position s to the position g. A piece can move on the grid horizontally and vertically, one square at a time. No step may be made into a forbidden shaded area.

(a) On the grid shown in Figure 4.12, number the nodes visited (in order) for a depth-first search from s to g, given that the order of the operators you will test is: up, left, right, then down. Assume there is a cycle check. A node is visited when it is taken off the frontier.

(b) For the same grid, number the nodes visited, in order, for a best-first search from s to g. Manhattan distance should be used as the evaluation function. The Manhattan distance between two points is the distance in the x-direction plus the distance in the y-direction. It corresponds to the distance traveled along city streets arranged in a grid. Assume that you have multiple-path pruning. What is the first path found?

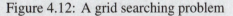

Figure 4.12: A grid searching problem

(c) On the same grid, number the nodes visited, in order, for a heuristic depth-first search from *s* to *g*, given Manhattan distance as the evaluation function. Assume a cycle check. What's the path found?

(d) Number the nodes in order for an A^* search for the same graph. What's the path found?

(e) Assume that you were to solve the same problem using dynamic programming. Give the *dist* value for each node, and show which path is found.

(f) Based on this experience, discuss which algorithms are best suited for this problem.

(g) Suppose that the graph extended infinitely in all directions. That is, there is no boundary, but *s*, *g*, and the blocks are in the same relative positions to each other. Which methods would no longer find a path? Which would be the best method, and why?

Exercise 4.5   Consider the following (real world) problem:

Due to Bill's excessive drinking, it has been decided to limit him to exactly one liter of beer a day. You have only a 17-liter jug and a 7-liter jug. You can (1) fill either jug from a keg, (2) drink the contents of a jug, or (3) pour beer from one jug to the other. Unfortunately, you are hopeless at guessing quantities. Given only these three operations, how can you restrict Bill's beer drinking to exactly one liter a day? (That is, how can you get one liter of beer in a jug if you only can perform the above three actions?)

The solution to this problem can be reduced to a graph-searching algorithm.

(a) Give the structure of a node in the graph.

(b) Define the neighbors of an arbitrary node.

(c) Draw the first two levels of a breadth-first search from the initial state of both jugs being empty.

(d) What's a suitable search strategy to use for this graph? Why?

Exercise 4.6 Draw two different graphs, indicating start and goal nodes, for which forward search is best in one and backward search is better in the other.

Exercise 4.7 Implement iterative-deepening A^*. This should be based on the iterative deepening searcher of Figure 4.7.

Exercise 4.8 Suppose that your goal was, rather than find the optimal path from the start to a goal, to find a path that wasn't more than, say 10% longer than the shortest path. Suggest an alternative to iterative-deepening A^* search that would let you do this. Why might this be advantageous to iterative-deepening A^* search?

Exercise 4.9 Priority queues are important to implement search algorithms. The use of a sorted list is very inefficient. This exercise is to implement a more efficient version of priority queues. The functionality you will need to provide is

- insertion of an element into a priority queue
- removal of the smallest element from the priority queue

Removal should fail if the queue is empty.

Insertion and deletion must both be done in $\log n$ time, where n is the number of elements in the priority queue.

Hint: Represent a priority queue as a binary tree with the following properties:

- The smallest element of the tree rooted at any node is at that node.
- At every node, the left subtree either has the same number of elements or has exactly one more element than the right subtree. To maintain this invariant upon insertion, insert a new element into the right subtree and swap the left and right subtrees. Do a similar operation on deletion.

Test this program by using it for heap sort, a sorting algorithm that involves putting all of the elements to be sorted onto a priority queue, and removing them in order.

Exercise 4.10 Consider the crossword puzzle shown in Figure 4.13. You must find six three-letter words: three words read across ($A1$, $A2$, and $A3$) and three words read down ($D1$, $D2$, and $D3$). Each word must be chosen from the list of 40 possible words shown. Try

A1,D1	D2	D3
A2		
A3		

Word list:
add, ado, age, ago, aid,
ail, aim, air, and, any,
ape, apt, arc, are, ark,
arm, art, ash, ask, auk,
awe, awl, aye, bad, bag,
ban, bat, bee, boa, ear,
eel, eft, far, fat, fit,
lee, oaf, rat, tar, tie.

Figure 4.13: A crossword puzzle to be solved with six words

to solve it yourself, first by intuition, then by hand using first domain consistency and then arc consistency. Using the implementation of Section C.3 (page 507), show how it can be represented and solved as a CSP.

Exercise 4.11 There are at least two ways to represent the crossword puzzle shown in Figure 4.13 as a constraint satisfaction problem.

The first is to represent the word positions ($A1$, $A2$, $A3$, $D1$, $D2$, and $D3$) as variables, with the set words as possible values. The constraints are that where the words intersect the letter is the same. This is represented in Section C.3 (page 507).

The second is to represent the nine squares as variables. The domain of each variable is the set of letters of the alphabet, $\{a, b, \ldots, z\}$. The constraints are that there is a word in the word list that contains the corresponding letters. For example, the top-left square and the center-top square cannot both have the value a, because there is no word starting with aa.

(a) Give an example of pruning due to domain consistency, using the first representation (if one exists).

(b) Give an example of pruning due to arc consistency, using the first representation (if one exists).

(c) Are domain consistency plus arc consistency adequate to solve this problem using the first representation? Explain.

(d) Give an example of pruning due to domain consistency, using the second representation (if one exists).

(e) Give an example of pruning due to arc consistency, using the second representation (if one exists).

(f) Are domain consistency plus arc consistency adequate to solve this problem using the second representation?

(g) Which representation leads to a more efficient solution using consistency-based techniques? Give the evidence on which you are basing your answer.

Exercise 4.12 Pose the 4-queens problem as a CSP and solve it using the CSP solver *csp*. A solution to the *n*-queens problem requires the placement of *n* queens on an $n \times n$ chess board so that no queen is on the same row, column, or diagonal as any other.

Exercise 4.13 Write a logic program that will generate the CSP query for the *n*-queens problem for any positive integer *n* and then solve it using *csp*.

Exercise 4.14 Compare the performance of three approaches to solving a CSP: generate-and-test, the backtrack method, and the CSP solver on the scheduling and crossword examples.

Exercise 4.15 Pose and solve the crypt-arithmetic problem *SEND + MORE = MONEY* as a CSP. In a crypt-arithmetic problem, each letter represents a different digit, the leftmost digit can't be zero (because then it wouldn't be there), and the sum must be correct considering each sequence of letters as a numeral, base ten. In this example, you know that $Y = (D + E)$ mod 10 and that $E = (N + R + ((D + E) \div 10))$ mod 10, etc.

Exercise 4.16 For any problem that's a candidate for the CSP approach there is a fundamental knowledge representation issue: What are the appropriate variables and constraints? Those decisions can have a major impact on the viability and efficiency of a proposed solution. For example, often the roles of the variables and the constraints in a CSP can be interchanged to get a *dual* form of the CSP. Consider the crossword puzzle example. Instead of thinking of the words as the variables and the letter at each intersection as a constraint, consider each square as a variable that can have as its value any of the 26 letters $\{a, b, c, \ldots, z\}$, and the constraint on each set of squares corresponding to a word location is that the appropriate letter values must form a word in the word list. Does this lead to a more efficient solution? Justify your answer with analysis, experimental results or both.

Chapter 5

Representing Knowledge

5.1 Introduction

This chapter explores the general principles behind using a computer to solve problems or carry out tasks. Chapters 2 and 3 introduced a particular representation and reasoning system and showed how it can be used. Chapter 4 presented various methods that can be used to solve problems once they are cast as graph searching. This chapter discusses refining problem specifications, choosing representation languages, mapping problems into representations, and choosing algorithms to apply. Much of the discussion is applicable to general problem solving with computers. We discuss the connections with software engineering where appropriate.

Typically a problem to solve or a task to carry out, as well as what constitutes a solution, is only given informally such as "deliver parcels promptly when they arrive" or "fix whatever is wrong with the electrical system of the house." To solve a problem, you need to flesh out the task and determine what constitutes a solution. You then need to represent the problem. Only then can a computer be used to solve the problem.

The general framework for solving problems by computer is given in Figure 5.1. In order to solve a problem, you have to represent the problem so an algorithm can be used to compute an answer. The output of the computation, the answer, either can be interpreted by someone as a solution or can be commands to actuators to actually carry out the derived activity.

Some of the questions that need to be considered when given a problem or a task are:

- What is a solution to the problem?
- What do you need in the language to represent the problem?

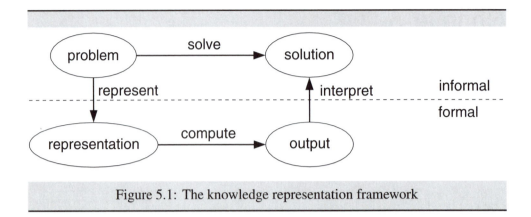

Figure 5.1: The knowledge representation framework

- How can you map from the informal problem description to a representation of the problem?
- What distinctions in the world are important to solve the problem?
- What knowledge is required?
- What level of detail is required?
- What reasoning strategies are appropriate?
- Is worst-case performance or average-case performance the critical time to minimize?
- Is it important for a human to understand how the answer was derived?
- How can you acquire the knowledge from experts or from experience?
- How can the knowledge be debugged, maintained, and improved?

This chapter considers questions such as these.

5.2   Defining a Solution

Given an informal description of a problem, and before even considering a computer, you need to determine what would constitute a solution. This question arises not only in AI but in any software design. Much of **software engineering** involves refining the specification of the problem.

Typically, problems are not well-specified. Not only is there usually much left unspecified, but the unspecified parts can't be filled in arbitrarily. For example, if you ask the infobot to find out all the information about cheating by AI students, you don't want the infobot to give back the whole information source, even though all of the information you requested is in there. Moreover, returning all of the information may be the only way for the infobot to guarantee that all of the requested information is there. Similarly, you don't want the delivery robot, when asked to take all of the trash to the garbage can, to take everything to the garbage can. Much work in AI is motivated by **common-sense reasoning**; you want the computer to be able to make common-sense conclusions about the unstated assumptions. Section 9.3 (page 323) discusses ways to specify defaults to enable the computer to draw common-sense conclusions.

Quality of Solutions

Once you have a well-defined problem, you can ask whether it matters if the answer returned is wrong. For example, if the specification asks for all instances, does it matter if some are missing? Does it matter if there are some extra instances? Often you want not only any solution but the best solution according to some criteria. There are four different classes of solutions:

Optimal solution   An optimal solution to a problem is one that's the best solution to the problem according some measure of solution quality. This measure is typically specified as an ordinal, where only the order matters. However, in some situations, such as when reasoning under uncertainty, or when combining multiple criteria, you need a cardinal measure, where the relative magnitudes also matter. For example, you may want the delivery robot to take all of the trash to the garbage can, minimizing its distance traveled, and explicitly specify a tradeoff between the effort required and the proportion of the trash taken out. It may be better to miss some trash than to waste too much time. One general cardinal measure of desirability, known as *utility*, is used in decision theory (page 392).

Satisficing solution   Often you don't need the best solution to a problem, but you just need some solution. A satisficing solution is one that's good enough. For this to make sense you need a Boolean description of which solutions are adequate. For example, you may tell the robot that it must take all of trash out, or tell it that you want three items of trash taken out.

Approximately optimal solution   One of the advantages of a cardinal measure of success is that you can find an approximately optimal solution. An approximately optimal solution is one whose measure of quality is close to the best that could theoretically be obtained. Typically, you don't need optimal solutions to problems, but only need to get close enough. For some problems, it's much easier computationally to get an approximately optimal solution; however, for some problems it's (asymptotically) just as difficult to find an approximately optimal solution as it is to find an optimal solution. Some approximation algorithms guarantee that a solution is within some range of optimal, but for some algorithms no guarantees are available. For example, you may not need the robot to do the optimal distance traveled to take out the trash but may only need it within, say, 10% of optimal.

Probable solution   A probable solution is one that, even though it may not actually be a solution to the problem, is likely to be a solution. This is one way to approximate, in a precise manner, a satisficing solution. For example, in the case where the delivery robot could drop the trash or fail to pick it up when it attempts to, you may need the robot to be 80% sure that it has picked up three items of trash. One question that can be asked of a solution is, if the computer got it right, say, 90% of the time, how good would this be. Often you want to distinguish the **false-positive** rate, the proportion of the answers given by the computer that are not correct, and the **false-negative** rate, which is the proportion of those answers not given by the computer that are indeed correct. Some applications are much more tolerant of one of these errors than the other.

These categories are not exclusive. A form of learning known as PAC learning considers probably learning an approximately correct concept (page 421).

Decisions and Outcomes

Sometimes unexpected things happen. Sometimes these are fortuitous and sometimes they are disastrous. Just because something fortuitous occurred does not mean that the agent made the right decision. For example, it may not be a good decision to have the robot travel fast near the top of the stairs, because it might fall down the stairs, with disastrous consequences. You may have a good outcome, where the agent doesn't slip near the stairs, but this doesn't mean that the decision that was made before you knew whether the agent would slip was a good decision.

You need to distinguish between a good answer or decision and a good outcome.

A good **decision** is a choice that seems to be good based on the information available to the agent. A good **outcome** is the result of a choice that happens to turn out well.

Example 5.1

If the infobot does not know whether some knowledge source has the information it needs, the infobot needs to make a good decision of whether to access the knowledge source based on the information it has available. If it accesses the knowledge base and the information is there, then there may be a good outcome. If it accesses the knowledge base, and the information isn't there, there may be a bad outcome.

A good decision is one that generally leads to a good outcome. However, a good decision sometimes leads to a bad outcome, and a bad decision sometimes leads to a good outcome. Relating decisions to outcomes is discussed in Section 10.4 (page 381).

Information Availability and Solution Quality

Once you start to be concerned about solution quality, you find that a number of activities an agent performs have costs associated with them. You need to determine whether these costs are worth the benefits.

Typically all of the information needed by an agent isn't immediately available to it. For example, the infobot may need to access a remote knowledge source in order to answer a question. Similarly, the delivery robot may need to go to the pick-up room to determine if there are any parcels to deliver. Information is valuable in that it, typically, leads to better solutions; but the actions needed to obtain information have costs associated with them.

When defining a solution, you have to be concerned about what information is available, how information can be obtained, and the costs and benefits associated with actions used to obtain the information. Explicit evaluation of the value of information is possible (page 391).

Computation Cost and Solution Quality

What's a good solution may depend on inference time. For example, in crossing a road, you have to be able to reason quickly so that another car hasn't arrived by the time you have derived that it's safe to cross the road. You don't want the delivery robot to spend an hour computing the optimal solution when that optimal solution may only save 5 minutes over a solution that can be obtained quickly. An **anytime algorithm** is an algorithm whose solution quality improves with time. In particular, it's one which can produce its current best solution at any time, but given more time it could produce even better solutions. The hill-climbing algorithms (page 156) can often be treated as anytime algorithms, particularly when you have had some experience in a domain with how fast they converge to optimal.

Example 5.2	Figure 5.2 shows how the computation time of an anytime algorithm can affect the solution quality. The absolute solution quality, had the action been carried out at the appropriate time, is improving. However, there is a penalty associated with taking time for computation. The penalty is proportional to the time taken to compute. These two values can be added to get the time-dependent value of computation. For the example of Figure 5.2, you should compute for about 7 time units. If you let the computation last for longer than 20 time units, the resulting solution quality is worse than if the algorithm just output the initial guess it can produce (virtually) without any computation.

When choosing a representation, you don't only need to be concerned about computing the right answer, but also computing it in a timely manner, and, as the anytime algorithm

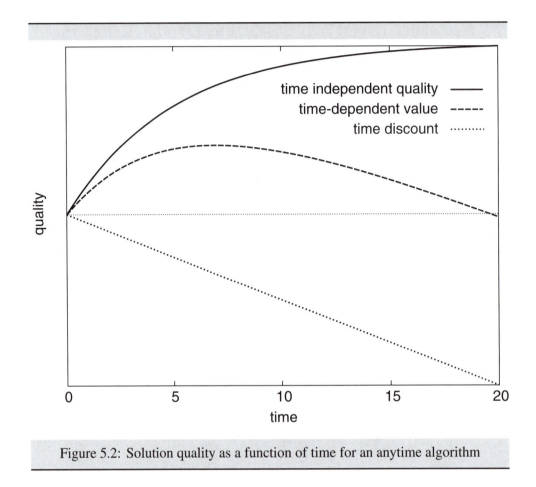

Figure 5.2: Solution quality as a function of time for an anytime algorithm

idea shows, you may need to find a compromise between finding the best answer and finding an answer quickly.

5.3   Choosing a Representation Language

Once you have some idea as to what a solution looks like, you need to choose a representation language in which to state the problem. You need some language in which to state the problem, whether it's a low-level programming language like C, the definite clause logic, or natural language. If you can't convey the problem to the computer, it can't be expected to solve it.

This chapter considers finding a good representation separately from how to derive an answer. In practice, however, as described in Section 5.2, the computational mechanisms available may influence the choice of representations.

A representation independent of a computational mechanism is called a logic.

A **logic** is a language together with a specification of what follows from a set of sentences in the language.

Section 2.5 (page 31) specified what follows from a set of sentences by the use of a semantics. This was convenient because it allowed you to give symbols meaning. In general a logic allows any specification of the functional relationship between the inputs and the outputs. In some cases, such as for learning in neural networks (page 408), you don't want to give an *a priori* semantics to the components of the network, as this is part of what is being learned.

Expressiveness and Complexity

To choose a language, you need some criteria for comparing different languages or logics. You can compare logics on a number of dimensions.

The first dimension is expressiveness:

Logic L_1 is **less expressive** than logic L_2 if every problem that can be expressed in logic L_1 can also be expressed in logic L_2. Logic L_1 is **equally as expressive** as logic L_2 if L_1 is less expressive than L_2 and L_2 is less expressive than L_1. Logic L_1 is **strictly less expressive** than logic L_2 if L_1 is less expressive than L_2 and L_2 is not less expressive than L_1.

Logic L_1 is a **sublogic** of logic L_2 if every sentence of L_1 is a sentence of L_2; and if S is a set of sentences in L_1, then what follows from S in L_1 is the same as what follows from S in L_2.

Figure 5.3 shows a network of logics with directed arcs between sublogics. For example, *datalog* is a sublogic of *definite clauses*.

It's easy to see that if L_1 is a sublanguage of L_2 then L_1 is less expressive than L_2. However two languages can be equally expressive, even if one is a strict sublogic of the other. In particular, each of the logics above the undecidable line are all Turing equivalent. They can compute any computable function, and so they are all equally expressive. The idea that all Turing equivalent languages are equally expressive is known as the **Turing tar-pit**. You need more than the notion of expressiveness to distinguish them.

You can also compare logics on worst-case complexity, which is the inherent complexity of the most difficult problem that can be represented in the logic. For example, the definite clause language with function symbols (page 58) is undecidable; you can't even guarantee that the programs will halt. The ground definite clause language has complexity linear in the number of clauses; the bottom-up procedure (page 47) computes consequences in time linear in the number of clauses. Datalog (page 29) is decidable, but determining consequences is NP-hard. (See page 50 for a description of

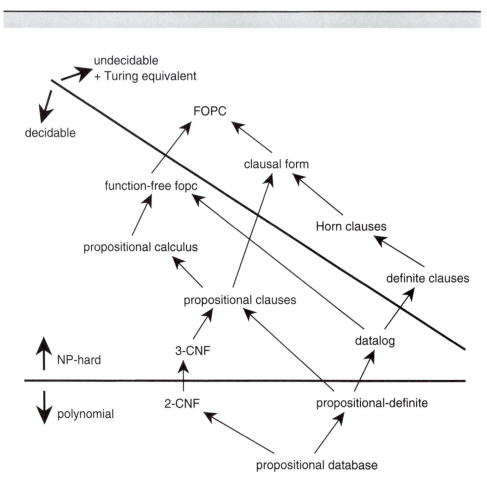

FOPC: first-order predicate calculus (page 270); **clausal form:** conjunctions of disjunctions of literals (CNF) (page 256); **Horn clauses:** definite clauses and integrity constraints (page 241); **definite clauses:** clauses with function symbols (page 58); **function-free FOPC:** FOPC without functions or existentially quantified variables in the scope of universally quantified variables; **datalog:** definite clauses without function symbols (page 29); **propositional calculus:** FOPC without variables or function symbols; **propositional clauses:** clausal form without variables or function symbols; **3-CNF:** propositional clauses with at most 3 disjuncts in each clause; **2-CNF:** propositional clauses with at most 2 disjuncts in each clause; **propositional definite:** definite clauses without variables or function symbols; **propositional database:** facts without variables or function symbols (no rules or disjunctions).

Figure 5.3: Lattice of sublogics

NP, NP-hard and NP-complete.) Finding consequences of the propositional calculus (page 270) is co-NP-complete.

If logic L_1 is less expressive than logic L_2 then the worst-case complexity of L_1 is either less or equal to that of L_2. Thus increasing the expressiveness can't reduce the worst-case complexity.

The inherent worst-case complexity of a representation may not matter very much if you have a particular problem you want to solve. The worst case for the representation may not coincide with the problem you are interested in. If you have a particular set of problems, you can't use a representation that has an inherent complexity lower than that of your problem class without blowing up the size of the representation.

Another dimension in which to compare logic is in the notion of **naturalness**. You want the representation of a problem to form a natural specification of the problem. At one extreme, the representation may be just a concise specification of the problem as one human may tell it to another, and at the other extreme, it may be very difficult to see how the problem and the representation mesh together. You would like to have a natural representation which facilitates both (a) the mapping from the problem to the representation of the problem so that it's straightforward to represent a problem; and (b) the mapping from the representation to the problem so that it's easy to see what the representation is saying. This is important for ease of knowledge acquisition; you don't want to have to hire skilled programmers to represent (program) every problem. It's also important for correctness. If you can see what the representation says, then it's easy to check to see whether it matches the problem. It's also important for maintainability. You want a small change in the problem to result in a small change to the representation of the problem.

It's this issue of naturalness that distinguishes AI representations from representations such as used by software engineers. Rather than having a machine-oriented programming language like C, you would rather have a high-level programming language that's closer to the specification to the problem, even if it's more difficult to see exactly what a computer would do to execute the problem. Note that you still may have to implement the representation languages in lower-level languages such as Prolog or C.

Levels of Representations

You can view representations at various abstraction levels, from high-level languages to low-level languages. Suppose you treat the mapping from representation to output in Figure 5.1 (page 170) as another problem to be solved. You can translate the representation into a lower-level representation so that computation at the lower-level will implement the high-level computation.

An example hierarchy of abstraction levels is shown in Figure 5.4. This hierarchy is the traditional software engineering view of the world. Given a problem, first give

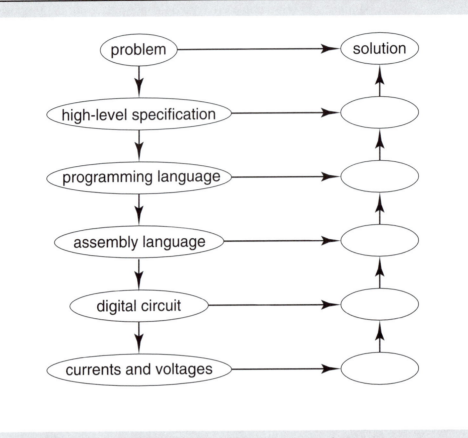

Figure 5.4: A hierarchy of representations

a high-level specification of the problem. This specification assumes an appropriate inference mechanism to specify what should be computed. Second you can map the high-level specification into a programming language (e.g., Prolog, or C), which can then be run to solve the problem. The programming language can in turn be mapped into an assembly language which in turn can be mapped into a digital circuit. Each time you map the problem into a lower-level representation it becomes harder to relate the representation to the problem, but it becomes easier to map to actual hardware. At lower levels you need to make more commitments, for example, at the highest level you may not be concerned with the actual ordering of computation steps, nor the details of the data structures. At the lower levels you must commit to the ordering of the steps and the details of the physical representation of the data structures. Thus specifying at the higher levels is easier as you don't have to commit to choices that you don't care about. This also potentially makes higher-level languages more efficient, as you can choose the ordering and the data structures to optimize performance.

Under this computer science view, mapping from the problem to the high-level specification is "problem specification." Mapping from the specification to the programming language is called "programming." Mapping from the programming language to the assembly language is called "compiling," and so on. Software engineering currently assumes that the first two mappings are done by humans, with only the third and lower ones being automated. In the 1950s and 1960s there were many claims made that high-level languages were inherently inefficient, so that the third arrow will always need to be done by humans. One distinguishing feature of computational intelligence from more traditional computer science is that we want as much of this automated as possible; in particular, we want the second mapping automated as well. This does not necessarily mean that you need automated programming, but it does mean that you want very high-level programming languages that are as close as possible to the problem. For most cases, computers can produce better code than humans because they can be optimized for the features of particular architectures, where a human doesn't want to do this. Programming parallel machines is particularly difficult. The high-level specification should be independent of whether a single machine or multiple machines are being used at lower levels.

Biological systems as well as computers can be described at multiple levels of abstraction. At successively lower levels are the neuron level, the biochemical level (what chemicals and what electrical potentials are being transmitted), the chemical level (what chemical reactions are being carried out), and the physics level (in terms of forces on atoms and quantum phenomena). What levels above the neuron level are needed to account for intelligence is still an open question. Note that these levels of description are echoed in the structure of science itself, where there are divisions of scientists into physicists, chemists, biologists, psychologists, anthropologists, and so on. While no level of description is more important than the other, we conjecture that you don't have to emulate every level to build an artificial computational intelligence, but can rather emulate the top levels and build them upon a new foundation.

Two levels that seem to be common among biological and computational entities are:

Knowledge level The knowledge level is a level of abstraction where you consider what an agent knows and what an agent's goals are, without necessarily knowing anything about how the agent reasons. For example, you can describe the delivery agent's behavior in terms of whether it knows that a parcel has arrived or not, and whether it knows where a particular person is or not. Both human or robotic agents can be described in this way. At this level you don't specify how the solution will be computed, or even which of the many possible strategies available to the agent will be used.

Symbol level The symbol level is a level of description of an agent in terms of what symbols it is manipulating. In order to implement the knowledge level, you must specify what symbols are involved and how they are manipulated in order to produce answers. Many cognitive science experiments are designed to determine what symbol manipulation occurs during reasoning. Note that while the knowledge level is about the external world to the agent (i.e., about what the agent believes about the external world and what its goals are in terms of the outside world), the symbol level is about what goes on inside an agent in order to reason about the external world.

This book uses logic both for the knowledge level to axiomatize a world using semantics (page 31) and for the symbol level to describe search strategies in logic. Chapter 6 discusses meta-interpreters which provide mechanisms for symbol-level reasoning to deliver knowledge-level functionality.

Some areas of computational intelligence, namely neural networks (page 408), are based on the belief that to automate the functionality of intelligence you need to also automate some abstraction of the mechanism of the brain. This hypothesis is being tested by concerted efforts to build intelligence both with and without automating the mechanism of the brain.

5.4   Mapping from Problem to Representation

Given a representation language, you have to consider how to map from a problem to a representation of the problem in that language. A number of issues arise in this context:

- What level of abstraction of the problem do you want to have to represent?
- What objects and relations in the world do you want to represent?
- How can you represent the knowledge to ensure that the representation is natural, modular, and maintainable?

Level of Abstraction

As in Figure 5.4, you can describe human and artificial reasoning systems at a number of levels of abstraction. You can describe virtually any situation at multiple levels of abstraction.

Example
5.3

In the delivery robot, you can model the environment at a high level of abstraction ignoring distances, the size of the robot, the steering angles needed, the slippage of the wheels, the weight of parcels, perception, obstacles, power for the robot, the political situation in Canada, and virtually everything else. You can model the

Qualitative Versus Quantitative Representations

Much of science and engineering considers quantitative reasoning with numerical quantities, using differential and integral calculus as the main tools rather than logic as presented here. Using logic doesn't preclude us from quantitative reasoning; calculus equations can be written in logic. Logical systems are quite capable of numerical reasoning, and many are even optimized for this. Logic and symbolic reasoning do, however, allow qualitative reasoning. This is reasoning about qualitative distinctions rather than numerical values for given parameters.

Qualitative reasoning use qualitative values that can take a number of forms:

- Landmarks are quantitative values that make qualitative distinctions in the object being modeled. When reasoning qualitatively, you may only need to reason about these landmark values and not the quantitative values. This is especially true when you can perceive when these landmarks are reached. For example, consider having the delivery robot pour and deliver coffee. Some important qualitative distinctions are whether the coffee cup is empty, partially full, or full. These landmark values are all that is needed to predict what happens if the cup is tipped upside down. It also lets you predict what happens when coffee is poured into the cup: An empty cup becomes partially full, a partially full cup stays partially full or becomes full, and a full cup overflows. Similarly, there is a qualitative difference between the pot being tipped enough for coffee to pour out and where it is not tipped enough.

- Orders of magnitude reasoning involves approximate reasoning that ignores minor distinctions. For example, whether the coffee cup is overfull, full enough to deliver, half empty, or empty may determine what actions the robot should carry out. These fuzzy terms have ill-defined borders between them, yet there should be some mapping between the actual amount of coffee in the cup and the qualitative description of the level of the coffee.

- Qualitative derivatives, namely whether some value is increasing, decreasing, or staying the same, can be seen as landmarks but in the derivative of the quantity. The derivative of the amount of coffee in the cup determines the possible transitions between landmark values or fuzzy values.

Qualitative reasoning is important for a number of reasons:

- You may not know the quantitative values, nor may you need quantitative answers. For example, the optimal level for the coffee cup may be 1.2 cm from the top, but neither you nor the robot may be able to sense that accurately.

- You may want the reasoning to be applicable no matter what the quantitative values are. For example, you may want a strategy for the robot that works regardless of what loads are placed on the robot, how slippery the floors are, or the actual charge on the batteries as long as they are within some normal operating ranges.

- You need to do qualitative reasoning to determine which quantitative laws are applicable. For example, if the delivery robot is filling a coffee cup, different quantitative formulae are appropriate to determine where the coffee goes when the coffee pot is not tilted enough for coffee to come out, when coffee comes out into a nonfull cup, and when the coffee cup is full and the coffee is soaking into the carpet.

robot at lower levels of abstraction by not ignoring some of these details. Some of these details may be irrelevant for the successful implementation of the robot, but some may be crucial for the robot to succeed. For example, in some situations the size of the robot and the steering angles may be crucial for not getting stuck around a particular corner. In other situations, if the robot stays close to the center of the corridor, you may not need to model the width of the robot or the steering angles.

Choosing an appropriate level of abstraction is difficult because:

- A high-level description is easier for a human to specify and understand.

- A low-level description can be more accurate and more predictive. Often high-level descriptions abstract away details that may be important for actually solving the problem.

- The lower the level, the more difficult it is to reason with. This is because a solution at a lower level of detail involves more steps, and so the search complexity to find a solution is much worse.

- You may not know the information needed for a low-level description. For example, the delivery robot may not know what obstacles it will encounter or how slippery the floor will be at the time that it needs to decide what to do.

It's often a good idea to model an environment at multiple levels of abstraction. Given multiple levels of abstraction, first solve the problem at a high level of abstraction (where it's easier to solve the problem), and then refine the solution for the lower level of abstraction. Unfortunately, you usually can't guarantee that the high-level solution can be refined to be a low-level solution.

Choosing Objects and Relations

Given a logic and a world, you have to choose what in the world you want to refer to; you have to choose what objects and relations there are. Thus you must commit to an **ontology** of the domain. It may seem that you can just refer to the objects and relations that exist in the world. However, the world does not determine what objects there are in a domain. How the world is divided into objects is invented by whoever is modeling the world. It's you that divides the world up into things, so that you can refer to parts of the world that make sense to you.

Example
5.4

It may seem as though "*red*" is a reasonable property to ascribe to things in the world. You may do this because you want to tell the delivery robot to go and get the red parcel. In the world, there are objects that absorb some frequencies and reflect other frequencies of light. Some user may have decided that, for some application, some particular set of reflectance properties should be called "red." Some other modeler of the domain might decide on another mapping of the spectrum and use the terms *"pink,"* *"puce,"* *"ruby"* and *"crimson."*

Just as you invent what are the objects, you also invent what are the relations in the world. There are, however, some guiding principles that can be used for choosing relations and individuals. These will be demonstrated through a sequence of examples.

Example 5.5	Suppose you have decided that "red" is an appropriate category for classifying objects. You can treat the name *"red"* as a unary relation and write that parcel a is red:

$red(a)$.

If you represent the color information in this way, then you can easily ask what's red:

$?red(X)$.

The X returned are the objects that are red.

With this representation, it's hard to ask the question, "What color is parcel a?" In the syntax of definite clauses, you can't ask

$?X(a)$.

because, according to the syntax, predicate names can't be variables. Moreover, this would return any property of a, and not just its color.

There are alternative representations that allow you to ask about the color of parcel a. There is nothing in the world that forces you to make *red* a predicate. You could just as easily say that colors are individuals too, and you could use the constant *red* to denote the color red. Given that *red* is a constant, you can use the predicate *color(Obj, Val)*, to mean that physical object *Obj* has color *Val*. "Parcel a is red" can now be written as

$color(a, red)$.

What you have done is reconceive the world: the world now consists of colors as individuals which you can name. You now have a new binary relation *color* between physical objects and colors. Under this new representation you can ask "What color is block a?" with the query

$?color(a, C)$.

It seems as though there is no disadvantage to the new representation of colors; everything that could be done before can be done now. It's not much more difficult to write $color(X, red)$ than $red(X)$, but you can now ask about the color of things. So the question arises as to whether you can do this to every relation, and what do you end up with.

Example 5.6

> You can do a similar analysis to the *color* predicate as for the *red* predicate in Example 5.5. The representation with *color* as a predicate doesn't allow you to ask the question, "Which property of parcel *a* has value *red*?," where the appropriate answer is "color." Carrying out a similar transformation to that of Example 5.5, you can view properties such as *color* as individuals, and you can invent a relation *prop* and write "object *a* has the *color* of *red*" as:
>
> > $prop(a, color, red)$.

You don't need to do this again, because you can write everything in terms of the *prop* relation.

The **object-attribute-value** representation is in terms of a single relation *prop* where

$$prop(Obj, Att, Val)$$

means that object *Obj* has value *Val* for attribute *Att*. The **domain** of an attribute is the set of possible values that the attribute can take on.

There are some predicates that seem to be too simple for this representation:

Example 5.7

> To transform *parcel(a)*, there does not seem to be appropriate attributes or values. There are two possible ways to transform this into the object-attribute-value representation: the first is to reify the concept parcel and to say that *a* is a parcel by
>
> > $prop(a, is\_a, parcel)$.
>
> Here *is_a* is a special attribute. The second is to make being a parcel a property of individuals and write "*a* is a parcel" as
>
> > $prop(a, parcel, true)$.
>
> In this representation, *parcel* is what's called a Boolean attribute.

A **Boolean** attribute is an attribute with domain {*true, false*}. That it, its possible values are *true* or *false*, where *true* and *false* are constant symbols in the language.

There are some predicates that may seem to be too complicated for this representation, but they can be put into this representation by inventing new individuals:

Example 5.8

> Suppose you want the infobot to be able to represent the relation
>
> > $scheduled(C, S, T, R)$,
>
> which is to mean that section *S* of course *C* is scheduled at time *T* in room *R*. For example, "section 2 of course *cs*422 is scheduled at 10:30 in room *cc*208," is written as
>
> > $scheduled(cs422, 2, 1030, cc208)$.

To represent this in the object-attribute-value representation, you need to invent a new individual, a booking. A booking has a number of properties, namely a course, a section, a time, and a room. To represent "section 2 of course *cs*422 is scheduled at 10:30 in room *cc*208," you name the room booking, say, the constant *b*123 and write

>   *prop*(*b*123, *course*, *cs*422).
>
>   *prop*(*b*123, *section*, 2).
>
>   *prop*(*b*123, *time*, 1030).
>
>   *prop*(*b*123, *room*, *cc*208).

This new representation has a number of advantages. The most important is that it's modular; which values go with which attributes can easily be seen. It's easy to add new attributes. With *scheduled* as a predicate, it was very difficult to add a new attribute—for example, the instructor of the section of the course. With the new representation it's easy to add that "Craig is teaching section 2 of course *cs*422, scheduled at 10:30 in room *cc*208":

>   *prop*(*b*123, *instructor*, *craig*).

Building Flexible Representations

Once you have just a single predicate symbol, you can omit it without any loss of information. You can interpret the *prop* relation in terms of a graph, where the relation:

>   *prop*(*Obj*, *Att*, *Val*)

is depicted with *Obj* and *Val* as nodes with an arc labeled with *Att* between them. Such a graph is called a **semantic network**.

| Example 5.9 | Figure 5.5 shows a semantic network for the delivery robot showing the sort of knowledge that the robot may have about a particular computer. Some of the knowledge represented in the network is |

>   *prop*(*comp*_2347, *owned_by*, *craig*).
>
>   *prop*(*comp*_2347, *deliver_to*, *ming*).
>
>   *prop*(*comp*_2347, *model*, *lemon_laptop*_10000).
>
>   *prop*(*comp*_2347, *brand*, *lemon_computer*).
>
>   *prop*(*comp*_2347, *logo*, *lemon_disc*).
>
>   *prop*(*comp*_2347, *color*, *brown*).
>
>   *prop*(*craig*, *room*, *r*107).
>
>   *prop*(*r*107, *building*, *comp_sci*).

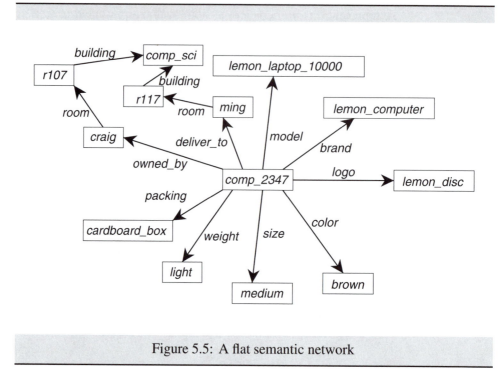

Figure 5.5: A flat semantic network

The network also shows how the knowledge is structured. For example, it's easy to see that computer number 2347 is owned by someone (Craig) whose office (r107) is in the *comp_sci* building. The direct indexing evident in the graph can be used by humans and machines.

This graphical notation has a number of advantages:

- It's easy for a human to see the relationships without needing to learn the syntax of a particular logic. The graphical notation helps the builders of knowledge bases to organize their knowledge.

- You don't have to label nodes which just have meaningless names (e.g., the name $b123$ in Example 5.8).

An alternative representation based on the object-attribute-value representation is to group all of the attribute-value pairs for a single object together into what is called a **frame**. This is useful as it brings all of the information about a object together. It also means that you don't have to create artificial constants. Within a frame, the attributes are often called **slots** and the values are called **fillers**.

Example 5.10

The frame for the scheduled relation of Example 5.8 can be written as

$$[course = cs422,$$

$$section = 2,$$
$$time = 1030,$$
$$room = cc208]$$

One of the advantages of the frame representation is that you can see what information you have about an individual.

Given such a frame representation you can define a **frame template** for each type of individual. A frame template specifies which slots are necessary and which are optional for individuals of that type. This helps in building a knowledge base, as it tells the providers of information exactly what knowledge is needed. It's no use providing a blank slate and asking someone to tell you everything that's true; they would not know where to start or what level of detail is required.

This frame for *scheduled* is like the original relational notation, but where the arguments are not defined by position but by naming them. In the relational definition, you had to decide initially exactly how many arguments there are, what they are, and what order they are in. With the frame representation, you don't have to decide on these at any time. If some particular program is only interested in some of the arguments, then it shouldn't have to care about the others. It should be able to name the ones it's interested in and not care about how many other properties there are.

This more flexible way of handling predicates and functions is known as a **unification grammar**. In a unification grammar, functions and relations are not specified by giving their arguments positions, but by giving them names, as in Example 5.10. This means that you can build procedures to manipulate functions without knowing the order or how many arguments there are. Unification in this framework works not by unifying position by position, but by matching values for the same attributes.

Example 5.11

Suppose you have a course schedule with frame

$$[course = cs422, section = 2, time = 1030, room = cc208]$$

and one with frame

$$[time = 1030, instructor = robin, course = cs422]$$

These two structures can unify, resulting in frame

$$[course = cs422, section = 2, time = 1030,$$
$$room = cc208, instructor = robin].$$

To represent this in the more traditional notation, you would need to decide on a positional notation that anticipates all possible attributes, such as the template:

$$schedule(Course, Section, Time, Room, Instructor, Duration, TA).$$

The above frame unification would be equivalent to the more traditional unification of

$$schedule(cs422, 2, 1030, cc208, Instructor, Duration, TA)$$
$$schedule(cs422, Section, 1030, Room, robin, Duration, TA)$$

resulting in

$$schedule(cs422, 2, 1030, cc208, robin, Duration, TA).$$

Unification grammars provide a very flexible way to represent functions and relations. They have been used mostly for natural language understanding (hence the name), where each part of the sentence provides for more constraints on the resulting term that gives the meaning. They allow you to build a parser without knowing all of the possible attributes that may be part of the sentence.

Primitive Versus Derived Relations

Typically you know more about a domain than a database of facts; you know general rules from which other facts can be derived. Which facts are explicitly given and which are derived is a choice to be made when building a knowledge base.

Primitive knowledge is that which is defined explicitly by facts. **Derived knowledge** is knowledge defined by rules.

The use of rules allows for a more compact representation of knowledge. Derived relations allow for conclusions to be drawn from observations of the domain. This is important because you don't directly observe everything about a domain. Much of what's known about a domain is inferred from the observations and more general knowledge.

Example
5.12

Example 5.9 explicitly specified that the logo for computer *comp_2347* was a lemon disc. You may, however, know that all Lemon-brand computers have this logo. An alternative representation is to associate the logo with *lemon_computer*, and derive the logo of *comp_2347*. The advantage of this representation is that if you find another Lemon-brand computer, you can infer its logo.

When expanding the semantic network to allow for deriving information, you distinguish classes from individuals. In Figure 5.6 the classes are the shaded boxes. Arcs from the classes are not the properties of the class, but are properties of the members of the class. In Figure 5.6, the semantic network has the logo associated with the brand. This says that all Lemon-brand computers have the lemon disc as the logo.

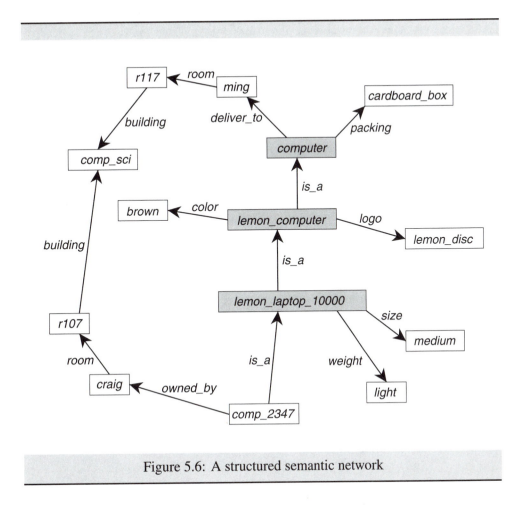

Figure 5.6: A structured semantic network

Note that you could also have properties of the class of Lemon-brand computers—for example, that it contains less than 50,000 elements. This is an attribute of the class of Lemon computers and not an attribute of any particular Lemon computer. We won't represent such information here, although it's not difficult to do so.

You can treat *is_a* as a special attribute that allows **property inheritance**. The attribute *is_a* between an individual and a class means that the individual is a member of the class. The attribute *is_a* between classes means that one class is a subclass of the other. The general idea is that an arc from a class means that the corresponding attribute has this value for every individual that's a member of the class. Thus, an arc labeled with p from class node c to node n is a representation of the clause

$$prop(Obj, p, n) \leftarrow$$
$$prop(Obj, is\_a, c).$$

Example | Given the translation, the following clauses can be directly read off the semantic
5.13 | network of Figure 5.6:

$prop(comp\_2347, is\_a, lemon\_laptop\_10000).$

$prop(comp\_2347, owned\_by, craig).$

$prop(X, weight, light) \leftarrow$
$\quad prop(X, is\_a, lemon\_laptop\_10000).$

$prop(X, size, medium) \leftarrow$
$\quad prop(X, is\_a, lemon\_laptop\_10000).$

$prop(X, is\_a, lemon\_computer) \leftarrow$
$\quad prop(X, is\_a, lemon\_laptop\_10000).$

$prop(X, is\_a, computer) \leftarrow$
$\quad prop(X, is\_a, lemon\_computer).$

$prop(X, logo, lemon\_disc) \leftarrow$
$\quad prop(X, is\_a, lemon\_computer).$

$prop(X, color, brown) \leftarrow$
$\quad prop(X, is\_a, lemon\_computer).$

$prop(X, packing, cardboard\_box) \leftarrow$
$\quad prop(X, is\_a, computer).$

$prop(X, deliver\_to, ming) \leftarrow$
$\quad prop(X, is\_a, computer).$

$prop(ming, room, r117).$

$prop(craig, room, r107).$

$prop(r117, building, comp\_sci).$

$prop(r107, building, comp\_sci).$

The *prop* relations that can be derived from these clauses are essentially the same as can be derived from the flat semantic network of Figure 5.5. With the structured representation, to incorporate a new Lemon Laptop 10000, you only need to add the arcs that are particular to it (e.g., that it's a Lemon Laptop 10000 and who its owner is). Other generic properties can be derived through inheritance.

There are some general guidelines that can be used for deciding what should be primitive and what should be derived:

- When associating an attribute value with an individual, choose the most general class that the individual is in with that attribute value, and associate the attribute value with that class. Inheritance can be used to derive the attribute value for the individual. This representation methodology tends to make knowledge bases more concise, and it means that it's easier to incorporate new individuals as they automatically inherit the attribute value if they are a member of the class.

- Don't associate contingent properties of a class with the class. For example, it may be true of the current computer environment that all of the computers come in brown boxes. However, it may not be a good idea to put that as a property of the *computer* class, as it would not be expected to be true as other computers are bought.

- Axiomatize in the **causal** direction. It's often the case that there is a choice a direction to make an implication. For example, consider two propositions a and b, both of which are true. You can write "$a \wedge b$," treating both a and b as primitive. You can write "b and $a \leftarrow b$," where a is primitive and b is derived. Alternatively, you can write "b and $a \leftarrow b$," where b is primitive and a is derived. These representations are logically equivalent; they can't be distinguished logically. A good way to test which representation to choose is to use implication in the causal direction: Suppose b were to be changed for some reason other than by changing a; if a would change as a side effect, write $a \leftarrow b$. Such a causal axiomatization is better because it's more stable with respect to changes in the knowledge base. This makes it easier to maintain such a knowledge base.

Example 5.14 | As an example of axiomatizing in the causal direction consider the electrical domain depicted in Figure 3.1 (page 71). In this domain, switch $s3$ is up and light $l2$ is lit. To consider the causal direction, suppose someone intervenes and puts $s3$ down. You would expect that light $l2$ would no longer be lit as a consequence. If someone intervened to make $l2$ no longer lit, you would not expect this to magically make the switch go down. Thus it seems reasonable to axiomatize in the causal direction and to make an implication from the switch being up to the light being lit, as was done in Section 3.2 (page 70). There is a justification as to why the implication direction of Section 3.2 seems natural.

For more discussion on causal reasoning see Section 9.5 (page 335) and the box on page 367.

Acquiring and Debugging a Knowledge Base

Acquiring and debugging knowledge bases are important activities. Much of Chapter 6 is devoted to these topics.

Given that you want to be able to interact with a computer at a very high level, you would like to acquire and debug knowledge without knowing internals of the compu-

tation or the inference procedure. You would like to acquire and debug knowledge at
the knowledge level:

Knowledge-level acquisition and debugging is where the knowledge base is
acquired and debugged given only the meaning of the symbols, without reference to
how answers are computed or to the internal representation of the knowledge.

Knowledge-level acquisition is explored in Section 6.4 (page 212) and knowledge-
level debugging is developed in Section 6.6 (page 221).

Symbol-level debugging is debugging of a knowledge base based on how the
symbols are being manipulated.

In symbol-level debugging you can take into account the search strategy and the
order in which goals are produced. The standard debugging tools for Prolog are
symbol-level debugging tools that rely on knowing the search strategy and the evaluation ordering. In essence they let you trace the symbol manipulation.

Suppose you have derived an answer from a knowledge base that's wrong. When
debugging a knowledge base (either automatically or manually) you need to distinguish
two different reasons for debugging:

- Knowledge (or belief) **update** is when the knowledge base is incorrect because
  the world has changed.

- Knowledge (or belief) **revision** is where the world has not changed, but you
  have determined that there is a mistake in the knowledge base.

It's important to distinguish these, because they have different characteristics. For
knowledge update, you want to determine what in the world has changed, and then
update the knowledge base based on the action. For knowledge revision, you expect
that there is one or a few things wrong and want to determine not an action but a fact
that's wrong.

5.5   Choosing an Inference Procedure

So far in this chapter you have only been considering one part of knowledge
representation, namely what should be computed, without taking into consideration
how it's computed.

You can partition what is required of intelligence into the **epistemological** component or just the logic, which provides a specification of *what* to compute, and the
heuristic component or control, which specifies *how* to compute. A representation
and reasoning system (page 23) contains both components.

A **representation and reasoning system** is a logic together with an inference
mechanism. Thus, as well as having a language and a specification of what an answer
is, an RRS also specifies a mechanism for computing answers.

You can judge an RRS on two different grounds:

- An RRS is **epistemologically adequate** if it can express the concepts and relations needed to solve the problem.

- An RRS is **heuristically adequate** if it can use the information expressed within reasonable computational resources.

It seems reasonable that you need a representation and reasoning system that's both capable of representing the problem (epistemological adequacy) and able to find a solution (heuristic adequacy). If one can't state a problem, then computation will not help because the computer can't determine what to compute. Similarly, it's not much use being able to state a problem if it can't be solved. Unfortunately, these two criteria are often at odds; epistemological adequacy tends to lead to richer representations that allow for the subtlety of expression that the problem may demand, however, the richer the representation, the more difficult it tends to be to compute with it.

You can compare RRSs on efficiency:

RRS_1 is **more efficient** than RRS_2 if, for each of the problems that both can represent, there is a representation of the problem in RRS_1 such that solving the problem using RRS_1 is faster than or equally as fast as solving the problem using any representation in RRS_2.

Note that where you could compare logics with respect to expressiveness (page 175) and complexity, you can't compare logics with respect to efficiency. Efficiency is with respect to an inference mechanism, whereas complexity has to do with a class of problems that can be solved rather than how efficiently they can be solved.

Just as being more expressive leads to a worse (or equal) complexity (page 175), more expressive RRSs lead to worse (or equal) efficiency. However, efficiency and complexity don't necessarily divide up the space of RRSs in the same way. For example, definite clauses and the clausal form of the first-order predicate calculus have the same complexity, but definite clause RRS can be more efficient than a clausal form RRS as it needs less overhead. (See Section 7.5 (page 256) for a description of a clausal form RRS.) At the other extreme, the definite clause language and datalog have different complexity, but, if using a top-down interpreter for both, they may have the same efficiency because the overhead of needing to be able to handle function symbols may have no cost if function symbols are not used.

Let's consider what issues affect efficiency of computing a solution to a problem. Because many of the algorithms used in AI are search algorithms, the efficiency is affected mainly by the size of the search space, the branching factor of the search, and the depth of solutions. Next is the existence of good heuristics. Good heuristics have no local minima that are not global minima and have no plateaus. The next most important issue is the algorithm used and whether it can take into account the heuristic information available. Other issues such as the level of detail affect the efficiency by changing the search space.

It needs to be emphasized that the representation chosen affects the search space and thus affects efficiency. Some representations lead to more efficient problem solving than others. Often a clever encoding of a problem can make the problem much easier to solve. Automating the clever encoding of problems is one of the goals of computational intelligence, but one that is not well developed.

Probably the least important factor affecting efficiency is the choice of programming language. It is probably better to use a programming language that is designed for the sorts of problems you want to solve rather than expecting that some low-level general-purpose programming language will be more efficient. You will probably have to implement all of the functionality of the high-level programming language, but will not have the time to optimize the code as would someone who is trying to develop an efficient programming language.

There are a number of general principles that can be used to help compute efficiently:

- Compile into another language for efficiency. Often the most appropriate high-level language isn't available. It's often useful to design the right high-level language for your application. This will be the language that has the appropriate constructs that you want to use to concisely specify the knowledge. For example, Chapter 6 defines a number of languages that are appropriate for different representation and reasoning applications. In that chapter interpreters for the languages are developed. Often the most efficient way to define a knowledge representation is to design an appropriate high-level language and then translate, or compile, that language into another, more efficient language rather than force your application into the available representations. In this way the translation can be tuned for efficiency independently of the particular knowledge base.

- Exploit restrictions in the language for efficiency. Often, by providing a language with restricted capabilities, you can implement it a lot faster. For example, if you are to implement a datalog program, you may be able to exploit the fact that every relation can be specified as a finite set of facts, as the extensional definition of the relation. By forward chaining on the rules you produce extensional definitions of derived relations. This does not work for the full definite clause language because you can't guarantee that the extensional definition of relations are finite.

- Exploit the structure of the problem to derive special purpose inference procedures. One example is exploiting properties of time. When building a robot controller (Chapter 12), you can exploit the fact that time marches forward and that a robot or any other agent can only observe its present time and can't directly observe its past or future.

- Don't commit to a choice unless forced to. This is the idea of *least commitment*. It's often best to not resolve a choice when it is encountered, but to delay making the choice. It is quite likely that some of the alternatives will be removed, in which case you don't have to make the choice at all. You only make the choice when there is nothing else to do. This was the basis of the efficient constraint satisfaction algorithms (page 147). This is also the basis of partial order planning (page 309), where you don't want to commit to an ordering of actions unless you are forced to.

- Cache results judiciously. If you have computed a result, then it's sometimes a good idea to save it so that it can be retrieved when needed. This can often save a great deal of time, as long as the relevant knowledge can be retrieved quickly. Unfortunately, if done naively, caching buys little and wastes a lot of space; when considering iterative deepening (page 140) we showed that in some cases it's better to recompute a value rather than to store it.

5.6   References and Further Reading

Brachman & Levesque (1985) present many classic knowledge representation papers. Davis (1990) is an accessible introduction to a wealth of knowledge representation issues in common-sense reasoning.

The use of anytime algorithms is due to Horvitz (1989) and Boddy & Dean (1994). See Dean & Wellman (1991, Chapter 8), Zilberstein (1996), and Russell (1997) for introductions to time-bounded computation.

The epistemological and heuristic distinction is due to McCarthy & Hayes (1969). A similar distinction between logic and control was advocated by Kowalski (1979a). The distinction between updating revising a knowledge base is described in Katsuno & Mendelzon (1991).

Semantic networks are analyzed in Woods (1975) and Brachman (1979). Frames are advocated in Minsky (1975). Scripts are temporal frames and are presented in Schank & Abelson (1977). See Bobrow & Winograd (1977) for an overview of a sophisticated frame-based knowledge representation language. For a critique of frames from a logical perspective see Hayes (1979). The use of the *prop* relation is based on the property lists of Lisp.

Qualitative reasoning is described by Forbus (1988) and Kuipers (1994). Weld & de Kleer (1990) contains many seminal papers on qualitative reasoning. See also Weld (1992) and related discussion in the same issue.

5.7   Exercises

Exercise 5.1
There are many possible kinship relationships you could imagine like mother, father, great-aunt, second-cousin-twice-removed, and natural-paternal-uncle. Some of these can be defined in terms of the others, for example:

$$brother(X, Y) \leftarrow father(X, Z) \wedge natural\_paternal\_uncle(Y, Z).$$

$$sister(X, Y) \leftarrow parent(Z, X) \wedge parent(Z, Y) \wedge$$
$$female(X) \wedge different(X, Y).$$

Give two quite different representations for kinship relationships based on different relations being primitive.

Consider representing the primitive kinship relationship using relation

$$children(Mother, Father, List\_of\_children)$$

What advantages or disadvantages may this representation have compared to the two you designed above?

Exercise 5.2
Tic-tac-toe is a game played on a 3×3 grid. Two players, X and O, alternately place their mark in an unoccupied position. X wins if, at its turn, it can place an X in an unoccupied position to make three X's in a row. For example, from the state of the game

X can win by moving into the left middle position.

Fred, Jane, Harold, and Jennifer have all written programs to determine if, given a state of the game, X can win in the next move. Each of them has decided on different representation of the state of the game. The aim of this question is to compare their representations.

Fred decided to represent a state of the game as a list of three rows, where each row was a list containing three elements, either x, o, or b (for blank). Fred represents the above state as the list

$$[[x, o, o], [b, x, b], [x, b, o]].$$

Jane decided that each position on the square could be described by two numbers, the position across and the position up. The top left X is in position $pos(1, 3)$, the bottom left X is in position $pos(1, 1)$, and so forth. She then decided to represent the state of the game as a pair $ttt(XPs, OPs)$ where XPs is the list of X's positions and OPs is the list of O's positions. Thus Jane represented the above state as

$$ttt([pos(1, 3), pos(2, 2), pos(1, 1)], [pos(2, 3), pos(3, 3), pos(3, 1)]).$$

Harold and Jennifer both realized that the positions on the tic-tac-toe board could be represented in terms of a so-called magic square:

6	7	2
1	5	9
8	3	4

Based on this representation, the game is transformed into one where the two players alternately select a digit. No digit can be selected twice, and the player who first selects three digits summing to 15 wins.

Harold decides to represent a state of game as a list of nine elements, each of which is x, o, or b, depending on whether the corresponding position in the magic square is controlled by X, controlled by O, or is blank. Thus Harold represents the game state above as the list:

$$[b, o, b, o, x, x, o, x, b].$$

Jennifer decides to represent the game as a pair consisting of the list of digits selected by X and the list of digits selected by O. She represented the state of the game above as

$$magic([6, 5, 8], [7, 2, 4]).$$

(a) For each of the four representations, write the relation $x\_can\_win(State)$, with the intended interpretation that X can win in the next move from the state of the game $State$.

(b) Which representation is easier to understand? Explain why.

(c) Which representation is more efficient to determine a win?

(d) Which representation results in the simplest algorithm to make sure a player doesn't lose on the next move, if such a move exists?

(e) Which do you think is the best representation? Explain why. Can you suggest a better representation?

Exercise 5.3 Implement unification of frames as in Example 5.11. You can use normal unification except when the two terms are lists, in which case you should assume they are frames. Note that the values of the attributes can also be frames. If an attribute appears in both frames, the values should unify. Unification should fail if the frames don't unify.

Exercise 5.4 Consider unifying the two frames

$$[type = mammal, blood = warm, legs = 4],$$

$$[type = dog, ears = 2, owner = kelly].$$

Our ordinary use of unification would fail in this case, as we couldn't make $type = dog$ match with $type = mammal$. However, Mary claims that these should unify with the result:

$$[type = dog, blood = warm, legs = 4, ears = 2, owner = kelly]$$

Explain why Mary might think these should unify. What is the general principle she is using? Describe how frame unification could be extended to create the new kind of result. Implement it.

Exercise 5.5   The generalization of the conception of the property *red* from an initial conception as *red(Obj)* to *color(Obj, Val)* is argued to provide extra flexibility, as it supports questions about the color of an object in terms of the query

$$?color(a, X).$$

Give at least one argument against this kind of generalization.

Exercise 5.6   Fred has proposed that any *n*-ary relation $P(X_1, X_2, X_3, ..., X_n)$ can be reexpressed as $n - 1$ binary relations—for example,

$$P_1(X_1, X_2).$$
$$P_2(X_2, X_3).$$
$$P_3(X_3, X_4).$$
$$\vdots$$
$$P_{n-1}(X_{n-1}, X_n).$$

Explain to Fred why this may not be such a good idea. What problems would arise if Fred tried to do this?

Chapter 6

Knowledge Engineering

6.1 Introduction

So far we have examined one representation and reasoning system and some different control structures. Before examining more representation and reasoning systems, it's important to look at another aspect of an RRS that is important for practical use of the technology, namely *knowledge acquisition* and *debugging*. This chapter considers the problem of acquiring knowledge from humans who possess the knowledge, and Chapter 11 considers acquiring knowledge from data. In order to build appropriate tools, this chapter also shows how to implement enhancements to the initial representation and reasoning system and how to build new RRSs.

Knowledge acquisition is an important issue in the construction of what are known as **expert systems**. Expert systems are knowledge-based systems where the knowledge base is a representation of an expert's knowledge of a domain. Expert systems can be built on various representation and reasoning systems, but there is a core of knowledge engineering tools that are common.

This chapter starts by considering the interaction between various users of a knowledge-based system. Different users of the system have different needs in terms of the tools that will be useful to them. Section 6.3 presents a general way to add features to a representation and reasoning system, in terms of a meta-interpreter that can be augmented to produce acquisition, debugging and other facilities. We then show how an initial *vanilla* meta-interpreter can be used to expand the representational capabilities of an RRS (page 207), to implement different search strategies, such as depth-bounded search (page 209) and other search strategies (page 226), and to collect assumptions (page 210).

We discuss interacting with users of a knowledge-based system, as well as how the user's view of semantics (page 37) can be used to explain to a user how the system

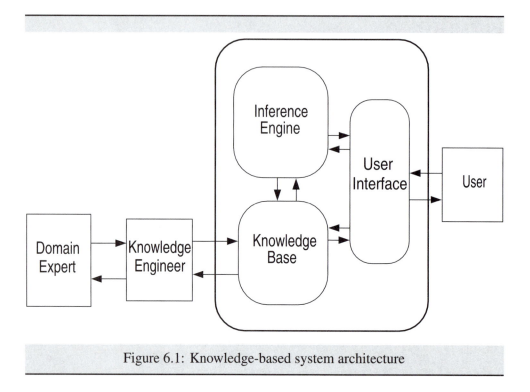

Figure 6.1: Knowledge-based system architecture

reached a conclusion. The same tools allow for a form of debugging that does not assume that the person debugging the system has any knowledge of how the system works. They need only know about the domain and what the symbols mean. We show how these facilities can be implemented in terms of meta-interpreters.

6.2   Knowledge-Based System Architecture

Different people are involved in the construction and use of a knowledge-based system. They have different skills, different knowledge, and different needs. It is important to distinguish the various roles in order to understand how to make the knowledge-based system most useful. Figure 6.1 shows the architecture of a knowledge-based system. The main components of the system are the control or inference engine, the knowledge base, and the user interface.

There are four major roles for people involved with a knowledge-based system:

Software engineers Those who build the inference engine and user interface. They typically know nothing about the contents of the knowledge base. They need not be experts in the use of the expert system they implement. For example, they may be experts in the use of a programming language like Prolog or Lisp or C, rather than in the language of the system they are designing.

Knowledge engineers Those who design, build, and debug the knowledge base in consultation with domain experts. They know about the details of the system and know about the domain through the domain expert. They know nothing about any particular case.

Domain experts Those whose expertise is to be captured within a knowledge-based system. They interact with knowledge engineers to define the substance of their expertise. They know about the domain, but typically know nothing about the particular case that may be under consideration (e.g., they know about diseases, but not about a particular patient). They typically do know about the details of how the expert system works. Often they have only a semantic view of the knowledge (page 37) and have no notion of the proof procedures used by the inference engine. It is unreasonable to expect that domain experts could debug a knowledge base by presenting them with traces of how an answer was produced. Thus, it's not appropriate to have debugging tools for domain experts that merely trace the execution of a program.

Users Those who have the need for expertise. They have information for individual problems which they can present to the knowledge-based system. They know about the particular case, but not about the domain of expertise of the knowledge base. They often do not know what information is needed by the system. Thus it's unreasonable to expect them to volunteer the information about a particular case. A simple and natural interface must be provided because users do not typically understand the internal structure of the system.

These should be seen as separate roles, even though the people in these roles may overlap. For example, the domain expert may act as a user in order to test or debug the system. The knowledge engineer may be the same person as the domain expert or the software engineer.

Each of the roles has different requirements for the tools they need. We do not consider the problem of software engineering, not because it's not important but because it's a field in its own right.

Section 6.3 introduces meta-interpreters that provide facilities for building the knowledge engineering tools. With this as background, let's present the tools and at the same time show how they can be implemented.

6.3   Meta-Interpreters

This chapter considers various tools that are needed by the people involved with a knowledge-based system. As well as discussing what tools are needed, we show how they can be implemented. The tools will be constructed by augmenting an initial meta-interpreter. This meta-interpreter is an interpreter for the simple representation and reasoning system within the same representation and reasoning system. As well as providing a basis for adding tools such as explanation and debugging facilities, this meta-interpreter is also useful for building more expressive representation and reasoning systems in later chapters.

When you implement one language inside another, the language being implemented is the **base language**, sometimes called the *object language*, and the language in which you implement it is the **metalanguage**. You say that expressions in the base language are at the base level, and expressions in the metalanguage are at the meta-level.

The base and metalanguages could be the same language, or different languages. In this chapter these are initially the same language, but soon diverge. Adding and modifying rules at the meta-level, without changing the metalanguage, can change the base language.

We define a sequence of more detailed interpreters. The first set of interpreters lets the meta-level do the searching and the unification. Later (page 226) we give an interpreter that specifies the searching but not the unification, and then we give a method that allows for more sophisticated treatment of variables (page 230) and show how to implement unification. Each of the interpreters can be extended to provide more features.

Base Languages and Metalanguages

To write an interpreter in the metalanguage, you need a representation of the base language; you need a representation of the base-level expressions that can be manipulated by the interpreter to produce answers. Recall (page 29) that the definite clause language is made up of terms, atoms, bodies, and clauses. In the metalanguage you need to be able to refer to these syntactic elements of the base language. When we introduced the semantics of the language (page 31), we said that the symbols could denote anything. Here the meta-level symbols will denote base-level terms, atoms, bodies and clauses.

In this section we adopt the **non-ground representation** for the base level, where you represent base-level terms as the same term in the metalanguage. This doesn't have to be done. In Section 6.8 we use the ground representation where you represent base language variables as constants in the metalanguage.

In this chapter we use the standard definite clause language (page 29) for the meta-level. In particular, "∧" is the meta-level conjunction and "←" is the meta-level implication.

Within the metalanguage you need to be able to represent all of the base-level constructs. In general, you represent base-level terms, atoms, and bodies as meta-level terms, and represent base-level clauses as meta-level facts. In the non-ground representation, you represent base-level variables, constants, and function symbols as the corresponding meta-level variables, constants, and function symbols. Thus, all terms in the base level are represented by the same term in the meta-level. A base-level predicate symbol p is represented by the corresponding meta-level function symbol p. Thus the base-level atom $p(t_1, \ldots, t_k)$ is represented as the meta-level term $p(t_1, \ldots, t_k)$.

Example 6.1

The base-level atom $foo(a, f(b), X)$ will be represented as the meta-level term $foo(a, f(b), X)$. In this expression the base-level predicate symbol foo is represented as the meta-level function symbol foo.

You represent base-level bodies as meta-level terms and represent base-level clauses as meta-level atoms. If e_1 and e_2 are meta-level terms that denote base-level atoms or conjunctions, then let's write the meta-level term $oand(e_1, e_2)$ to denote the base-level conjunction of e_1 and e_2. Thus $oand$ is a meta-level function symbol that denotes base-level conjunction. Let's represent the base-level empty body as the meta-level constant *true*.

If h is a base-level atom and b is a meta-level term that represents a base-level body, let's use the meta-level atom $clause(h, b)$ that is true if "h if b" is a base-level clause. A base-level fact a is represented as the meta-level atom $clause(a, true)$.

Example 6.2

The base-level clauses from Section 3.2 (page 70)

$$connected\_to(l_1, w_0).$$

$$connected\_to(w_0, w_1) \leftarrow up(s_2).$$

$$lit(L) \leftarrow light(L) \wedge ok(L) \wedge live(L).$$

can be represented as the meta-level facts

$$clause(connected\_to(l_1, w_0), true).$$

$$clause(connected\_to(w_0, w_1), up(s_2)).$$

$$clause(lit(L), oand(light(L), oand(ok(L), live(L)))).$$

Syntactic construct		Meta-level representation of the syntactic construct	
variable	X	variable	X
constant	c	constant	c
function symbol	f	function symbol	f
predicate symbol	p	function symbol	p
"and" operator	$\wedge$	function symbol	$\&$
"if" operator	$\leftarrow$	predicate symbol	$\Leftarrow$
clause	$h \leftarrow a_1 \wedge \cdots \wedge a_n.$	atom	$h \Leftarrow a_1 \& \cdots \& a_n.$
clause	$h.$	atom	$h \Leftarrow true.$

Figure 6.2: The non-ground representation for the base language

In order to make the base-level more readable, let's use the infix function symbol "&" rather than *oand*. Thus, instead of writing $oand(e_1, e_2)$, you write $e_1 \& e_2$. When you use the conjunction symbol "&", it's an infix function symbol of the metalanguage that denotes an operator, between atoms, of the base language. This is just a syntactic variant of the "*oand*" representation. It just makes it easier to read base-level formulae.

Instead of writing $clause(h, b)$ you can write $h \Leftarrow b$, where $\Leftarrow$ is an infix meta-level predicate symbol. Thus the base-level clause "$h \leftarrow a_1 \wedge \cdots \wedge a_n$" is represented as the meta-level atom

$$h \Leftarrow a_1 \& \cdots \& a_n.$$

This meta-level atom is true if the corresponding base-level clause is part of the base-level knowledge base. In the meta-level, this atom can be used like any other atom. It just happens to denote an base-level clause.

Figure 6.2 summarizes how the base language is represented in the meta-level.

Example 6.3

Using the infix notation, the base-level clauses of Example 6.2 can be represented as the meta-level facts

$$connected\_to(l_1, w_0) \Leftarrow true.$$

$$connected\_to(w_0, w_1) \Leftarrow up(s_2).$$

$$lit(L) \Leftarrow light(L) \& ok(L) \& live(L).$$

This notation is easier for humans to read than that of Example 6.2, but as far as the computer is concerned, it's essentially the same.

Note that the meta-level function symbol "&" and the meta-level predicate symbol "$\Leftarrow$" are not predefined symbols of the meta-level. You could have used any other symbols. We just assume they are written in infix notation for readability.

% *prove*(*G*) is true if base-level body *G* is a logical consequence of the base-level
% clauses that are defined using the predicate symbol "⇐".

> *prove*(*true*).
>
> *prove*((*A* & *B*)) ←
>> *prove*(*A*) ∧
>> *prove*(*B*).
>
> *prove*(*H*) ←
>> (*H* ⇐ *B*) ∧
>> *prove*(*B*).

Figure 6.3: The vanilla definite clause meta-interpreter

A Vanilla Meta-Interpreter

In this section we build a very simple vanilla meta-interpreter for the definite clause language in the definite clause language. In subsequent sections we will augment this meta-interpreter to provide various knowledge engineering tools. It is important to understand what is going on in the simple case presented here in order to make it easier to understand the more sophisticated meta-interpreters of later sections.

Figure 6.3 defines a meta-interpreter for the definite clause language. This is an axiomatization of the relation *prove*(*G*) that is true when base-level body *G* is a logical consequence of the base-level clauses.

As when you are axiomatizing any other relation, you write down the clauses that are true, and you make sure that they cover all of the cases and that there is some simplification through recursion. This meta-interpreter essentially covers each of the cases that are allowed in the body of a clause or in a query, and it specifies how to solve each case. A body is either empty, a conjunction, or an atom. The empty base-level body *true* is trivially proved. To prove the base-level conjunction *A* & *B*, you can prove *A* and prove *B*. To prove the atom *H*, you find a base-level clause with *H* as the head, and prove the body of the clause.

Example 6.4	Consider the meta-level representation of the base-level knowledge base in Figure 6.4. This knowledge base is adapted from Section 3.2 (page 70). Note that this knowledge base consists of meta-level atoms, all with the same predicate symbol, namely "⇐".

To prove the base-level goal *live*(w_5) you would issue the query

> ?*prove*(*live*(w_5)).

$lit(L) \Leftarrow$
  $light(L)$ &
  $ok(L)$ &
  $live(L)$.
$live(W) \Leftarrow$
  $connected\_to(W, W_1)$ &
  $live(W_1)$.
$live(outside) \Leftarrow true$.
$light(l_1) \Leftarrow true$.
$light(l_2) \Leftarrow true$.
$down(s_1) \Leftarrow true$.
$up(s_2) \Leftarrow true$.
$up(s_3) \Leftarrow true$.
$connected\_to(l_1, w_0) \Leftarrow true$.
$connected\_to(w_0, w_1) \Leftarrow up(s_2)$ & $ok(s_2)$.
$connected\_to(w_0, w_2) \Leftarrow down(s_2)$ & $ok(s_2)$.
$connected\_to(w_1, w_3) \Leftarrow up(s_1)$ & $ok(s_1)$.
$connected\_to(w_2, w_3) \Leftarrow down(s_1)$ & $ok(s_1)$.
$connected\_to(l_2, w_4) \Leftarrow true$.
$connected\_to(w_4, w_3) \Leftarrow up(s_3)$ & $ok(s_3)$.
$connected\_to(p_1, w_3) \Leftarrow true$.
$connected\_to(w_3, w_5) \Leftarrow ok(cb_1)$.
$connected\_to(p_2, w_6) \Leftarrow true$.
$connected\_to(w_6, w_5) \Leftarrow ok(cb_2)$.
$connected\_to(w_5, outside) \Leftarrow true$.
$ok(l_1) \Leftarrow true$.
$ok(l_2) \Leftarrow true$.
$ok(cb_1) \Leftarrow true$.
$ok(cb_1) \Leftarrow true$.

Figure 6.4: A base-level knowledge base for house wiring

The third clause of *prove* is the only clause that matches this query. You then look for a clause of the form *live*(w_5) $\Leftarrow$ *B* and find

$$live(W) \Leftarrow connected\_to(W, W_1) \,\&\, live(W_1).$$

where *W* unifies with w_5, and *B* unifies with *connected_to*(w_5, W_1) & *live*(W_1). You then try to prove

$$prove((connected\_to(w_5, W_1) \,\&\, live(W_1))).$$

The second clause for *prove* is applicable. You then try to prove

$$prove(connected\_to(w_5, W_1)).$$

Using the third clause for *prove*, you look for an atom that can unify with

$$connected\_to(w_5, W_1) \Leftarrow B,$$

and find *connected_to*(w_5, *outside*) $\Leftarrow$ *true*, binding W_1 to *outside*. You then try to prove *prove*(*true*) which succeeds using the first clause for *prove*.

The second half of the conjunction, *prove*(*live*(W_1)) with $W_1 = outside$, reduces to *prove*(*true*) which is again immediately solved.

Expanding the Base Language

Without changing the metalanguage you can change the base language by modifying the meta-interpreter. By adding clauses, you can increase what can be proved, and by adding extra conditions to the meta-interpreter rules, you can restrict what can be proved.

There are a number of refinements that can be made to the meta-interpreter. In all practical systems, not every predicate is defined by clauses. For example, it would be impractical to axiomatize arithmetic on current machines that can do arithmetic quickly. There are some predicates that you do not want to axiomatize, but you would rather just call the underlying system directly. Assume that the meta-level has the procedure *call*(*G*) that evaluates *G* directly in the meta-level. Note that writing *call*(*p*(*X*)) is the same as writing *p*(*X*). You need *call* because the language doesn't allow free variables as atoms.

You can evaluate built-in procedures at the base level by defining the meta-level relation *built_in*(*X*) that is true if all instances of *X* are to be evaluated directly. *X* is a meta-level variable that must denote a base-level atom. Don't assume that "*built_in*" is built-in. It can be axiomatized like any other relation.

You can also expand the base language to allow for disjunction (inclusive or) in the body of a clause.

The **disjunction**, $A \vee B$, is true in an interpretation *I* when either *A* is true in *I* or *B* is true in *I* (or both are true in *I*).

% *prove*(*G*) is true if base-level body *G* is a logical consequence of the base-level
% knowledge base.

> *prove*(*true*).
>
> *prove*((*A* & *B*)) ←
> > *prove*(*A*) ∧
> > *prove*(*B*).
>
> *prove*((*A* ∨ *B*)) ←
> > *prove*(*A*).
>
> *prove*((*A* ∨ *B*)) ←
> > *prove*(*B*).
>
> *prove*(*H*) ←
> > *built_in*(*H*) ∧
> > *call*(*H*).
>
> *prove*(*H*) ←
> > (*H* ⇐ *B*) ∧
> > *prove*(*B*).

Figure 6.5: A meta-interpreter that uses built-in calls and disjunction

---

You can add disjunction to the base-level without needing disjunction in the metalanguage. Note that this only allows disjunction in the body of rules. (We will not add disjunction to the head of rules until Section 7.5 (page 256).)

Example 6.5

An example of the kind of base-level rule you can now interpret is:

> *can_see* ⇐ *eyes_open* & (*lit*(*l*$_1$) ∨ *lit*(*l*$_2$)),

which says that *can_see* is true if *eyes_open* is true and either *lit*(*l*$_1$) or *lit*(*l*$_2$) is true (or both).

Figure 6.5 shows a meta-interpreter that allows built-in procedures to be evaluated directly, as well as allowing disjunction in the bodies of rules. This assumes that there is a database of built-in assertions and that *call*(*G*) is a way to prove *G* in the meta-level.

Once you have such an interpreter, the meta-level and the base level are different languages. The base level allows built-in predicates and allows disjunction in the body. The meta-level may or may not have such facilities, but it doesn't need them to provide

% *bprove*(*G, D*) is true if *G* can be proved with a proof tree of depth less than or equal
% to number *D*.

> *bprove*(*true, D*).
> *bprove*((*A* & *B*), *D*) ←
> > *bprove*(*A, D*) ∧
> > *bprove*(*B, D*).
> *bprove*(*H, D*) ←
> > *D* ≥ 0 ∧
> > D_1 is *D* − 1 ∧
> > (*H* ⇐ *B*) ∧
> > *bprove*(*B*, D_1).

Figure 6.6: A meta-interpreter for depth-bounded search

these facilities for the base language. The meta-level needs a way to interpret *call*(*G*),
which the base level can't handle. You can, however, add the facility for the base level
to interpret the command *call*(*G*) by adding the meta-level clause:

> *prove*(*call*(*G*)) ←
> > *prove*(*G*).

Depth-Bounded Search

The previous section showed how adding extra meta-level clauses can be used to
expand the base language. This section shows how adding extra conditions to the
meta-level clauses can reduce what can be proved.

A useful meta-interpreter is one that implements depth-bounded search. This can
be used to look for short proofs, or as part of an iterative-deepening searcher (page
140) which carries out repeated depth-bounded, depth-first searches, increasing the
bound at each stage.

Figure 6.6 gives an axiomatization of the relation:

> *bprove*(*G, D*)

that is true if *G* can be proved with a proof tree of depth less than or equal to the non-
negative integer *D*. In this figure, "is" is an infix predicate symbol, where "*V* is *E*"
is true if *V* is the numerical value of expression *E*. Within the expression "−" is the

infix subtraction function symbol. Thus "D_1 is $D-1$" is true if D_1 is one less than the number D.

One aspect of this meta-interpreter is that if D is bound to a number in the query, this meta-interpreter will never go into an infinite loop. It will miss proofs whose depth is greater than D. Thus this interpreter is incomplete when D is set to a fixed number. However, every proof that can be found for the *prove* meta-interpreter can be found for this meta-interpreter if the value D is set large enough. The idea behind an iterative-deepening (page 140) searcher is to exploit this fact by carrying out repeated depth-bounded searches, each time increasing the depth-bound. Sometimes the depth-bounded meta-interpreter can find proofs that *prove* can't. This occurs when the *prove* meta-interpreter goes into an infinite loop before exploring all proofs.

This isn't the only way to build a bounded meta-interpreter. Alternative measure of the size of proof trees could also be given. For example, you could use the number nodes in the tree as opposed to the maximum depth of the proof tree. You could also make conjunction incur a cost by changing the second rule. (See Exercise 6.1.)

See page 502 for code that uses *bprove* to implement an iterative deepening meta-interpreter.

Delaying Goals

One of the most useful abilities of a meta-level is to delay goals. Some goals, rather than being proved, can be collected in a list. At the end of the proof you have derived the implication that if the delayed goals were all true then the computer answer would be true.

There are a number of reasons for providing a facility for collecting goals that should be delayed, including:

- To delay subgoals with variables, in the hope that subsequent calls will ground the variables (page 255).

- To delay assumptions, so that you can collect assumptions that are needed to prove a goal (this is the primary mechanism of the algorithms in Chapter 9).

- To create new rules that leave out intermediate steps. For example, if the delayed goals are to be asked of a user or queried from a database. This is called *partial evaluation* and is used for explanation-based learning (page 433).

Figure 6.7 gives a meta-interpreter that provides delaying. A base-level atom G can be declared to be delayable using the meta-level fact $delay(G)$. The delayable atoms can be collected into a list without being proved.

Suppose you can prove $dprove(G, [\,], D)$. Let D' be the conjunction of base-level atoms in D. Then you know that the implication $G \leftarrow D'$ is a logical consequence of the clauses, and $delay(d)$ is true for all $d \in D$.

% $dprove(G, D_0, D_1)$ is true if D_0 is an ending of D_1 and G logically follows from the
% conjunction of the delayable atoms in D_1.

$dprove(true, D, D)$.

$dprove((A \& B), D_1, D_3) \leftarrow$
$\qquad dprove(A, D_1, D_2) \wedge$
$\qquad dprove(B, D_2, D_3)$.

$dprove(G, D, [G|D]) \leftarrow$
$\qquad delay(G)$.

$dprove(H, D_1, D_2) \leftarrow$
$\qquad (H \Leftarrow B) \wedge$
$\qquad dprove(B, D_1, D_2)$.

Figure 6.7: A meta-interpreter that collects delayed goals

Example 6.6

As an example of the second reason for using delaying, consider the base-level knowledge base of Figure 6.4 (page 206), but without the rules for ok. Suppose, instead, that $ok(G)$ is delayable. This is represented as the meta-level fact

$delay(ok(G))$.

The query

$?dprove(live(p_1), [\,], D)$.

has one answer, namely, $D = [ok(cb_1)]$. If $ok(cb_1)$ were true, then $live(p_1)$ would be true.

This idea of hypothesizing causes that can be used to prove a goal is a simple form of what is called **abduction** (page 332).

Example 6.7

The query

$?dprove((lit(l_2) \& live(p_1)), [\,], D)$.

has the answer $D = [ok(cb_1), ok(cb_1), ok(s_3)]$. If cb_1 and s_3 are ok, then l_2 will be lit and p_1 will be live.

Note that $ok(cb_1)$ is an element of this list twice. $dprove$ doesn't check for this condition. It's possible to write a version of $dprove$ that doesn't add delayables that are already there. (See Exercise 6.2.)

6.4   Querying the User

Users are one class of people who use a knowledge-based system (page 200). Users are not expected to know anything about the internals of the system and are not experts in the domain of the knowledge base. They do, however, know about a particular case, and so you need to be able to extract that knowledge from the user. Users typically don't know what is relevant or how their knowledge should be used, and so they can't be expected to axiomatize their knowledge. One aspect of the problem of knowledge acquisition is how to extract knowledge from a user most effectively.

Typically there is no information in the knowledge base about particular cases. The user must be asked to provide such information. For example, a medical-diagnosis program may know about particular diseases, but not what particular symptoms that a particular patient manifests. You would not expect that the user would want to, or even be able to, volunteer all of the information about a particular case, as often the user doesn't know what information is relevant or know the syntax of the representation language. They would prefer to answer explicit questions put to them in a more natural language. The idea of *querying the user* is that the system can treat the user as a source of information and ask the user specific questions about a particular case under consideration. The system, embodying the experts' knowledge, should know what information is relevant or important to a case. It should ask the user for the information it needs. When it queries the user it should only ask for relevant information that the user may have.

If you think of the problem solver in a goal-directed top-down manner, one way to solve a goal is to ask the user for the relevant information. There are three types of goals to be solved:

- Goals for which the user isn't expected to know the answer, so the system never asks.

- Goals for which the user should know the answer, and for which they have not already provided an answer. In this case they should be asked for the answer and the answer should be recorded (so they don't get asked again).

- Goals for which the user has already provided an answer. In this case that answer should be used, and the user should not be asked again about this query.

It is important to note that there is a symmetry between roles of the user and the system. They can both ask questions and give answers. Questions take the form of queries with free variables of goals to be solved. Answers (page 40) are instances of the question that are true. The system and the user may give multiple answers for a query, which correspond to different instances that can be derived. The user at the top level asks the system a question, and at each step the system asks a question which is answered

% *aprove*(*G*) is true if base-level body *G* is a logical consequence of the base-level
% knowledge base, and the answers provided by asking the user yes/no questions.

$$aprove(true).$$

$$aprove((A \; \& \; B)) \leftarrow$$
$$\quad aprove(A) \wedge$$
$$\quad aprove(B).$$

$$aprove(H) \leftarrow$$
$$\quad askable(H) \wedge$$
$$\quad answered(H, yes).$$

$$aprove(H) \leftarrow$$
$$\quad askable(H) \wedge$$
$$\quad unanswered(H) \wedge$$
$$\quad ask(H, Ans) \wedge$$
$$\quad record(answered(H, Ans)) \wedge$$
$$\quad Ans = yes.$$

$$aprove(H) \leftarrow$$
$$\quad (H \Leftarrow B) \wedge$$
$$\quad aprove(B).$$

Figure 6.8: An ask-the-user interpreter in pseudo-code

either by finding the relevant clauses or by asking the user. The whole system can be characterized by a protocol of questions and answers.

Yes/No Questions

The simplest form of question is a ground query (one that contains no free variables). It requires just a yes/no answer from the user. The system should ask a question only once. Thus the user is only asked a question if

- the question is askable, and
- the user hasn't previously answered the question.

When the user has answered a question, the answer needs to be recorded (in order to satisfy the second point).

Figure 6.8 gives a pseudo-code interpreter that incorporates querying the user. This interpreter assumes that there is some extralogical external database that records answers to queries. The meta-level knowledge base is changing as queries are answered. In particular, meta-level facts of the form *answered*(*H*, *Ans*) are being added. *ask*(*H*, *Ans*) is true if *H* is asked of the user; and *Ans*, either *yes* or *no*, is given by the user as a reply. *unanswered*(*H*) means there is nothing in the database of the form *answered*(*H*, *Ans*) for any *Ans*. Note that the intended meaning of the fourth clause is that it succeeds only if the answer is *yes*, but the answer gets recorded whether the user answered *yes* or *no*. (See Section C.2 (page 504) for an implementation in Prolog using Prolog's *asserta*.)

Example 6.8	Consider the knowledge base of Figure 6.4 (page 206), but without the rules for *up* or *down*. To make the status of the switches askable, you can add the meta-level assertions

$$askable(up(S)).$$

$$askable(down(S)).$$

The following dialogue is for the query ?*aprove*(*lit*(*L*)), with user responses in bold:

Is *up*(s_2) true? **yes.**
Is *up*(s_1) true? **no.**
Is *down*(s_2) true? **no.**
Is *up*(s_3) true? **yes.**
Answer: $L = l_2$

Note that the intended interpretation for the symbols allows the user to understand isolated queries. This use of semantics is very important for making the system intelligible.

Functional Relations

There is one case where you may know the answer to a question, even though it hasn't been answered. This occurs when you know a relation is a function. Relation $r(X, Y)$ is functional if for every X there is a unique Y such that $r(X, Y)$ is true. If you have already found one Y for a particular X for which $r(X, Y)$ is true, you should not re-ask for more instances.

Example 6.9	In the example above, it's redundant to ask the question "Is *down*(s_2) true?" when the user has already told the system that *up*(s_2) is true. It may be better to have a relation *pos*(*Sw*, *Pos*) where the position is a function of the switch. Once you have determined a position of the switch, you don't need to ask any more.

It is usually very inefficient to ask yes/no questions to a functional relation. Rather than enumerating every possible position of the switch and asking whether the switch is in that position or not, it may be better to ask once for the position of the switch and not ask this question again. If the system wants to know the age of a person, it would be better to ask outright for the age of the person rather than enumerate all of the ages and ask true/false questions.

Asking these kinds of more general questions probably would not be appropriate for nonfunctional relations. For example, if the switch could be in many positions at once, then it would probably be better to ask for each position that is needed.

There is one more complication that arises, and that has to do with the vocabulary of answers. The user may not know the vocabulary that is expected by the knowledge engineer. There are two sensible solutions to this problem:

- The system designer provides a menu of items from which the user has to select the best fit.

- The system designer provides a large dictionary to anticipate all possible answers, and it maps these into the internal forms expected by the system. There would be an assumption that the vocabulary used by the user would be normal language. Thus the user may not expect the system to understand such terms as "giganormous" (meaning "very big," of course).

More General Questions

Yes/no questions and functional relations do not cover all of the cases of a query. The general form of a query occurs when there are free variables in a query that is to be asked of the user.

Example 6.10

For the subgoal $p(a, X, f(Z))$ the user should be asked something like

for which X, Z is $p(a, X, f(Z))$ true?

The user would then be expected to give bindings for X and Z that make the subgoal true, or reply *no*, meaning that there are no more instances. This follows the spirit of the query protocol you already have for asking the system questions.

A number of issues arise:

- Should users be expected to give all instances which are true, or should they give the instances one at a time, with the system prompting for new instances? One of the major criteria for acceptance of expert systems is in how sensible they appear. Asking questions in a natural, logical way is a big part of this. For this reason it may be better to get one instance and then prompt when another instance is needed. In this way, the system can, for example, probe into one individual in depth before considering the next individual. This protocol would then allow the in-depth analysis of one instance case before asking for new instances. Of course, how well this works depends on the structure of the knowledge base as well.

- When should you not ask/re-ask a question? For example, consider the question

  For which X is $p(X)$ true?

  to which the user replies $X = f(Z)$. In other words, the user replied that "for all values of Z, $p(f(Z))$ is true." In this case the user should not be asked about whether $p(f(b))$ is true or whether $p(f(h(W)))$ is true, but should be asked about whether $p(a)$ or even whether $p(X)$ is true (asking for a different instance, of course). If the user subsequently replies *no* to the question

  For which X is $p(X)$ true?

  they mean that there are no more answers to this question, rather than that $p(X)$ is false for all X (as you have already been told that $X = f(b)$ is a true instance).

  The general rule is: Don't ask a question that is more specific than a query to which either a positive answer has already been given or the user has replied *no*.

- Should you ask the question as soon as it's encountered, or should you delay the goal until more variables are bound? There may be some goals which will subsequently fail no matter what answer the user gives. For these goals it may be better to find this out rather than asking the user at all. There may also be free variables in a query that subsequently will become bound. Rather than asking the user to enumerate the possibilities until she stumbles on the instance that doesn't fail, it may be better to delay the goal, using, for example, the delaying meta-interpreter of Figure 6.7 (page 211).

  Consider the query $?p(X) \& q(X)$, where $p(X)$ is askable. If there is only one instance of q that is true, say $q(k)$, it may be better to delay asking the $p(X)$ question until X is bound to k. Then the system can ask $p(k)$ directly instead of asking the user to enumerate all of the p's until they stumble on $p(k)$ being true. However, if there is a large database of instances of $q(X)$, it may be better to ask for instances of $p(X)$ and then check them with the database rather than asking a yes/no question for each element of the database.

These pragmatic questions are very important if you want to have a user-friendly interface to a KB system.

6.5 Explanation

To make the system usable by people, it can't just give an answer and expect the user to believe it. Consider the case of a system advising doctors who are legally responsible for the treatment that they carry out based on the diagnosis. They need to be convinced that the diagnosis is appropriate. The system must be able to justify and argue that its answer is correct. The same feature can be used to explain how the system found a result and to allow debugging of the knowledge base.

Two complementary means of interrogation, namely "HOW" and "WHY," can be used to justify answers and why questions are asked. HOW is used when the system has returned an answer. It allows the user to ask how the answer was derived. The system provides the instances of the clause used to deduce the answer. For any atom in the body of the clause, the user can ask HOW the system derived that atom, and thus traverse the proof tree. The user can ask WHY in response to being asked a question. The system justifies the question as reasonable by giving the rule which produced the question. This allows you to move up the proof tree. Together these rules allow you to traverse the proof tree in either direction.

How Did the System Prove a Goal?

The first explanation procedure allows the user to ask HOW a goal was derived. If g was derived as an answer, there must have been a rule with g as the head where each element of the body has been derived. This rule will be an instance of a rule in the knowledge base. Suppose it is

$$g \Leftarrow a_1 \& \ldots \& a_k.$$

If the user asks HOW in response to g being derived, the system can display this rule which shows one step of the proof of g. In response to this, the user can ask

    HOW i

which will give the rule that was used to prove a_i. The user can continue using the HOW command to move down the proof tree.

A **proof tree** for derived atom g is a tree with g at the root, such that the children of g are labeled with atoms $a_1, \ldots, a_k$ where $g \Leftarrow a_1 \& \ldots \& a_k$ is the instance of the rules used to prove g, and the subtree rooted at a_i is a proof tree for atom a_i.

In order to implement the HOW command, you need to maintain a representation of the proof tree for a derived answer. Our meta-interpreter can be extended to build a proof tree. Figure 6.9 gives a meta-interpreter that incorporates built-in predicates and builds a representation of a proof tree. This proof tree can be traversed in order to implement HOW questions.

% *hprove*(*G*, *T*) is true if base-level body *G* is a logical consequence of the base-level
% knowledge base, and *T* is a representation of the proof tree for the corresponding
% proof.

> *hprove*(*true*, *true*).
>
> *hprove*((*A* & *B*), (*L* & *R*)) ←
> > *hprove*(*A*, *L*) ∧
> > *hprove*(*B*, *R*).
>
> *hprove*(*H*, *if* (*H*, *built_in*)) ←
> > *built_in*(*H*) ∧
> > *call*(*H*).
>
> *hprove*(*H*, *if* (*H*, *T*)) ←
> > (*H* ⇐ *B*) ∧
> > *hprove*(*B*, *T*).

Figure 6.9: A meta-interpreter that builds a proof tree

Example 6.11

Consider the base-level clauses for the wiring domain (page 70) and the base-level query ?*lit*(*L*). There is one answer, namely $L = l_2$. The meta-level query ?*hprove*(*lit*(*L*), *T*) returns the answer $L = l_2$ and the tree

$T = if (lit(l_2),$
    $if (light(l_2), true)$ &
    $if (ok(l_2), true)$ &
    $if (live(l_2),$
        $if (connected\_to(l_2, w_4), true)$ &
        $if (live(w_4),$
            $if (connected\_to(w_4, w_3),$
                $if (up(s_3), true))$ &
            $if (live(w_3),$
                $if (connected\_to(w_3, w_5),$
                    $if (ok(cb_1), true))$ &
                $if (live(w_5),$
                    $if (connected\_to(w_5, outside), true)$ &
                    $if (live(outside), true))))))$

While this tree can be understood if properly formatted, it requires a skilled user to understand it. An implementation of a tree traversal algorithm that uses HOW to make the tree more comprehensible is included in Appendix C (page 498).

The answer to the query $?lit(L)$ is $l = l_2$. In response to the user question HOW, the system can reply

$lit(l_2) \Leftarrow$
$\qquad light(l_2)$ &
$\qquad ok(l_2)$ &
$\qquad live(l_2)$.

Which was the instance of the top-level rule used to prove $lit(l_2)$. To find out how $live(l_2)$ was proved the user can ask

HOW 3

the rule being

$live(l_2) \Leftarrow$
$\qquad connected\_to(l_2, w_4)$ &
$\qquad live(w_4)$.

To find how second conjunct was proved, the user can issue the command

HOW 2

The second conjunct was proved using the rule instance

$live(w_4) \Leftarrow$
$\qquad connected\_to(w_4, w_3)$ &
$\qquad live(w_3)$.

To find how first conjunct was proved, the user can issue the command

HOW 1

The first conjunct was proved using the rule

$connected\_to(w_4, w_3) \Leftarrow$
$\qquad up(s_3)$.

To find how first conjunct was proved, the user can issue the command

HOW 1

and the system will report that $up(s_3)$ is a fact.

Any part of the tree that the user is unsure about can be queried in this manner.

Why Did the System Ask a Question?

Another useful explanation is to find out why a question was asked. This is useful because:

- You want the system to appear intelligent. Knowing why a question was asked will increase the user's confidence that the system is working sensibly.

- One of the main measures of complexity of an interactive system is the number of questions asked of a user; you want to keep this to a minimum. Knowing why a question was asked will help the knowledge engineer optimize this complexity.

- An irrelevant question is usually a symptom of a deeper problem.

- The user may learn something from the system by knowing why the system is doing something. This learning is much like an apprentice asking a master why she is doing something.

You want to implement the user command

WHY

which is read as "Why did you ask me that question?" The answer should be the rule which needed the question answered. The question will be one of the elements of the body of the rule, and the head of the rule will be something that the system wants to find out. If you ask WHY again, then you should find out why the question at the head of the rule was asked, and so forth. Repeatedly asking WHY will eventually give us the path of subgoals to the top-level question. If all of these rules are reasonable, this justifies why the question is reasonable.

Figure 6.10 gives a meta-interpreter that can be used to find the list of ancestor rules for a WHY question. The second argument to *wprove* is a list of clauses of the form $(H \Leftarrow B)$ for each head of a rule in the current part of the proof tree. This meta-interpreter needs to be combined with the meta-interpreter of Figure 6.8 that actually asks questions. When the user is asked a question and responds with "WHY," the list can be used to give the set of rules that were used to generate the current subgoal. The WHY described above can be implemented by stepping through the list, and thus up the proof tree, one step at a time.

Example 6.12	Consider the dialog of Example 6.8 (page 214). The following shows how repeated use of WHY can move up the proof tree. The following dialogue is for the query $?aprove(lit(L))$, with user responses in bold:

Is $up(s_2)$ true? **why.**
I used the rule: $connected\_to(w_0, w_1) \Leftarrow up(s_2)$. **why.**
I used the rule: $live(w_0) \Leftarrow connected\_to(w_0, w_1) \& live(w_1)$. **why.**
I used the rule: $live(l_1) \Leftarrow connected\_to(l_1, w_0) \& live(w_0)$. **why.**
I used the rule: $lit(l_1) \Leftarrow light(l_1) \& ok(l_1) \& live(l_1)$. **why.**
Because that is what you asked me!

% *wprove*(*G*, *A*) is true if base-level body *G* is a logical consequence of the base-level
% knowledge base, and *A* is a list of ancestors for *G* in the proof tree for the original
% query.

> *wprove*(*true*, *Anc*).
>
> *wprove*((*A* & *B*), *Anc*) ←
> > *wprove*(*A*, *Anc*) ∧
> > *wprove*(*B*, *Anc*).
>
> *wprove*(*H*, *Anc*) ←
> > (*H* ⇐ *B*) ∧
> > *wprove*(*B*, [(*H* ⇐ *B*)|*Anc*]).

Figure 6.10: A meta-interpreter that collects ancestor rules for WHY questions

Typically, HOW and WHY are used together, HOW moving from the root to the leaves
of the tree, and WHY moving from the leaves to the root of the tree. Together they let
the user traverse the tree.

Example 6.13

As an example of the need to combine HOW and WHY, consider the case of a user
asking WHY subgoal $live(w_0)$ was asked, and the system giving the rule:

$$live(l_1) \Leftarrow connected\_to(l_1, w_0) \ \& \ live(w_0)$$

meaning that $live(w_0)$ is asked because the system wants to know $live(l_1)$ and is
using this rule to try to prove $live(l_1)$. The user may think it's reasonable that the
system wants to know $live(l_1)$, but may think it inappropriate that $live(w_0)$ be asked
because the user doesn't know why $connected\_to(l_1, w_0)$ should have succeeded.
In this case it's useful for the user to ask HOW $connected\_to(l_1, w_0)$ was derived.

6.6   Debugging Knowledge Bases

Just as in other software, there can be errors and omissions in knowledge bases, and
you need to be able to debug a knowledge base as well as add knowledge. In knowledge-
based systems, debugging is difficult because the domain experts and users who have
the domain knowledge required to detect a bug don't necessarily know anything about

the internal working of the system, nor do they want to. Standard debugging tools such as providing traces of the execution are useless because they require a knowledge of the mechanism by which the answer was produced. In this section we show how the idea of semantics (page 31) can be exploited to provide powerful debugging facilities for knowledge-based systems. Whoever is debugging the system need only know the meaning of the symbols and whether specific facts are true or not. This is important as this is the kind of knowledge that a domain expert and a user may have.

There are four types of nonsyntactic errors that can arise in rule-based systems:

- An incorrect answer is produced; that is, some atom that is false in the intended interpretation was derived.

- Some answer wasn't produced; that is, the proof failed when it should have succeeded, or some particular true atom wasn't derived.

- The program gets into an infinite loop.

- The system asks irrelevant questions.

These are the only four nonsyntactic errors that can occur. Ways to debug each of the other errors are examined in turn.

Incorrect Answers

Assume that whoever is debugging the knowledge base, such as a domain expert or a user, knows the intended interpretation of the symbols of the language, and can determine whether a particular fact is true or false in the intended interpretation. This should not be an invalid assumption if the person building the knowledge base is debugging the knowledge base, and they followed the methodology advocated in Chapter 2 (page 37). Moreover, someone would not be able to tell there is a bug unless they knew what at least some of the symbols are intended to mean and at least something about what is true in the world.

An **incorrect answer** is a derived answer which is false in the intended interpretation. An incorrect answer can only be produced if there is an incorrect rule that was used in the proof. Otherwise you have a contradiction to the soundness of the rule of inference. You need only to examine the proof tree; one of the instances of the rules used in the proof tree must be wrong.

If there is a subgoal in the proof tree which is false in the intended interpretation, then either:

- some child of the subgoal is false in the intended interpretation (in which case you can determine how *it* was proved), or

- all children of the subgoal are true in the intended interpretation. In this case the subgoal and its children are instances of a clause in the knowledge base, and you

debug(*g*) such that *g* is false in the intended interpretation and *g* is proved from the knowledge base:

> Find the clause used to prove *g*, say $g \Leftarrow a_1 \& \ldots \& a_k$.
> If there is some a_i that is false in the intended interpretation
> > then *debug*(a_i)
> > else return $g \Leftarrow a_1 \& \ldots \& a_k$ as the buggy rule instance.

Figure 6.11: An algorithm for debugging incorrect answers

have shown an instance of the clause that is false in the intended interpretation, therefore that clause must be incorrect.

This leads to an algorithm, presented in Figure 6.11, to debug a knowledge base when an atom that is false in the intended interpretation is derived. This procedure can easily be carried out by the use of the HOW command (page 217). Given a proof for *g* that is false in the intended interpretation, you can ask HOW that atom was proved. This will return an instance of a rule which was used in the proof. You use HOW to ask about an atom in the body which was false in the intended interpretation. This will return the rule that was used to prove that atom. Repeat this until you find a clause where the head of the clause is false and all of the elements of the body are true. This is the incorrect clause. All you have assumed is that you can determine whether an atom is true or false in the intended interpretation.

Example 6.14	Let's continue the example of the electrical domain (page 70) but assume there is a bug in the program. Suppose that the domain expert or user had inadvertently said that whether w_1 is connected to w_3 depends on the status of s_3 instead of s_1 (see Figure 3.1 on page 71). Thus the clauses about *connected* include the following incorrect rule:

$$connected\_to(w_1, w_3) \Leftarrow up(s_3).$$

All of the other axioms are the same as in Figure 6.4. Given this axiom set, the atom $lit(l_1)$ can be derived, which is false in the intended interpretation. Consider how a domain expert or user would go about finding this incorrect clause when they detected this incorrect answer.

First they can ask how $lit(l_1)$ was derived, which will give the rule

$$lit(l_1) \Leftarrow light(l_1) \& ok(l_1) \& live(l_1).$$

Then, given that $lit(l_1)$ is false in the intended interpretation, they know that either $light(l_1)$ is false in the intended interpretation, or $ok(l_1)$ is false in the intended interpretation, or $live(l_1)$ is false in the intended interpretation, or this rule is buggy.

They know that $live(l_1)$ is false in the intended interpretation so they ask

> HOW 3

which elicits the rule

$$live(l_1) \Leftarrow connected\_to(l_1, w_0) \,\&\, live(w_0).$$

$live(w_0)$ is false in the intended interpretation so they ask

> HOW 2

which elicits the rule

$$live(w_0) \Leftarrow connected\_to(w_0, w_1) \,\&\, live(w_1).$$

$live(w_1)$ is false in the intended interpretation so they ask

> HOW 2

which elicits the rule

$$live(w_1) \Leftarrow connected\_to(w_1, w_3) \,\&\, live(w_3).$$

$connected\_to(w_1, w_3)$ is false in the intended interpretation so they ask

> HOW 1

which elicits the rule

$$connected\_to(w_1, w_3) \Leftarrow up(s_3)$$

$up(s_3)$ is true in the intended interpretation so this is the buggy rule.

Note how the user or domain expert can find the buggy clause without needing to know the internal workings of the system or how the proof was found. They only need knowledge about the intended interpretation and the disciplined use of HOW. Again notice how semantics is playing a crucial role in debugging the knowledge base.

Missing Answers

The second type of error occurs when an expected answer isn't produced. This manifests itself by a failure when an answer is expected, or when an expected answer isn't among the answers produced.

The preceding algorithm doesn't work in this case. What is in the proof tree is irrelevant; the rules *not* in the proof tree are what you are concerned about. When a query fails, there is no proof tree.

An appropriate answer isn't produced only if there is some part of the search space where an appropriate rule is missing, or there is an incorrect answer rule. By knowing the intended interpretation of the symbols and by knowing what queries should succeed, you can debug a missing answer. Given an atom g that failed when it

debug_failure(*g*), where *g* is an atom which should have a proof, but which fails:

Suppose one of the bodies of a rule with *g* as head fails when it should have succeeded. Note that *all* of the bodies of the rules for *g* failed; that is why *g* failed. Either

- one of the atoms in the body gives an answer that is false, in which case HOW can be used to debug the program,

- there is an instance of atom *a* that failed that should have succeeded, in which case you call *debug_failure*(*a*), or

- you have found where the extra knowledge needs to be added, as there is no clause relevant to proving *g*.

Figure 6.12: An algorithm for debugging missing answers

should have succeeded, you can find either an incorrect rule or a missing clause using the algorithm described in Figure 6.12.

Example 6.15   Suppose that for the axiomatization of the electrical domain (page 70), the world of Figure 3.1 (page 71) actually has s_2 down. The axiomatization of Section 3.2 fails to prove $lit(l_1)$ when it should succeed. Let's find the bug.

$lit(l_1)$ failed, so you need to find all of the rules instances for $lit(l_1)$. There is one such rule:

$lit(l_1) \Leftarrow$
  $light(l_1)$ &
  $ok(l_1)$ &
  $live(l_1)$.

$light(l_1)$ can be derived and is true in the intended interpretation (there is no bug here). $ok(l_1)$ can be derived and is true in the intended interpretation. $live(l_1)$ fails, but is true in the intended interpretation. You need to debug the failure of $live(l_1)$. The applicable rule is

$live(l_1) \Leftarrow$
  $connected\_to(l_1, W0)$ &
  $live(W0)$.

$connected\_to(l_1, W0)$ correctly succeeds with substitution $\{W0/w_0\}$. $live(w_0)$ failed but should have succeeded. The applicable rule is

$live(w_0) \Leftarrow$

$$connected\_to(w_0, W_1) \wedge$$
$$live(W_1).$$

$connected\_to(w_0, W_1)$ succeeds with substitution $\{W_1/w_1\}$ and $live(w_1)$ failed and should have failed. There are two bugs here. The first is that $connected\_to(w_0, w_1)$ succeeded and should have failed. This can be debugged using HOW as in the preceding section. The problem clause is $up(s_2)$. When this bug is corrected by removing the incorrect clause, $lit(l_1)$ still fails. The second bug that the procedure finds is that $connected\_to(w_0, w_2)$ failed but should have succeeded. The applicable rule is

$$connected\_to(w_0, w_2) \Leftarrow down(s_2).$$

$down(s_2)$ failed but should have succeeded. There is no rule whose head unifies with $down(s_2)$. You have found the bug.

Infinite Loops

The third problem is that the program could get into an infinite loop. In general there is no way of automatically diagnosing such errors, since this is equivalent to the halting problem. If you could do this, you would have a decision procedure for undecidable logic. Therefore, any solution is necessarily heuristic.

The first thing to notice is that if some node in the search tree is identical to one of its ancestors, you have a loop. Identical means that all variables are bound to each other (not just unifiable, but identical). This test can be automated.

The second possibility is to assume that when the program was written, there was some well-ordered set which was supposed to be reducing at each iteration. For example, in the electrical domain the number of steps away from the outside is meant to be reducing by one each time through the loop. If you can find a rule which causes the ordering to increase, or a loop through which it's not reducing, then you have a candidate for why the program is looping. Disciplined and explicit use of a well-founded ordering can be used to detect infinite loops.

6.7   A Meta-Interpreter with Search

The vanilla meta-interpreter of Figure 6.3 (page 205) and the meta-interpreters based on it, each inherit the search strategy of the meta-level. The only control you had over the search strategy was to fail earlier than normal. If you want to be able to use some other search strategies such as breadth-first, A^*, or some stochastic strategy, you need another sort of meta-interpreter that provides control over the search alternatives.

To do this, you can axiomatize the abstract definition of SLD resolution (page 56). By implementing this procedure, you can build a meta-interpreter for the language that is at a different level than the meta-interpreters presented so far in this chapter.

The idea of this meta-interpreter is that you explicitly maintain a list of all of the generalized answer clauses (page 56). Instead of letting the underlying interpreter do the search for us, you can maintain all potential alternate proofs in the meta-interpreter. This is best viewed as treating finding proofs as a graph-searching problem where a node corresponds to a generalized answer clause, and the neighbors of a node correspond to each of the possible resolutions for a selected subgoal.

In terms of the graph, you can let a node be of the form $goal(Ans, S)$ where Ans is the form of an answer such as $yes(t_1, \ldots, t_k)$ (page 56), and S is a list of atoms, the subgoals, representing their conjunct.

You have found an answer when S is the empty conjunction

   $is\_goal(goal(Ans, [\,]))$

with Ans corresponding to an answer.

The neighbors of node $goal(Ans, S)$, where $S = Atom \wedge Rest$ ($Atom$ is the atom selected and $Rest$ is the conjunction of the remaining atoms), could be defined in pseudocode as:

   $neighbors(goal(Ans, Atom \,\&\, Rest),$

      $\{goal(Ans, Body \,\&\, Rest) : (Atom \Leftarrow Body)\}).$

This is pseudocode as it hasn't committed to a representation of the conjuncts, and it requires that you find all clauses whose head unifies with $Atom$.

The goal node has no neighbors:

   $neighbors(goal(Ans, [\,]), \{\}).$

An implementation of the meta-interpreter in pseudocode is given in Figure 6.13. This procedure incorporates SLD resolution (page 49) into the generic search algorithm (page 119). The predicate $add\_to\_frontier$ and $select\_query$ corresponds to the corresponding frontier operations in the generic search algorithm. The predicate $select\_atom$ corresponds to the selection of which atom to resolve against in SLD resolution.

To use this meta-interpreter to find all solutions to the base-level query

   $?a_1 \,\&\, a_2 \,\&\, \ldots \,\&\, a_n$

with free variables $V_1, \ldots, V_m$, you issue the query

   $?prove\_all([goal(yes(V_1, \ldots, V_m), [a_1, a_2, \ldots, a_n]), Ans)$

% *prove_all*(*Qs*, *As*) that is true if *Qs* is a list of generalized answer clauses, and *As* is
% the list of all answers.

> *prove_all*([ ], [ ]).
> *prove_all*(*Qs*, [*A*|*Ans*]) ←
>> *select_query*(*goal*(*A*, [ ]), *Qs*, *RQs*) ∧
>> *prove_all*(*RQs*, *Ans*).
> *prove_all*(*Qs*, *Ans*) ←
>> *select_query*(*goal*(*Ans*, *Conj*), *Qs*, *RQs*) ∧
>> *select_atom*(*Atom*, *Conj*, *Rest*) ∧
>> *NS* = {*goal*(*Ans*, *Body* & *Rest*) : (*Atom* ⇐ *Body*)} ∧
>> *add_to_frontier*(*NS*, *RQs*, *ND*) ∧
>> *prove_all*(*ND*, *Ans*).

Figure 6.13: Meta-interpreter with search

Ans becomes the list of all instances of *yes*(V_1, ..., V_n) that are a logical consequence of the clauses.

This meta-interpreter never backtracks, and if it halts, it finds all of the solutions. You can get different behavior by varying *select_query*, *add_to_frontier*, or *select_atom*. With *select_query* and *add_to_frontier* acting as a stack you get depth-first search. A breadth-first meta-interpreter is obtained by *select_query* and *add_to_frontier* acting as a queue. You can use leftmost selection criteria with *select_atom* always selecting the first element of the list, and you can delay goals or cycle through all of the goals by changing the definition of *select_atom*.

Example 6.16

Suppose you have the base-level program, where *app*(*A*, *B*, *C*) means that appending list *A* to *B* results in *C*:

> *app*([ ], *L*, *L*) ⇐ *true*.
> *app*([*A*|*X*], *Y*, [*A*|*Z*]) ⇐ *app*(*X*, *Y*, *Z*).

sublist(*S*, *C*) is true if list *S* is in the middle of list *C*

> *sublist*(*S*, *L*) ⇐ *app*(_, *B*, *L*) & *app*(*S*, _, *B*).

Suppose you are interested in the base-level query

> ?*sublist*([*X*, *Y*], [*a*, *b*, *c*, *d*, *e*]).

where you are interested in the values of X and Y. The corresponding meta-level query is

$$prove\_all([goal(yes(X, Y), [sublist([X, Y], [a, b, c, d, e])])], Ans).$$

The resulting solution is

$$Ans = [yes(a, b), yes(b, c), yes(c, d), yes(d, e)].$$

Let's consider the trace of this in more detail. The calls to *prove_all* have first argument given below:

$$[goal(yes(V_0, V_1), [sublist([V_0, V_1], [a, b, c, d, e])])].$$

You substitute the body of definition of *sublist* for the call to *sublist*, rename the variables, and produce the conjunctive goal:

$$[goal(yes(V_4, V_5), [app(V_2, V_3, [a, b, c, d, e]), app([V_4, V_5], V_6, V_3)])].$$

Suppose you select the first atom in the conjunction to resolve against. There are two clauses whose head unifies with the selected atom. For the first clause, you unify the head of the clause with the selected atom in a copy of the goal and then replace that atom with the empty body of that clause. For the second applicable clause, you unify the head of the clause with the selected atom in another copy of the goal and then replace the atom with the body of the clause. Let's call this new goal α. The resulting first argument to *prove_all* is

$$[goal(yes(V_6, V_7), [app([V_6, V_7], V_8, [a, b, c, d, e])]), \alpha]$$

where

$$\alpha = goal(yes(V_{11}, V_{12}), [app(V_9, V_{10}, [b, c, d, e]),$$
$$app([V_{11}, V_{12}], V_{13}, V_{10})]).$$

Only the second clause for *app* is applicable to this goal, and you substitute the body for it into the goal:

$$[goal(yes(a, V_{13}), [app([V_{13}], V_{14}, [b, c, d, e])]), \alpha].$$

Again the only applicable clause is the second one:

$$[goal(yes(a, b), [app([\,], V_{15}, [c, d, e])]), \alpha]$$

Now the only applicable clause is the first, which makes $V_{15} = [c, d, e]$ and results in an empty conjunct:

$$[goal(yes(a, b), [\,]), \alpha].$$

You now have an empty conjunct, and so one answer is found. You then backtrack to the last choice point to find other answers:

$$[\alpha].$$

Continue as before until you have found the two other answers and eventually end up with the empty disjunct signifying there are no more answers:

> [].

An implementation of this meta-interpreter in Prolog is given on page 506.

6.8   Unification

The meta-interpreters given assumed that the unification was handled by the underlying system. This works fine if all of the base-level variables are universally quantified, as are the variables in the definite clause RRS (page 41). Not all variables you will be interested in will be universally quantified; you may not be able to use just any instance of a variable in a proof. If you want to handle other quantification of variables or you want to have different unification than is provided by the meta-level, you can't rely on all the unification happening at the meta-level. You need to use a different representation for the object level.

The **ground representation** for the object level represents all base-level terms, including variables, as meta-level ground terms.

Unlike the non-ground representation (page 202), where meta-level variables represent variables at the base-level, the ground representation uses meta-level constants to represent base-level variables.

You need to axiomatize a meta-level predicate $variable(T)$ that is true if term T is a base-level variable, and a meta-level predicate $non\_variable(T)$ that is true if term T isn't a base-level variable.

Example 6.17

To represent the base-level using the ground representation, you first declare which meta-level constants denote base-level variables, such as

> $variable(w)$.
>
> $variable(x)$.
>
> $variable(y)$.
>
> $variable(z)$.
>
> $non\_variable(a)$.
>
> $non\_variable(b)$.
>
> $non\_variable(c)$.

You can represent the base-level term $f(X, a, g(Z))$ in the meta-level as the ground term $f(x, a, g(z))$. What is a base-level variable is defined not by the case of its starting letter but by explicit declaration.

To use the ground representation you need to build an even more explicit meta-interpreter where you also axiomatize the unification. In this section we give an implementation of unification that can be used in such an interpreter.

Recall (page 52) that a unifier of two atoms or terms is a substitution that implies the equality of the expressions. This section defines a **unification algorithm** that finds a most general unifier (mgu) of two expressions.

Figure 6.14 gives pseudocode for the unification assuming the ground representation. Substitutions are represented as lists of elements of the form V/T, where V is a variable and T is a term. Note that this is only pseudologic code because the last clause in *unif*1 can be seen as a schema; there is an instance of this clause for each function and predicate symbol. You need to have either (a) one instance of that rule for each predicate and function symbol, or (b) a procedure that can take apart the structure of terms.

Note that if the chosen equality is of the form $X = t$ where t is a term in which X occurs and isn't X, the unification fails. You need this because you can't guarantee that, in the intended interpretation, there is a value which is a fixed point of the functions involved. This step in the unification algorithm is known as the **occurs check**. The occurs check is necessary for soundness as the following example shows.

Example 6.18	Suppose you want to unify X and $s(X)$. You can't ensure that the intended interpretation isn't the domain of natural numbers where the function symbol s denotes the successor function, where $s(n)$ is the number after n. In this interpretation, there is no number X which is equal to its successor. Without the occurs check, X and $s(X)$ unify. This makes the proof procedure unsound.

Consider the theory with only one axiom $lt(X, s(X))$ with the intended interpretation of the domain of integers with lt meaning "less than" and $s(X)$ the integer after X. The query $?lt(X, X)$ should fail because it's false in the intended interpretation since there is no number less than itself. However, without the occurs check, this query would succeed, and so the proof procedure would be unsound as something could be derived that is false in a model of the axioms.

The following examples demonstrate how the unification algorithm works:

Example 6.19	Consider unifying $p(x, a, q(x))$ with $p(z, z, w)$, where x, z, and w are base-level variables (i.e., *variable*(x), *variable*(z) and *variable*(w) are meta-level facts). The following is a sequence of calls to *unif*1:

$$unif1(p(x, a, q(x)), p(z, z, w), U).$$
$$unif1([p(x, a, q(x)) = p(z, z, w)], [\,], U).$$
$$unif1([x = z, a = z, q(x) = w], [\,], U).$$

Eventually you derive the answer: $U = [x/a, z/a, w/q(a)]$.

% *unify*$(T1, T2, U)$ is true if U is the most general unifier of terms $T1$ and $T2$:

> *unify*$(T1, T2, U) \leftarrow$
>> *unif* $1([T1 = T2], [\,], U)$.

% *unif* $1(TU, U_1, U_2)$ is true if TU is a list of equalities and U_1 is a substitution, and
% U_2 is a substitution made of U_1, together with bindings to unify all of the equalities
% in TU.

> *unif* $1([\,], U, U)$.
> *unif* $1([T = T|R], U_1, U_2) \leftarrow$
>> *unif* $1(R, U_1, U_2)$.
> *unif* $1([V = T|R], U_1, U_3) \leftarrow$
>> *variable*$(V) \wedge$
>> *not_appears_in*$(V, T) \wedge$
>> *replace_all*$(V, T, R, R1) \wedge$
>> *replace_all*$(V, T, U_1, U_2) \wedge$
>> *unif* $1(R1, [V/T|U_2], U_3)$.
> *unif* $1([T = V|R], U_1, U_2) \leftarrow$
>> *variable*$(V) \wedge$
>> *non_variable*$(T) \wedge$
>> *unif* $1([V = T|R], U_1, U_2)$.
> *unif* $1([f(S_1, \ldots, S_n) = f(T_1, \ldots, T_n)|R], U_1, U_2) \leftarrow$
>> *unif* $1([S_1 = T_1, \ldots, S_n = T_n|R], U_1, U_2)$.

% *not_appears_in*(V, T) is true if variable V doesn't appear in term T.

% *replace_all*(V, T, F_1, F_2) is true if expression F_2 is an instance of expression F_1,
% with all occurrences of expression V replaced by term T.

Figure 6.14: Pseudocode for unification using the ground representation

Example 6.20	If you try to unify $p(x, a, q(x))$ with $p(w, z, z)$, you derive the equality $q(w) = a$, and so the unification fails. Note that this fails not because these are necessarily different in the intended interpretation, but rather that there are interpretations in which they denote different individuals so you can't use their equality to build a proof.

Example 6.21	If you try to unify $p(x, a, q(x))$ with $p(z, w, z)$, you derive $x = q(x)$, and so by the occurs check the unification fails, as you can't guarantee that there is a fixed point of q (i.e., you can't guarantee that in the intended interpretation there is an individual x such that $x = q(x)$).

6.9   References and Further Reading

The ask-the-user facility and HOW and WHY questions are from the medical diagnosis program Mycin (Buchanann & Shortliffe, 1984). Mycin used a very different logic than the one presented here. The ask-the-user facility for logic programs was first suggested by Sergot (1983).

The algorithmic debugging is due to Shapiro (1983). Shapiro's algorithm is more sophisticated than the one here in that it tries to ask the minimum number of questions. He also considers learning a logic program as starting with the empty program and debugging it.

The collection by Abramson & Rogers (1989) contains many papers on the use of meta-interpreters. Bowen (1985) discusses knowledge representation and meta-interpreters. Forbus & de Kleer (1993) show how many different problem solvers can be implemented.

6.10   Exercises

Exercise 6.1 Modify the depth-bound meta-interpreter of Figure 6.6 (page 209) so that

(a) The bound is on the total length of the proof, where the length is the total number of instances of base-level atoms that appear in the proof.

(b) Different base-level atoms can incur different costs on the bound. For example, most atoms could have zero cost, and some atoms could incur a positive code.

Discuss why either of these may be better or worse than using the depth of the tree.

What conditions on the atom costs would guarantee that when a positive bound is given, the proof procedure doesn't go into an infinite loop?

Exercise 6.2 The program of Figure 6.7 (page 211) allows duplicate delayed goals. Write a version of *dprove* that returns minimal sets of delayed goals, in their simplest forms.

Exercise 6.3 Explain the difference between using a symbolic debugger for a programming language like C and using the explanation facility of an expert system. Discuss what needs to be known in order to use either debugging tool.

Exercise 6.4 Write a meta-interpreter that allows for asking the users yes/no questions. Make sure it doesn't ask questions that it already knows the answer to.

Exercise 6.5 Extend the ask-the-user meta-interpreter to have questions that ask for instances. You should be able to interpret declarations that say that a predicate is functional and respond accordingly.

Exercise 6.6 Write a meta-interpreter that allows both HOW and WHY questions. In particular, it should allow the user to ask HOW questions about a goal that has been proved after a WHY question. Explain how such a program may be useful.

Exercise 6.7 Write a meta-interpreter for definite clauses that does iterative deepening search. Make sure that you only get one answer for each proof and that the system says *no* whenever the depth-first searcher says *no*. This should be an expansion of the depth-bounded prover rather than the meta-interpreter with search. *Hint:* You need to distinguish between failing naturally and failing unnaturally (hitting the depth bound). You only want to increase the depth bound when the system fails unnaturally. Also keep track of the depth of a proof found so that you can determine that it would have failed unnaturally on the previous iteration.

Exercise 6.8 Using the ideas of meta-interpreter with search (page 226), write a breadth-first interpreter for definite clauses. Why may this be better than an iterative-deepening search? Why may it be worse? Demonstrate each with an example using your implementation.

Chapter 7

Beyond Definite Knowledge

7.1 Introduction

This chapter shows how to extend the definite clause representation and reasoning system in several ways. Each extension is motivated by some desirable feature that is beyond the logic presented so far. We show how each extension can be implemented by extending our meta-interpreter.

First, we add equality to the language. We then add negation in a simple form by allowing rules that imply *false*, which is an atom that is, by definition, false in all interpretations. A second form of negation incorporates a complete knowledge assumption, where we assume that the clauses defining a predicate are complete so that an atom is false if all of the bodies of the clauses defining that atom are false. In Section 7.5 we depart from complete knowledge and allow disjunctive assertions. The language is then expanded to include explicit quantification of variables. We then define the *first-order predicate calculus* and show how it can be transformed to the previous forms. Finally, we present a way to use modal knowledge.

7.2 Equality

Sometimes it's useful to use more than one term to name a single individual. For example, the terms 4×4, 2^4, $273 - 257$, and 16 all may denote the same number. Sometimes you want to have each name refer to a different individual. For example, you may want a separate name for different courses in a university. Sometimes

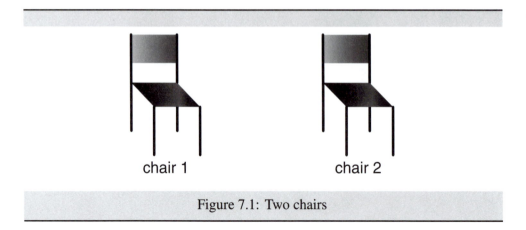

Figure 7.1: Two chairs

you don't know whether two names denote the same individual or not—for example, whether the 8 a.m. delivery person is the same as the 1 p.m. delivery person.

This section considers the role of equality, which allows us to represent whether or not two terms denote the same individual. Note that in the RRS presented in Chapter 2, all of the answers were valid whether or not the terms denoted the same individuals.

Equality is a special predicate symbol that has an accepted domain-independent intended interpretation:

Term t_1 **equals** term t_2, written $t_1 = t_2$, is true in interpretation I if t_1 and t_2 denote the same individual in I.

Equality doesn't mean similarity. If a and b are constants and $a = b$, it's not the case that there are two things that are similar or even identical. Rather it means there is one thing with two names.

Example
7.1

Consider the world of two chairs given in Figure 7.1. In this world it's not true that $chair1 = chair2$, even though the two chairs may be identical in all respects; without representing the exact position of the chairs, they can't be distinguished. It may be the case that $chairOnRight = chair2$. It is not the case that the chair on the right is *similar* to chair2. It *is* chair2.

Allowing Equality Assertions

If you don't allow equality assertions, the only thing that is equal to a term is itself. This can be captured as though you had the assertion $X = X$. This means that for any ground term, t, t denotes the same individual as t.

If you want to permit equality assertions (e.g., stating that $chairOnRight = chair2$), the RRS needs to be able to derive what follows from a knowledge base that includes clauses with equality in the head. There are two major ways of doing this. The first is to axiomatize equality like any other predicate. The second is to build special-purpose inference machinery for equality. Both of these ways are considered here.

Axiomatizing Equality

Equality can be axiomatized like any other relation. The axioms are as follows. The first three axioms state that equality is reflexive, symmetric, and transitive:

$$X = X.$$
$$X = Y \leftarrow Y = X.$$
$$X = Z \leftarrow X = Y \wedge Y = Z.$$

The other axioms depend on the set of function and relation symbols in your language, thus they form what is called an **axiom schema**. The general idea is that you can substitute a term by an equal term in functions and in relations. For each n-ary function symbol f there is a rule of the form

$$f(X_1, \ldots, X_n) = f(Y_1, \ldots, Y_n) \leftarrow X_1 = Y_1 \wedge \cdots \wedge X_n = Y_n.$$

For each n-ary predicate symbol p, there is a rule of the form

$$p(X_1, \ldots, X_n) \leftarrow p(Y_1, \ldots, Y_n) \wedge X_1 = Y_1 \wedge \cdots \wedge X_n = Y_n.$$

Example 7.2

The binary function $cons(X, Y)$ needs the axiom

$$cons(X_1, X_2) = cons(Y_1, Y_2) \leftarrow X_1 = Y_1 \wedge X_2 = Y_2.$$

The ternary relationship $prop(O, A, V)$ needs the axiom

$$prop(O_1, A_1, V_1) \leftarrow prop(O_2, A_2, V_2) \wedge O_1 = O_2 \wedge A_1 = A_2 \wedge V_1 = V_2.$$

Having these axioms explicit as part of the knowledge base turns out to be very inefficient. The use of these rules isn't guaranteed to halt using a top-down depth-first interpreter. For example, the symmetric axiom will cause an infinite loop unless identical subgoals are noticed.

Special-Purpose Equality Reasoning

One way to augment the proof procedure to implement equality is called **paramodulation**.

The general idea is that if you have $t_1 = t_2$, then you can replace any occurrence of t_1 by t_2. You thus treat equality as a *rewrite rule*, substituting equals for equals. This works best if you can select a canonical representation for each individual and rewrite all other representations into that representation. One classic example of this is in the representation of numbers. There are many terms that represent the same number, e.g.,

4×4, $13 + 3$, $273 - 257$, 2^4, 4^2, 16, but typically you treat the sequence of digits as the canonical representation of the number. The reason that universities invented student numbers is to provide a canonical representation for each student; different students with the same name can be distinguished and different names for the same individual (e.g., Randolph G. Goebel or Randy Goebel) can be mapped to the same individual.

Unique Names Assumption

Instead of being agnostic about the equality of each term and expecting the user to axiomatize exactly which names denote the same individual and which denote different individuals, it's often easier to have the convention that different ground terms denote different individuals.

Example 7.3   In the student database example of Section 3.6 (page 86) a student had to have two courses as science electives. If a student has passed *math*302 and *psyc*303, then you only know whether they have passed two courses if you know *math*302 $\neq$ *psyc*303. That is, the constants *math*302 and *psyc*303 denote different courses. Thus you need to know which course numbers denote different courses. Rather than writing $n \times (n-1)$ inequality axioms for n objects, it may be better to have the convention that every course number denotes a different course and thus avoid the use of any inequality axioms.

This approach to handling equality is known as the unique names assumption.

The **unique names assumption (UNA)** is the assumption that distinct ground terms denote different individuals. That is, for every pair of distinct ground terms t_1 and t_2, assume $t_1 \neq t_2$, where "$\neq$" means "not equal to."

Note that this doesn't follow from the semantics for the definite clause language (page 31). As far as that RRS was concerned, distinct ground terms t_1 and t_2 could denote the same individual or could denote different individuals.

In the logic presented so far, the unique names assumption only matters if you have explicit inequalities in the bodies of clauses or equality in the head of clauses. With the unique names assumption, you don't use equality in the head of rules other than those defining equality above. These clauses implying equality will either be tautologies or inconsistent with the unique name axioms.

The unique names assumption can be axiomatized with the following axiom schema for inequality, which consists of the axiom schema for equality (page 237) together with the axiom schema:

- $c \neq c'$ for any distinct constants c and c'.

- $f(X_1, \ldots, X_n) \neq g(Y_1, \ldots, Y_m)$ for any distinct function symbols f and g.

- $f(X_1, \ldots, X_n) \neq f(Y_1, \ldots, Y_n) \leftarrow X_i \neq Y_i$, for any function symbol f. There are n instances of this schema for every n-ary function symbol f (one for each i such that $1 \leq i \leq n$).

- $f(X_1, \ldots, X_n) \neq c$ for any function symbol f and constant c.

- $t \neq X$ for any term t in which X appears (where t is not the term X).

With this axiomatization, two ground terms are not equal if and only if they don't unify, as ground terms are identical if and only if they unify. This isn't the case for non-ground terms. For example, $a \neq X$ has some instances which are true, for example, when X has value b and an instance which is false, namely when X has value a.

The unique names assumption is very useful for database applications, where you don't, for example, want to have to state that $alan \neq david$ and $alan \neq randy$ and $randy \neq david$. The definite clause logic presented in Chapter 2 doesn't care whether or not the constants $alan$ and $david$ denote the same individual. The unique names assumption allows us to use the convention that each name denotes a different individual.

Sometimes the unique names assumption is inappropriate, as $2 + 2 \neq 4$ is wrong, nor may it be the case that $clark\_kent \neq superman$.

Top-Down Procedure for the Unique Names Assumption

The top-down procedure incorporating the unique names assumption can't treat inequality as just another predicate, mainly because, if there is a function symbol, there are an infinite number of axioms defining inequality. Searching through the infinite set of axioms takes too long.

If you have a subgoal $t_1 \neq t_2$, for terms t_1 and t_2 there are three cases:

- t_1 and t_2 don't unify. In this case, $t_1 \neq t_2$ succeeds.

    For example, the inequality $f(X, a, g(X)) \neq f(t(X), X, b)$ succeeds as the two terms don't unify.

- t_1 and t_2 are identical including having the same variables in the same positions. In this case, $t_1 \neq t_2$ fails.

    For example, $f(X, a, g(X)) \neq f(X, a, g(X))$ fails.

    Note that for any pair of ground terms either case 1 or case 2 occurs.

- Otherwise, there are instances of $t_1 \neq t_2$ that succeed and instances of $t_1 \neq t_2$ that fail.

    For example, consider the subgoal $f(W, a, g(Z)) \neq f(t(X), X, Y)$. The most general unifier of $f(W, a, g(Z))$ and $f(t(X), X, Y)$ is $\{X/a, W/t(a), Y/g(Z)\}$. Some instances of the inequality, such as the ground instances consistent with the unifier, should fail. Any instance that isn't consistent with the unifier should succeed. Unlike other goals, you don't want to enumerate every instance that succeeds because that would mean unifying X with every function and constant different to a, as well as enumerating every pair of values for Y and Z where Y is different to $g(Z)$.

You need a slightly more complicated procedure for the unique names assumption with free variables in the queries. If you have an inequality subgoal of the third type, you can delay (page 210) the inequality goals until subsequent goals unify variables so that one of the first two cases occur. If the inequality goals can't be delayed any more (this is, if there are no more nondelayed subgoals) and neither of the first two cases is applicable, the goal should succeed as there always is an instance of the inequality that succeeds, namely the instance where every variable gets a different constant that doesn't appear anywhere else. When this occurs, you have to be careful to interpret the free variables in the answer; they don't mean that the answer is true for every instance of the free variables, but rather that it's true for some.

Example 7.4

Consider the rules adapted from Section 3.6 (page 86):

$$passed\_two\_courses(S) \leftarrow$$
$$\quad C_1 \neq C_2 \wedge$$
$$\quad passed(S, C_1) \wedge$$
$$\quad passed(S, C_2).$$
$$passed(S, C) \leftarrow$$
$$\quad grade(S, C, M) \wedge$$
$$\quad M \geq 50.$$
$$grade(mike, engl101, 87).$$
$$grade(mike, phys101, 89).$$

For the query

$$?passed\_two\_courses(mike),$$

you first try to prove $C_1 \neq C_2$, which can't be resolved and so needs to be delayed. You then call $passed(mike, C_1)$ which binds $engl101$ to C_1. You then call $passed(mike, C_2)$ which in turn calls $grade(mike, C_2, M)$, which can succeed with substitution $\{C_2/engl101, M/87\}$. At this stage the variables for the delayed inequality are bound enough to determine that the inequality should fail. Another clause can be chosen for $grade(mike, C_2, M)$, returning substitution $\{C_2/phys101, M/89\}$. The variables in the delayed inequality are bound enough to test the inequality, and this time the inequality succeeds. You can then go on to prove that $89 > 50$, and the goal succeeds.

One question that may arise from the above example is why not simply make the inequality the last call, because then it doesn't need to be delayed. There are two reasons for this. First it may be more efficient to delay. In this example, the delayed inequality can be tested before you have checked whether $87 > 50$. While this particular inequality test may be fast, in many cases substantial computation can be avoided by noticing violated inequalities as soon as possible. Second, if you were to return one of the values before it's bound, you need to remember the inequality constraint, so that any future unification that violates the constraint can fail.

7.3 Integrity Constraints

In the electrical wiring domain (page 71), it's useful to be able to say that some prediction, such as that light l_2 is on, isn't true. This will enable diagnostic reasoning to deduce that some switches, lights, or circuit breakers are broken. In adding knowledge to the university database (page 86), there are certain constraints that, if violated, indicate a problem with the database. For example, the constraint that a student can't have two different grades for the same course in the same term should be enforced.

You can expand the definite clause language to allow such reasoning by allowing rules with the special atom *false*, which is false in all interpretations, at the head of rules.

An **integrity constraint** is a clause of the form

$$false \leftarrow a_1 \wedge \ldots \wedge a_k$$

where the a_i are atoms and *false* is a special atom that is false in all interpretations.

A **Horn clause** is either a definite clause (page 29) or an integrity constraint. That is, Horn clauses have either *false* or a normal atom on the left-hand side of the arrow.

Integrity constraints allow us to prove that some conjunction of atoms is false in all models of the axioms (i.e., to prove disjunctions of negations of atoms).

We replace the Assumption DK (page 28) by the Assumption HK (named after the logician Horn):

Assumption   **(Horn Knowledge)** An agent's knowledge of the world can be described in
HK   terms of Horn clauses.

Example   Consider the clauses T_1:
7.5
$$false \leftarrow a \wedge b.$$
$$a \leftarrow c.$$
$$b \leftarrow c.$$

You can conclude that c is false in all models of T_1. If c were true in model I of T_1, then a and b would both be true in I (otherwise I would not be a model of T_1). As *false* is false in I and a and b are true in I, the first clause is false in I, a contradiction to I being a model of T. Thus c is false in all models of T_1.

If α is a formula, the **negation** of α, written $\neg\alpha$, read "not α," is a formula that is true in interpretation I if α is false in I, and is false in interpretation I if α is true in I.

Example 7.6

Using the negation symbol in Example 7.5, we can write

$$T_1 \models \neg c$$

which means that $\neg c$ is true in all models of T_1, or equivalently that c is false in all models of T_1.

Example 7.7

Consider the clauses T_2:

$$false \leftarrow a \wedge b.$$
$$a \leftarrow c.$$
$$b \leftarrow d.$$
$$b \leftarrow e.$$

Either c is false or d is false in every model of T_2. If they were both true in some model of T_2, you would get a contradiction. Similarly, either c is false or e is false in every model of T_2.

The **disjunction** of formulae α and β, written $\alpha \vee \beta$ and read "α or β," is true in interpretation I if α is true in I or if β is true in I (or both), and is false in I if both α and β are false in I.

Example 7.8

Given the above notation, the conclusion of Example 7.7 can be written as

$$T_2 \models \neg c \vee \neg d \text{ and}$$
$$T_2 \models \neg c \vee \neg e.$$

Although the language doesn't let you state disjunctions and negations, you can derive disjunctions of negations of atoms.

An **assumable** is an atom whose negation you are prepared to accept as part of a (disjunctive) answer. The assumables are those atoms you assume in a proof by contradiction and whose negation you eventually prove when you reach a contradiction.

If T is a set of clauses, a **conflict** of T is a set of assumables that, given T imply *false*. That is, $C = \{c_1, \ldots, c_r\}$ is a conflict of T if

$$T \models false \leftarrow c_1 \wedge \ldots \wedge c_r.$$

In this case, you know as an **answer**:

$$T \models \neg c_1 \vee \ldots \vee \neg c_r.$$

A **minimal conflict** is a conflict such that no strict subset is also a conflict.

Example
7.9

> Continuing Example 7.7, if $\{c, d, e, f, g, h\}$ is the set of assumables, then $\{c, d\}$ and $\{c, e\}$ are minimal conflicts of T_2. $\{c, d, e, h\}$ is also a conflict, but not a minimal conflict.

A set of clauses is **unsatisfiable** if it has no models. A set of clauses is logically **inconsistent** with respect to a proof procedure if *false* can be derived from the clauses using that proof procedure. If a proof procedure is sound and complete, a set of clauses is inconsistent if and only if it's unsatisfiable.

The use of *false* in the head of clauses allows the RRS to go beyond what could be done with just definite clauses. Now that the RRS language has integrity constraints, there is the possibility of a set of clauses being unsatisfiable.

Example
7.10

> The set of clauses $\{a, false \leftarrow a\}$ is unsatisfiable. There is no interpretation which satisfies both clauses. Both a and *false* $\leftarrow a$ can't be true in any interpretation.

It is always possible to find a model for a set of definite clauses. The interpretation with all atoms true is a model of any set of definite clauses. Thus, a set of definite clauses can never be proven inconsistent. However, with integrity constraints inconsistency can occur.

Applications of Integrity Constraints

While it may seem that the use of integrity constraints has limited usefulness, it turns out to be a very powerful tool. For many activities it's useful to know that some combination of truths are incompatible. We give two specific examples, one in diagnosis and one in databases. As other examples, it's useful in planning to know that some combination of actions you are trying to do is impossible. It's useful in design to know that some combination of components can't work if the laws of physics hold.

Consistency-Based Diagnosis

The extra power provided by extending the representation and reasoning system with integrity constraints can be put to good use in diagnostic programs. Consider having a system, together with a description of how it's supposed to work. If the system doesn't work according to its specification, then you can prove not only that it's not working, but you can actually identify which components could be faulty.

Example
7.11

> Consider the house wiring example depicted in Figure 3.1 (page 71) and axiomatized in Figure 6.4 (page 206). In this example we will show how that axiomatization can be adapted for consistency-based diagnosis.
>
> Assume that the user is able to observe whether a light is lit or dark and whether a power outlet is dead or live. A light can't be both lit and dark, and a power outlet

can't be both dead and live. This knowledge can be stated in the following integrity constraints:

$$false \leftarrow dark(L) \land lit(L).$$
$$false \leftarrow dead(L) \land live(L).$$

To use this knowledge base for diagnosis, remove all rules and facts with ok in the head and make $ok(X)$ assumable.

Suppose the three switches s_1, s_2, and s_3 are all up:

$$up(s_1).$$
$$up(s_2).$$
$$up(s_3).$$

Suppose the user has observed that l_1 and l_2 are both dark. This is represented as the facts:

$$dark(l_1).$$
$$dark(l_2).$$

Given this set of axioms, there are two minimal conflicts, namely

$$\{ok(cb_1), ok(s_1), ok(s_2), ok(l_1)\} \text{ and}$$
$$\{ok(cb_1), ok(s_3), ok(l_2)\}.$$

Thus you can derive:

$$\neg ok(cb_1) \lor \neg ok(s_1) \lor \neg ok(s_2) \lor \neg ok(l_1)$$

which means that at least one of the components cb_1, s_1, s_2, or l_1 must be not ok. Also you can derive

$$\neg ok(cb_1) \lor \neg ok(s_3) \lor \neg ok(l_2).$$

Because these are both logical consequences of the clauses, their conjunction

$$(\neg ok(cb_1) \lor \neg ok(s_1) \lor \neg ok(s_2) \lor \neg ok(l_1))$$
$$\land (\neg ok(cb_1) \lor \neg ok(s_3) \lor \neg ok(l_2))$$

follows from the clauses. This conjunction of disjunctions can be distributed into **disjunctive normal form (DNF)**, a disjunction of conjunctions of literals:

$$\neg ok(cb_1) \lor$$
$$(\neg ok(s_1) \land \neg ok(s_3)) \lor (\neg ok(s_1) \land \neg ok(l_2)) \lor$$
$$(\neg ok(s_2) \land \neg ok(s_3)) \lor (\neg ok(s_2) \land \neg ok(l_2)) \lor$$
$$(\neg ok(l_1) \land \neg ok(s_3)) \lor (\neg ok(l_1) \land \neg ok(l_2)).$$

Thus you know that either cb_1 is broken, or else there is at least one of six double faults.

Integrity Constraints in Databases

In databases, there are often constraints that the designer of a database knows should never be violated. These are known as integrity constraints. If a database ever violates an integrity constraint, an error message should be displayed, and the offending clauses should be identified.

There are two types of integrity constraints. The first type specifies that some conjunction of atoms should never be true of a database, for example, that no student should have two different marks for the same course in the same term. The other type specifies that some atom should always be provable when another is, for example, that a mark must appear for a student in a course if they are enrolled in the course, and the course is finished. We will only consider the first type.

You can use integrity constraints to detect inconsistent databases, as in the following example:

Example 7.12

The following integrity constraint declares that a student can't have two grades for the same course.

$$false \leftarrow grade(St, Course, Gr_1) \land grade(St, Course, Gr_2) \land Gr_1 \neq Gr_2.$$

If this rule is added to the database of Figure 3.3 (page 87), an inconsistency can be derived, with $Course = math302$. Unfortunately, such integrity constraints only allow us to conclude that the database is inconsistent, not why it's inconsistent. To determine why it's inconsistent, you can use the debugging tools for incorrect answers (page 222). Such debugging tools could also be used whenever *false* is derived, because you know that the atom *false* is always false in the intended interpretation.

If you want to be more sophisticated, you can add a justification to each fact or rule which may be contentious. A **justification** is an atom, possibly containing free variables, which encodes information about why a clause is in the knowledge base. This atom is added to the body of the appropriate clauses, and made assumable. When *false* is derived, the corresponding conflict will encode the reason why the fault occurred.

Reasoning with Integrity Constraints

This section presents a bottom-up implementation and a top-down implementation for finding conflicts, and thus answers, in Horn clause knowledge bases.

Bottom-Up Implementation

The bottom-up implementation is an augmented version of the bottom-up algorithm for definite clauses presented in Section 2.7 (page 47).

$C := \{\langle a, \{a\}\rangle : a$ is assumable $\}$;
repeat
      select clause "$h \leftarrow b_1 \wedge \ldots \wedge b_m$" in T such that
          $\langle b_i, A_i \rangle \in C$ for all i and
          $\langle h, A \rangle \notin C$ where $A = A_1 \cup \ldots \cup A_m$;
    $C := C \cup \{\langle h, A \rangle\}$
until no more selections are possible

Figure 7.2: Bottom-up proof procedure for computing conflicts of T

The modification to that algorithm is that the conclusions are pairs $\langle a, A \rangle$, where a is an atom and A is a set of assumables that imply a in the context of T.

Initially the conclusion set C is $\{\langle a, \{a\}\rangle : a$ is assumable$\}$. Rules can be used to form new conclusions. If there is a rule $h \leftarrow b_1 \wedge \ldots \wedge b_m$ such that for each b_i there is some A_i such that $\langle b_i, A_i \rangle \in C$, then $\langle h, A_1 \cup \ldots \cup A_m \rangle$ can be added to C. Figure 7.2 gives code for the algorithm. (See page 495 for a Prolog implementation.)

When the pair $\langle false, A \rangle$ is generated, you know that the assumptions in A form a conflict. Thus if $A = \{a_1, \ldots, a_k\}$, then you know that

$$T \models \neg a_1 \vee \ldots \vee \neg a_k.$$

This is essentially proof by contradiction. You assume a set of propositions. If you show that some subset of the assumables leads to a contradiction, then in every model of T, one element of that subset must be false.

One refinement of this program is to prune supersets of assumptions. If $\langle a, A_1 \rangle$ and $\langle a, A_2 \rangle$ are in C, where $A_1 \subset A_2$, then $\langle a, A_2 \rangle$ can be removed from C (or not added if you were about to add it). There is no need to make the extra assumptions to imply a. Similarly, if $\langle false, A_1 \rangle$ and $\langle a, A_2 \rangle$ are in C, where $A_1 \subseteq A_2$, then $\langle a, A_2 \rangle$ can be removed from C as you already know that A_1 and any superset—including A_2—is inconsistent with the clauses given, and so nothing more can be learned from considering such sets of assumables.

Example 7.13

Consider the axiomatization of the electrical domain Figure 6.4 (page 206), but with no rules for ok and with $ok(X)$ assumable for each light, switch, and circuit breaker X.

Initially, in the algorithm of Figure 7.2, C has the value

$$\{\langle ok(l_1), \{ok(l_1)\}\rangle , \langle ok(l_2), \{ok(l_2)\}\rangle , \langle ok(s_1), \{ok(s_1)\}\rangle , \langle ok(s_2), \{ok(s_2)\}\rangle ,$$
$$\langle ok(s_3), \{ok(s_3)\}\rangle , \langle ok(cb_1), \{ok(cb_1)\}\rangle , \langle ok(cb_2), \{ok(cb_2)\}\rangle\}$$

The following shows a sequence of values added to C under one sequence of selections:

$\langle live(outside), \{\} \rangle$

$\langle connected\_to(w5, outside), \{\} \rangle$

$\langle live(w5), \{\} \rangle$

$\langle connected\_to(w3, w5), \{ok(cb_1)\} \rangle$

$\langle live(w3), \{ok(cb_1)\} \rangle$

$\langle up(s_3), \{\} \rangle$

$\langle connected\_to(w4, w3), \{ok(s_3)\} \rangle$

$\langle live(w4), \{ok(cb_1), ok(s_3)\} \rangle$

$\langle connected\_to(l_2, w4), \{\} \rangle$

$\langle live(l_2), \{ok(cb_1), ok(s_3)\} \rangle$

$\langle light(l_2), \{\} \rangle$

$\langle lit(l_2), \{ok(cb_1), ok(s_3), ok(l_2)\} \rangle$

$\langle dark(l_2), \{\} \rangle$

$\langle false, \{ok(cb_1), ok(s_3), ok(l_2)\} \rangle$.

Thus you can conclude

$$\neg ok(cb_1) \vee \neg ok(s_3) \vee \neg ok(l_2).$$

Top-Down Implementation

The top-down implementation is a direct application of the delaying prover presented in Section 6.3 (page 210). Figure 7.3, based on Figure 6.7 (page 211), shows the top-down implementation that collects assumables that can be used to prove a goal.

$dprove(G, D_1, D_2)$ means that G can be proven from the clauses in the knowledge base, given D_1 as the set of initial assumptions, forming D_2 as a set of assumptions that includes D_1 and also implies G. Thus is it true when $D_1 \subseteq D_2$ and $KB \models D_2 \rightarrow G$.

$conflict(C)$ is true if C is a conflict. Note that C need not be a minimal conflict, as there may be another proof that returns a smaller conflict. However, the minimal conflicts will be among those returned by a call to $conflict$.

Example 7.14 Consider the representation of the circuit in Example 7.11 (page 243). You can use the above meta-interpreter by adding the assertion

$assumable(ok(G))$.

When you issue the query ?$conflict(C)$, there are two answers

$C = [ok(cb_1), ok(s_1), ok(s_2), ok(l_1)]$,

$C = [ok(cb_1), ok(s_3), ok(l_2)]$.

% $dprove(G, D_0, D_1)$ is true if list D_0 is an ending of list D_1 such that assuming the
% elements of D_1 lets you derive G. Procedurally, D_0 is the assumptions made before
% proving G, and D_1 contains D_0 as well as any extra assumptions needed to prove G.

> $dprove(true, D, D)$.
> $dprove((A \ \& \ B), D_1, D_3) \leftarrow$
> > $dprove(A, D_1, D_2) \wedge$
> > $dprove(B, D_2, D_3)$.
>
> $dprove(G, D, D) \leftarrow$
> > $member(G, D)$.
>
> $dprove(G, D, [G|D]) \leftarrow$
> > $notin(G, D) \wedge$
> > $assumable(G)$.
>
> $dprove(H, D_1, D_2) \leftarrow$
> > $(H \Leftarrow B) \wedge$
> > $dprove(B, D_1, D_2)$.
>
> $conflict(C) \leftarrow$
> > $dprove(false, [\ ], C)$.

Figure 7.3: A meta-interpreter to find conflicts

Note that the answer extraction in this case works differently to that for definite
clauses (page 43). Rather than knowing which instances of a query follow from a
knowledge base, you want to know which instances of assumables are inconsistent
with the knowledge base. Answer extraction in this case is done by collecting these
instances that imply *false*.

7.4   Complete Knowledge Assumption

When you have a database of facts, often you want to assume that these facts are
complete in the sense that anything not stated as a fact is false.

Example
7.15

You may want to conclude that a database of students enrolled in a class is complete,
so that if someone isn't listed there, you know they are not enrolled. Under this

> assumption you may conclude that a class is empty if there is no one who is listed as being enrolled in it.

In Section 3.7 (page 103), we showed how the initial RRS doesn't embody this complete knowledge assumption. The lack of complete knowledge means you can't conclude anything from the lack of clauses. The initial definite clause RRS is **monotonic** in the sense that anything that could be concluded before a fact or rule is added can still be concluded after it's added; adding knowledge can't reduce the set of everything that can be concluded. In this section we present a formalism that is based on the assumption that a knowledge base is complete. For example, you can conclude that no one is enrolled in a course if you can't find anyone who is enrolled in it. The resulting formalism is thus **nonmonotonic**: Some conclusions can be invalidated by adding more knowledge. For example, the conclusion that a course is empty can be made invalid by adding the fact that someone is enrolled in the course.

Assuming that the knowledge about a predicate is complete is quite different than stating integrity constraints as in Section 7.3. Under the completeness assumption, an atom is false if none of the bodies defining the atom are true. This is the Assumption CK:

Assumption CK   (**Complete Knowledge**) The agent's knowledge of the world is *complete*.

This assumption requires that everything relevant about the world be known to the agent. For example, this **Closed World Assumption** allows the agent to assume safely that a fact is false if it can't infer that it is true.

For the ground case, if you have the rules for some atom a as

$$a \leftarrow b_1$$
$$\vdots$$
$$a \leftarrow b_n,$$

then the Complete Knowledge Assumption says that if a is true in some interpretation then one of the b_i must be true in that interpretation. Thus you implicitly have

$$a \rightarrow b_1 \vee \ldots \vee b_n.$$

Since the clauses defining a are equivalent to

$$a \leftarrow b_1 \vee \ldots \vee b_n,$$

the meaning of the clauses can be seen as the conjunction of these two formulae, namely, the definition

$$a \leftrightarrow b_1 \vee \ldots \vee b_n$$

where $\leftrightarrow$ is read as "if and only if." This form is the ground form of **Clark's completion** of the clauses for a.

For the non-ground case you have to be more careful.

Example
7.16

Suppose you have a student relation that is defined by

$student(mary).$

$student(john).$

$student(ying).$

then the complete knowledge assumption would say that these three are the only students, that is,

$student(X) \leftrightarrow X = mary \lor X = john \lor X = ying.$

That is, if X is *mary*, *john*, or *ying*, then X is a student; and if X is a student, then X must be one of these three. Thus you have

$\neg student(X) \leftarrow X \neq mary \land X \neq john \land X \neq ying.$

In order to conclude $\neg student(alan)$, you have to be able to prove $alan \neq mary \land alan \neq john \land alan \neq ying$. In order to derive the inequalities, you need the unique names assumption (page 238).

For this section we adopt the unique names assumption and assume the axioms for equality (page 237) and inequality (page 238).

The **Clark normal form** of the clause:

$p(t_1, \ldots, t_k) \leftarrow B$

is the clause

$p(V_1, \ldots, V_k) \leftarrow \exists W_1 \ldots \exists W_m \; V_1 = t_1 \land \ldots \land V_k = t_k \land B,$

where $V_1, \ldots, V_k$ are k different variables that did not appear in the original clause, and $W_1, \ldots, W_m$ are the original variables in the clause. "$\exists$" means "there exists" (page 268). When the clause is a fact (page 30), B is *true*.

Suppose you put all of the clauses for p into Clark normal form, with the same set of introduced variables, giving

$p(V_1, \ldots, V_k) \leftarrow B_1$

$\qquad \vdots$

$p(V_1, \ldots, V_k) \leftarrow B_n$

then you can rewrite the clauses for p in terms of a disjunct of the created bodies:

$p(V_1, \ldots, V_k) \leftarrow B_1 \lor \ldots \lor B_n.$

This implication is logically equivalent to the set of original clauses.

Clark's completion of p is the equivalence

$p(V_1, \ldots, V_k) \leftrightarrow B_1 \lor \ldots \lor B_n,$

which means that $p(V_1, \ldots, V_k)$ is true if and only if at least one B_i is true.

Clark's completion of a knowledge base consists of the completion of every predicate symbol, along with the axioms for equality (page 237) and inequality (page 238).

Example 7.17

For the clauses

$$student(mary).$$
$$student(john).$$
$$student(ying).$$

the Clark normal form is

$$student(V) \leftarrow V = mary.$$
$$student(V) \leftarrow V = john.$$
$$student(V) \leftarrow V = ying.$$

which is equivalent to

$$student(V) \leftarrow V = mary \lor V = john \lor V = ying.$$

The completion of the *student* predicate is

$$student(V) \leftrightarrow V = mary \lor V = john \lor V = ying.$$

Example 7.18

Consider the following recursive definition from Section 3.6 (page 86):

$$passed\_each([\ ], St, MinPass).$$
$$passed\_each([C|R], St, MinPass) \leftarrow$$
$$\quad passed(St, C, MinPass) \land$$
$$\quad passed\_each(R, St, MinPass).$$

In Clark normal form this can be written as

$$passed\_each(L, S, M) \leftarrow L = [\ ].$$
$$passed\_each(L, S, M) \leftarrow$$
$$\quad L = [C|R] \land$$
$$\quad passed(S, C, M) \land$$
$$\quad passed\_each(R, S, M).$$

Here we have removed the equalities that specify renaming of variables and have renamed the variables by hand. Thus Clark's completion of *passed_each* is

$$passed\_each(L, S, M) \leftrightarrow L = [\ ] \lor$$
$$\quad (L = [C|R] \land$$
$$\quad passed(S, C, M) \land$$
$$\quad passed\_each(R, S, M)).$$

If you have an atom p which unifies with the head no clauses in the knowledge base, the completion is $p \leftrightarrow false$. In other words $\neg p$.

Once you have the completion you can interpret negations in the bodies of clauses.

The formula $\sim p$ means that p is false under the Complete Knowledge Assumption. This is called **negation as failure**. That is, p is false in all models of the completion of the program. We use a different symbol to the previous negation where $\neg p$ is true in an interpretation if p is false in the interpretation, since that symbol doesn't incorporate the Complete Knowledge Assumption. Instead, we have $T \models \sim p$ iff $T' \models \neg p$, where T' is the completion of T.

| Example 7.19 | Consider the clauses |

$$p \leftarrow q \wedge \sim r.$$
$$p \leftarrow s.$$
$$q \leftarrow \sim s.$$
$$r \leftarrow \sim t.$$
$$t.$$
$$s \leftarrow w.$$

The completion of this program is

$$p \leftrightarrow (q \wedge \neg r) \vee s.$$
$$q \leftrightarrow \neg s.$$
$$r \leftrightarrow \neg t.$$
$$t \leftrightarrow true.$$
$$s \leftrightarrow w.$$
$$w \leftrightarrow false.$$

From which it follows that w is false (in all models of the completion), so s is false, q is true, t is true, r is false, and p is true.

Under this assumption, you can now define relations that you couldn't define using only definite clauses.

| Example 7.20 | In Section 3.7 (page 103) we showed that for the initial RRS, given a database of *course*(C) that is true if C is a course, and *enrolled*(S, C) which means that student S is enrolled in course C, you couldn't define *empty_course*(C) that is true if there are no students enrolled in course C. Using negation as failure, this can be defined by: |

$$empty\_course(C) \leftarrow course(C) \wedge \sim has\_Enrollment(C).$$
$$has\_Enrollment(C) \leftarrow enrolled(S, C).$$

As a word of caution: You have to be very careful when you include free variables within negation as failure. They usually don't mean what you think they might. We introduced the predicate *has_Enrollment* to avoid having a free variable within a negation as failure. Consider what would have happened if you had not done this.

Example
7.21

One may be tempted to define *empty_course* in the following manner:

$$empty\_course(C) \leftarrow course(C) \wedge \sim enrolled(S, C)$$

This isn't correct. Given the clauses

course(*cs*422).
course(*cs*486).
enrolled(*mary*, *cs*422).
enrolled(*sally*, *cs*486).

the clause

$$empty\_course(cs422) \leftarrow course(cs422) \wedge \sim enrolled(sally, cs422)$$

is an instance of the above clause for which the body is true, and the head is false, as *cs*422 isn't an empty course. So this is a contradiction to the truth of the above clause.

Proof Procedures

Bottom-Up Procedure

The bottom-up procedure is a simple modification of the bottom-up procedure for definite clauses (page 47). The only difference is that you will be able to add literals of the form $\sim p$ to the set C of consequences. $\sim p$ is added to C when you have derived that p must fail. Failure can be defined recursively. p **fails** when every body with p as the head fails. A body fails if one of the literals in the body fails. An atom b_i in a body fails if $\sim b_i$ has been derived (is in C). A negation $\sim b_i$ in a body fails if b_i has been derived.

Figure 7.4 gives a bottom-up negation as failure interpreter for computing consequents of a ground *KB*. (Prolog code is given on page 493.) This is similar to the bottom-up definite clause interpreter of Figure 2.4 (page 47), but where we add the negation of atoms which have no clauses that could possibly succeed.

$C := \{\};$
repeat
    either
        select $r \in KB$ such that
            r is "$h \leftarrow b_1 \wedge \ldots \wedge b_m$"
            $b_i \in C$ for all i, and
            $h \notin C$;
        $C := C \cup \{h\}$
    or
        select h such that for every rule "$h \leftarrow b_1 \wedge \ldots \wedge b_m$" $\in KB$
            either for some b_i, $\sim b_i \in C$
            or some $b_i = \sim g$ and $g \in C$
        $C := C \cup \{\sim h\}$
until no more selections are possible

Figure 7.4: Bottom-up negation as failure proof procedure

Example 7.22

Consider the clauses

$p \leftarrow q \wedge \sim r.$

$p \leftarrow s.$

$q \leftarrow \sim s.$

$r \leftarrow \sim t.$

$t.$

$s \leftarrow w.$

The following is a sequence of atoms added to C:

t

$\sim r$

$\sim w$

$\sim s$

q

$p.$

t is derived trivially as it's a fact. $\sim r$ is derived because $\sim t$ finitely fails as $t \in C$. $\sim w$ is derived as there are no clauses for w, and so the conditions for finite failure trivially hold. $\sim s$ is derived as its only body finitely fails as $\sim w \in C$. q and p are derived as the bodies are all proved.

Top-Down Procedure

The top-down procedure for the complete knowledge assumption proceeds by **negation as failure**. If the proof for a fails, you can conclude $\sim a$. Failure can be defined recursively. Suppose you have rules for atom a:

$$a \leftarrow b_1$$
$$\vdots$$
$$a \leftarrow b_n$$

If each body b_i fails, a fails. A body fails if one of the conjuncts in the body fails.

Note that you require *finite* failure. Just because you have the rule $p \leftarrow p$ and can argue that there will never be a proof, you can't conclude that p is false. Just as a backward chaining proof procedure never halts for this program, the completion, $p \leftrightarrow p$, gives no information.

Example 7.23

Consider the clauses

$$p \leftarrow q \wedge \sim r.$$
$$p \leftarrow s.$$
$$q \leftarrow \sim s.$$
$$r \leftarrow \sim t.$$
$$t.$$
$$s \leftarrow w.$$

Suppose the query is ?p. When you try to prove p, you try and prove q, and then try to fail to prove r (i.e., you try to prove r, and the proof must fail before you can conclude $\sim r$). When trying to prove q, you try and fail to prove s. To prove s, you try to prove w. There are no clauses for w, so the proof for w fails, and so the proof for s fails, and so you can conclude $\sim s$. As you can prove $\sim s$, you can use the third rule to prove q.

When you try to prove r, you try to prove t. The proof for t succeeds, so $\sim t$ fails, and so the proof for r fails, and so the proof for $\sim r$ succeeds. You have thus proven the body of the first clause, and so you can conclude p.

As in the case of the unique names assumption, a problem arises when you have free variables in negated goals.

Example 7.24

Consider the clauses

$$p(X) \leftarrow \sim q(X) \wedge r(X).$$
$$q(a).$$
$$q(b).$$
$$r(d).$$

According to the semantics, there is only one answer to the query $?p(X)$, which is $X = d$. As $r(d)$ follows, you can derive $\sim q(d)$ and so $p(d)$. However, in a top-down proof procedure, you don't want to encounter $\sim q(X)$, try to prove $q(X)$ which succeeds (with substitution $\{X/a\}$), and thus fail $\sim q(X)$, and hence fail the subgoal $p(X)$. In this case the proof procedure would be incomplete. As with the unique names assumption (Section 7.2), you may need to delay the negated subgoal until the free variable is bound.

You can see that you need a slightly more complicated procedure when you have calls to negation as failure with free variables:

- You need to delay (page 210) negation as failure goals that contain free variables until the variables become bound.

- If the variables never become bound, the goal **flounders**. In this case you can't conclude anything about the goal. The following example shows that you need to do something more sophisticated for the case of floundering goals.

Example 7.25

Consider the clauses:

$$p(X) \leftarrow \sim q(X)$$
$$q(X) \leftarrow \sim r(X)$$
$$r(a)$$

and the query

$$?p(X).$$

The completion of the knowledge base is:

$$p(X) \leftrightarrow \neg q(X),$$
$$q(X) \leftrightarrow \neg r(X),$$
$$r(X) \leftrightarrow X = a.$$

Substituting $X = a$ for r gives $q(X) \leftrightarrow \neg X = a$, and so $p(X) \leftrightarrow X = a$. Thus there is one answer, namely $X = a$, but delaying the goal will not help find it. You need to analyze the cases for which the goal failed in order to derive this answer. Such a procedure is, however, beyond the scope of this book.

7.5   Disjunctive Knowledge

All of the knowledge you have been able to represent so far is definite. You have not been able to represent vague knowledge. The complete knowledge assumption (page 248) strengthened the assumptions behind the definite clause language so that you can represent knowledge that is definite and complete. In this section we weaken the assumption and allow vague knowledge.

Some examples of the need for vague knowledge include the following:

- In the course domain, you may not know when a course is to be scheduled, and a student may want to show they can enroll in the course no matter which time it's scheduled.

- For the electrical trouble-shooting domain, with just definite clauses you can't state the common single-fault assumption that only one component can be broken.

- The delivery robot may not know which of three possible offices Craig is in, but may have to develop a plan that will work no matter which office he is in.

In order to represent such vague knowledge, we remove the Assumption HK that all the knowledge about the world is Horn. We replace HK with the following:

Assumption
DNK
(Disjunctive and Negative Knowledge) Allow the representation of *disjunction* in the head of clauses and allow atoms to be *negated*.

The form of negation you need is different from negation as failure. You can now explicitly state negative knowledge, rather than assuming that if something can't be proven, then it's false. Note that having disjunctive knowledge is incompatible with the notion of complete knowledge embodied in negation as failure. Once you have a disjunction, say $a \lor b$, you can't prove a, and you can't prove b, but to assume both negatives would be incompatible with the given knowledge that one of a or b is true in every model.

We add to the language the ability to state explicit negation and the ability to state explicit disjunctions. This will be incorporated as an extension to the clausal language.

Allowing disjunction is closely linked to allowing negation, since "a or b," written $a \lor b$, is equivalent to "a if not b," written $a \leftarrow \neg b$.

Syntax

We now present the syntax for an extension of the language given in Chapter 2 that will incorporate disjunctions and definite clauses as special cases. We use the same notion of *variable*, *term*, *atomic formula (atom)* as in the definite clause language (page 29) and add the following:

A **literal** is an atom or the negation of an atom. The negation of atom a, read "not a," is written $\neg a$.

A **general clause** is of the form

$$L_1 \lor \cdots \lor L_k \leftarrow L_{k+1} \land \cdots \land L_n$$

where the L_i are literals. If $k = 1$ and all of the literals are atoms (i.e., none of the L_i are negated) you have a definite clause (as in Section 2.4). If $k = n$, you have a disjunction of literals, in which case the "$\leftarrow$" can be omitted.

Note that the precedence of the operators means that the above clause is grouped as

$$(L_1 \vee \cdots \vee L_k) \leftarrow (L_{k+1} \wedge \cdots \wedge L_n)$$

Since conjunction and disjunction are commutative and associative, as far as the logic is concerned, the order of the literals in the left and right parts of a clause is irrelevant. However, in some proof procedures the order affects the efficiency or the completeness, so some ordering may be more appropriate than another under a particular proof procedure.

The notion of a query is similar to that of definite clauses (page 40). The query

$$?L_1 \wedge \cdots \wedge L_n$$

is defined to mean the following: Add to the current set of clauses the clause

$$yes(\overline{X}) \leftarrow L_1 \wedge \cdots \wedge L_n$$

where "yes" is a special predicate symbol and $\overline{X}$ are the free variables in the query, then ask the question, "For what values of $\overline{X}$ is $yes(\overline{X})$ a logical consequence of the knowledge base?"

Semantics

The meaning of a set of clauses C is given by the normal model-theoretic semantics (page 32). An **interpretation** specifies the domain, specifies the denotation of constants and function symbols, and specifies which instances of atoms are true and false. A **model** of a set of clauses is an interpretation in which all of the clauses are true. The only difference is that you need to give an enhanced account of how the truth of a composite formula is built from the truth of its components. This is given by the truth table of Figure 7.5. Using this truth table, a ground clause

$$L_1 \vee \ldots \vee L_k \leftarrow L_{k+1} \wedge \ldots \wedge L_n$$

is false in some interpretation I if and only if all of the $L_1, \ldots, L_k$ are all false in I, and $L_{k+1}, \ldots, L_n$ are all true in I.

One difference between the forms of rules here and for definite clauses is that here there is no privileged version of the rules (unlike definite clauses where there was always a unique head of the clause). As an example of this, $a \leftarrow b \wedge \neg c$ is equivalent to $a \vee c \leftarrow b$, which is in turn equivalent to $c \leftarrow b \wedge \neg a$ and also to $\neg b \vee c \leftarrow \neg a$.

The following proposition trivially follows from the semantics:

Proposition 7.1   You can swap a literal from one side of the "$\leftarrow$" to the other if you negate it.

p	q	$\neg p$	$p \wedge q$	$p \vee q$	$p \leftarrow q$
true	true	false	true	true	true
true	false	false	false	true	true
false	true	true	false	true	false
false	false	true	false	false	true

Figure 7.5: Truth table for the clausal operators

This gives us a method for transforming clauses into equivalent clauses. Some of these clauses are of a form similar to definite clauses, but with literals rather than just atoms allowed in the head and the body.

A **contrapositive** form of a clause is an equivalent clause (resulting from applying Proposition 7.1) with exactly one literal on the left-hand side of the "←." If there are n literals in a clause, there are n contrapositive forms of that clause.

The **normal form** of a clause c is an equivalent clause without any literals on the right-hand side of the "←." Any literal on the right-hand side of the arrow is negated and moved to the left side of the arrow. The arrow can then be omitted. Any clause has a unique normal form up to associativity and commutativity of disjunction.

Example
7.26

The clause $a \vee b \leftarrow c \wedge \neg d$ has the contrapositive forms

$$a \leftarrow \neg b \wedge c \wedge \neg d$$

$$b \leftarrow \neg a \wedge c \wedge \neg d$$

$$\neg c \leftarrow \neg a \wedge \neg b \wedge \neg d$$

$$d \leftarrow \neg a \wedge \neg b \wedge c$$

and the normal form

$$a \vee b \vee \neg c \vee d.$$

Queries and Answers

You use the notion of a query and an answer in a similar way to the definite clause RRS. A variable-free query, assuming that a proof procedure halts, either has the answer *yes* (there is a proof) or *no* (there is no proof).

For a query with variables, an answer can be more complicated.

Example
7.27

Consider the following clauses and query:

$$p(X) \leftarrow q(X).$$

$$q(a) \vee q(b).$$

$$?p(Y).$$

In every model of the clauses there is an instance of Y such that $p(Y)$ is true in that interpretation. In some of the models the instance is $\{Y/a\}$ and in other models the instance is $\{Y/b\}$. There is no value for Y that's true in every model of the clauses. The statement "there is some Y such that $p(Y)$" is true in every model of the clauses; thus you would expect a reply of *yes*, but the answer will be more complicated. In some of the models $p(a)$ is true, and in the other models $p(b)$ is true. Thus in the intended interpretation, there is some Y such that $p(Y)$ is true. Either $p(a)$ or $p(b)$ is true in the intended interpretation. It is just that you don't know which is true. Thus the answer to the query should be: "yes, $Y = a$ or $Y = b$."

In general an **answer** to a query with free variables is a disjunction of instances of the query that follows from the knowledge base. A minimal answer is one such that no smaller disjunct also follows from the knowledge base.

Proof Procedures

Bottom-Up Procedure

This section gives a procedure for ground (variable-free) sets of clauses. Assume all clauses are in normal form. You treat a clause as the set of literals (implicitly disjoined) in that clause. The set representation means that duplicate elements are removed, and allows you to use the subset relationship.

The bottom-up proof procedure consists of finding all of the prime implicates of a theory (set of clauses).

An **implicate** of a theory is a clause that logically follows from a theory. Clause C is a **minimal implicate** if C is an implicate and no strict subset of C is an implicate. Some of the minimal implicates of a theory will be trivial implicates of the form $a \vee \neg a$ for some atom a. A **prime implicate** of a theory is a minimal implicate that isn't of the form $a \vee \neg a$.

Clause $L_1 \vee \ldots \vee L_k$ follows from a theory if and only if either:

- there are some L_i and L_j such that $L_i = \neg L_j$, or
- some subset of $\{L_1, \ldots, L_k\}$ forms a prime implicate of the theory.

So, if you can compute all of the prime implicates, then you can perform deduction by table lookup, which can be very fast.

The basic step of the algorithm is binary resolution:

The **binary resolution** of $\{L\} \cup R$ and $\{\neg L\} \cup S$ is $R \cup S$. In clause notation, the binary resolution of $R \vee L$ and $S \vee \neg L$ is $R \vee S$.

The binary resolution of two clauses is a logical consequence of the clauses. This is because in every model I of the clauses, literal L is either true or false in I. If L is

---

$I := KB$;
repeat

      select $\{L\} \cup R \in I$ and $\{\neg L\} \cup S \in I$ such that

                       there is no atom A such that $\{A, \neg A\} \subseteq R \cup S$ and

                       there is no $C \in I$ such that $C \subseteq R \cup S$;

      remove all C from I for which $R \cup S \subset C$;

      $I := I \cup \{R \cup S\}$

until no more selections are possible

Figure 7.6: Bottom-up proof procedure for computing prime implicates of KB

---

true in I, then S must be true in I; if L is false in I, then R is true in I. In either case, $R \vee S$ is true in I.

Note that the representation of clauses as sets means that you automatically collapse a disjunct of the form $A \vee A$ into the set $\{A\}$. Without this treatment, you would need another inference rule called **factoring**, which says that from $A \vee A \vee L$, infer $A \vee L$.

Figure 7.6 gives an algorithm to compute prime implicates of a set of clauses KB. It assumes that each clause is represented as the set of literals in the normal form of the clause. This algorithm does repeated binary resolution until no new clauses are formed.

Example 7.28

Consider the clauses

$$a \vee \neg b \leftarrow c.$$
$$\neg e \leftarrow \neg c.$$
$$b \vee d.$$
$$a \vee b \leftarrow d.$$
$$e \leftarrow \neg a.$$

In set notation these are given as

$$\{\{a, \neg b, \neg c\}, \{\neg e, c\}, \{b, d\}, \{a, b, \neg d\}, \{e, a\}\}.$$

You can resolve the first two sets to derive $\{a, \neg b, \neg e\}$. This can be resolved with the last clause to derive $\{a, \neg b\}$. At this point the first clause in the above set can be removed, as it's subsumed by the new clause. This resolution and removal of subsumed clauses can be repeated until you stop. The prime implicates of these clauses derived are:

$$a,$$
$$b \vee d,$$

$\neg e \vee c.$

You can now quickly answer queries such as $?a \vee \neg b$; the answer is *yes*, as a is a prime implicate. The query $?\neg e \vee \neg b$ has answer *no*, as no subset forms a prime implicate. The query $?b \vee \neg e \vee \neg b$ is trivially answered *yes*, as it has complementary literals.

A Top-Down Proof Procedure

There are two main problems you have to consider in extending the top-down definite clause proof procedure to work for general clauses. The first is that there is no privileged form for rules (there are many contrapositive forms of a clause). The second is that the simple rule of reasoning-by-cases:

$$prove(A \vee B) \leftarrow prove(A).$$
$$prove(A \vee B) \leftarrow prove(B).$$

which was given in Figure 6.5 (page 208), is sound but not complete for general clauses. For example, given $a \vee b$, you can't prove a nor can you prove b, but you still want to be able to prove $a \vee b$. This is important because the rules $c \leftarrow a$ and $c \leftarrow b$ are equivalent to $c \leftarrow a \vee b$. You can't use each rule in isolation because you may need both rules in order to infer c.

The following example demonstrates the sort of reasoning you need:

Example 7.29

Consider the following clauses and query:

$a \vee b.$

$c \leftarrow a.$

$c \leftarrow b.$

$?c.$

In this example the answer is *yes*. Semantically, in every model I of the clauses either a or b is true in I, and c is true in I regardless of which one of them is true.

To prove c, you use the following reasoning: By the second clause, c is true if a is true. By the first clause, a is true if b is false. By the third clause, b is false if c is false, as $c \leftarrow b$ is equivalent to $\neg b \leftarrow \neg c$. So if c is false, you can derive c is true. Since assuming $\neg c$ gives a contradiction, c must be true. This corresponds to *reductio ad absurdum* or proof by contradiction.

In this example we have expanded the definite clause proof procedure in two ways:

- You can use any contrapositive form of each clause in the clause set C.

% *gprove*(*G*, *A*) is true if $G \leftarrow A$ follows from *KB*

> *gprove*(*true*, *A*).
> *gprove*((*G* & *H*), *A*) ←
>> *gprove*(*G*, *A*) ∧
>> *gprove*(*H*, *A*).
> *gprove*(*G*, *A*) ←
>> *member*(*G*, *A*).
> *gprove*(*G*, *A*) ←
>> *clause*(*G*, *B*) ∧
>> *neg*(*G*, *NG*) ∧
>> *gprove*(*B*, [*NG*|*A*]).

Figure 7.7: Meta-interpreter for general clauses

- You need to carry out an **ancestor search**. This is a search up the proof tree (page 217) looking for the negation of the current subgoal. The idea is that you can assume the descendant subgoal and discharge the assumption when you use it to prove the ancestor. This is the same as the idea of proving c by assuming $\neg c$ and deriving a contradiction. Here you can get a contradiction by actually deriving c.

The clausal proof procedure is defined as follows: A subgoal g is proven if

- there is a contrapositive form of an input clause that unifies with g, such that all of the literals on the right-hand side of the contrapositive form are proven, or
- g unifies with the negation of an ancestor literal such that all substitutions are consistent. This is called the **negative ancestor rule**.

Let's express this procedure in terms of a *gprove* relation in Figure 7.7, with the following intended interpretations:

- *gprove*(*G*, *A*), interpreted procedurally, is a request to prove *G*, where *A* is the list of negated ancestors of *G*. Interpreted semantically, *gprove*(*G*, *A*) is true if $G \leftarrow A$ is a logical consequence of the clausal knowledge base.
- *neg*(*X*, *Y*) is true if *X* is the negative of *Y*, both in their simplest form.
- *clause*(*H*, *B*) is true if there is the contrapositive form of an input clause such that *H* is the head, and *B* is the body. In particular, you know ($H \leftarrow B$) is true in all models of the knowledge base.

- $n(a)$ is a term in the meta-level that denotes the negation of base-level atom a thus the meta-level function symbol n denotes the base-level negation. (See page 202 for the difference between the base language and metalanguage).

Example 7.30

The clause $a \vee \neg b \leftarrow c$, is represented as:

$\quad$ *clause*$(a, (b \,\&\, c))$.

$\quad$ *clause*$(n(b), (n(a) \,\&\, c))$.

$\quad$ *clause*$(n(c), (n(a) \,\&\, b))$.

In a way similar to how we represented the base-level definite clauses for the vanilla meta-interpreter (page 205), the base-level fact d is represented as the meta-level fact *clause*$(d, true)$.

Example 7.31

As an example of the procedure of Figure 7.7, consider the clauses from Example 7.29:

$\quad a \vee b.$

$\quad c \leftarrow a.$

$\quad c \leftarrow b.$

$\quad ?c.$

These are represented by the meta-level facts

$\quad$ *clause*$(a, n(b))$.

$\quad$ *clause*$(b, n(a))$.

$\quad$ *clause*(c, a).

$\quad$ *clause*$(n(a), n(c))$.

$\quad$ *clause*(c, b).

$\quad$ *clause*$(n(b), n(c))$.

$\quad$ *clause*(yes, c).

$\quad$ *clause*$(n(c), n(yes))$.

The following shows the sequence of calls to *gprove*:

$\quad$ *gprove*$(yes, [\,])$

$\quad$ *gprove*$(c, [n(yes)])$

$\quad$ *gprove*$(a, [n(c), n(yes)])$

$\quad$ *gprove*$(n(b), [n(a), n(c), n(yes)])$

$\quad$ *gprove*$(n(c), [b, n(a), n(c), n(yes)])$

at which stage the proof succeeds by the negative ancestor rule.

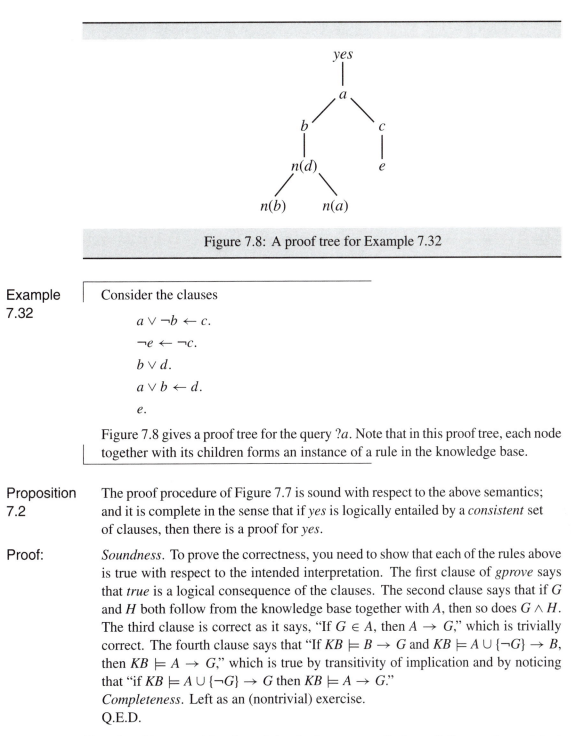

Figure 7.8: A proof tree for Example 7.32

Example 7.32

Consider the clauses

$$a \vee \neg b \leftarrow c.$$
$$\neg e \leftarrow \neg c.$$
$$b \vee d.$$
$$a \vee b \leftarrow d.$$
$$e.$$

Figure 7.8 gives a proof tree for the query ?a. Note that in this proof tree, each node together with its children forms an instance of a rule in the knowledge base.

Proposition 7.2

The proof procedure of Figure 7.7 is sound with respect to the above semantics; and it is complete in the sense that if *yes* is logically entailed by a *consistent* set of clauses, then there is a proof for *yes*.

Proof:

Soundness. To prove the correctness, you need to show that each of the rules above is true with respect to the intended interpretation. The first clause of *gprove* says that *true* is a logical consequence of the clauses. The second clause says that if G and H both follow from the knowledge base together with A, then so does $G \wedge H$. The third clause is correct as it says, "If $G \in A$, then $A \rightarrow G$," which is trivially correct. The fourth clause says that "If $KB \models B \rightarrow G$ and $KB \models A \cup \{\neg G\} \rightarrow B$, then $KB \models A \rightarrow G$," which is true by transitivity of implication and by noticing that "if $KB \models A \cup \{\neg G\} \rightarrow G$ then $KB \models A \rightarrow G$."
Completeness. Left as an (nontrivial) exercise.
Q.E.D.

Note that this proposition doesn't imply that you can find proofs from an inconsistent set of clauses. For example, b is a logical consequence of the set of two clauses $\{a, \neg a\}$.

These have no model, and so b is true in all of the models of $\{a, \neg a\}$. The procedure of Figure 7.7 can't derive b from this set of clauses.

Answer Extraction

When dealing with definite clauses, the answer you wanted was the value of the variable substitutions you used when you proved the query. As described in Section 7.5, you may need to derive disjunctive answers. This section shows how the meta-level theorem prover can be used to derive disjunctive answers.

Example 7.33

Consider the clauses and query of Example 7.27 (page 259). To get the appropriate answer you can assume $\neg yes(b)$ to prove $yes(a)$. You have proved $\neg yes(b) \rightarrow yes(a)$ which is equivalent to $yes(b) \vee yes(a)$.

A way to add answer extraction to the prover of Figure 7.7 is to delay (page 210) subgoals of the form $\neg yes(X)$. Assuming some instances of $\neg yes(X)$, you can go on to prove an instance of $yes(Y)$. At the end you will have proven something like:

$$\neg yes(a_1) \wedge \ldots \wedge \neg yes(a_n) \rightarrow yes(a),$$

which is equivalent to

$$yes(a_1) \vee \ldots \vee yes(a_n) \vee yes(a).$$

Thus you have derived a disjunctive answer.

Figure 7.9 gives a theorem prover that extracts answers. This is based on the theorem prover in Figure 7.7 (page 263), as well as on the meta-interpreter that delays subgoals (page 210).

Procedurally, $gaprove(G, Anc, Yes_1, Yes_2)$ means that G with ancestors Anc can be proved, where Yes_1 is the disjunct of answers before G is proven and Yes_2 is the resulting disjunct of answers after G is proven.

The meta-level query to extract answers is of the form

$$?gaprove(yes(X), [\,], yes(X), Answers),$$

where $Answers$ provides a disjunct of answers to the query.

For example, to get the disjunctive answer above, you can give the above query and obtain the answer $Answers = yes(a) \vee yes(b)$.

If you have a query $?p(X)$, the clause $yes(X) \leftarrow p(X)$ must be added to the knowledge base before the proof is commenced. Answer extraction needs the contrapositive of this clause, namely $\neg p(X) \leftarrow \neg yes(X)$.

% *gaprove*(*G*, *Anc*, *Yes*$_1$, *Yes*$_2$) is true if *G*∨*Yes*$_2$ ← *Anc*∨*Yes*$_1$ is a logical consequence
% of the knowledge base, where *Yes*$_1$ and *Yes*$_2$ are disjuncts of atoms of the form *yes*(*X*).
% *Yes*$_2$ extends *Yes*1 to also include the answers that must be true if *G* isn't true.

> *gaprove*(*true*, *Anc*, *Yes*, *Yes*).
>
> *gaprove*((*G* ∧ *H*), *Anc*, *Yes*$_1$, *Yes*$_3$) ←
>> *gaprove*(*G*, *Anc*, *Yes*$_1$, *Yes*$_2$) ∧
>> *gaprove*(*H*, *Anc*, *Yes*$_2$, *Yes*$_3$).
>
> *gaprove*(*G*, *Anc*, *Yes*, *Yes*) ←
>> *member*(*G*, *Anc*).
>
> *gaprove*(¬*yes*(*X*), *Anc*, *Yes*, (*yes*(*X*) ∨ *Yes*)).
>
> *gaprove*(*G*, *Anc*, *Yes*$_1$, *Yes*$_2$) ←
>> *clause*(*G*, *B*) ∧
>> *neg*(*G*, *NG*) ∧
>> *gaprove*(*B*, [*NG*|*A*], *Yes*$_1$, *Yes*$_2$).

Figure 7.9: Theorem prover with answer extraction

Application Examples

In the electrical wiring example of Example 7.11 (page 243), there are some things you can't say with integrity constraints alone. One is the *single-fault assumption* (or the *n*-fault assumption for some *n*). This is the assumption that there is only a single fault in the system. Note that this assumption isn't generally true and may remove some diagnoses which may indeed be correct. The following clause defines the single fault assumption:

> *ok*(*G*1) ←
>> ¬*ok*(*G*2) ∧
>> *G*1 ≠ *G*2.

This is equivalent to the disjunction

> *ok*(*G*1) ∨ *ok*(*G*2) ∨ *G*1 = *G*2.

The single fault assumption is interesting, as often single faults are more plausible than double faults. (Comparisons of diagnoses will be explored when we consider probabilities in Chapter 10.) If the single-fault assumption is inconsistent, then you can deduce that there are no single faults that are consistent with the observations, so there must be at least two faults. If the single fault assumption is inconsistent, you can deduce its negation, namely some instance of

$$\neg ok(G1) \wedge \neg ok(G2) \wedge G1 \neq G2.$$

If the single fault assumption is consistent, then you can add it and deduce single faults. If you add this clause representing the single fault assumption to the clauses of Example 7.11, you can derive $\neg ok(cb_1)$.

Other uses for disjunction include the "limited failure assumption." This is the assumption that components can only break in certain ways. For example, a light can only break by being permanently dark, but a switch can be broken so that a current flows even when it's off; however, none of these can break so that a current flows out when no current flows in. To axiomatize this, you can have as a clause the possible faults disjoined with the "ok" state for each component. You then need to axiomatize what follows from the faulty states.

7.6   Explicit Quantification

The final idea that you need before you have the full power of first order logic is to consider explicit quantification of variables. So far all of the variables have been universally quantified at the level of the clause. If a clause with universally quantified variables is true in an interpretation, any substitution of values for the variables will also be true in that interpretation. The clause is true for all values of a variable.

We discuss two ways to quantify variables and how they interact. The first is where a formula is true for all values of a variable (universal quantification), and the second is where a formula is true for some values for the variable (existential quantification). Note that there are many other possible quantifications, some of which are easy to define in terms of these predicates (e.g., there exists a unique value) and some of which are more difficult to model (e.g., most, or nearly all).

We will expand the language to have two forms of quantification. We will allow expressions of the following form to be formulae:

- $\forall X \ w$, where w is a formula in which variable X appears free. X is said to be a **universally quantified** variable. Formula $\forall X \ w$ is true in an interpretation I if w would be true in that interpretation if X were considered as a constant denoting any individual.

- $\exists X \ w$, where w is a formula in which variable X appears free. X is said to be an **existentially quantified** variable. Formula $\exists X \ w$ is true in an interpretation if there is some individual such that if X were considered as a constant denoting that individual, w would be true in that interpretation.

Let's write $w[X]$ to mean formula w with X appearing free, and write $w[t]$ to mean w with t uniformly substituted for X in w.

One way to view existential quantification is as an infinite disjunction, $\exists X \; w[X]$ meaning the disjunction:

$$w[c_1] \lor w[c_2] \lor w[c_3] \lor \dots$$

for each ground term $c_1, c_2, c_3, \dots$

Similarly, universal quantification can be viewed as an infinite conjunction, with $\forall X \; w[X]$ meaning the conjunction

$$w[c_1] \land w[c_2] \land w[c_3] \land \dots$$

Quantification becomes more tricky when an existentially quantified variable is in the scope of a universally quantified variables—for example, as in

$$\forall V_1 \forall V_2 \dots \forall V_n \exists X \; w[V_1, V_2, \dots, V_n, X],$$

where $w[V_1, V_2, \dots, V_n, X]$ is a formula with free variables $V_1, V_2, \dots, V_n, X$. The value of X that exists may depend on the values of the V_i.

For example, in mathematics the following statement is true:

$$\forall N \exists M \; (M = N + 1 \leftarrow integer(N)).$$

The value of M that exists obviously depends on the value of N.

This is very different from the statement

$$\exists M \forall N \; (M = N + 1 \leftarrow integer(N)).$$

where M doesn't depend on N. This isn't true for the integers: No such M exists that is one more than every integer.

To handle existentially quantified variables, you carry out a process known as **Skolemization**. This process essentially gives a name to the individuals which are said to exist.

Suppose you have an occurrence of $(\exists X \; w)$, which is in the scope of universally quantified variables $V_1, \dots, V_n$

$$\forall V_1 \forall V_2 \dots \forall V_n \exists X \; w[V_1, V_2, \dots, V_n, X].$$

To **Skolemize** variable X there are two cases:

- $n = 0$; that is, X isn't in the scope of any universally quantified variable. Replace X by a new constant symbol c. This new constant is called a **Skolem constant**. The resulting formula is

$$\forall V_1 \forall V_2 \dots \forall V_n \; w[V_1, V_2, \dots, V_n, c].$$

- $n > 0$. Find a new function symbol, f, and replace all occurrences of X in w by $f(V_1, ..., V_n)$. This new function symbol is called a **Skolem function**. The resulting formula is

$$\forall V_1 \forall V_2 \ldots \forall V_n\ w[V_1, V_2, \ldots, V_n, f(V_1, V_2, \ldots, V_n)].$$

A formula is **Skolemized** by Skolemizing all of the existentially quantified variables in the formula.

Example 7.34

The formula

$$\forall X \exists Y\ w(X, Y)$$

has Skolemized form $w(X, f(X))$, where f is a new unary function symbol.

$$\exists Y \forall X\ w(X, Y)$$

has Skolemized form $w(X, c)$, where c is a new constant.

$$\exists V \forall W \exists X \forall Y \exists Z\ p(V, W, X, Y, Z)$$

has Skolemized form $p(c_1, W, f_1(W), Y, f_2(W, Y))$, where c_1 is a new constant, and f_1 and f_2 are new function symbols. The value of variable V doesn't depend on the value of any other variable. The value of X depends on the value of W. The value of Z depends on both W and X.

7.7 First-Order Predicate Calculus

So far we have considered definite clauses and general clauses. In this section we present a language that allows essentially arbitrary mixes of conjunctions, disjunctions, implications, equivalences, and quantifiers in nested form. The use of all of these elements allows for a less restrained expression of knowledge. It turns out that you have enough tools to be able to make the extension to arbitrary formulae just an exercise in making the language more flexible rather than needing extra power. The language which incorporates all of these extensions is called the **first-order predicate calculus**, defined as follows:

A first-order predicate calculus **well-formed formula** (wff) is an atom (page 29) or is of the form $(u \lor w)$, $(u \land w)$, $(u \rightarrow w)$, $(u \leftarrow w)$, $(u \leftrightarrow w)$, $\neg w$, $\forall V\ w$, or $\exists V\ w$, where u and w are wffs and V is a variable.

The restriction of this language to the variable-free case is known as the **propositional calculus** or the zeroth-order predicate calculus.

In order to understand the meaning of quantified formula, it's important to understand how variables are scoped: A variable that appears in a formula is either free or bound. A free variable is defined as follows:

Zeroth-, First-, Second-, Third-, and Higher-Order Logics

The language presented in Section 7.7 is called the first-order predicate calculus. One question that arises is, Why is it "first"-order logic? Are there other-order logics? What is a second-order logic? What is third or fourth-order logics? What about zeroth-order logic? Is there a logic that is the limit of k-order logic as $k \to \infty$?

First-order logic is first-order because you can only quantify over individuals in the domain. All of the predicates are true or false of individuals in the domain. In first-order logic you can't have predicates as variables and you can't quantify over predicates.

In second-order logic you can quantify over first-order relations and define predicates whose arguments are first-order relations. These are second-order relations. For example, the second-order logic formula:

$$\forall R \; symmetric(R) \leftrightarrow (\forall X \forall Y \; R(X, Y) \to R(Y, X))$$

defines the second-order relation *symmetric*.

In third-order logic you can quantify over second-order relations (as well as first-order relations and individuals), as well as having predicates over these relations and individuals. Similarly, in fourth-order logic you can quantify over third-order relations.

Zeroth-order logic is logic without any quantification. Zeroth-order predicate calculus is called the *propositional calculus.*

Second order logic seems necessary for many applications as transitive closure isn't first-order definable. For example, suppose you want *lt* to be the transitive closure of *next* where $next(X, s(X))$ is true. Think of *next* meaning the "next integer" and *lt* denoting "less than," or *next* meaning "next floor down," and *lt* denoting "lower than." You may think of writing the formula that is Clark's completion of a recursive definition of *lt*:

$$\forall X \forall Y \; lt(X, Y) \leftrightarrow Y = s(X) \vee lt(s(X), Y).$$

This doesn't accurately capture the definition, because, for example,

$$\forall X \forall Y \; lt(X, Y) \to \exists W \; Y = s(W)$$

doesn't logically follow from the above formula. This is because there are nonstandard models of the above formula with Y denoting *infinity*. In order to capture the transitive closure, you need a formula stating that *lt* is the minimal predicate that satisfies the completion formula. This can be done using second-order logic. Using second-order logic to state that the only values that satisfy a formula are those that are known to satisfy the formula is the basis of the nonmonotonic reasoning formalism Circumscription (McCarthy, 1986). See Chapter 9.

p	q	$\neg p$	$p \wedge q$	$p \vee q$	$p \leftarrow q$	$p \rightarrow q$	$p \leftrightarrow q$
true	true	false	true	true	true	true	true
true	false	false	false	true	true	false	false
false	true	true	false	true	false	true	false
false	false	true	false	false	true	true	true

Figure 7.10: Truth table for predicate calculus operators

- a variable that appears in an atom is **free**
- a variable that appears free in part of a composite formula that doesn't involve quantification is free in the composite formula
- if V is free in w, then V is **bound** and not free in $\forall V\ w$ or $\exists V\ w$
- all other variables that are free in w are free in $\forall V\ w$ and in $\exists V\ w$

A **closed formula** is one in which no variable is free. A variable V that appears outside of $\forall V\ w$ or $\exists V\ w$ should be regarded as a different variable to the bound variable V. One of these should be renamed to avoid confusion. The following semantics assumes this renaming is done.

The truth of ground atoms in an interpretation is as in the formal semantics of definite clauses (page 32). The truth of arbitrary ground formulae in an interpretation is defined by the truth tables of Figure 7.10. $(\exists V\ w)$ is true in an interpretation if there exists an individual in the domain such that if V denotes that individual then w is true. $(\forall V\ w)$ is true in an interpretation if and only if $\neg(\exists V\ \neg w)$, that is, if and only if there is no individual for which w is false.

Truth in an interpretation is only defined for closed formulae.

Proof Procedure

A standard proof procedure for the first-order predicate calculus is to convert an arbitrary formula into clausal form and then use a clausal theorem prover.

Converting to Negation Normal Form

You can convert an arbitrary formula into a normal form called **negation normal form** (NNF). Negation normal form is a formula which has arbitrary conjunctions and disjunctions of literals but no other operators.

You can convert a formula to NNF by doing the following:

1. Transform the implication operators into conjunctions, disjunctions and negations as follows: Implication $u \rightarrow w$ becomes $\neg u \vee w$. Implication $u \leftarrow w$ becomes $u \vee \neg w$. Logical equivalence $u \leftrightarrow w$ becomes either a conjunction $(\neg u \vee w) \wedge (u \vee \neg w)$ or a disjunction $(u \wedge w) \vee (\neg u \wedge \neg w)$.

2. Repeatedly apply De Morgan's law to move all of the negations in, and remove repeated negations: negated disjunction $\neg(u \vee w)$ becomes $(\neg u \wedge \neg w)$; negated conjunction $\neg(u \wedge w)$ becomes $(\neg u \vee \neg w)$; double negation $\neg\neg w$ becomes w. The quantifiers are duals of each other: $\neg(\exists V\ w)$ becomes $(\forall V\ \neg w)$ and $\neg(\forall V\ w)$ becomes $(\exists V\ \neg w)$.

The resulting formula, made up of conjunctions and disjunctions of literals, is then in negation normal form.

Example 7.35

Given the formula

$$\neg((a \leftrightarrow b) \vee (c \rightarrow \neg(d \wedge (e \leftarrow f))))$$

you can replace the implications by the more primitive operators forming

$$\neg(((a \wedge b) \vee (\neg a \wedge \neg b)) \vee (\neg c \vee \neg(d \wedge (e \vee \neg f)))),$$

which, by moving negation in, becomes in NNF:

$$(((\neg a \vee \neg b) \wedge (a \vee b)) \wedge (c \wedge (d \wedge (e \vee \neg f)))).$$

Skolemization

The next step is to remove existentially quantified variables by Skolemization, as described in Section 7.6.

Converting Negation Normal Form to Clausal Form

This section shows how to transform a wff in NNF into conjunctive normal form (CNF, also called clausal form). A formula is in **conjunctive normal form** (CNF) when it is a conjunction of disjunctions of literals. CNF is thus equivalent to a set of clauses. Each clause is a representation of a disjunction of literals. These clauses are conjoined.

Assume f is a formula in NNF. A formula in negation normal form is in clausal form when there are no conjuncts in the scope of a disjunct. That is, if f contains something of the form (up to associativity and commutativity of conjunction and disjunction)

$$(\alpha \vee (\beta \wedge \gamma)),$$

then it's not in conjunctive normal form (clausal form).

One way to convert it into CNF is to distribute into the logically equivalent:

$$((\alpha \vee \beta) \wedge (\alpha \vee \gamma)).$$

This reduces the number of $\wedge$'s within the scope of $\vee$'s by one each time. Repeated distribution will convert any NNF formula to clausal form.

This distribution forms two copies of the sub-formula α. This may cause a problem since a large number of such transformations produces an exponential growth of subterms.

An alternate way to convert to clausal form is to form f^0 which is f with $(\alpha \vee (\beta \wedge \gamma))$ replaced by $(\alpha \vee \delta)$, where δ is an atom of the form $p(X_1, ..., X_n)$, where the X_i are the free variables in $\beta \wedge \gamma$, and where p is a unique predicate symbol. Now form $f' = f^0 \wedge (\neg \delta \vee \beta) \wedge (\neg \delta \vee \gamma)$. Repeated use of this transformation from f to f' will convert a formula from NNF to CNF because:

- The only way an NNF formula will not be in CNF is if it has a sub-expression of the form $(\alpha \vee (\beta \wedge \gamma))$ in which case the transformation can be repeated.

- The number of $\wedge$'s within the scope of $\vee$'s is reduced by at least one each time, thus repeated use of the transformation will terminate.

- This transformation preserves satisfiability of the set of all clauses.

Example 7.36

Consider the case where there are no axioms and you want to prove

$$(\exists Y \forall X\ p(X, Y)) \rightarrow (\forall X \exists Y\ p(X, Y)),$$

which has no free variables. You first make it so that this form implies *yes*:

$$yes \leftarrow ((\exists Y \forall X\ p(X, Y)) \rightarrow (\forall X \exists Y\ p(X, Y))).$$

You then map out the definitions of $\leftarrow$, and move negations in:

$$yes \vee ((\exists Y \forall X\ p(X, Y)) \wedge (\exists X \forall Y\ \neg p(X, Y)))$$

which can be Skolemized to

$$yes \vee (p(X, c) \wedge \neg p(d, Y))$$

and then distributed into clausal form

$$yes \leftarrow \neg p(X, c).$$
$$\neg p(d, Y) \leftarrow \neg yes.$$

for which there is a very simple proof given the proof procedure (binding X to d and Y to c).

Example 7.37

Consider a proof for

$$(\forall X \exists Y\ p(X, Y)) \rightarrow (\exists Y \forall X\ p(X, Y)),$$

which is Skolemized into:

$$(p(X, f(X)) \wedge \neg p(g(Y), Y)) \vee yes$$

which is then converted into clausal form:

$$yes \leftarrow \neg p(X, f(X)).$$
$$\neg p(g(Y), Y) \leftarrow \neg yes.$$

for which there is no proof of *yes*, due to the occurs check, as there is no value for X such that $X = g(f(X))$.

7.8 Modal Logic

One of the things that you may want to be able to do is to refer to a formula inside an operator—for example, to be able to say that a formula is "necessary" or that a formula is "believed" or a formula is "known." In this section, we give an overview of propositional modal logic (i.e., without variables).

Some of the uses for modal logic include the following:

- You may want to prove properties about the delivery robot, for example, that it never gets stuck in a room under a certain strategy. Thus, you would like to prove that it's necessary (in all states of the robot) that it's possible to get to a particular point. Such statements are best given in terms of modal logic.

- Our infobot may need to reason about the belief and knowledge of other agents. An atom of the agent may be the statement that another agent believes or knows some formula. Similar ideas can be used by an agent to reason about their own knowledge.

The syntax we consider is the propositional language above, augmented with a **modal operator** L. If α is a formula, then $L\alpha$ is an atom. For example, $a \rightarrow b$ is a formula, and $L(a \rightarrow b)$ is an atom. Note that this modal operator is different from a predicate symbol, because it takes a formula as an argument, not a term as an argument.

You also define a dual operator M to L, defined by $M = \neg L \neg$ (i.e., $M\alpha$ means $\neg L \neg \alpha$). If you read L as "necessarily," then M can be read as "possibly."

Possible Worlds Semantics

The semantics for modal logic is defined in terms of **possible worlds.** Possible worlds are like interpretations, in that they define a truth assignment to propositions, but they also have some structure. Thus different possible worlds can assign the same truth values to propositions, but different interpretations can't assign the same truth values to all propositions (or else they would be the same interpretation). What is

Modal Logic	Constraints on R	Axioms in Proof Theory
K	no constraint	A2
T	reflexive	A1, A2
S4	reflexive, transitive	A1,A2,A3
S5	reflexive, transitive, symmetric	A1, A2, A3, A4
Weak S4	transitive	A2, A3
Weak S5	transitive, Euclidean	A2, A3, A4

Figure 7.11: Modal logic, their constraints on R, and their axioms

important about possible worlds is that there can be relations between the worlds. In this section we consider a binary accessibility relation between possible worlds and later we consider a semantics for probability (page 349) where there is a number (measure) associated with each possible world.

Consider a binary **accessibility relation** R on worlds. If world w_1 is related to w_2 by relation R, you write $w_1 R w_2$ (rather than the more traditional $R(w_1, w_2)$).

The notion of truth of a formula in a possible world is defined exactly as if the possible world were an interpretation (i.e., the possible world assigns truth values to the atoms, and you can use the truth table to determine the values of a compound formula), except that you have to specify how L is assigned a value. This is done as follows: $L\alpha$ is true in world w if α is true in all worlds w' such that wRw'.

Different logics can be obtained by placing constraints on the relation R. The following constraints are often used:

Reflexive R is **reflexive** if wRw for all worlds w.

Symmetric R is **symmetric** if $w_1 R w_2$ implies $w_2 R w_1$.

Transitive R is **transitive** if $w_1 R w_2$ and $w_2 R w_3$ implies $w_1 R w_3$.

Euclidean R is **Euclidean** if $w_1 R w_2$ and $w_1 R w_3$ implies $w_2 R w_3$.

Figure 7.11 shows different logics that are obtained by giving different constraints on the accessibility relation.

Intuitively, possible worlds are like possible states of affairs. The accessibility relation R defines how such states of affairs are related. This provides the basis for interpreting the L and M operators in terms of what is true in related states of affairs.

Name	Axiom
A1 Necessity	$L\alpha \rightarrow \alpha$
A2 Distribution	$L(\alpha \rightarrow \beta) \rightarrow (L\alpha \rightarrow L\beta)$
A3 Positive Introspection	$L\alpha \rightarrow LL\alpha$
A4 Negative Introspection	$\neg L\alpha \rightarrow L\neg L\alpha$

Figure 7.12: Axioms for L

Proof Procedures

You can also give proof procedures for the modal logics by adding to the proof procedure for the propositional logic the following rule:

if $\vdash \alpha$, then $\vdash L\alpha$;

that is, if you can derive α, you can infer $L\alpha$.

You can also impose some new axioms for L. These are given in Figure 7.12.

For each logic there is a possible worlds semantics, along with a sound and complete proof procedure. The proof procedures are specified by stating which axioms for L can be used. This is done in Figure 7.11

Such modal logics have been suggested for reasoning about necessity and possibility. They have also been suggested for use as the logic of knowledge (particularly **S4** and **S5**), as well as for logics of belief (particularly **weak S4** and **weak S5**). One of the major problems for using modal logics as a representation for human or robot belief or knowledge is the problem of *logical omniscience*—namely, that in modal logics, you know or believe all tautologies and all consequences of your beliefs. This is clearly unreasonable for a model of knowledge or belief. There have, however, been suggestions as to how to overcome this problem.

7.9 References and Further Reading

Resolution was invented by Robinson (1965). The top-down proof procedure given in Section 7.5 is based on the MESON proof procedure (Loveland, 1978). For a good introduction to theorem proving see Chang & Lee (1973).

The diagnostic application (page 243) is based on Genesereth (1984), de Kleer & Williams (1987), and Reiter (1987). See Hamscher, Console & de Kleer (1992) for classic papers on model-based diagnosis, and see Davis & Hamscher (1988) for a review.

The bottom-up integrity constraint algorithm (page 245) is based on the assumption-based truth maintenance system (ATMS) of de Kleer (1986), but the ATMS is also concerned about incremental addition of rules and assumptions. The bottom-up negation as failure interpreter of Section 7.4 are based on the justification-based truth maintenance system (JTMS) justification-based truth maintenance system of Doyle (1979), where positive atoms in the body of rules are known as IN, and negated atoms are OUT. Again, incremental addition of rules is stressed in the JTMS.

The work on negation as failure (page 248), as well as the unique names assumption (page 238), is based on the work of Clark (1978). See the book by Lloyd (1987) for a formal treatment of logic programming in general and negation as failure in particular. Apt & Bol (1994) provide a more recent survey of different techniques for handling negation as failure.

The modal logic section is based on Hughes & Cresswell (1968). For applications to reasoning about knowledge see Halpern & Moses (1985) and Fagin, Halpern, Moses & Vardi (1994).

7.10   Exercises

Exercise 7.1   Consider the following clauses and integrity constraints:

$$false \leftarrow a \wedge b.$$
$$false \leftarrow c.$$
$$a \leftarrow d.$$
$$a \leftarrow e.$$
$$b \leftarrow d.$$
$$b \leftarrow g.$$
$$b \leftarrow h.$$
$$c \leftarrow h.$$

Given that $\{d, e, f, g, h, i\}$ is the set of assumables, what is the set of minimal conflicts of the above clauses?

Exercise 7.2   Write a bottom-up proof procedure for definite clauses with integrity constraints that allows for incremental addition of clauses. When a clause is added, its effect should be propagated through the set of conclusions.

Exercise 7.3   Give Clark's completion of the knowledge base:

$$g \leftarrow d \wedge \sim c.$$
$$a \leftarrow b \wedge \sim c.$$

$$a \leftarrow d.$$
$$b \leftarrow e.$$
$$d \leftarrow e.$$
$$d \leftarrow f.$$
$$f.$$
$$c \leftarrow h.$$

What literals are derived by a bottom-up proof procedure run on the above clauses?

Exercise 7.4 Give Clark's completion of the knowledge base:

$$member(E, [E|L]).$$
$$member(E, [H|T]) \leftarrow member(E, T).$$
$$disjoint([], L).$$
$$disjoint([H|T], L) \leftarrow$$
$$\qquad \sim member(H, L) \wedge$$
$$\qquad disjoint(T, L).$$

Show a bottom-up derivation that is able to answer the query:

$$?disjoint([a, d], [e, f]).$$

Exercise 7.5 Write a bottom-up proof procedure for definite clauses with negation as failure that allows for incremental addition and removal of clauses. When a clause is added, its effect should be propagated through the set of conclusions.

Exercise 7.6 Define a depth-bounded meta-interpreter that incorporates the complete knowledge assumption—that, is with negation as failure. You need to distinguish failing "naturally," without reaching the depth bound, and failing because the depth bound has been reached. Only in the first case can you conclude the negation.

Extend this to an iterative deepening meta-interpreter that includes negation as failure.

Exercise 7.7 Prove that both ways to convert a formula to clausal form (page 273) are correct. That is, show that one formula is unsatisfiable if and only if the other is.

Exercise 7.8 One of the ways of converting an equivalence to negation normal form may be more efficient than another. Give some characterization of when one is better than the other.

Chapter 8

Actions and Planning

8.1   Introduction

So far we have only considered axioms about a static world. You wrote axioms that were true in a single intended interpretation. But the world changes over time. So you want to consider actions in the world, including actions that your agent performs, actions that other agents perform, and actions that nature performs.

There are two complementary aspects to agents and time:

- Agents reason in time
- Agents reason about time

If you have an agent that acts in an environment, it must act and reason in time. Time passes as it's acting and reasoning. If the agent has sensors, the values its sensors get from the environment change in time. The issues related to reasoning in time are explored in Chapter 12.

An agent that reasons in time doesn't necessarily have to reason about time. An agent whose actions are a simple function of its inputs is called a *reactive agent*. For example, a burglar alarm may turn on when motion is detected in a certain area. A thermostat is another example of a reactive agent. While reactive systems are very useful, they typically need to be hard-wired for the behaviors they exhibit. They can't adapt to changing goals or to changing circumstances.

Reasoning about time is most important when you want to give your agent goals and want it to perform actions to achieve those goals. What an agent should do depends on its goals, what it knows about the world, and what it intends to do in the future. For example, a delivery robot may have a goal to deliver a parcel to a particular room. It may know that the door to the room is locked; and it may reason that it should go and get the key, then go to the door, and then unlock it. Reasoning about time is essential

281

for building intelligent agents. To solve a goal intelligently, an agent needs to think not only about what it will do now, but also what it will do in the future.

This chapter first presents some general issues about representing time. It then gives a detailed example based on the delivery robot (page 12). It then presents three representations for reasoning about time. The chapter ends with a number of algorithms for solving goals based on a description of the possible actions and a model of the world. Chapter 12 discusses the complementary issues involved in reasoning in time, and it shows how the kinds of plans produced by the planners in this chapter can be used in a robot.

Representing Time

To properly study reasoning about time, you have to consider the nature of time. While there is much philosophic debate about the nature of time, you can take a very pragmatic view and consider the characterization of time that's most useful for your purposes. Time can be viewed in a number of ways:

Discrete time Time can be modeled as being discrete, as jumping from one time point to another. For example, you can model some situation from millisecond to millisecond or from day to day, treating each millisecond or day as a different time point.

Continuous time You can model time as being continuous. In this way there are no gaps in time. This is most useful when considering continuous quantities—for example, the orientation of the robot.

Event-based time Time steps don't have to be uniform; you can consider the time steps between interesting events, such as the time a package is delivered, and the time the robot picks up the package.

State space Rather than considering time explicitly, you can consider actions as mapping from one state to another. For example, you can consider the action "pick up parcel a" as a mapping from a state to where parcel a is on the ground to a state where the agent is holding parcel a and a is no longer on the table.

Each of these views may be the right one depending on the domain being modeled, the level of abstraction, and what you are trying to do.

There are a number of ways that you can add time to the RRSs presented:

- Actions and time can be external to the logic. You can axiomatize a single state in the logic, and you can use other statements outside of the logic to specify how an action maps from one state description to the next. The STRIPS representation (page 288) is an example of such a framework.

- Time or the state can be **reified**. This means that you consider time to be another object that you can talk about in the logic. You can add an extra parameter to say when some property is true. For example, you can treat the term $pm(2, 30)$ to denote a particular time, and you can say that David teaches the course cs502 at 2:30 p.m. using the atom $teaches(david, cs502, pm(2, 30))$. The situation calculus (page 290) is a representation that uses the state-space view of time and reifies the state.

- The logic can be at the meta-level, where sentences in the logic are about what's true at what time. For example, you can have a predicate $holds(r, t)$ to mean that relation r is true at time t. The *holds* predicate is much like the *prove* predicate of Chapter 6, but it says not only that something is true, but when it's true. This is closely related to the previous case; it's not very different to state $holds(at(david, office), am(11, 30))$ or $at(david, office, am(11, 30))$. The use of *holds* enables us to quantify over relations (by having a variable in the first argument to *holds*). This is similar to the development of the *prop* predicate (page 182).

You can also consider whether to represent time in terms of **points** or in terms of **intervals**. It might be most useful to consider all events as occurring over an interval and to consider all propositions as being true or false over an interval. An interval can be easily described in terms of its end points, but you have to be careful as to whether the end points are part of the interval or not. When considering the mapping between intervals and time points, you have to distinguish between pointwise properties and gestalt properties of intervals.

Pointwise properties of intervals are those properties that are only true of an interval when they are true of every point in the interval. Pointwise properties of intervals are true of every subinterval. **Gestalt properties** are those that are true of an interval but may not true of any subinterval.

Example 8.1	The property of the robot "delivering Craig's parcel" may only be true of an interval if it's true of every point in the interval. If the robot was delivering Craig's parcel from 2:45 to 2:55 p.m., then it was delivering it from 2:51 to 2:53 p.m. It is thus a pointwise property. "The robot did all of its daily chores" may be true of the interval from 8:30 to 10:00 a.m., but not true of any subinterval if it actually took all of the hour and a half to do the chores. It is thus a gestalt property.

When modeling relations, you distinguish two basic types:

- **Static relations** are those relations whose value does not depend on time.
- **Dynamic relations** are relations whose truth values depends on time. Furthermore, the dynamic relations can be divided into **derived relations**, whose definition can be derived from other relations for each time, and **primitive relations**, whose truth value can be determine by considering previous times.

Which relations are derived or primitive is not a property of the domain, but is a modeling assumption. Often the world can be modeled in many different ways with different relations being primitive or derived. Examples of static and dynamic relations are given in the following section.

The Delivery Robot World

This chapter uses the example of the delivery robot world (page 12), but with parcels to deliver, doors that can be locked, and keys that can be used to unlock doors. The domain is shown in Figure 8.1. This domain is quite simple, yet is rich enough to demonstrate many of the problems in representing actions and in planning. For this domain you will consider a number of different action representations and a number of different planning algorithms. Note that we go beyond the path planning of Example 4.6 (page 121), in that we allow for more complex actions than just moving the robot.

To define the domain, you need to specify the individuals in the domain, the relations on those individuals, and the possible actions.

The individuals are rooms, doors, keys, parcels, and the delivery robot.

The actions available to the robot are to move from room to room, to pick up and put down objects (either keys or packages), and to unlock doors (if it's carrying the appropriate keys). In a more realistic domain, you would also expect the robot to be able to sense its environment (when a package arrives, when something is blocking its path) and to receive goals and requests from users.

The robot needs relations to represent its position, to represent the position of packages and keys and locked doors, and to represent what it's holding.

Relations

The relations that are true of situations or states are as follows:

- $at(Obj, Loc)$ is true in a situation if object Obj is at location Loc in the situation.

- $carrying(Ag, Obj)$ is true in a situation if agent Ag is carrying Obj in that situation.

- $sitting\_at(Obj, Loc)$ is true in a situation if object Obj is sitting on the ground (not being carried) at location Loc in the situation.

- $unlocked(Door)$ is true in a situation if door $Door$ is unlocked in the situation.

- $autonomous(Ag)$ is true if agent Ag can move autonomously. We assume that this is static in this domain.

- $opens(Key, Door)$ is true if key Key opens door $Door$. This is assumed to be static.

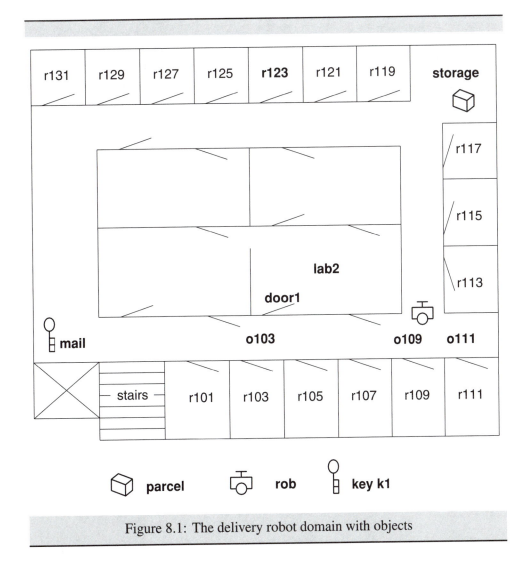

Figure 8.1: The delivery robot domain with objects

- *adjacent*(Pos_1, Pos_2) is true if position Pos_1 is adjacent to position Pos_2 so that the robot can move from Pos_1 to Pos_2 in one step.

- *between*($Door$, Pos_1, Pos_2) is true if $Door$ is between position Pos_1 and position Pos_2. If the door is unlocked , the two positions are adjacent. *between* is assumed to be static; you won't create new doors.

Actions

The actions are:

- *move*(Ag, $From$, To) is the action of agent Ag moving from location $From$ to location To. The agent can only carry out this action if it's sitting at location $From$ and location To is an adjacent location.

- *pickup*(Ag, Obj) is the action of agent Ag picking up object Obj. The agent can only carry out this action if it's at the same location that Obj is sitting.

- *putdown*(Ag, Obj) is the action of agent Ag putting down object Obj. It can only carry out this action if it's holding Obj.

- *unlock*(Ag, $Door$) is the action of agent Ag unlocking door $Door$. It can only carry out this action if it's outside the door and is carrying the key to the door.

One could imagine many other predicates, actions, and variants on these (e.g., that the agent has a maximum carrying capacity), but these are enough to demonstrate the problems of planning.

Initial World Description

Consider the initial world depicted in Figure 8.1. This can be described by

sitting_at(rob, o109).

sitting_at(parcel, storage).

sitting_at(k1, mail).

We assume the following static facts about the domain:

between(door1, o103, lab2).

opens(k1, door1).

autonomous(rob).

Derived Relations

As well as the above descriptions, the following clauses are true of every state. This is a simplified version of the domain depicted in Figure 4.1 (page 115):

adjacent(o109, o103).

$adjacent(o103, o109)$.

$adjacent(o109, storage)$.

$adjacent(storage, o109)$.

$adjacent(o109, o111)$.

$adjacent(o111, o109)$.

$adjacent(o103, mail)$.

$adjacent(mail, o103)$.

$adjacent(lab2, o109)$.

$adjacent(P_1, P_2) \leftarrow$
$\quad between(Door, P_1, P_2) \wedge$
$\quad unlocked(Door)$.

While the clauses are true of every state, what's unlocked can change from state to state, and so what's *adjacent* depends on the state. Thus *adjacent* is a derived relation. The *between* relation does not depend on the state; it's a *static* relation.

Assume that *at* is a derived relation defined as

$at(Obj, Pos) \leftarrow sitting\_at(Obj, Pos)$.

$at(Obj, Pos) \leftarrow carrying(Ag, Obj) \wedge at(Ag, Pos)$.

where $sitting\_at(Obj, Pos)$ is true if object *Obj* is at position *Pos* and is not being carried by an agent. This allows us to model the fact that when an agent moves, everything that it's carrying also moves. It also allows us to have a hierarchy of agents carrying each other; for example, the robot could be carrying a box that's in turn carrying an object. Wherever the robot is so is the box and the object.

8.2   Representations of Actions and Change

This section considers three different representations of actions and change. These can be fitted into the distinctions outlined in the introduction. These representations provide different views of the dynamics of the world. The first two, the STRIPS representation and the situation calculus, both give a state-based view, where actions map one state into another. In the STRIPS representation, the actions are external to the logic. In the situation calculus, the state is reified (see Section 8.1). These two representations have different representational capabilities; the situation calculus is more expressive, but the restrictions imposed by the STRIPS representation can be exploited in the planning algorithms. In contrast, the third representation, the event calculus, considers time explicitly (either continuous or discrete) and lets you axiomatize concurrent actions and events that are external to the agents as well as the agent's actions. This event calculus is used in Chapter 12 for building reactive robots.

The STRIPS Representation

One popular state-action representation with the actions external to the logic is the STRIPS representation. STRIPS, which stands for "STanford Research Institute Problem Solver," was the problem solver used in Shakey, one of the first robots built using AI technology. In this chapter we consider both the representation used for actions, the STRIPS representation, and the search strategy used for finding plans, the STRIPS planner (page 301). These should not be confused. You can use the STRIPS representation with other planners, and you can use the STRIPS planner with other representations.

The representation is used for specifying the following problem: Given a state and an action, determine whether the action can be carried out in that state and, if it can, determine what is true in the state resulting from carrying out the action.

First, divide the dynamic relations that describe the world into primitive and derived relations. You use normal clauses to determine the truth of derived relations from the true values of other relations in any given state. The STRIPS representation is used to determine the truth values of primitive relations based on the previous state and the action that changed the previous state into the current state.

The STRIPS representation is based on the idea that most things are not affected by a single action. For each action you state when an action is possible and what primitive relations are affected by the action. This incorporates the **STRIPS assumption**: All of the primitive relations not mentioned in the description of the action stay unchanged.

The **STRIPS representation** for an action consists of:

preconditions   A list of atomic formulae that need to be true for the action to occur

delete list   A list of those primitive relations no longer true after the action is performed

add list   A list of the primitive relations made true by the action

Example 8.2

In the delivery robot domain, the action *pickup(Ag, Obj)* can be defined by:

preconditions $[autonomous(Ag), Ag \neq Obj, at(Ag, Pos), sitting\_at(Obj, Pos)]$
delete list $[sitting\_at(Obj, Pos)]$
add list $[carrying(Ag, Obj)]$

Thus after the action the agent *Ag* is still *at* the position *Pos*, but is now carrying object *Obj*. Because of the definition of *at*, object *Obj* is still *at* position *Pos*, but is no longer *sitting_at* position *Pos* because it's being carried.

Example 8.3

The action *move*(*Ag*, *Pos*$_1$, *Pos*$_2$) can be defined by:

preconditions [*autonomous*(*Ag*), *adjacent*(*Pos*$_1$, *Pos*$_2$, *S*), *sitting_at*(*Ag*, *Pos*$_1$)]

delete list [*sitting_at*(*Ag*, *Pos*$_1$)]

add list [*sitting_at*(*Ag*, *Pos*$_2$)]

Note that the agent not only must be *at* position *Pos*$_1$, but also must be *sitting_at* position *Pos*$_1$, and thus not being carried. Because the *move* action does not affect the *carrying* relation or any other *sitting_at* relations, nothing else moves except those objects that the agent is carrying; they also end up at position *Pos*$_2$.

Example 8.4

Consider the initial world of Figure 8.1 described by

> *sitting_at*(*rob*, *o*109).
>
> *sitting_at*(*parcel*, *storage*).
>
> *sitting_at*(*k*1, *mail*).

In this world the action *pickup*(*rob*, *parcel*) is not possible as its preconditions don't hold.

The action *move*(*rob*, *o*109, *storage*) is possible. The world resulting from carrying out this action is described by

> *sitting_at*(*rob*, *storage*).
>
> *sitting_at*(*parcel*, *storage*).
>
> *sitting_at*(*k*1, *mail*).

In this world, the action *pickup*(*rob*, *parcel*) is possible. The world resulting from carrying out this action is described by:

> *sitting_at*(*rob*, *storage*).
>
> *carrying*(*rob*, *parcel*).
>
> *sitting_at*(*k*1, *mail*).

In this world, *move*(*rob*, *storage*, *o*109) is possible and the resulting world is

> *sitting_at*(*rob*, *o*109).
>
> *carrying*(*rob*, *parcel*).
>
> *sitting_at*(*k*1, *mail*).

Note that we only need to describe the world in terms of the primitive relations. Derived relations, such as *at* can be inferred from the primitive relations. For example, in this world *at*(*parcel*, *o*109) is true, thus capturing the intuition that the parcel moves where Rob moves when it is being carried.

One way to simplify the STRIPS representation is for any action A to always have a single precondition $poss(A)$, where $poss$ is a derived relation. You can then axiomatize $poss$ as you would any other relation.

Example
8.5

Suppose you decided to use the $poss$ relation, for the $move(Ag, Pos_1, Pos_2)$ relation. The action would have one precondition, namely $poss(move(Ag, Pos_1, Pos_2))$, and you would have the following axiom:

$$poss(move(Ag, Pos_1, Pos_2)) \leftarrow$$
$$autonomous(Ag) \wedge$$
$$adjacent(Pos, Pos_1, S) \wedge$$
$$sitting\_at(Ag, Pos_1).$$

This axiom is to be interpreted in each state, because whether $sitting\_at(Ag, Pos_1)$ is true depends on the state.

The Situation Calculus

The situation calculus is a state-action oriented representation in which the states and actions are reified.

You use constants and terms to denote particular states. There are two different ways to refer to states:

- $init$ is the initial state

- $do(A, S)$ is the state resulting from doing action A in state S, if it is possible to do action A in state S.

Example
8.6

If the initial situation depicted in Figure 8.1 (page 285) is state $init$, then

$$do(move(rob, o109, o103), init)$$

is the state resulting from Rob moving from position $o109$ in state $init$ to position $o103$. In this state, Rob is at $o103$, the key $k1$ is still at $mail$, and the package is at $storage$.

The situation

$$do(move(rob, o103, mail),$$
$$do(move(rob, o109, o103),$$
$$init)).$$

is one where the robot has moved from position $o109$ to $o103$ to $mail$, and is currently at mail. The situation

$$do(pickup(rob, k1),$$
$$do(move(rob, o103, mail),$$
$$do(move(rob, o109, o103),$$
$$init))).$$

is a situation in which Rob is at position *mail* carrying the key $k1$.

Note that it may not be possible to do action A in state S; its preconditions may not be satisfied; so $do(A, S)$ need not be a legal state. In other words, $do(A, S)$ is a partial function.

Example 8.7

The term $do(unlock(rob, door1), init)$ does not denote a state at all, as it's not possible for Rob to unlock the door when it is not at the door and doesn't have the key.

You can represent predicates either at the meta-level (using a $holds(R, S)$ relation to state that relation R holds in state S) or by adding an extra argument to each predicate stating when it's true. For different uses, either of these may be the more useful representation. Let's use the latter.

Example 8.8

The atom

$$at(rob, o109, init)$$

is true if the robot *rob* is at position $o109$ in the initial situation. The atom

$$at(rob, o103, do(move(rob, o109, o103), init))$$

is true if robot *rob* is at position $o103$ in the situation resulting from *rob* moving from position $o109$ to position $o103$ from the initial situation. The atom

$$at(k1, mail, do(move(rob, o109, o103), init))$$

is true if $k1$ is at position *mail* in the situation resulting from *rob* moving from position $o109$ to position $o103$ from the initial situation.

You can axiomatize a dynamic relation by specifying in which states it's true in. This is done inductively in terms of the structure of states.

- You specify what is true in the initial state using axioms with *init* as the situation parameter.
- Primitive relations are defined by specifying which instances are true in situations of the form $do(A, S)$ in terms of what holds in situation S.
- Derived relations are defined using clauses with a variable in the situation argument, as their truth in a situation depends on what else is true in the situation.
- Static relations can be defined without reference to the situation.

Example
8.9

The following axioms describe the initial situation of Figure 8.1 (page 285)

> $at(rob, o109, init)$.
>
> $at(parcel, storage, init)$.
>
> $at(k1, mail, init)$.

The *adjacent* relation is a dynamic, derived relation defined as

> $adjacent(o109, o103, S)$.
>
> $adjacent(o103, o109, S)$.
>
> $adjacent(o109, storage, S)$.
>
> $adjacent(storage, o109, S)$.
>
> $adjacent(o109, o111, S)$.
>
> $adjacent(o111, o109, S)$.
>
> $adjacent(o103, mail, S)$.
>
> $adjacent(mail, o103, S)$.
>
> $adjacent(lab2, o109, S)$.
>
> $adjacent(P_1, P_2, S) \leftarrow$
>
> > $between(Door, P_1, P_2) \wedge$
> >
> > $unlocked(Door, S)$.

Notice the free S variable; these clauses are true for all states. We can't omit the S because what rooms are adjacent depends on the state, because whether a door is unlocked depends on the state.

The *between* relation is static and does not need a state variable:

> $between(door1, o103, lab2)$.

You need to also axiomatize when actions are possible (in a similar way to the preconditions of the STRIPS representation). The relation $poss(A, S)$ is true if action A is possible in state S.

Example
8.10

An agent can always put down an object it's carrying:

> $poss(putdown(Ag, Obj), S) \leftarrow$
>
> > $carrying(Ag, Obj, S)$.

Example
8.11

For the *move* action, autonomous agents can move from a position where they are sitting to an adjacent position:

> $poss(move(Ag, Pos_1, Pos_2), S) \leftarrow$

$$autonomous(Ag) \land$$
$$adjacent(Pos_1, Pos_2, S) \land$$
$$sitting\_at(Ag, Pos_1, S).$$

Example 8.12

The most complicated precondition for an action is for the unlock action where the agent must be at the correct side of the door carrying the appropriate key:

$$poss(unlock(Ag, Door), S) \leftarrow$$
$$autonomous(Ag) \land$$
$$between(Door, P_1, P_2) \land$$
$$at(Ag, P_1, S) \land$$
$$opens(Key, Door) \land$$
$$carrying(Ag, Key, S).$$

Note that the *between* relation encodes which is the correct side of the door.

You also axiomatize how what's true in a state depends on the previous state and what action occurred between the states.

Example 8.13

The primitive *unlocked* relation can be defined by specifying how different actions can affect it being true. The door is unlocked in the state resulting from an unlock action, as long as the unlock action was possible:

$$unlocked(Door, do(unlock(Ag, Door), S)) \leftarrow$$
$$poss(unlock(Ag, Door), S).$$

In our simplified domain, doors can't be relocked, so no actions make *unlocked* no longer true. Thus it's true in a situation following an action if it was true before, and the action was possible:

$$unlocked(Door, do(A, S)) \leftarrow$$
$$unlocked(Door, S) \land$$
$$poss(A, S).$$

Such axioms are called **frame axioms**. These specify what remains unchanged during an action.

Example 8.14

Similarly, the *carrying* predicate can be defined as follows: An agent is carrying an object after picking up the object:

$$carrying(Ag, Obj, do(pickup(Ag, Obj), S)) \leftarrow$$
$$poss(pickup(Ag, Obj), S).$$

The only action that undoes the carrying predicate is the *putdown* action. Thus carrying is true after an action if it was true before and the action was not the action to putdown the object; thus you have the following frame axiom:

$$carrying(Ag, Obj, do(A, S)) \leftarrow$$
$$carrying(Ag, Obj, S) \wedge$$
$$poss(A, S) \wedge$$
$$A \neq putdown(Ag, Obj).$$

Example 8.15

$sitting\_at(Obj, Pos, S_1)$ is true in a state S_1 resulting from object Obj moving to Pos, as long as the action was possible:

$$sitting\_at(Obj, Pos, do(move(Obj, Pos_0, Pos), S)) \leftarrow$$
$$poss(move(Obj, Pos_0, Pos), S).$$

The other action that makes *sitting_at* true is the *putdown* action. An object is sitting at the point where the agent who put it down was. Thus you have the following axiom:

$$sitting\_at(Obj, Pos, do(putdown(Ag, Obj), S)) \leftarrow$$
$$poss(putdown(Ag, Obj), S) \wedge$$
$$at(Ag, Pos, S).$$

The only other time that *sitting_at* is true in a (non-initial state) is when it was true in the previous state, and it was not undone by an action. The only actions that undo *sitting_at* is a *move* action or a *pickup* action. This can be specified by the following frame axiom:

$$sitting\_at(Obj, Pos, do(A, S)) \leftarrow$$
$$poss(A, S) \wedge$$
$$sitting\_at(Obj, Pos, S) \wedge$$
$$\forall Pos_1 \ A \neq move(Obj, Pos, Pos_1) \wedge$$
$$\forall Ag \ A \neq pickup(Ag, Obj).$$

The Event Calculus

The third representation is a temporal representation, where time is either continuous or discrete, called the **event calculus**. The event calculus models how the truth value of predicates change because of events that occur at certain times.

Event E occurring at time T is written as $event(E, T)$.

For event E, we specify what E makes true and what E makes no longer true:

- *initiates*(E, P, T) is true if event E makes predicate P true at time T.
- *terminates*(E, P, T) is true if event E makes predicate P no longer true at time T.

Time T is a parameter to *initiates* and *terminates* because the effect of an event depends on what else is true at the time (e.g., the effect of attempting to unlock a door depends on the position of the robot and whether it's carrying the appropriate key).

For each predicate, you want to determine whether it holds (is true) at a particular time. Predicate P holds at time T if there was an event that occurred before T that made P true, and there was no event between these that made P no longer true. This can be specified as the following rule:

$$holds(P, T) \leftarrow$$
$$\quad event(E, T_0) \wedge$$
$$\quad T_0 < T \wedge$$
$$\quad initiates(E, P, T_0) \wedge$$
$$\quad \sim clipped(P, T_0, T).$$
$$clipped(P, T_0, T) \leftarrow$$
$$\quad event(E_1, T_1) \wedge$$
$$\quad terminates(E_1, P, T_1) \wedge$$
$$\quad T_0 < T_1 \wedge$$
$$\quad T_1 < T.$$

where *clipped*(P, T_0, T) is true if there is an event between times T_0 and T that makes P no longer true. $T_0 < T_1$ is true if time T_0 is before time T_1. Here $\sim$ is negation as failure (page 248).

You can axiomatize actions in terms of what properties they initiate and terminate.

Example 8.16

The *pickup* action initiates a *carrying* relation as long as the preconditions for *pickup* are true. You could also axiomatize what happens if a *pickup* is attempted when the preconditions don't hold; for example, the robot may get stuck, or maybe nothing happens.

$$initiates(pickup(Ag, Ob), carrying(Ag, Ob), T) \leftarrow$$
$$\quad poss(pickup(Ag, Ob), T).$$
$$terminates(pickup(Ag, Ob), sitting\_at(Ob, Pos), T) \leftarrow$$
$$\quad poss(pickup(Ag, Ob), T).$$
$$poss(pickup(Ag, Ob), T) \leftarrow$$
$$\quad autonomous(Ag) \wedge$$
$$\quad Ag \neq Obj \wedge$$

$holds(at(Ag, Pos), T) \wedge$

$holds(sitting\_at(Obj, Pos), T).$

The event calculus is used in Chapter 12 to represent agents that carry out actions and maintain state (page 452).

Comparing the Representations

The situation calculus and the STRIPS representation are similar in their adoption of the state space approach. The situation calculus is strictly more powerful than the STRIPS representation in that anything that can be stated in the STRIPS representation can be stated in the situation calculus, but not vice versa.

It's instructive to see how STRIPS can be axiomatized in the situation calculus. Here we'll use the $holds(C, S)$ relation to mean that C is true in situation S. The following is a representation of STRIPS in the situation calculus, specifying when primitive condition C is true immediately after action A is performed in situation W:

$holds(C, do(A, W)) \leftarrow$

$\quad preconditions(A, P) \wedge$

$\quad holdsall(P, W) \wedge$

$\quad add\_list(A, AL) \wedge$

$\quad member(C, AL).$

$holds(C, do(A, W)) \leftarrow$

$\quad preconditions(A, P) \wedge$

$\quad holdsall(P, W) \wedge$

$\quad delete\_list(A, DL) \wedge$

$\quad notin(C, DL) \wedge$

$\quad holds(C, W).$

where

- $preconditions(A, P)$ means that P is the list of preconditions of action A

- $add\_list(A, L)$ is true if L is the list of primitive predicates added by action A

- $delete\_list(A, L)$ means L is the list of primitive predicates deleted by action A

- $notin(C, DL)$ is true if C is not a member of list DL

- $holdsall(L, W)$ means all the conditions in the list L hold in world W

holdsall can be defined by:

$$holdsall([C|L], W) \leftarrow$$
$$holds(C, W) \wedge$$
$$holdsall(L, W).$$
$$holdsall([\ ]).$$

As previously mentioned, the STRIPS representation is not as powerful as the situation calculus. Consider the following example:

Example 8.17

Consider the *drop_everything* action where an agent drops everything it's carrying. This cannot be represented in STRIPS with the relation *sitting_at* as primitive (or with *at* as primitive), because the instances of the *sitting_at* relation to be deleted depend on what was being carried (and so cannot be specified in one step).

In the situation calculus, the following axiom can be added to the definition of *sitting_at* to say that everything the agent was carrying is now on the ground:

$$sitting\_at(Obj, Pos, do(drop\_everything(Ag), S)) \leftarrow$$
$$poss(drop\_everything(Ag), S) \wedge$$
$$at(Ag, Pos, S) \wedge$$
$$carrying(Ag, Obj, S).$$

You also need to add the exception to the frame axiom for *carrying* so that an agent is not carrying an object after a *drop_everything* action.

$$carrying(Ag, Obj, do(A, S)) \leftarrow$$
$$poss(A, S) \wedge$$
$$carrying(Ag, Obj, S) \wedge$$
$$A \neq drop\_everything(Ag) \wedge$$
$$A \neq putdown(Ag, Obj).$$

The event calculus is different from the other two representations in that it's based on a temporal representation rather than a state-based representation; the T argument to the predicates is a time and not a state. This means that you can reason about discrete or continuous time, and also reason about multiple agents carrying out actions in time. You specify when the different events by the different agents occurred. In the situation calculus and STRIPS you have to interleave the actions by different agents. The event calculus also lends itself to the case where times have durations. You can specify the exact times that events occur and measure the distance between times. Planning in the situation calculus can be done by a constructive proof of the existence of a situation (see the next section), whereas in the event calculus you need to hypothesize the occurrence of events to make the goal true (see abduction described on page 332).

8.3   Reasoning with World Representations

A planner is a problem solver that can produce **plans** (sequences of actions) to achieve some **goal**. The input to such a system is, typically, an initial world description, a specification of possible actions, and a goal description. The planner's task is to find a sequence of actions that will transform the initial world into one in which the goal description is true.

In this section we consider the problem of finding plans to satisfy a goal where the actions are described using the situation calculus or STRIPS.

The specification of the planning problem can be posed to a knowledge base consisting of situation calculus axioms defining the actions and the initial situation by issuing a query consisting of the goal with the situation term written as a free variable. When the goal is proved, answer extraction (see Section 2.6 on page 43) can be used to return a situation in which the goal will be true.

Example 8.18

If you want a plan for Rob to be carrying key $k1$, you can give the query

$$?carrying(rob, k1, S).$$

If you can find a proof for this, then answer extraction can be used to determine the plan. For example, given the initial situation of Figure 8.1, the above query has an answer

$$S = do(pickup(rob, k1),$$
$$\quad do(move(rob, o103, mail),$$
$$\quad\quad do(move(rob, o109, o103), init))).$$

This says that in the state resulting from Rob starting at state *init*, moving from $o109$ to $o103$ and then to *mail* and then picking up $k1$ is a state with Rob carrying $k1$.

Example 8.19

If you want a plan to achieve Rob holding the key $k1$ and being at $o103$, you can issue the query

$$?carrying(rob, k1, S) \wedge at(rob, o103, S).$$

This has an answer

$$S = do(move(rob, mail, o103),$$
$$\quad do(pickup(rob, k1),$$
$$\quad\quad do(move(rob, o103, mail),$$
$$\quad\quad\quad do(move(rob, o109, o103), init)))).$$

Note that when there is a solution to the query with $S = do(A, S_1)$, action A is the last action in the plan. If you were in state S_1 and you carried out action A, then you would be in state S.

Forward Planning

One of the simplest planning strategies is to treat the planning problem as a path planning problem in the **state-space graph**. In a state-space graph, nodes represent states, and arcs correspond to action from one state to another. The arcs coming out of a state correspond to all of the legal actions that can be carried out in that state. Thus the neighbors of a node correspond to the states that are reachable by a single action. A plan is a path from the initial state to a state that satisfies the goal condition.

In a *forward planner*, you search the state-space graph from the initial state looking for a state that satisfies a goal description. You can use any of the search strategies described in Chapter 4. This strategy produces one of the conceptually simplest planners.

A complete search strategy, such as A* or iterative deepening, is guaranteed to find a solution. The complexity of the search space is defined by the branching factor (page 116) of the search. The branching factor is the set of all possible actions at any state, which may be quite large. For the simple robot delivery domain, the branching factor is three for the initial situation and is up to six for other situations. When the domain becomes bigger, the branching factor increases and so the search space explodes. This complexity may be reduced by finding good heuristics (see Exercise 8.4), but the heuristics have to be very good to overcome the combinatorial explosion.

There is a certain amount of flexibility in your choice of a state representation; you can represent a state in terms of either:

(a) *A complete world description*, in terms of a set of axioms which imply everything which holds in the world. This could be in terms of primitive relations and general rules of how to compute nonprimitive predicates from primitive ones or as a database of all the facts true in a situation.

(b) *A path from an initial state*, for example, as the situation calculus name of the state. In this case the relations that hold in a certain state can be deduced from axioms that specify the effects of actions.

The difference between (a) and (b) amounts to the difference between computing a whole new world description for each world created, or by calculating what holds in a world as necessary. Alternative (b) may save on space (particularly if there is a complex world description) and will allow faster creation of a new node, but it will be slower to determine what actually holds in any given world. Another difficulty with option (b) is that the comparison of two world states is expensive.

Example
8.20

Consider how you might verify that the two worlds named by

$$do(move(rob, o103, o109), do(move(rob, o109, o103), init))$$

and

$$do(move(rob, storage, o109), do(move(rob, o109, storage), init))$$

are identical in that what's true in both worlds is the same. One method is to check every proposition that is true in the first world and to verify that it's true in the second, and vice versa. A second method is to determine the cumulative additions and deletions in the STRIPS representations. (See Exercise 8.5.)

Determining equality of states, as needed, for example, in multiple path checking to avoid looping, is more difficult for the second representation than for the first.

It's often more efficient to search a different space, namely the problem space, and to backward chain from the goal to be solved, as the branching factor in this space may be lower. This idea is pursued in the following sections. Note that the forward chaining approach is not simply extensible to backward planning because you don't typically have complete world descriptions either as the goal to solve or as the precondition to some action. Thus you cannot search the state space backwards.

Planning as Resolution

A second approach is to backward chain on the situation calculus rules or the situation calculus axiomatization of STRIPS, using the SLD resolution (page 49) and, say, depth-first search. This is analogous to using the situation calculus clauses as a logic program. Given a goal to achieve, the answer extracted for the situation term will correspond to a plan to achieve the goal.

The problem with this approach is that backward chaining on situation calculus rules typically never halts. You may always be able to use a frame rule by choosing an action that maintains the truth of a goal. Such a planner will never backtrack, and so will never find a solution unless entirely by luck.

Example
8.21

Suppose the subgoal is to achieve $carrying(rob, k1, S)$. The frame axiom for *carrying* (page 294), with the action $move(rob, o109, o103)$, could be used to derive $carrying(rob, k1, S)$. This axiom cannot be pruned *a priori*, as the action of moving Rob from $o103$ to $o109$ may be an appropriate last action—for example, when the agent also has to achieve the subgoal $at(parcel, o109, S)$. Similarly, the second last action could be $move(rob, o109, o103)$, with the same frame axiom. Unless there is some kind of loop detection, you will get into an infinite loop, repeatedly using frame axioms on these two actions. Even with a loop detection, the

> branching factor is large, as the frame axiom for *carrying* is applicable for every action except *putdown*(*rob*, *k*1).

A third approach is to consider a search method other than depth-first search—for example, breadth-first search or iterative deepening search. Then the problem is that the search is largely unconstrained by the goal. The use of frame axioms means that you have to consider the last action to be any of the actions that does not undo the goal. The branching factor of the search tree is thus very large. This is, however, an easy method to implement and is useful for debugging situation calculus axioms.

Intuitively you would like to consider only those actions that actually achieve a goal. The problem is that you cannot consider conjoined subgoals separately, as the following example shows.

Example 8.22

> Consider solving the the goal
>
> $$carrying(rob, parcel, S) \wedge in(rob, lab2, S).$$
>
> The S that achieves $carrying(rob, parcel, S)$ must also achieve $in(rob, lab2, S)$. If you were to solve the left conjunct without taking into account the fact that you also must achieve the right conjunct, the last actions to solve $carrying(rob, parcel, S)$ may need to be irrelevant to $carrying(rob, parcel, S)$ to also solve $in(rob, lab2, S)$. They may just use frame axioms.

Thus straightforward use of the generic SLD resolution on the goals is not such a good idea. The branching factor for this search is about the same as the branching factor for the forward search, because for each subgoal you have to consider every possible action that does not undo the subgoal.

The three planners in the following sections consider only those actions that actually achieve a goal. The result is a much smaller search space.

The STRIPS Planner

The basic idea behind the STRIPS planner is divide and conquer: to create a plan to achieve a conjunction of goals, create a plan to achieve one goal, and then create a plan to achieve the rest of the goals.

The idea behind the STRIPS planner is simple. To achieve a list of goals choose one of them to achieve. If it is not already achieved, choose an action that makes the goal true, achieve the preconditions of the action, carry out the action, and then achieve the rest of the goals.

Figure 8.2 gives a specification of the STRIPS planning algorithm. Suppose you want to achieve a list of goals from world W_0, where the list represents the conjunction of the goals. If the list is empty, then it is trivially solved, with the resulting world W_0.

% *achieve_all*(Gs, W_1, W_2) is true if W_2 is the resulting world after achieving every
% element of the list Gs of goals from the world W_1.

$achieve\_all$([ ], W_0, W_0).

$achieve\_all$($Goals$, W_0, W_2) $\leftarrow$
  $remove$(G, $Goals$, Rem_Gs) $\wedge$
  $achieve$(G, W_0, W_1) $\wedge$
  $achieve\_all$(Rem_Gs, W_1, W_2).

% *achieve*(G, W_0, W_1) is true if W_1 is the resulting world after achieving goal G from
% the world W_0.

$achieve$(G, W, W) $\leftarrow$
  $holds$(G, W).

$achieve$(G, W_0, W_1) $\leftarrow$
  $clause$(G, B) $\wedge$
  $achieve\_all$(B, W_0, W_1).

$achieve$(G, W_0, do($Action$, W_1)) $\leftarrow$
  $achieves$($Action$, G) $\wedge$
  $preconditions$($Action$, Pre) $\wedge$
  $achieve\_all$(Pre, W_0, W_1).

Figure 8.2: A simple STRIPS planner that uses STRIPS representation

If the list of goals is nonempty, first choose a goal G to achieve. Suppose achieving G
results in world W_1. You then achieve the rest of the goals from world W_1.

Suppose you want to achieve a goal G from world W_0. If G holds in world W_0, it is
trivially achieved and the resulting world is W_0. If G is a derived relation, find a clause
with G in the head and achieve all of the elements of the body. If G is primitive and does
not hold in W_0, choose an action A that makes G true and achieve the preconditions of
A. If achieving the preconditions of A results in world W_1, the resulting world which
achieves G is do(A, W_1), the world resulting from doing action A in world W_1.

In the algorithm of Figure 8.2:

- *achieve_all*($Goals$, $World_1$, $World_2$) means that $World_2$ is the resulting world
  after achieving each goal in the list of $Goals$ from $World_1$.

- *achieve*($Goal$, $World_1$, $World_2$) means that $World_2$ is the resulting world after
  achieving $Goal$ from $World_1$. The first clause is for goals that already hold in
  the world W. The second clause is for derived relations, and the third clause is

for primitive relations.

- *holds(Goal, World)* is true if *Goal* holds in *World* (defined on page 296).
- *achieves(Action, Goal)* means *Goal* is one of the elements on the add-list of *Action*.
- *clause(G, B)* is true if there is a clause in the knowledge base with derived relation *G* in the head such that *B* is the list of atoms in the body of the clause.
- *preconditions(Action, Pre)* is true if *Pre* is the list of preconditions of *Action*.

The first clause for *achieve* covers the case where *G* already holds. The second clause is for derived predicates, and the third clause is for primitive predicates.

Example 8.23

Consider the planning problem of Figure 8.1. The problem is to achieve a world in which *carrying(rob, parcel)* and *sitting_at(rob, lab2)* are true from the initial world depicted in Figure 8.1 (page 285). This can be done with the query:

$$?achieve\_all([carrying(rob, parcel), sitting\_at(rob, lab2)], init, S).$$

Suppose it chooses the first element of the list to achieve. First the algorithm tries to achieve *carrying(rob, parcel)*, which doesn't hold initially. Then it looks for an action to achieve *carrying(rob, parcel)* and finds the action *pickup(rob, parcel)*. It then tries to achieve the precondition of *pickup(rob, parcel)*, namely,

$$[autonomous(rob), sitting\_at(parcel, Pos), at(rob, Pos)].$$

The first condition holds initially and the second holds initially if you bind *Pos* to *storage*. So then you must achieve *at(rob, storage)*. This can be achieved by the action *move(rob, Pos$_1$, storage)*, with preconditions

$$[autonomous(rob), adjacent(Pos_1, storage), sitting\_at(rob, Pos_1)]$$

all of which are true initially if *Pos$_1$ = o109*. The algorithm then must try to achieve *sitting_at(rob, lab2)* from the state

$$do(pickup(rob, parcel, storage), do(move(rob, o109, storage), init)).$$

The final state that achieves both goals is

$$do(move(rob, o103, lab2),$$
$$do(unlock(rob, door1),$$
$$do(move(rob, mail, o103),$$
$$do(pickup(rob, k1, mail),$$
$$do(move(rob, o103, mail),$$
$$do(move(rob, o109, o103),$$
$$do(move(rob, storage, o109),$$
$$do(pickup(rob, parcel, storage),$$
$$do(move(rob, o109, storage),$$
$$init))))))))).$$

To read this as a plan, you should start at the end, and read the actions in reverse order: the first action is *move(rob, o109, storage)*, the second action is *do(pickup(rob, parcel, storage)*, and so on.

There is a major problem that can arise with the algorithm given above, namely that it's possible that subsequent actions undo previously achieved goals.

Example 8.24

If you had tried to achieve [*sitting_at(rob, lab2), carrying(rob, parcel)*], this algorithm, as presented, gets the wrong answer. First it achieves *sitting_at(rob, lab2)*, then it achieves *carrying(rob, parcel)*. But it can only achieve the second goal by undoing the first goal.

Two solutions have been proposed:

- You can protect subgoals, so that subsequent actions cannot undo them before they are needed, where subgoals are needed either because they are part of the top-level goal, such as *sitting_at(rob, lab2)*, or are needed as the precondition of an action. You may need to try different permutations of the subgoals to determine some order in which they can be carried out. The algorithm of Figure 8.2 does this if *remove* is implemented as a choice that can remove different elements. If this was done with the goal [*sitting_at(rob, lab2), carrying(rob, parcel)*], removing the first element of the list would lead to a protection violation, and so the algorithm would remove the second element, and the desired path would be found.

- An alternative method is to reachieve subgoals if they have been undone. In this example, with the goal [*sitting_at(rob, lab2), carrying(rob, parcel)*], you would first get *rob* into *lab2*, then move *rob* to *storage* to pick up the parcel, then move *rob* back into *lab2*.

It seems as though re-achieving subgoals wastes a lot of effort, so it is of interest to ask whether the first method can always work. Can you always reorder subgoals so that a plan can be found? The answer to this question is no, as the following example demonstrates:

Example 8.25

Consider a slight variation of the delivery robot example where the robot can only be carrying one item at a time (it cannot carry both the key and the parcel at the same time), and consider the goal

$$sitting\_at(rob, lab2) \land carrying(rob, parcel)$$

No matter which is achieved first, in order to achieve the other subgoal the first achieved goal must be undone. There is a neat solution (namely to unlock the door, then pickup the parcel then enter lab2), but neither protecting subgoals nor reachieving subgoals will allow this solution to be found.

The following section presents a method that can find the shortest plan in this example.

Regression

An approach to dealing with interacting goals is called **regression**. A regression planner keeps track of all of the subgoals that must be solved. Dependent subgoals are not attempted independently as they are in STRIPS.

A regression planner can best be seen in terms of graph searching. In this directed graph, nodes are labeled with sets of goals. Arcs correspond to actions.

A node labeled with goal set G has a neighboring node for each action A that achieves one of the goals in G. The neighbor corresponding to action A is labeled with the goals G_A that must be true immediately before the action A so that all of the goals in G are true immediately after A. The goals G_A are called the **weakest preconditions** for action A and goal list G. Let $wp(A, GL, WP)$ be true if WP is the weakest precondition that must occur immediately before the action A so that every element of the goal list GL is true immediately after the action. It's weakest in the sense that any other precondition must imply it.

For the STRIPS representation with all predicates primitive, $wp(A, GL, WP)$ is as follows:

- It is *false* if any element of GL is on delete list of action A. In this case it's impossible for every element of GL to be true immediately after action A.

- Otherwise the value of WP is

  $$preconds(A) \cup \{G \in GL : G \notin add\_list(A)\}.$$

  where, as before, $preconds(A)$ is the list of preconditions of action A and $add\_list(A)$ is the set of conditions added by action A (i.e., the add list of action A).

wp is defined in Figure 8.3. This is simplified in two ways. First, it doesn't take derived relations into account. Second, there can be repeated elements of the weakest precondition list.

Example 8.26

The weakest precondition for $[sitting\_at(rob, lab2), carrying(rob, parcel)]$ to be true after $move(rob, Pos, lab2)$ is that

$$[autonomous(rob), adjacent(Pos, lab2), sitting\_at(rob, Pos),$$
$$carrying(rob, parcel)]$$

is true immediately before the action. Thus the query

$$wp(move(rob, Pos, lab2),$$
$$[sitting\_at(rob, lab2), carrying(rob, parcel)], WP).$$

has answer $WP = [autonomous(rob), adjacent(Pos, lab2), sitting\_at(rob, Pos),$
$carrying(rob, parcel)]$.

% *wp*(*A*, *GL*, *WP*) is true if *WP* is the weakest condition that must hold immediately
% before action *A*, so that *A* is possible and every element of the goal list *GL* is true
% immediately after *A*.

    wp(*A*, [ ], *P*) ←
        preconditions(*A*, *P*).
    wp(*A*, [*G*|*R*], P_1) ←
        wp(*A*, *R*, P_0) ∧
        regress(*G*, *A*, P_0, P_1).

% *regress*(*G*, *A*, P_0, P_1) is true if P_1 is a list of conditions that extends P_0 to include
% enough conditions that must hold immediately before action *A* to guarantee that *G*
% is true immediately after *A*.

    regress(*G*, *A*, *P*, *P*) ←
        achieves(*A*, *G*).
    regress(*G*, *A*, *P*, [*G*|*P*]) ←
        not_on_add_list(*A*, *G*) ∧
        not_on_delete_list(*A*, *G*).

Figure 8.3: Finding weakest preconditions using the STRIPS representation

The situation calculus axioms can be seen as explicit regression rules, since each rule
says what must be true in a previous world for an action to take place.

A planning problem is posed to a regression planner by giving it the set of goals to
be true in the final state. It has solved the planning problem when it has a set of goals,
all of which are true in the initial state.

A regression planner works by keeping a set of goals to be achieved at any time.
The idea is to choose an action to occur last, then find the weakest precondition to
hold before that action, so that the set of goals will hold after the action. To do this
you *regress* the goals through the actions. Then you make a plan to solve the new set
of goals, and you continue until you reach goals that are true in the start state.

The algorithm is given in Figure 8.4, where the predicates have the following
meaning:

- *holdsall*(*GL*, *init*) means every element of the goal list *GL* hold in the initial
  world *init*.

- *consistent*(*GL*) means the goals *GL* are self-consistent (you don't want to try to
  achieve *GL* if they cannot be true in any world).

% *solve*(*GL*, *W*) is true if every element of goal list *GL* is true in world *W*.

$\quad$ *solve*(*GoalSet*, *init*) ←
$\qquad$ *holdsall*(*GoalSet*, *init*).
$\quad$ *solve*(*GoalSet*, *do*(*A*, *W*)) ←
$\qquad$ *consistent*(*GoalSet*) ∧
$\qquad$ *choose_goal*(*Goal*, *GoalSet*) ∧
$\qquad$ *choose_action*(*Action*, *Goal*) ∧
$\qquad$ *wp*(*Action*, *GoalSet*, *NewGoalSet*) ∧
$\qquad$ *solve*(*NewGoalSet*, *W*).

Figure 8.4: A regression planner

- *choose_goal*(*G*, *GL*) is true if *G* is an element of goal list *GL*.
- *choose_action*(*A*, *G*) is true in action *A* achieves goal *G* (e.g., for the STRIPS representation, *G* is on the add list of action *A*).
- *wp*(*A*, *GL*, *WP*) is true *WP* is the weakest precondition for goal list *GL* to be true immediately after action *A*.

A problem with the regression planner that you did not have with the previous planners is that you are dealing with conjunctions of goals, and these may be unachievable. The problem of deciding when a set of goals is not consistent (unachievable) is often not one which you can easily derive from the actions and their effects. For example, you may be required to know that an object cannot be at two different places at the same time, or that something cannot be both sitting at a place and being held. In order to do such pruning, this method requires some domain knowledge.

Loop detection can be easily incorporated into a regression planner. For the regression planner you don't need to visit exactly the same node in order to prune the search; you can prune the search if an ancestor of a node is a subset of the node, in which case it's simpler to solve the higher-level goal than the current goal.

Example 8.27

Consider the delivery robot example (page 284) and the goal:

$\quad$ [*carrying*(*rob*, *parcel*), *sitting_at*(*rob*, *lab2*)].

The possible actions that achieve one of the elements of this goal are

- *pickup*(*rob*, *parcel*), which achieves *carrying*(*rob*, *parcel*)
- *move*(*rob*, *P*, *lab2*), which achieves *sitting_at*(*rob*, *lab2*)

The last action in your plan should be one of these. If you have a plan for the above goal with any other action last, then it can be removed, and you still must have a legal plan for the goal, as the goal must be true before that action.

Consider that the last action is *move*(*rob*, *P*, *lab*2). For this to be the last action, immediately before this action:

$$[sitting\_at(rob, P), adjacent(P, lab2), carrying(rob, parcel)]$$

must all be true. By backward chaining on the derived relation *adjacent*(*P*, *lab*2), you can determine that there is only one room adjacent to *lab*2, and this needs *door*1 unlocked. Thus this goal can be reduced to

$$[sitting\_at(rob, o103), unlocked(door1), carrying(rob, parcel)]$$

all of which must be true immediately before Rob moves to lab2.

You can then consider the second last action, which must achieve one of these subgoals. If the chosen action is *unlock*(*door*1), regressing the goals gives the subgoals:

$$[at(rob, o103), carrying(rob, k1), carrying(rob, parcel)],$$

which must all be true immediately before *unlock*(*door*1).

Suppose that as the second last action you had chosen to achieve *at*(*rob*, *o*103) by carrying out the action *move*(*rob*, *P*, *o*103). Immediately before this action you would need the following subgoals to hold:

$$[sitting\_at(rob, P), adjacent(P, o103), unlocked(door1),$$
$$carrying(rob, parcel)]$$

This can also lead to a solution (e.g., where Rob unlocks the door, then goes to pick up the parcel), which may be an appropriate plan if Rob can only carry one item at a time (as in Example 8.25).

In making regression practical it's important to ensure that you do not waste time searching for a solution to an unachievable set of goals.

Example 8.28   Suppose you had the subgoal

$$[at(rob, storage), unlocked(door1)].$$

The action *unlock*(*door*1) achieves *unlocked*(*door*1), and the weakest precondition for the subgoal to hold after the action is that

$$[at(rob, storage), carrying(rob, k1), at(rob, o103)].$$

From the specification of the actions and the initial situation, it's nontrivial to prove that *rob* can only be at one position at any time (the action specifications alone don't imply this; it's only because Rob only starts at one position and every time a *sitting_at* relation is added, another is removed). If extra knowledge can be added to say that a subgoal in which Rob is at two different positions is unachievable, then you don't need to waste time searching for actions to this goal.

In regression, as in the STRIPS planner, you must commit to a particular total ordering of actions, even if there is no particular reason for one ordering over another. This commitment to a total ordering tends to increase the complexity of the search space if the actions don't interact much. For example, you must test each permutation of the actions, when it may be possible to show that all orderings don't succeed. In the next section we consider an algorithm that only adds ordering as required.

Partial-Order Planning

All the above planners enforce a total ordering on actions at all stages of the planning process. This means that they have to commit to an ordering of actions when adding them to a partial plan, even if there is no particular reason to put one action before another.

The idea of a **partial-order planner** is to have a partial ordering between actions, and only commit to an ordering between actions when forced to. This is sometimes also called a **nonlinear planner**, but this is a misnomer as such planners often produce a linear plan.

A partial ordering is a less-than relation that's transitive and asymmetric. A **partial-order plan** is a set of actions together with a partial ordering, representing a "before" relation on actions, such that any total ordering of the actions, consistent with the partial ordering, will solve the goal from the initial state.

Write $A_0 < A_1$ if action A_0 is before action A_1 in the partial order. This means that action A_0 must occur before action A_1.

For uniformity, let's treat *start* as an action that achieves the relations that are true in the initial state, and treat *finish* as an action whose preconditions are the goal to be solved. The pseudo-action *start* is before every other action, and *finish* is after every other action. The use of these as actions will mean that you don't have special cases for the initial situation and for the goals.

An action, other than *start* or *finish*, will be in a partial-order plan either to achieve one of the conjuncts in the final goal or to achieve one of the preconditions of an action in the plan. Because of the introduction of *finish*, in the shortest plan every action is there to achieve one of the preconditions of an action (possibly the *finish* action). Moreover, each precondition of an action in the plan will either be true in the initial state, and so achieved by *start*, or there will be an action in the plan that achieves it.

Let us put the previous paragraph more concretely. Each precondition P of an action A_1 in a plan will have an action A_0 associated with it such that A_0 achieves precondition P for A_1. This forms what's called a *causal link*. In the partial ordering you need action A_0 to be before action A_1, written $A_0 < A_1$. You must achieve the precondition for action A_1 before you carry out action A_1. Any other action A that deletes P—makes P false—must either be before A_0 or after A_1.

Informally a partial-order planner works as follows: Begin with the actions *start* and *finish* and the partial order *start* < *finish*. You maintain an agenda which is the list of preconditions of actions in the plan that need to be achieved. The elements of the agenda are called subgoals. Initially the agenda is the list of preconditions for *finish*, namely the final goal to be achieved. At each stage in the planning process, a subgoal P is selected from the agenda. P is the precondition for some action, say A_1. You then choose an action, A_0, to achieve P. That action is either already in the plan—it could be the *start* action, for example—or is a new action that's added to the plan. Action A_0 must be made to go before A_1 in the partial order. You then add a causal link that records that A_0 achieves P for action A_1. Any action in the plan that deletes P must go either before A_0 or after A_1. If A_0 is a new action, its preconditions are added to the agenda, and the process continues until the agenda is empty.

This is a nondeterministic procedure. There are two choice points over which you have to search:

- Which action is selected to achieve P
- Whether an action that deletes P goes before A_0 or after A_1

The following definitions will be used to make the above algorithm more concrete:

A **causal link** is a term of the form $cl(A_0, P, A_1)$, where A_0 and A_1 are actions and P is a proposition. P is a precondition of action A_1. Action A_0 achieves P. The causal link means that action A_0 achieves precondition P for A_1. P is said to be **supported** by action A_0.

A **partial plan** is a term of the form $plan(As, Os, Ls)$, where As is a list of actions, Os is a partial ordering on actions, and Ls is a list of causal links.

$plan(As_1, Os_1, Ls_1)$ is an **extension** of $plan(As_2, Os_2, Ls_2)$ if $As_2 \subseteq As_1$, $Os_2 \subseteq Os_1$, and $Ls_2 \subseteq Ls_1$.

Action A **threatens** causal link $cl(A_0, P, A_1)$ if action A deletes proposition P. $plan(As, Os, Ls)$ is **safe** if whenever action $A \in As$ threatens $cl(A_0, P, A_1) \in Ls$, the partial-ordering Os entails either $A < A_0$ or $A_1 < A$.

Associated with a partial plan is an **agenda** which is the set of subgoals for each unsupported precondition for each action in As. A **subgoal** is a term of the form $goal(P, A_1)$ where P is an atomic proposition that's a precondition for action A_1.

A **complete plan** is a safe partial plan with an empty agenda. A complete plan corresponds to a partial order plan as in Definition 8.3.

Figure 8.5 gives the top-level description of a partial-order planner. The general idea is that you maintain a safe partial plan. You have finished when the agenda is empty. If the agenda is not empty (there are unsupported preconditions for actions), you extend the current partial plan by removing a subgoal from the agenda, finding an

% *pop*(*CPlan*, *Agenda*, *FPlan*) is true if *CPlan* is a safe partial plan, with agenda
% *Agenda* and *FPlan* is a complete plan that's an extension of *CPlan*.

pop(*Plan*, [ ], *Plan*).

pop(*CPlan*, *Agenda*, *FPlan*) ←
    select(*Goal*, *Agenda*, $Agenda_1$) ∧
    solve_goal(*Goal*, *CPlan*, *NPlan*, $Agenda_1$, *NAgenda*) ∧
    pop(*NPlan*, *NAgenda*, *FPlan*).

% *select*(*Goal*, *Agenda*, *NewAgenda*) is true if *Goal* is selected from *Agenda* with
% *NewAgenda* the remaining elements. Here is one of many possible selection algo-
% rithms (see Exercise 8.8):

select(*G*, [*G*|*A*], *A*).

% *solve_goal*(*Goal*, *CPlan*, *NPlan*, *CAgenda*, *NAgenda*) chooses an action to solve
% *Goal*, updating plan *CPlan* to *NPlan* and agenda *CAgenda* to *NAgenda*.

solve_goal(*goal*(*P*, A_1), *plan*(*As*, *Os*, *Ls*),
        plan(*As*, *NOs*, [*cl*(A_0, *P*, A_1)|*Ls*]), *Ag*, *Ag*) ←
    member(A_0, *As*) ∧
    achieves(A_0, *P*) ∧
    add_constraint($A_0 < A_1$, *Os*, Os_1) ∧
    incorporate_causal_link(*cl*(A_0, *P*, A_1), *As*, Os_1, *NOs*).

solve_goal(*goal*(*P*, A_1), *plan*(*As*, *Os*, *Ls*),
        plan([A_0|*As*], *NOs*, [*cl*(A_0, *P*, A_1)|*Ls*]), *Ag*, *NAg*) ←
    achieves(A_0, *P*) ∧
    add_constraint($A_0 < A_1$, *Os*, Os_1) ∧
    add_constraint(*start* $< A_0$, Os_1, Os_2) ∧
    incorporate_action(A_0, *Ls*, Os_2, Os_3) ∧
    incorporate_causal_link(*cl*(A_0, *P*, A_1), *As*, Os_3, *NOs*) ∧
    add_preconds(A_0, *Ag*, *NAg*).

Figure 8.5: Partial-order planner

action to achieve this subgoal, and ensuring that the resulting partial plan is safe. An action to achieve the subgoal is either an existing action (the first clause of *solve_goal*) or a new action (the second clause of *solve_goal*).

To handle threats you need to enforce one of the two orderings in order to keep the partial plan safe. That is, if action A threatens causal link $cl(A_0, P, A_1)$, nondeterministically make either $A < A_0$ or $A_1 < A$.

Assume the following relations:

- *add_constraint*$(A_0 < A_1, Os, Os_1)$ is true if Os_1 is the minimal partial ordering made up of the ordering in Os together with $A_0 < A_1$. This fails if $A_0 < A_1$ is incompatible with Os. There are many ways this can be implemented. See Exercise 8.7.

- *achieves*(A, P) is true if action A makes predicate P true.

- *deletes*(A, P) is true if action A makes predicate P not true.

- *add_preconds*(A, Ag, NAg) is true if NAg is the agenda made up of the element of Ag together with element of the form $goal(P, A)$ for every precondition P of action A.

Example 8.29

Consider the goal

$$carrying(rob, parcel) \wedge sitting\_at(rob, lab2)$$

To solve this goal, invent pseudo-actions *start*, such that the initial conditions are achieved by *start*, and *finish* with preconditions corresponding to the goal to be solved, and call

$?pop(plan([start, finish], [start < finish], [\,]),$
$[goal(carrying(rob, parcel), finish),$
$goal(sitting\_at(rob, lab2), finish)], P).$

There are two subgoals on the agenda. Suppose that

$goal(carrying(rob, parcel), finish)$

is selected from the agenda. If you choose action *pickup*$(rob, parcel, P)$ to solve this goal, the resulting partial plan is

$plan([pickup(rob, parcel, P), start, finish],$
$[start < finish,$
$pickup(rob, parcel, P) < finish,$
$start < pickup(rob, parcel, P)],$
$[cl(pickup(rob, parcel, P), carrying(rob, parcel), finish)])$

% *incorporate_causal_link*(*CL*, *As*, *Os*, *NOs*) is true if partial-order *NOs* extends
% partial-ordering *Os* to ensure no action in *As* interferes with causal link *CL*.

    incorporate_causal_link(_, [ ], *Os*, *Os*).
    incorporate_causal_link(*CL*, [*A*|*RAs*], *Os*, *NOs*) ←
        protect(*CL*, *A*, *Os*, *Os*$_1$) ∧
        incorporate_causal_link(*CL*, *RAs*, *Os*$_1$, *NOs*).

% *incorporate_action*(*A*, *CLs*, *Os*, *NOs*) is true if partial-order *NOs* extends partial-
% ordering *Os* so action *A* doesn't interfere with any causal link in *CLs*.

    incorporate_action(_, [ ], *Os*, *Os*).
    incorporate_action(*A*, [*CL*|*Ls*], *Os*, *NOs*) ←
        protect(*CL*, *A*, *Os*, *Os*$_1$) ∧
        incorporate_action(*A*, *Ls*, *Os*$_1$, *NOs*).

% *protect*(*CL*, *Action*, *Os*, *NOs*) is true if *NOs* is a partial-ordering that extends partial
% ordering *Os* such action *A* doesn't interfere causal link *CL*.

    protect(*cl*(*A*$_0$, *P*, *A*$_1$), *Act*, *Os*, *Os*$_1$) ←
        threatens(*Act*, *cl*(*A*$_0$, *P*, *A*$_1$)) ∧
        enforce_order(*A*$_0$, *Act*, *A*$_1$, *Os*, *Os*$_1$).
    protect(*CL*, *Act*, *Os*, *Os*) ←
        ∼*threatens*(*Act*, *CL*).

% *enforce_order*(*A*$_0$, *A*, *A*$_1$, *Os*, *Os*$_1$) is true if *Os*$_1$ is a partial-ordering that extends
% partial ordering *Os* to make action *A* either before action *A*$_0$ or after action *A*$_1$.

    enforce_order(*A*$_0$, *A*, _, *Os*, *Os*$_1$) ←
        add_constraint(*A* < *A*$_0$, *Os*, *Os*$_1$).
    enforce_order(_, *A*, *A*$_1$, *Os*, *Os*$_1$) ←
        add_constraint(*A*$_1$ < *A*, *Os*, *Os*$_1$).

% *threatens*(*Act*, *CL*) is true if action *Act* threatens causal link *CL*.

    threatens(*Act*, *cl*(*A*$_0$, *P*, *A*$_1$)) ←
        Act ≠ *A*$_0$ ∧ *Act* ≠ *A*$_1$ ∧ *deletes*(*Act*, *P*).

Figure 8.6: Threat handler for the partial-order planner

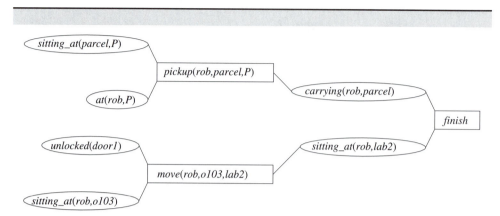

Rectangles represent actions and ovals represent conditions. The ovals joined to the left of an action are the preconditions of the action. The rectangle joined to the left of a condition represent an action that achieves the condition. Each condition together with the action joined to its left and the action joined to its right corresponds to a causal link.

Figure 8.7: Partial elaboration of the partial-ordered planning example

and the corresponding agenda is

> [goal(*sitting_at*(*rob*, *lab2*), *finish*),
>   goal(*sitting_at*(*parcel*, *P*), *pickup*(*rob*, *parcel*, *P*)),
>   goal(*at*(*rob*, *P*), *pickup*(*rob*, *parcel*, *P*))].

Suppose next that goal(*sitting_at*(*rob*, *lab2*), *finish*) is selected from the agenda and solved. The resulting partial plan is

> plan([*move*(*rob*, *o*103, *lab2*), *pickup*(*rob*, *parcel*, *P*), *start*, *finish*],
>   [*start* < *finish*,
>   *pickup*(*rob*, *parcel*, *P*) < *finish*,
>   *start* < *pickup*(*rob*, *parcel*, *P*),
>   *move*(*rob*, *o*103, *lab2*) < *finish*,
>   *start* < *move*(*rob*, *o*103, *lab2*)],
>   [*cl*(*move*(*rob*, *o*103, *lab2*), *sitting_at*(*rob*, *lab2*), *finish*),
>   *cl*(*pickup*(*rob*, *parcel*, *P*), *carrying*(*rob*, *parcel*), *finish*)])

with the corresponding agenda

> [goal(*sitting_at*(*parcel*, *P*), *pickup*(*rob*, *parcel*, *P*)),

$$goal(at(rob, P), pickup(rob, parcel, P)),$$
$$goal(unlocked(door1), move(rob, o103, lab2)),$$
$$goal(sitting\_at(rob, o103), move(rob, o103, lab2))].$$

This stage is depicted in Figure 8.7. Note that there is no ordering constraint between the two actions.

When $goal(sitting\_at(parcel, P), pickup(rob, parcel, P))$ is chosen from the agenda, the pseudo-action *start* that's already in the plan can be used to solve this goal, with $P = storage$. This would mean that the causal link

$$cl(start, sitting\_at(parcel, P), pickup(rob, parcel, P))$$

is added to the plan.

When $goal(at(rob, storage), pickup(rob, parcel, storage))$ is chosen from the agenda, the action $move(rob, P_1, storage)$ is added to the plan along with the causal link

$$cl(move(rob, P_1, storage), at(rob, storage), pickup(rob, parcel, storage))$$

This causal link means that any action that deletes $at(rob, storage)$ must come either before $move(rob, P_1, storage)$ or after $pickup(rob, parcel, storage)$.

The above algorithm has glossed over one important detail that you need to build a real planner. It's sometimes necessary to do some action twice. The above algorithm will not work in this case, as it will try to find a partial ordering with both instances of the action occurring at the same time. You have to be more careful about what's being ordered. The ordering should be between action instances, and not actions themselves. To handle this, you can assign an index to each instance of an action in your plan, and the ordering is on the action instance indexes, and not the actions themselves. Appendix C.5 gives code that incorporates instances of actions and uses depth-bounded search (where the depth is in terms of the number of action instances in the final plan).

8.4   References and Further Reading

Both the STRIPS representation and the STRIPS algorithm are due to Fikes & Nilsson (1971).

The situation calculus is due to McCarthy & Hayes (1969). The form of the frame axioms presented here can be traced back to Kowalski (1979b), Schubert (1990), and Reiter (1991). There have been many other suggestions about how to solve the frame problem, which is the problem of concisely specifying what does not change during an action. For a critique of these see Hanks & McDermott (1987). Shanahan (1997) provides an excellent introduction to the issues involved in representing change and to the frame problem in particular.

The event calculus is due to Kowalski & Sergot (1986). Shanahan (1997) discusses more recent advances as well as a comparison with the situation calculus.

There is much ongoing research into how to plan sequences of actions. Yang (1997) presents a textbook overview of planning. For a collection of classic papers see Allen, Hendler & Tate (1990).

Forward planning has been used successfully for planning in the blocks world, where some good heuristics have been identified (Bacchus & Kabanza, 1996). (See Exercise 8.4).

Regression is due to Waldinger (1977). The use of weakest preconditions is due to Dijkstra (1976), where it was used to define the semantics of imperative programming languages. This should not be too surprising because the commands of an imperative language are actions that change the state of the computer.

Partial-order planning was introduced in Sacerdoti's (1975) NOAH and followed up in Tate's (1977) NONLIN system, Chapman's (1987) TWEAK algorithm, and McAllester & Rosenblitt's (1991) SNLP algorithm. See Weld (1994) for an overview of partial-order planning and see Kambhampati, Knoblock & Yang (1995) for a comparison of the algorithms. The version presented here is basically SNLP (but see Exercise 8.9).

See Wilkins (1988) for a discussion on practical issues in planning and see McDermott & Hendler (1995) and associated papers for a recent overview. Some of the most efficient current planners are UCPOP (Weld, 1994), Graphplan (Blum & Furst, 1995), and the use of GSAT (page 160) and variants for planning (Kautz & Selman, 1996).

8.5   Exercises

Exercise 8.1   Write a complete description of the limited robot delivery world, and then draw a state-space representation that includes at least two instances of each of the blocks-world actions discussed in this chapter. Notice that the number of different arcs depends on the numbers of instances of actions.

Exercise 8.2   Change the axiomatization for the delivery robot world (page 284) so that

(a) the agent can only carry one object at a time;

(b) the agent can carry a box in which it can place objects (so it can carry the box and the box can carry other objects).

Using this new axiomatization try to solve the goal of the parcel sitting at lab2, assuming that the box starts at room $o111$.

Exercise 8.3	For the other actions, complete the axiomatization of the delivery robot world (page 8.1) using the situation calculus.

Using a depth-bounded meta-interpreter, run the axioms as a logic program. How does the running time grow with the size of the plan?

Exercise 8.4	Can you think up a good heuristic for a forward planner to use in the robot delivery domain? Implement it. How well does it work?

Exercise 8.5	Suppose you have a STRIPS representation for actions a_1 and a_2, and you want to define the STRIPS representation for the composite action $a_1; a_2$, which means that you do a_1 then do a_2.

(a) What is the add list for this composite action?

(b) What is the delete list?

(c) What are the preconditions for this composite action?

(d) Using the delivery robot domain, give the STRIPS representation for the composite action $move(Ag, Pos_1, Pos_2); pickup(Ag, Obj)$.

(e) Give the STRIPS representation for the composite action $move(Ag, Pos_1, Pos_2); move(Ag, Pos_2, Pos_1)$.

(f) Give the STRIPS representation for the composite action $move(Ag, Pos_1, Pos_2); pickup(Ag, Obj); move(Ag, Pos_2, Pos_1)$ make up of three primitive actions.

Exercise 8.6	The STRIPS representation allows the definition of an algorithm for detecting cycles in a forward planner. See if you can think of what it is. *Hint:* Consider the composite action (Exercise 8.5) consisting of the last k actions at any stage.

Exercise 8.7	In order to implement the relation $add\_constraint(A_0 < A_1, Os, Os_1)$ used in the partial-order planner, you have to choose a representation for a partial ordering. Implement the following as different representations for a partial ordering:

(a) Represent a partial ordering as a set of less-than relations that entail the ordering—for example, as the list $[1 < 2, 2 < 4, 1 < 3, 3 < 4, 4 < 5]$.

(b) Represent a partial ordering as the set of all the less-than relations entailed by the ordering—for example, as the list $[1 < 2, 2 < 4, 1 < 4, 1 < 3, 3 < 4, 1 < 5, 2 < 5, 3 < 5, 4 < 5]$.

(c) Represent a partial ordering as a list of terms of the form $after(E, L)$, where E is an element in the partial ordering and L is the list of all elements that are after E in the partial ordering. For every E there is a unique term of the form $after(E, L)$. An example of such a representation is $[after(1, [2, 3, 4, 5]), after(2, [4, 5]), after(3, [4, 5]), after(4, [5]), after(5, [ ])]$.

For each of these representations, how big can the partial ordering be? How easy is it to check for consistency of a new ordering? How easy is it to add a new less-than ordering constraint? Which do you think would be the most efficient representation? Can you think of a better representation?

Exercise 8.8   The selection algorithm used in the partial-order planner is not very sophisticated. It may be sensible to order the subgoals selected. For example, in the robot world, the robot should try to achieve a *carrying* subgoal before an *at* subgoal because it may be sensible for the robot to try to be carrying an object as soon as it knows that it needs to be carrying it. However, the robot does not want to necessarily move to a particular place unless it's carrying everything it needs to carry. Implement a selection algorithm that incorporates such a heuristic. Does this selection heuristic actually work better than selecting the last added subgoal? Can you think of a general selection algorithm that does not require each pair of subgoals to be ordered by the knowledge engineer?

Exercise 8.9   The SNLP algorithm is the same as the partial-order planner presented here but defines a threat in a different way:

$$threatens(Act, cl(A_0, P, A_1)) \leftarrow$$

$$Act \neq A_0 \wedge$$

$$Act \neq A_1 \wedge$$

$$deletes(Act, P).$$

$$threatens(Act, cl(A_0, P, A_1)) \leftarrow$$

$$Act \neq A_0 \wedge$$

$$Act \neq A_1 \wedge$$

$$achieves(Act, P).$$

This enforces *systematicity*, which means that for every linear plan there is a unique partial-ordered plan. Explain why systematicity may or may not be a good thing (e.g., discuss how it changes the branching factor or reduces the search space). Test the different algorithms on different domains.

Chapter 9

Assumption-Based Reasoning

9.1   Introduction

So far we have examined reasoning strategies which treat reasoning as determining what logically follows from the knowledge base. This chapter considers another paradigm for reasoning, namely, assumption-based reasoning. Rather that just doing deduction from facts, you can reason from the facts together with a set of assumptions you are prepared to make. There are four main applications of this idea:

- In default reasoning, the user provides defaults which are assumptions of normality that can be used to predict what is true. For example the default "Mary is interested in articles about AI" can be used to predict that Mary will be interested in a particular article. There may be exceptions to this general rule, but you want to make the prediction in the absence of knowing that the particular article is uninteresting to Mary.

- In model-based **diagnosis** and **recognition**, through what is called abduction, you hypothesize causes to explain observations about the world. For example, in diagnosis, given the symptoms of a patient, you hypothesize diseases that the person may have that would account for the symptoms.

- In **design**, you hypothesize components that provably fulfill some design goals and are feasible. Given a design goal, you can select a design that is possible and meets the design goal. For example, the robot can design a path that will get it from one place to another; there may be many possible paths, but the robot can select whichever it likes. It could select the shortest one, or the one that provides the most information. The infobot could design a presentation of the material it has available; it needs to construct a display of the information that will convey the appropriate information, and it is free to select a display that best suits its needs.

- In inductive learning, you generate hypotheses to account for a set of training examples or previous experiences. For example, from observing a user selecting articles based on keywords, the infobot can hypothesize what sort of article the user is interested in. Learning is covered in detail in Chapter 11.

Two different tasks use assumption-based reasoning:

Design   In a design task the aim is to design an artifact or plan. The designer can select whichever design they like the best that satisfies the design criteria.

Recognition   In a recognition, prediction, or diagnosis task, the aim is to find out what is true based on observations. If there are a number of possibilities, it's not up to the recognizer to select the one they like best. The underlying reality is fixed, and the aim is to find out what it is.

These distinctions are highlighted in the following examples:

Example 9.1

For the diagnostic assistant, there may be many consistent assumptions about what is wrong with a patient or what is wrong with a device. The diagnostician doesn't get to select what disease the patient has, or what is wrong with the device. They can observe the symptoms and hypothesize what may be wrong. Unless there is enough information to rule out all but one hypothesis, the diagnostician at best can conclude a disjunction of possible diseases or faults (perhaps with a probability distribution over them; see Chapter 10).

What the diagnostician gets to select is what treatment to prescribe. In particular, they get to design a treatment strategy. This strategy can include tests as well as actions that are contingent on the outcome of tests. The diagnostician can't select a disjunction of treatments to administer; they can't give the patient drug A or drug B without actually selecting one of the drugs to give the patient. An agent can't do a disjunction of actions without actually doing one of them. This act of selecting a treatment is an instance of design.

Example 9.2

You may want the infobot to design or recognize meeting times based on its knowledge of the constraints of the participants.

Suppose the possible meeting times based on everyone's constraints are 9:00 a.m., 10:00 a.m., or 12:00 noon. Consider the different tasks for this example:

Design   In a design task, there are three possible meeting times. It is up to the designer to select whichever time it likes. The designer can't select the disjunction 9:00 a.m. or 10:00 a.m., as the meeting must happen at one of these times. At some stage the choice has to be made, and the agent or mechanism that makes the choice is the designer.

Recognition   In a recognition task, you can't arbitrarily select the meeting time. All you can conclude is that it is at one of the three times; the meeting is at 9:00 or 10:00 or 12:00. Some of these may be more plausible that others, but unless you are the designer of the meeting time you can't arbitrarily select the meeting time.

The difference between design and recognition can often be understood by considering the question, "Who selects the assumptions?" There is a difference between a design task, where the designers can select whichever design they like, and a recognition task where the aim is to recognize the underlying reality. Recognizers don't get to select, for example, what disease a patient has. In the latter case you can think of either the worst case, where it is as though an adversary gets to select the assumptions, or an average case, where nature gets to select the assumptions; all you can do is make educated guesses at what is true in the world.

This chapter considers both design and recognition in terms of assumption-based reasoning. First we define a general framework for assumption-based reasoning, then we discuss two strategies for using this framework, namely, default reasoning and abduction.

9.2   An Assumption-Based Reasoning Framework

Logic tells us the consequences of some premises. Up to now, this book has considered cases where the premises are axioms given as true. This chapter considers premises to be facts plus assumptions. The basic property that you want assumptions to have is that you don't want to make assumptions that you know are wrong. You know the assumptions are wrong if they are inconsistent with the facts.

In the framework presented in this chapter, the system is provided with a set of facts and a set of possible hypotheses. Instances of the possible hypotheses can be assumed if they are consistent with the facts. This is sometimes called *hypothetical reasoning*.

More formally, the representation and reasoning system is given two sets of formulae:

- F is a set of closed formula called the **facts**. These are formulae that are given as true in the world. Assume the facts are consistent. Here we consider the facts to be Horn clauses.

- H is a set of formulae called the **possible hypotheses** or **assumables**. Ground instance of the possible hypotheses can be assumed if consistent. A ground instance of a hypothesis is obtained by substituting a ground term for each of the free variables appearing in it.

A **scenario** of $\langle F, H \rangle$ is a set D of ground instances of elements of H such that $F \cup D$ is satisfiable. The elements of D are implicitly conjoined. D is impossible given F if $F \cup D \models \textit{false}$ or equivalently if $F \models \neg D$. $F \cup D$ is **satisfiable** if there is a model in which F and D are both true. This means that $F \not\models \neg D$, which is read as "F doesn't entail not D."

An **explanation** of g from $\langle F, H \rangle$ is a scenario that, together with F, implies g.

That is, an explanation of formula g is a set D of ground instances of elements of H such that

$$F \cup D \models g \text{ and}$$

$$F \not\models \neg D$$

A **minimal explanation** of g from $\langle F, H \rangle$ is an explanation of g from $\langle F, H \rangle$ such that no strict subset is also an explanation of g from $\langle F, H \rangle$. That is, D is a minimal explanation of g if it is an explanation of g, and for no $D' \subset D$ is D' an explanation for g.

Another way to characterize assumption-based reasoning is to consider making as many assumptions as possible without making them inconsistent:

An **extension** of $\langle F, H \rangle$ is the set of logical consequences of F together with a maximal scenario of $\langle F, H \rangle$. A scenario D is maximal if there is no scenario D' that is a strict superset of D.

It is easy to see the following:

Proposition 9.1   Formula g can be explained if and only if g is in an extension.

One important distinction is between whether the g is known to be true, or whether it is to be determined. This gives us two different strategies:

Default reasoning   Where the truth of g is unknown and is to be determined. The explanations in this case are of interest only because of what they explain. In this case you often speak in terms of arguments: An explanation for g corresponds to an argument for g.

Abduction   Where the g is given, either as an observation in a recognition task or as a design goal in a design task. In this case it is the explanations that are of interest.

These two strategies are presented in the following two sections, and ways they can be combined are presented in Section 9.5. Section 9.6 discusses how the strategies can be implemented.

9.3   Default Reasoning

Default reasoning is ubiquitous in common-sense reasoning. Virtually every state-ment that isn't a mathematical definition has exceptions (even this one). When telling a computer some information, you don't want to have to enumerate all of the exceptions, even if you could think of them all. It is more reasonable that you tell the computer general knowledge, and then modularly add exceptions. You want the computer to be able to use the general knowledge to make conclusions in the absence of knowledge that a particular case is exceptional.

This type of reasoning can't be done using classical logic (e.g., the definite clause language introduced in Chapter 2 or the first-order predicate calculus (page 270)) because classical logic is **monotonic**: If proposition g logically follows (page 36) from axioms A, then it also follows from any superset of A. However, default reasoning is **nonmonotonic**: when you add the fact that something is exceptional, you may no longer be able to conclude what you could before.

Making Normality Assumptions

The simplest way to use the assumption-based framework for default reasoning is to consider H as assumptions of normality. F specifies what follows from the normality assumptions and exceptions to the normality assumptions.

An explanation for g can be seen as an **argument** for g based on normality as-sumptions. The consistency requirement means that you don't want to use arguments that you know are wrong.

Example 9.3

Suppose you want the infobot to be able to represent preferences about users and want users to be able to "tell" the infobot about the kind of articles they are interested in. A user should be able to tell the infobot the general rules, and be able to modularly add exceptions and other rules. For this domain it is imperative that users not be forced to give all their rules at the start, because they typically don't know what kind of information is in an information space, and can't be expected to anticipate all eventualities.

Suppose a user wants to represent the default "I am generally interested in articles about AI." This is an assumption that an infobot can use for each article about AI. The infobot can assume that each article is interesting to the user, unless there is some reason the infobot knows that it isn't interesting. This is a default because the user may also, for some reason, tell the infobot that they are not interested in articles about formal logic, even though articles about formal logic may be about

AI. Suppose that the user also says that articles that contain introductory questions are not interesting. You can represent this using the possible hypothesis:

$$H = \{int\_ai(X)\},$$

where the intended interpretation of $int\_ai(X)$ is that X is an interesting article if it is about AI. This is just an assumption as some X's may be uninteresting even though they are about AI. $int\_ai(X)$ is false when X is about AI and not interesting. Making it a possible hypothesis means you can assume it is true unless it's inconsistent to do so.

The corresponding facts may be:

$$interesting(X) \leftarrow about\_ai(X) \wedge int\_ai(X).$$

$$about\_ai(X) \leftarrow about\_fl(X).$$

$$about\_ai(X) \leftarrow about\_ml(X).$$

$$false \leftarrow about\_fl(X) \wedge interesting(X).$$

$$false \leftarrow intro\_question(X) \wedge interesting(X).$$

where $about\_ai(X)$ is true if article X is about AI; $about\_fl(X)$ is true if article X is about formal logic; $about\_ml(X)$ is true if article X is about machine learning and so, by definition, about AI; $intro\_question(X)$ is true if article X is an introductory question.

The first rule is of most interest. It says that if an article is about AI, and the normality assumption is true of that article, then the article is interesting.

Suppose the infobot wants to reason about particular articles it encounters which have the following properties:

$$about\_ai(article\_23).$$

$$about\_fl(article\_77).$$

$$about\_ml(article\_34).$$

$$about\_ml(article\_99).$$

$$intro\_question(article\_99).$$

This example is depicted in Figure 9.1.

$interesting(article\_23)$ can be explained using $\{int\_ai(article\_23)\}$, which is consistent with the facts and, with F, implies $interesting(article\_23)$. You can also explain $interesting(article\_34)$, assuming $\{int\_ai(article\_34)\}$. You can't explain $interesting(article\_77)$, as articles about formal logic are not interesting; in particular, $F \models \neg int\_ai(article\_77)$. Similarly, $interesting(article\_99)$ can't be explained because introductory questions are not interesting; in particular $F \models \neg int\_ai(article\_99)$.

Notice that you conclude $article\_34$ is interesting, but don't conclude $article\_99$ is interesting, even though everything known about $article\_34$ is also known about

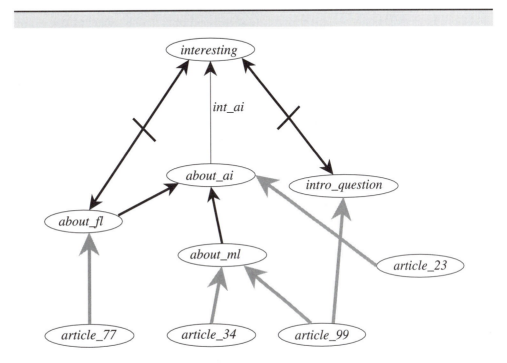

- Thick lines represent implication.
- Thin (labeled) lines represent defaults, which are rules that contain possible hypotheses. These possible hypotheses are given as the labels to the arcs.
- Light lines represent membership of an individual in a class.
- Crossed lines indicate that the two properties are disjoint; the truth of one entails the negation of the other.

Figure 9.1: Diagrammatic representation of the defaults from Example 9.3

article_99. This ability to no longer be able to derive a conclusion based on more evidence is why this is referred to as nonmonotonic reasoning. The set of conclusions doesn't grow monotonically with the knowledge as they do with deduction from a knowledge base.

Overriding Assumptions

When you have multiple defaults, it is often the case that there are multiple competing arguments. This section shows how arguments can be undermined to give preferences for some defaults over others.

Example 9.4

Suppose you change Example 9.3 to make it

- a default, as opposed to a fact, that articles about formal logic are not interesting, and

- a default that introductory questions are not interesting.

Let's axiomatize these defaults in the same way as the single default in Example 9.3. The hypotheses that can be used in an argument are

$$H = \{unint\_fl(X),$$
$$unint\_iq(X),$$
$$int\_ai(X)\},$$

where *unint_fl(X)* is the assumption that X is uninteresting if it is about formal logic, *unint_iq(X)* is the assumption that X is uninteresting if it is an introductory question, and *int_ai* is the same as in Example 9.3.

The corresponding facts are:

$$interesting(X) \leftarrow about\_ai(X) \wedge int\_ai(X).$$
$$about\_ai(X) \leftarrow about\_fl(X).$$
$$about\_ai(X) \leftarrow about\_ml(X).$$
$$false \leftarrow about\_fl(X) \wedge unint\_fl(X) \wedge interesting(X).$$
$$false \leftarrow intro\_question(X) \wedge unint\_iq(X) \wedge interesting(X).$$

Suppose you encounter the same articles as in Example 9.3.

There are two arguments about whether *article_77* is interesting:

- You can explain ¬*interesting(article_77)* using {*unint_fl(article_77)*}, which is consistent with the facts and which, along with the facts, implies that *article_77* isn't interesting. This is an argument that *article_77* isn't interesting because it is about formal logic and articles about formal logic are generally not interesting.

- You can explain *interesting(article_77)* using {*int_ai(article_77)*}. This is an argument that says that you expect *article_77* to be interesting because you know that *article_77* is about AI and that articles about AI are generally interesting.

This example shows us that there may be multiple explanations as to why you expect some outcome or its negation. This should be regarded as a natural consequence of using assumptions; there may be many assumptions that are applicable.

It seems that the second argument should not be applicable to *article_77* because the fact that *article_77* is about formal logic should undermine the argument that *article_77* is interesting because it is about AI. This is an instance of preference for **more specific** defaults (see Exercise 9.3). The argument that *article_77* is interesting because it is about AI can be undermined by adding the fact

$$false \;\leftarrow\; about\_fl(X) \wedge int\_ai(X)$$

to *F*. This is known as a **cancellation rule**. It cancels the *int_ai(X)* default when *about_fl(X)* is true. It has the effect of making the second explanation inconsistent. You can't use the "articles about AI are interesting" default for articles that are about formal logic.

Consider a case which is not an instance of specificity. There are two arguments about whether *article_99* is interesting:

- You can explain that *article_99* isn't interesting using {*unint_iq(article_99)*}. This is an argument that *article_99* isn't interesting because it is an introductory question and articles that are introductory questions are generally not interesting.

- You can explain *interesting(article_99)* using {*int_ai(article_99)*}. This is an argument that says that you expect *article_99* to be interesting because you know that *article_99* is about AI and that articles about AI are generally interesting.

Neither default used is more specific than the other. It should be up to the user to decide which default has preference when they compete. Suppose the user wants the introductory question default to win when it conflicts with the default about AI articles. This can be represented by allowing the fact that an article is an introductory question to undermine the argument that AI articles are interesting, by adding to *F* the fact

$$false \;\leftarrow\; intro\_question(X) \wedge int\_ai(X).$$

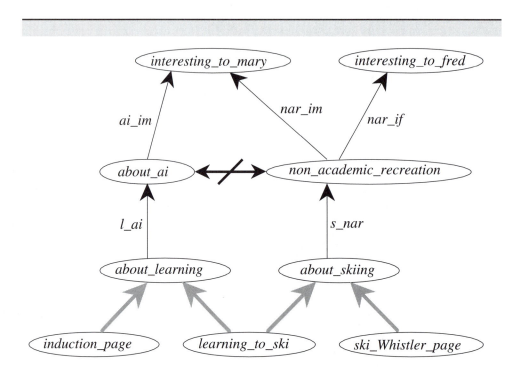

The crossed thick line represents the fact that the two properties are disjoint; the truth of one entails the negation of the other. Thin labeled lines represent defaults, where the label is a possible hypothesis used in the default. Light lines represent membership of an individual in a class.

Figure 9.2: A diagrammatic representation of Example 9.5

Resolving Competing Arguments

There are some cases where it is reasonable that there are two incompatible goals which can be explained and where there are no cancellation rules applicable. This problem is known as the **multiple extension problem**, as there must be incompatible extensions. This section considers what should be predicted based on a set of possibly contradictory defaults.

Example
9.5

Consider the example depicted in Figure 9.2 with two categories of articles: An article is about AI or is about nonacademic recreation. By definition an article can't be both about AI and about nonacademic recreation. Thus, an argument for an article being about AI is an argument against it being about nonacademic recreation, and vice versa.

Consider the following possible hypotheses:

$$H = \{l\_ai(X), s\_nar(X), ai\_im(X), nar\_im(X), nar\_if(X)\},$$

where

- $l\_ai(X)$ is the assumption that if X is about learning it is about AI.
- $s\_nar(X)$ is the assumption that if X is about skiing, it is about nonacademic recreation.
- $ai\_im(X)$ is the assumption that if X is about AI, it is interesting to Mary.
- $nar\_im(X)$ is the assumption that if X is about non-academic recreation, it is interesting to Mary.
- $nar\_if(X)$ is the assumption that if X is about non-academic recreation, it is interesting to Fred.

The corresponding facts are:

$$about\_ai(X) \leftarrow about\_learning(X) \wedge l\_ai(X).$$
$$non\_academic\_recreation(X) \leftarrow about\_skiing(X) \wedge s\_nar(X).$$
$$false \leftarrow about\_ai(X) \wedge non\_academic\_recreation(X).$$
$$interesting\_to\_mary(X) \leftarrow about\_ai(X) \wedge ai\_im(X).$$
$$interesting\_to\_mary(X) \leftarrow non\_academic\_recreation(X) \wedge nar\_im(X).$$
$$interesting\_to\_fred(X) \leftarrow non\_academic\_recreation(X) \wedge nar\_if(X).$$
$$about\_skiing(learning\_to\_ski).$$
$$about\_learning(learning\_to\_ski).$$
$$about\_skiing(ski\_Whistler\_page).$$
$$about\_learning(induction\_page).$$

Suppose that the user hasn't added any cancellation rules to say that some defaults override others.

Both $about\_ai(learning\_to\_ski)$ and $\neg about\_ai(learning\_to\_ski)$ can be explained:

- You can explain $about\_ai(learning\_to\_ski)$ using $\{l\_ai(learning\_to\_ski)\}$. That is, you can say "We expect $learning\_to\_ski$ to be about AI because it is about learning and articles about learning are, by default, about AI."
- You can explain $\neg about\_ai(learning\_to\_ski)$ using $\{s\_nar(learning\_to\_ski)\}$. That is, "$learning\_to\_ski$ is, by default about non-academic recreation because it is a about skiing; because it is about non-academic recreation it can't be about AI."

There is no semantic way to distinguish these arguments. So you would not expect to predict that *learning_to_ski* is about AI nor would you expect to predict that *learning_to_ski* isn't about AI.

Consider the case of whether you should predict

interesting_to_fred(*learning_to_ski*).

It seems obvious that you shouldn't predict *interesting_to_fred*(*learning_to_ski*), as it rests on *non_academic_recreation*(*learning_to_ski*), which, as discussed above, you don't predict.

It seems reasonable, however, to predict

interesting_to_mary(*learning_to_ski*)

based on the defaults given, as Mary would be interested in *learning_to_ski* whether it was about AI or about nonacademic recreation.

Defining Default Prediction

Based on the discussion of Example 9.5 above and considering the alternatives (page 331), a reasonable criteria for default prediction is to predict formula g if g can be explained from every set of consistent assumptions. That is, no matter what consistent assumptions are made, you can still derive g by making perhaps more consistent assumptions.

To put it into another form:

Predict g if g is in all extensions of $\langle F, H \rangle$.

This is a form of **skeptical** default reasoning. One way to think about this is that g is predicted even if an adversary gets to select the assumptions, as long as the adversary is forced to select something. You do not predict g if the adversary can pick assumptions from which g can't be explained.

Example 9.6

Using this definition of prediction, in Example 9.5 you would not predict

non_academic_recreation(*learning_to_ski*)

nor would you predict

interesting_to_fred(*learning_to_ski*)

because they can't be explained given {*l_ai*(*leaning_to_ski*)}.

You would, however, predict

interesting_to_mary(*learning_to_ski*)

because from every scenario you can also assume either *l_ai*(*leaning_to_ski*) or *s_nar*(*leaning_to_ski*). Whichever hypothesis you add to the scenario can be used to prove *interesting_to_mary*(*learning_to_ski*).

Alternate Definitions of Default Prediction

Consider the question of what should be predicted based on a set of, possibly conflicting, defaults. There are a number of alternate proposals for default prediction which we haven't adopted:

Proposal One. The first proposal is to predict g if g can be explained. Thus you would predict a proposition that is in any extension. Under this proposal, in Example 9.5 you would predict

> $non\_academic\_recreation(learning\_to\_ski)$

assuming $\{s\_nar(learning\_to\_ski)\}$ and predict

> $\neg non\_academic\_recreation(learning\_to\_ski)$

assuming $\{l\_ai(learning\_to\_ski)\}$. This seems like a very peculiar form of prediction, where you predict some proposition is true and predict it is false.

Proposal Two. The second proposal is to predict g if g can be explained and $\neg g$ can't be explained. This would ensure that you do not predict a proposition and also predict its negation. Under this proposal, in Example 9.5 you would not predict

> $non\_academic\_recreation(learning\_to\_ski)$

as you can explain its negation. You would predict

> $interesting\_to\_fred(learning\_to\_ski)$

as you can explain it, assuming $\{s\_nar(learning\_to\_ski), nar\_if(learning\_to\_ski)\}$, and can't explain its negation. This is rather peculiar as the only way to get to

> $interesting\_to\_fred(learning\_to\_ski)$

is from $non\_academic\_recreation(learning\_to\_ski)$, which you do not predict. Thus, it doesn't seem reasonable to predict $interesting\_to\_fred(learning\_to\_ski)$.

Proposal Three. The third proposal is to predict g if there is an explanation $D = \{d_1, \cdots, d_k\}$ such that for no i can $\neg d_i$ be explained. This is to say that there is an argument for g that can't be undermined by making different assumptions. Under this proposal, in Example 9.5 you would not predict

> $non\_academic\_recreation(learning\_to\_ski)$

nor would you predict either

> $interesting\_to\_fred(learning\_to\_ski)$ or
>
> $interesting\_to\_mary(learning\_to\_ski)$

which you would like to predict. The problem is that even though each argument for Mary being interested in the article can be undermined, each explanation needed to undermine an argument also leads to the same conclusion, namely that Mary is interested in the learning-to-ski article.

You also predict

$$non\_academic\_recreation(learning\_to\_ski) \lor about\_ai(learning\_to\_ski).$$

A Minimal Model Semantics for Default Prediction

The previous section presented a syntactic criterion for prediction. This section gives an equivalent semantic definition of prediction. In earlier chapters we concluded a goal if it was true in all models (page 32). You can generalize this idea to predict a goal if it is true in all *minimal* models: You want to consider only those models that violate as few assumptions as possible.

Suppose M_1 and M_2 are models of the facts. Each hypothesis is either true or false in each model. $M_1 <_H M_2$, read "M_1 is more normal than M_2 with respect to H," if the hypotheses violated by M_1 are a strict subset of the hypotheses violated by M_2. That is,

$$M_1 <_H M_2 \text{ if } \{h \in H' : h \text{ is false in } M_1\} \subset \{h \in H' : h \text{ is false in } M_2\},$$

where H' is the set of ground instances of elements of H.

M is a **minimal model** of F with respect to H if M is a model of F and there is no model M_1 of F such that $M_1 <_H M$.

g is **minimally entailed** from $\langle F, H \rangle$ if g is true in all minimal models of F with respect to H.

Proposition 9.2 g is minimally entailed from $\langle F, H \rangle$ if and only if g is in all extensions of $\langle F, H \rangle$.

This proposition can be proved by showing that the minimal models of F with respect to H are exactly the models of the extensions of $\langle F, H \rangle$. (See Exercise 9.4.)

This means that prediction can either be based on the membership in all extensions, or on truth in all minimal models. They both give the same answer. Note that this proposition corresponds to the soundness and completeness of membership in all extensions with respect to the semantic definition of truth in all minimal models.

9.4   Abduction

Another instance of assumption-based reasoning is where, instead of making predictions about what is true in the world, you observe some thing being true in the world, and want to conjecture what may have produced this observation. This kind of reasoning is known as **abduction**.

> ### Conditional Defaults

According to the definition of prediction, you predict what is in all minimal models. Suppose you have some background knowledge base, K, and some possible hypotheses H. If you are given some particular facts A about the world, you add these as facts and derive what is true in all the minimal models of $K \cup A$. Let's call these the A-minimal models, where K and H are implicit. If formula B is true in all of the A-minimal models, let's write:

$$A \models_{\langle K,H \rangle} B.$$

The symbol $\models_{\langle K,H \rangle}$ is said to be a **consequence relation**.

Another style of default reasoning can be obtained if, instead of specifying K and H, you write constraints on the consequence relation. Let's write $A \mathrel{|\!\sim} B$, if B is true in all A-minimal models, where you don't actually specify the K and the H.

You can use the consequence relation to specify constraints on K and H. For example, let's specify some defaults based on Example 9.3, but simplified to ignore variables:

> $about\_ai \mathrel{|\!\sim} interesting$
>
> $about\_fl \mathrel{|\!\sim} about\_ai$
>
> $about\_fl \mathrel{|\!\sim} \neg interesting$

Given these constraints on the consequence relation, you can derive what you should predict if you were given both $about\_fl$ and $about\_ai$:

> $about\_fl \wedge about\_ai \mathrel{|\!\sim} \neg interesting$

Notice how you automatically get preference for the more specific default.

Four sound and complete rules of inference can be used to derive valid new consequence relations:

> if $A \models B$, then $A \mathrel{|\!\sim} B$
>
> if $A \mathrel{|\!\sim} B$ and $A \mathrel{|\!\sim} C$, then $A \wedge B \mathrel{|\!\sim} C$
>
> if $A \mathrel{|\!\sim} B$ and $A \wedge B \mathrel{|\!\sim} C$, then $A \mathrel{|\!\sim} C$
>
> if $A \mathrel{|\!\sim} C$ and $B \mathrel{|\!\sim} C$, then $A \vee B \mathrel{|\!\sim} C$

The use of conditional logics for default reasoning was pioneered by Delgrande (1987). Others obtained essentially the same framework by considering minimal models (Kraus, Lehmann & Magidor, 1990), extreme probabilities in the form of ϵ-semantics (Pearl, 1988b), or by representing typicality in modal logic (Boutilier, 1994). See Pearl (1989) for a survey.

One problem with conditional logic is, unlike the case where K and H are specified, irrelevant facts are not ignored. For example, from the conditional defaults above, if you were given $about\_ai \wedge short$, you could not conclude $interesting$. As far as the logic is concerned, the short articles could be exceptional. Finding a framework with automatic preference for more specific defaults that ignores irrelevant facts is an open problem. See Lehmann (1989), Geffner & Pearl (1992), and Simari & Loui (1992) for some proposed solutions.

Instead of only having normality assumptions, as in default reasoning, you may also need other assumptions about what can occur, such as hypothesizing underlying causes that may be unusual. Indeed, you often use abduction in exactly the cases where the observation is unexpected.

Determining what is going on inside a system based on observations about the behavior is the problem of **diagnosis**. In **abductive diagnosis**, you hypothesize diseases and malfunctions as well as that some parts are working normally in order to explain the observed symptoms. This differs from consistency-based diagnosis (page 243) in that you model faulty behavior as well as normal behavior, and the observations are explained rather than added as facts. So abductive diagnosis provides a mechanism that allows for more detailed modeling and gives more detailed diagnoses. It also means that you can diagnose systems where there is no normal behavior. One example is for the infobot to try to determine why a user asked a question to a knowledge base. This can help the infobot to correct misconceptions that the user may have.

Abduction also provides a paradigm for *recognition*, where you hypothesize what is in a scene that could have produced what you observed. You develop a theory, or explanation, of what is in the world that accounts for what is observed.

The term *abduction* was coined by C.S. Peirce (1839–1914) to differentiate this type of reasoning from deduction, which is determining what logically follows from a set of axioms, and induction, which is inferring general relationships from examples. In abduction you hypothesize what may be true about an observed case.

| Example 9.7 | A user of the infobot may not want to have to program the infobot and tell it all of her preferences. It may be better to have the infobot observe the user and to hypothesize her beliefs and preferences. This is known as **user modeling**. Based on its theory about the beliefs and preferences of the user, the infobot can then suggest other pieces of information that may be useful to the user. Such proactive help is important in large information spaces, where the user may not suspect that some information even exists and thus can't be expected to ask for it. |

Suppose you want to be able to infer users' interests based on what articles they select to read.

You can hypothesize what agents are interested in:

$$H = \{interested\_in(Ag, Topic)\}.$$

The corresponding facts are

$selects(Ag, Art) \leftarrow$
    $about(Art, Topic) \wedge$
    $interested\_in(Ag, Topic).$
$about(article\_94, ai).$
$about(article\_94, information\_highway).$

about(*article_34*, *ai*).

about(*article_34*, *skiing*).

There are two minimal explanations of *selects*(*fred*, *article_94*):

{*interested_in*(*fred*, *ai*)}.

{*interested_in*(*fred*, *information_highway*)}.

Each explanation is consistent and, together with the facts, implies the observation.

If you observe *selects*(*fred*, *article_94*) ∧ *selects*(*fred*, *article_34*), there are two minimal explanations:

{*interested_in*(*fred*, *ai*)}.

{*interested_in*(*fred*, *information_highway*),

   interested_in(*fred*, *skiing*)}.

Note that there are other explanations for the observation, but these are all supersets of the minimal explanations.

9.5 Evidential and Causal Reasoning

Much reasoning in AI can be seen as **evidential reasoning**, going from observations to a theory about what is inside the system, followed by **causal reasoning**, going from a theory about the mechanism of a system to predicting output of the system.

Evidential reasoning is sometimes called **diagnosis**, where diseases or malfunctions are hypothesized to explain symptoms. These diagnoses make predictions which lead to tests being performed, which lead to further symptoms (test results being symptoms). Evidential reasoning is also called **perception**, where an agent hypothesizes what is in the world to account for what is perceived by the agent. Based on its hypothesis about what is in the world, the agent acts and then receives further percepts. One view of **science** is that given observations about the world, scientists create theories, which lead to predictions which lead to experiments to test the predictions, which in turn lead to further observations. This cycle from observation to theory to prediction to experiment to further observations is the **cycle of perception**.

There are essentially two ways to handle evidential and causal reasoning:

- Axiomatize only one direction, and use different mechanisms for evidential and causal reasoning. For example, you can write implications just in the causal direction, specifying how causes imply effects, and use abduction for evidential reasoning and deduction or default reasoning for causal reasoning.

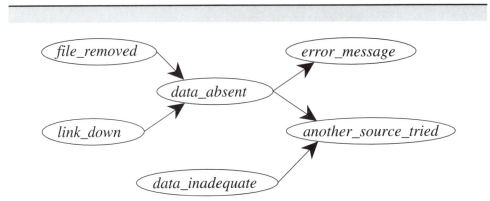

Nodes represent propositions, and arcs are drawn from possible causes to their effects.

Figure 9.3: Causal network for Example 9.8

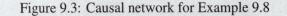

- Axiomatize both implications from effects to causes and from causes to effects, and use a single reasoning mechanism, such as deduction or default reasoning.

Note that, although presented in a logical formalism in this section, exactly the same reasoning occurs in other representations. For example, Bayesian networks (page 363) can be seen as axiomatizing in the causal direction. Probabilistic inference in Bayesian networks follows the two-phased reasoning: Given observations, first abduce to causes going backwards along the arrows, then predict going forward along the arrows. Neural networks (page 408) can be seen as modeling in the evidential direction. In a feed-forward network, you model how the classification of the network follows from the input. These networks can't be used for causal reasoning.

| Example 9.8 | Suppose that you need to reason about the performance of the infobot, and you want to know why it was trying another information source. There may be two reasons for trying another information source, namely that data were absent from the original source or that they were inadequate. Data could be absent because the link to the information provider was down, or because the file was removed. Data absent also results in an error message. This example is depicted in Figure 9.3.
    Causal and evidential reasoning can be tricky. Suppose it was observed that another source was tried. In this case, data absent is a reasonable hypothesis. This would predict an error message. However, if you instead observed that the data were inadequate, this would predict that another source was tried. This should not in turn let us predict that the data were absent. Observing that another source was tried and deducing that another source was tried need to produce different behavior, and thus need to be distinguished. |

Background knowledge is that knowledge that is possessed by an agent before it observes anything. **Observations** or **evidence** is the knowledge directly acquired through interacting with the world.

Typically, observations need to be explained in order to understand the world, whereas background knowledge can be just absorbed. This distinction is part of what is done in modeling the domain. When building a knowledge base, you could take into account a particular piece of information about the domain, in which case that piece of information would be background knowledge. Alternatively, the knowledge base could be built not taking into account that information, in which case the information would need to be considered as evidence.

This distinction arises in a number of different areas—for example, in decision trees (page 403), the trees themselves represent the background knowledge and the instance to be classified represents the observation. Similarly, in neural networks (page 408) the background knowledge is represented as the network and the particular cases classified are the evidence. This distinction also is important in probability theory (see page 355).

Abduction Followed by Default Reasoning

One way to allow causal and evidential reasoning is to axiomatize in the causal direction only. When an observation is received, reasoning occurs in two phases: First use abduction to determine the possible causes and then use deduction or default reasoning to predict what else will follow from the causes.

One important issue needs to be considered. When using default reasoning for prediction, you only want to use normality assumptions. In the abduction phase, when explaining the observations, you may need to assume that some things are abnormal in order to explain an observation that isn't expected. This can be seen best in diagnosis, where you may need to hypothesize diseases or faults that you don't want to assume without any evidence. This arises in all applications, not just diagnosis; the world will keep providing you with surprises that you have to make sense of.

In order to represent this two-phase reasoning, you can divide the hypothesis set H into H_{nor}, the set of normality assumptions, and H_{ab}, the set of abnormality assumptions. For default prediction, you use only H_{nor}; and for abduction, you use both H_{nor} and H_{ab}. You predict g if g is true in all extensions (using H_{nor}) of all explanations (using $H_{nor} \cup H_{ab}$ as possible hypotheses) of the observations.

Example 9.9

Example 9.8 can be represented with all of the arrows in Figure 9.3 (page 336) as defaults, and the three leftmost nodes in Figure 9.3 as abnormality hypotheses:

$$H_{nor} = \{ld\_da, fr\_da, da\_em, da\_ast, di\_ast\},$$
$$H_{ab} = \{file\_removed, link\_down, data\_inadequate\},$$

where *ld_da* means the data are absent because the link is down, *fr_da* means the data are absent because the file was removed, and so forth. Assume the following facts:

$$data\_absent \leftarrow file\_removed \land fr\_da.$$
$$data\_absent \leftarrow link\_down \land ld\_da.$$
$$error\_message \leftarrow data\_absent \land da\_em.$$
$$another\_source\_tried \leftarrow data\_absent \land da\_ast.$$
$$another\_source\_tried \leftarrow data\_inadequate \land di\_ast.$$

If *error_message* was observed, there are two minimal explanations:

{*file_removed, fr_da, da_em*},
{*link_down, ld_da, da_em*}.

So you predict *another_source_tried* as well as *file_removed* ∨ *link_down*.

If *another_source_tried* was observed instead, there are three minimal explanations:

{*file_removed, fr_da, da_ast*},
{*link_down, ld_da, da_ast*},
{*data_inadequate, di_ast*},

and so you predict *error_message* ∨ *data_inadequate*.

If *data_inadequate* was observed instead, there is one minimal explanation:

{*data_inadequate*}

and so you predict *another_source_tried*.

Note that when you conclude *another_source_tried* causally, you do not need to explain it. However, when you reached *data_absent* when *another_source_tried* was observed, you also needed to explain it.

You will see this kind of reasoning again when you consider belief networks (page 363).

Axiomatizing Both Causal and Evidential Directions

An alternative is to axiomatize both the causal and evidential directions and use a single mechanism, such as deduction or default reasoning.

The general idea is for an effect to imply the disjunction of its possible causes.

| Example 9.10 | You can give an alternative axiomatization to that of Example 9.9. In this axiomatization you have only the normality assumptions. That is, H_{nor} of Example 9.9 is the set of assumables. In addition to the facts of Example 9.9, suppose you also have the facts: |

$$(\textit{file\_removed} \wedge \textit{fr\_da}) \vee (\textit{link\_down} \wedge \textit{ld\_da}) \leftarrow \textit{data\_absent}.$$

$$\textit{data\_absent} \leftarrow \textit{error\_message}.$$

$$\textit{em\_da} \leftarrow \textit{error\_message}.$$

$$(\textit{data\_absent} \wedge \textit{da\_ast}) \vee (\textit{data\_inadequate} \wedge \textit{di\_ast}) \leftarrow$$
$$\textit{another\_source\_tried}.$$

This axiomatization implies exactly the same conclusions as Example 9.9, with each axiomatization using its appropriate reasoning strategy.

The mapping from the abductive axiomatization to the deductive or default reasoning axiomatization is very much like Clark's completion (page 248), at least for the propositional case. The main differences are that you complete only the atoms that are not hypotheses. For example, you don't conclude ¬*link_down* just because there are no rules for it. This equivalence holds when the causal structure is acyclic.

The disjunctive rules can be eliminated by appropriate encoding of defaults, noting that competing assumptions will lead us to predict the disjunction.

| Example 9.11 | To axiomatize the rule |

$$\textit{file\_removed} \vee \textit{link\_down} \leftarrow \textit{data\_absent}$$

without using disjunction, you can use the axioms

$$\textit{file\_removed} \leftarrow \textit{data\_absent} \wedge \textit{da\_fr}.$$

$$\textit{link\_down} \leftarrow \textit{data\_absent} \wedge \textit{da\_ld}.$$

$$\textit{false} \leftarrow \textit{da\_fr} \wedge \textit{da\_ld}.$$

where *da_fr* and *da_ld* are defaults. The third axiom is there to force the disjunction to be predicted. Without it, or something similar, you would be able to go from *file_removed* to *data_absent* using a causal rule, and from *data_absent* to *link_down* using an evidential rule.

9.6 Algorithms for Assumption-Based Reasoning

This section considers how to implement the general assumption-based framework of Section 9.2. You have already seen virtually all of the tools needed. This section assumes that F is restricted to Horn clauses (page 241). (See Exercise 9.6 to see how this restriction can be relaxed.)

A **subexplanation** of g from $\langle F, H \rangle$ is a set D of ground instances of elements of H such that

$$F \cup D \models g$$

An explanation of g is a subexplanation of g that isn't a subexplanation of *false*. Checking that a subexplanation of g isn't a subexplanation of *false* is known as a **consistency check**. The notion of a subexplanation is used to implement explanation and prediction. Subexplanations will be computed using a complete deduction method, either top-down or bottom-up.

To implement explanation, you can use the notion of a conflict. Recall (page 242) that a set N of ground instances of elements of H forms a conflict if $F \cup N \models \textit{false}$. A minimal conflict is a conflict such that no strict subset is also a conflict. A set D of ground hypotheses is consistent if and only if there is no minimal conflict N, such that $N \subseteq D$.

When implementing the system, you can either explicitly build the minimal conflict sets independently of any particular assumption set, or you can check the consistency of subexplanations on demand. A **nogood** is a conflict that has been found and stored. Each nogood is thus a conflict. The opposite isn't necessarily true; there may be some conflicts not yet discovered.

A top-down interpreter is a direct application of the delaying meta-interpreter (page 210). (See the algorithm given in Figure 7.3 on page 248.) You can also add a check that an assumable is consistent before adding it to the set of assumptions.

You have a choice as to how to implement the consistency check:

- You can precompute the conflicts and store them as nogoods. This can be done by using the algorithm of Figure 7.3 (page 248) to compute conflicts, store them, and then check the consistency of subexplanations against the stored nogoods. One problem, is that there may be infinitely many minimal conflicts. You could also implement this by iterative deepening, where you generate all conflicts of length k before you consider any sets of hypotheses of length k or more. Such an algorithm is a straightforward extension of the iterative deepening meta-interpreter (page 502).

- An alternative is to check consistency on demand. Suppose you know that set D of ground instances of assumables is consistent. In order to determine whether G can be consistently added to D, you can forward chain from G, perhaps using elements of D. To do this, suppose you had a base-level rule of the form

$$h \Leftarrow g \mathbin{\&} r.$$

To forward chain from g, first try to prove r, perhaps assuming elements of D. Proving r lets you derive h. Then continue to forward chain from h. [$G|D$] is consistent if you can't derive *false* by forward chaining from G.

- A third approach is, given a set D, to try to prove *false* in a top-down manner, allowing elements of D to be immediately proved. If *false* can be proved, then D is inconsistent with F. If *false* can't be proved, then D is consistent with F.

Which of these is more efficient depends on the number of assumptions, the number of facts they imply, and the number of rules that imply *false*. None of the strategies are guaranteed to halt in the general case as the problem of determining whether a set of formulae is consistent is, in general, undecidable.

Another way to implement Horn clause assumption-based reasoning is bottom-up. (See page 245 for a description and page 495 for an implementation).

The above discussion on computing explanations seems to be of little use in computing whether something should be predicted. Default predication is about membership in *all* extensions or about minimal models, both of which may seem to be very difficult to compute. A way to compute whether something should be predicted follows from the following result:

Proposition 9.3

g is in all extensions of $\langle F, H \rangle$ if and only if there is a set $\mathbf{D} = \{D_1, D_2, \ldots\}$ where each D_i is an explanation of g such that

$$\neg (D_1 \vee D_2 \vee \cdots)$$

can't be explained from $\langle F, H \rangle$.

Proof:

Suppose g isn't in extension E. Then E must be an explanation of $\neg D_i$ for each D_i. If E were not an explanation for $\neg D_i$, then $E \cup D_i$ is consistent and implies g, which is a contradiction to E being an extension that doesn't contain g. Thus E is an explanation of $\neg (D_1 \vee D_2 \vee \cdots)$.

Conversely, let $\mathbf{D}$ be the set of all explanations of g. If C was an explanation of $\neg (D_1 \vee D_2 \vee \cdots)$, then C is inconsistent with every explanation of g. Extend C to be an extension by adding as many assumables as possible. We have constructed an extension which doesn't contain g, so g isn't true in all extensions. Q.E.D.

This proposition can be used to implement prediction. To determine whether g is predicted, you find explanations of g and then try to find explanations for the negations of these explanations of g.

- If you can find a set $\mathbf{D}$ of explanations of g such that a complete proof procedure fails to find an explanation for the negations of the elements of $\mathbf{D}$, then g should be predicted. This is because there is at least one element of $\mathbf{D}$ in every extension, otherwise the extension would be an explanation for the negation of every element of $\mathbf{D}$.

- If there is an explanation S for the negations of every explanation of g, then g should not be predicted. If you can generate such an S, then g isn't in all extensions. In particular, it isn't in any extension that is a superset of S.

9.7   References and Further Reading

The assumption-based reasoning framework presented here is based on Theorist (Poole, Goebel & Aleliunas, 1987). The defaults presented here are equivalent to the prerequisite-free normal defaults of the default logic of Reiter (1980). The use of naming defaults is based on the abnormality predicate of McCarthy (1986). The minimal models semantics is based on that of Geffner & Pearl (1992) and is similar to that of circumscription (McCarthy, 1986). The collection by Ginsberg (1987) contains many classic papers on nonmonotonic reasoning.

The notion of default reasoning as forming arguments has been advocated by Loui (1987), Lin & Shoham (1989), Pollock (1994), Dung (1995), and Bondarenko, Dung, Kowalski & Toni (1997).

Abduction has been used for diagnosis (Peng & Reggia, 1990), natural language understanding (Hobbs, Stickel, Appelt & Martin, 1993) and temporal reasoning (Shanahan, 1989). See Kakas, Kowalski & Toni (1993) for a review of abductive reasoning from a logic programming perspective.

The distinction between causal and evidential reasoning was first pointed out by Pearl (1988a). Kautz (1991) proposed using default reasoning for axiomatizing both the causal and evidential directions and applied this technique to user modeling— in particular, recognizing the plans of users. The two-phase technique to combine abduction with default reasoning of Section 9.5 was advocated by Shanahan (1989) and Poole (1989; 1990). The comparison between axiomatizing in both directions and the two-phased approach is discussed in Console, Theseider Dupré & Torasso (1991), Konolige (1992), and Poole (1994).

The bottom-up Horn implementation for finding explanations is based on the ATMS (de Kleer, 1986). The ATMS is more sophisticated in that it considers the problem of incremental addition of facts and hypotheses, which we ignored (see Exercise 9.7). Przymusinski (1989), Ginsberg (1989), and Poole (1989) describe proof methods which are similar to that suggested by Proposition 9.3.

9.8 Exercises

Exercise 9.1 Suppose you have the following facts and possible hypotheses for an abductive reasoning system:

$$H = \{walking, hunting, robbing, banking\}$$
$$F = \{goto(forest) \leftarrow walking,$$
$$\quad get(gun) \leftarrow hunting,$$
$$\quad goto(forest) \leftarrow hunting,$$
$$\quad get(gun) \leftarrow robbing,$$
$$\quad goto(bank) \leftarrow robbing,$$
$$\quad goto(bank) \leftarrow banking,$$
$$\quad false \leftarrow puton(goodShoes) \wedge goto(forest)\}.$$

(a) Suppose you observe *get(gun)*. What are all of the minimal explanations for this observation?

(b) Suppose you observe *get(gun)* ∧ *goto(bank)*. What are all of the minimal explanations for this observation?

(c) Is there something that could be observed to remove one of these explanations?

Exercise 9.2 Select a domain and construct a natural example with two sets of instances of defaults that are consistent, but contradict each other. For the domain, which should be preferred?

Exercise 9.3 Why should you prefer more specific defaults over more general defaults? *Hint:* Suppose you don't; when could the more specific default get used?

Exercise 9.4 Prove Proposition 9.2 (page 332). As intermediate steps, prove the following two lemmas. First show that if M is a minimal model of F with respect to H, then it's a model of an extension of $\langle F, H \rangle$. Second, if M is a model of an extension of $\langle F, H \rangle$, it's a minimal model of F with respect to H.

Exercise 9.5 Build an iterative deepening assumption-based reasoning system. This can be based on the iterative deepening meta-interpreter (page 502), but the depth bound should be based on the number of elements of the set of hypotheses used in the proof. You should interleave proving *false*—consistency checking—with finding explanations for the goal.

Exercise 9.6 Write an interpreter for finding explanations that isn't restricted to Horn clauses. *Hint:* Combine the theorem proving interpreter in Chapter 7, with the assumption-based interpreter given in Section 9.6. When might such an interpreter be useful?

Exercise
9.7

Suppose you are implementing a bottom-up Horn explanation implementation, and you want to add new rules or new assumables. When a new rule is added, how are the minimal explanations affected? When a new assumable is added, how are the minimal explanations affected? What other information may make the incremental addition of clauses and assumables easier?

Chapter 10

Using Uncertain Knowledge

10.1 Introduction

Neither human nor computer agents are omniscient—they don't know everything that is true. However, they need to make decisions when they don't know the exact state of the world. They have to use whatever evidence they have. Even when an agent senses the world to find out more information, it doesn't find out the exact state of the world. This chapter considers such reasoning under uncertainty, where the agent must make decisions about what action to take even though it can't precisely predict the outcomes of its actions.

To make a good decision, an agent cannot simply assume what the world is like and act according to those assumptions. Consider the following example:

Example 10.1 | Many people consider it sensible to wear a seat belt when traveling in a car because, in an accident, wearing a seat belt reduces the risk of serious injury. However, consider an agent that commits to assumptions and bases its decision on these assumptions. If the agent assumes it will not have an accident, it will not bother with the inconvenience of wearing a seat belt. If it assumes it will have an accident, it won't go out. In neither case would it wear a seat belt! A more intelligent agent may wear a seat belt because the inconvenience of wearing a seat belt is far outweighed by the increased risk of injury or death if it has an accident. It doesn't stay at home too worried about an accident to go out; the benefits of being mobile, even with the risk of an accident, outweigh the benefits of the extremely cautious approach of never going out. These decisions depend on the likelihood of having an accident, how much a seat belt helps in an accident, the inconvenience of wearing a seat belt, and how important it is to go out. The various tradeoffs may be different

345

> for different agents. Some people don't wear seat belts, and some people don't go
> out because of the risk of accident.

The calculus used in such reasoning under uncertainty is that of probability and decision theory. Probability is the calculus of gambling. When an agent makes decisions and there are uncertainties involved about the outcomes of its action, it is gambling on the outcome. However, unlike a gambler at the casino, the agent can't opt out and decide not to gamble; whatever it does—including doing nothing—involves uncertainty and risk. If it doesn't take the probabilities and outcomes into account, it will lose at gambling to an agent who does. This doesn't mean, however, that making the best decision guarantees a win. (See the discussion on page 172 about the difference between a decision and an outcome.) Just as at the casino where gamblers can win occasionally but if they play enough they will always lose, an agent that makes many decisions based on probabilities can expect to be better off than if it didn't.

10.2   Probability

Many of us learn probability as the theory of tossing coins and rolling dice. Although this may be a good way to present probability theory, probability is applicable to a much richer set of applications than coins and dice. The calculus of probability is surprisingly simple (page 352). Some important issues involved in the meaning of probability need to be discussed first.

There are two main interpretations of probability theory:

- The **statistical** or **frequentist** view of probabilities is that probability is a measure of proportion of individuals, or the long-range frequency of a set of "events." Under this view, the probability of a bird flying is the proportion of birds that fly out of the set of all birds. Under this interpretation, you can't assign probability to a particular bird flying: It either flies or it doesn't. Probabilities are defined instead over the set of all individuals or events. Similarly, you don't refer to the probability of a "6" on a particular roll of a die, but instead to the probability of the event of a "6" appearing on a roll of a die. This probability is the long-range frequency of 6's that appear on rolls of dice.

- The **personal**, **subjective**, or **Bayesian** view is that probability is an agent's measure of belief in some proposition based on the agent's knowledge. Your probability of a bird flying is your measure of belief in the flying ability of an individual based only on the knowledge that the individual is a bird. Others may have different probabilities, as they may, for example, have had different experiences with birds or may have different knowledge about this particular bird.

An agent's belief in a bird's flying ability is affected by what the agent knows about that bird, such as that it is sick or an emu. Instead of using probability as a measure over populations (or sequence of events), an agent's probability can be viewed as a measure over all of the worlds that are possible given the agent's knowledge about the particular situation. In each possible world, the bird either flies or it doesn't. An agent needs probability theory because it doesn't know which of the possible worlds it is in.

The term *subjective* doesn't mean *arbitrary*, but rather it means "belonging to the subject." For example, suppose there are three agents A, B, and C, and one dice which has been tossed. Suppose A observes that the outcome is a "6" and tells B that the outcome is even, but C knows nothing about the outcome. In this case, A has a probability of 1 that the outcome is a "6," B has a probability of $\frac{1}{3}$ that it's a "6" (assuming B believes A and treats all of the even outcomes with equal probability), and C may have probability of $\frac{1}{6}$ that the outcome is a "6." They all have different probabilities because they all have different knowledge.

Both views of probability have the same calculus. The same probability theory can be used for both interpretations, even if the second interpretation allows for a more liberal interpretation of what can have a probability.

In this book we focus on the second view, the interpretation that probability is a measure of belief. This is because we are interested in making decisions, and decisions are always based on the particular case that the agent encounters. Agents don't encounter generic events: they instead have to make decisions in particular situations.

Probability theory can be defined as the study of *how knowledge affects belief.* Belief in some proposition, f, can be measured in terms of a number between 0 and 1. The probability f is 0 means that f is believed to be definitely false (no new evidence will shift that belief), and a probability of 1 means that f is believed to be definitely true. Using 0 and 1 is purely a convention.

Adopting the 'belief' view of probabilities doesn't mean that statistics are ignored. Statistics of what has happened in the past is knowledge that can be conditioned on and used to update belief. (See page 424 for a discussion of learning in the context of Bayesian probability.)

We are assuming that the uncertainty is **epistemological**—pertaining to your knowledge of the world—rather than **ontological**—how the world is. We are assuming that your knowledge of the truth of propositions is uncertain, not that there are degrees of truth. For example, if you are told someone is very tall, you know they have some height; you just have vague knowledge about the actual value of their height. If the probability of some f is greater than zero and less than one, this doesn't mean that f is true to some degree, but rather that you are ignorant of whether f is true or false. The probability reflects your ignorance.

Random Variables

It's often convenient to refer to a term that has some value, even if you don't know its value. We will call such terms random variables, as these will have associated probability distributions.

A **random variable** is a ground (variable free) term in our language. The **domain** of a random variable x, written $dom(x)$, is the set of possible values that x can take. The elements of a domain are exclusive and exhaustive: a random variable can't have more than one value from a domain, and a random variable must take on one value in the domain. We write the atomic proposition $x = v$ to mean that variable x has value $v \in dom(x)$.

This is related to the object-attribute-value representation (page 184), where a random variable corresponds to an object and an attribute. Because we want to refer to the object-attribute pair as a single entity, we will use the syntax that is standard in probability theory, rather than using the *prop* relation.

A **discrete** random variable is one whose domain is a finite or countably infinite set of values. You can also have variables that are not discrete; for example, a random variable whose domain corresponds to the real line is a **continuous** random variable.

Example 10.2

In the electrical domain depicted in Figure 3.1 (page 71),

- s_1_pos may be a discrete binary random variable denoting the position of switch s_1 with domain {up, $down$}, where $s_1\_pos = up$ means switch s_1 is up, and $s_1\_pos = down$ means switch s_1 is down.

- s_1_st may be a random variable denoting the state of switch s_1 with domain {ok, $upside\_down$, $short$, $intermittent$, $broken$}, where $s_1\_st = ok$ means switch s_1 is working normally, $s_1\_st = upside\_down$ means switch s_1 is installed upside down, $s_1\_st = short$ means switch s_1 is shorted and acting as a wire, $s_1\_st = intermittent$ means switch s_1 is working intermittently, and $s_1\_st = broken$ means switch s_1 is broken and doesn't allow electricity to flow.

- $number\_of\_broken\_switches$ may be an integer-valued random variable denoting the number of switches that are broken in our electrical domain.

- $current(w_1)$ may be a real-valued random variable denoting the current flowing through wire w_1. $current(w_1) = 1.3$ means there are 1.3 amps flowing through wire w_1. We also allow inequalities between variables and constants as atomic propositions. $current(w_1) \geq 1.3$ means there are at least 1.3 amps flowing through wire w_1.

A **Boolean** random variable is a random variable with domain {$true$, $false$}. Often $x = true$ is written simply as x, and $x = false$ as $\neg x$.

<table>
<tr><td>Example
10.3</td><td>For example, l_1_lit may be a Boolean random variable representing whether light l_1 is lit. "Light l_1 is lit" is represented as $l_1\_lit = true$, or simply as l_1_lit. "Light l_1 isn't lit" is represented as $l_1\_lit = false$, or simply as $\neg l_1\_lit$.</td></tr>
</table>

You can build propositions from atomic propositions using the normal logical connectives such as negation, conjunction, and disjunction. For example,

$$current(w_1) \geq 1.3 \vee (\neg l_1\_lit \wedge s_1\_st = upside\_down)$$

is a proposition.

There is nothing truly random about random variables. A random variable having a value is a proposition with the standard logical meaning. They are called random variables (apart from the historical reason) because typically agents don't know the value of the variables. For example, you may not know Eric's height; you may just know that Mary said he is tall. We introduce random variables so we can refer to a variable that has some value but we don't know what the value is. Most of what follows can be done without the use of random variables using our previous ideas of (ground) atoms and (ground) formulae, but the use of random variables should make some things clearer and is useful to define some concepts. (See, for example, independence of random variables on page 361.)

It's often useful to build complex variables from simpler ones. Suppose $V_1, \ldots, V_n$ are random variables. $\langle V_1, \ldots, V_n \rangle$ is a complex random variable. The domain of the complex random variable is the cross product of the domains of its components. That is,

$$\begin{aligned} dom(\langle V_1, \ldots, V_n \rangle) &= dom(V_1) \times \ldots \times dom(V_n) \\ &= \{\langle d_1, \ldots, d_n \rangle : d_i \in dom(V_i)\}. \end{aligned}$$

Variable $\langle V_1, \ldots, V_n \rangle$ has value $\langle d_1, \ldots, d_n \rangle$, written $\langle V_1, \ldots, V_n \rangle = \langle d_1, \ldots, d_n \rangle$, means that each V_i has value d_i.

<table>
<tr><td>Example
10.4</td><td>Suppose you are in a diagnostic setting where there are n possible diseases, each of which the patient either has or doesn't have. The random variable representing the patient's state assigns one of the 2^n disease complexes to the patient.</td></tr>
</table>

Semantics of Probability

Just as we have given a semantics for our other representation and reasoning systems, we give a possible world semantics for probability. The general idea is that the interpretations (or possible worlds) as well as specifying the truth of atomic formulae, also have a measure that specifies how likely the true state of affairs corresponds to the interpretation. Observations will rule out those possible worlds that are incompatible with the observations.

We assume a finite number of discrete random variables. Although probability theory can be defined for a more general class of random variables, this involves more sophisticated measure theory than we will use. (See the box on page 351.)

A **possible world** corresponds to an interpretation (page 32) but with more structure. When considering modal logic (page 275), we considered a binary accessibility relation between possible worlds. To define the semantics of probability, we instead define a measure on possible worlds. This measure is a number that represents the belief that the interpretation matches the true state of the world. By convention this number is in the range [0, 1]. Zero means that you know the possible world doesn't correspond to the true state of the world, and the measures are normalized so that the measures on the possible worlds sum to one. The probability of any proposition is the sum of the measures of the worlds in which the proposition is true. Using the number one is by convention. We could have just as easily defined the probability as summing to 100; however, the use of one as the sum makes some arithmetic easier.

Below is the standard definition of a possible world, cast in terms of random variables.

A **possible world** is an assignment of exactly one value to every random variable. Let Ω be the set of all possible worlds. If $\omega \in \Omega$ and f is a formula, f is true in ω (written $\omega \models f$) is defined inductively on the structure of f:

- $\omega \models x = v$ if ω assigns value v to x.
- $\omega \models f \wedge g$ if $\omega \models f$ and $\omega \models g$.
- $\omega \models f \vee g$ if $\omega \models f$ or $\omega \models g$ (or both).
- $\omega \models \neg f$ if $\omega \not\models f$.

A **tautology** is a formula that is true in all possible worlds. For example, $a \vee \neg a$ is a tautology.

Associated with each possible world is a **measure**. When there are only a finite number of possible worlds, the measure of world ω, written as $\mu(\omega)$, has the following properties:

- $0 \leq \mu(\omega)$ for all $\omega \in \Omega$.
- The measures of the possible worlds sum to one:

$$\sum_{\omega \in \Omega} \mu(\omega) = 1.$$

When there are more possible worlds, we need a more sophisticated measure theory. (See the box on page 351.)

The **probability** of formula f, written $P(f)$, is the sum of the measures of the possible worlds in which f is true. Formally,

$$P(f) = \sum_{\omega \models f} \mu(\omega).$$

Thus, $P(f)$ is the weighted proportion of the worlds in which f is true.

More Sophisticated Measures

The definition of probability can be made much more general than the one provided here. It can allow for both infinite sets of variables and for nondiscrete random variables—for example variables whose domain is some subset of the real line. In general, we have to use the integral rather than the sum, defining a measure over measurable or Borel subsets of possible worlds. The measure must be nonnegative and integrate to 1:

$$\int_\omega \mu(\omega) \, d\omega = 1.$$

To determine the probability of a formula f, we integrate over the worlds satisfying f

$$P(f) = \int_{\omega \models f} \mu(\omega) \, d\omega.$$

The only nondiscrete probability we will use is where the domain is the real line. In this case, we can consider a possible world for each point and a measure representing a **probability density function**, which we write as p. The probability that random variable x has value between a and b is given by

$$P(a \leq x \leq b) = \int_a^b p(x) \, dx.$$

The probability density function is a function from reals into nonnegative reals that integrates to 1. The most common way to specify such functions is as a **parametric** equation as a formula that specifies what each value should be.

A common parametric distribution is the **normal or Gaussian distribution** with mean μ and variance σ^2 defined by:

$$p(x) = \frac{1}{\sqrt{2\pi}\sigma} e^{-\frac{1}{2}((x-\mu)/\sigma)^2}$$

where σ is the standard deviation. The normal distribution is used for measurement errors, where there is an average value, given by μ, and a variation in values specified by σ. The **central limit theorem**, proved by Laplace (1812), specifies that accumulation of independent errors will approach the Gaussian distribution. This and its nice mathematical properties account for the widespread use of the normal distribution.

Another distribution discussed in the learning under uncertainty section (page 424) is the binomial distribution.

Axioms for Probability

The preceding section gave a semantic definition of probability. In this section we give an alternative axiomatic definition of probability that defines the same calculus.

Four axioms can be used to define what follows from a set of probabilities:

Axiom 1 $P(f) = P(g)$ if $f \leftrightarrow g$ is a tautology. That is, if f and g are logically equivalent, they have the same probability. This shouldn't be surprising because they select the same possible worlds as being true. We often say that probabilities are functions on the propositions, not on formulae, since what is important is the meaning of the formula, not the form in which it's stated.

Axiom 2 $0 \leq P(f)$ for any formula f. The belief in any proposition can't be negative.

Axiom 3 $P(\tau) = 1$ if τ is a tautology. That is, if τ is true in all possible worlds, its probability is 1.

Axiom 4 $P(f \vee g) = P(f) + P(g)$ if $\neg(f \wedge g)$ is a tautology. In other words, if f and g can't both be true—they are mutually exclusive—the probability of the disjunction can be obtained by summing the probability of the disjuncts.

These axioms can be justified by arguing proportions, but they can be seen as an alternative to the possible world definition of probability theory. If your measure of beliefs follows these intuitive axioms, they are covered by probability theory, whether or not they are derived from actual frequency counts. These axioms form a sound and complete axiomatization of the meaning of probability. A probability follows in the possible worlds semantics from a set of probabilities if and only if it can be derived from these axioms:

Proposition 10.1 If there are a finite number of finite discrete random variables, Axioms 1 through 4 are sound and complete with respect to the semantics.

The proof is left as an exercise. (See Exercise 10.1.)

Proposition 10.2 The following hold for all propositions f and g:

(a) Negation of a proposition:
$$P(\neg f) = 1 - P(f).$$

(b) Reasoning by cases:
$$P(f) = P(f \wedge g) + P(f \wedge \neg g).$$

(c) If v is a random variable with domain D, then for all formulae f,
$$P(f) = \sum_{d \in D} P(f \wedge v = d).$$

(d) Disjunction for nonexclusive disjuncts:
$$P(f \vee g) = P(f) + P(g) - P(f \wedge g).$$

The proof is left as an exercise (Exercise 10.2, page 395).

Conditional Probability

Typically you don't want to know only the prior probability of some formula, but rather would like to know the probability based on some evidence. Agents observe the world, and they update their beliefs based on this new knowledge.

The measure of belief in formula h based on formula e is called the **conditional probability** of h **given** e, written $P(h|e)$.

A formula e representing the conjunction of all of the agent's observations of the world is called **evidence**. Given evidence e, the conditional probability $P(h|e)$ is the agent's **posterior probability** of h. The probability $P(h)$ is the **prior probability** of h and is the same as $P(h|true)$.

Example 10.5	For the diagnostic assistant, the patient's symptoms will be the evidence. The prior probability distribution over possible diseases is used before you see, or know anything about, the patient. The posterior probability distribution is the probability that you use after you have gained some evidence. The posterior probability is a measure of your knowledge. When you acquire new evidence through discussions with the patient, observing symptoms, or the results of lab tests, you must update the posterior probability to reflect the new evidence.

Other Possible Measures of Belief

Justifying other measures of belief is difficult. Consider, for example, the proposal that the belief in $f \wedge g$ is some function of the belief of f and the belief in g. Such a measure of belief is called **compositional**. To see why this isn't sensible, consider the single toss of a fair coin. Compare the case where f_1 is "the coin will land heads" and g_1 is "the coin will land tails" with the case where f_2 is "the coin will land heads" and g_2 is "the coin will land heads." For these two cases, the belief in f_1 would seem to be the same as the belief in f_2, and the belief in g_1 would be the same as the belief in g_2. But the belief in $f_1 \wedge g_1$, which is impossible, is very different from the belief in $f_2 \wedge g_2$, which is the same as f_2.

Note also that the conditional probability $P(f|e)$ is very different from the probability of the implication $P(e \rightarrow f)$. The latter is the same as $P(\neg e \vee f)$, which is the measure of the interpretations for which f is true or e is false. For example, suppose you have a domain where birds are relatively rare, and nonflying birds are a small proportion of the birds. Here $P(\neg flies|bird)$ would be the proportion of birds that don't fly, which would be low. $P(bird \rightarrow \neg flies)$ is the same as $P(\neg bird \vee \neg flies)$, which would be dominated by nonbirds and so would be high. Similarly, $P(bird \rightarrow flies)$ would also be high, the probability also being dominated by the nonbirds. It's difficult to imagine a situation where the probability of an implication is the kind of knowledge that is appropriate or useful.

Example
10.6

> The information that the delivery robot receives from its sensors is its evidence. When sensors can be noisy, the evidence is what is known, such as the particular pattern received by the sensor, not that there is a person in front of the robot. It could be mistaken about what is in the world, but it knows what information it received.

Remember that the posterior probability involves conditioning on *everything* the agent knows about a particular situation. All evidence must be conditioned on to obtain the correct posterior probability.

Semantics of Conditional Probability

Evidence e will rule out all possible worlds that are incompatible with e. Like the definition of logical consequence, the given formula e selects the possible worlds in which e is true. Evidence e induces a new measure, μ_e, over possible worlds where all worlds in which e is false have measure 0, and the remaining worlds are normalized so that the sum of the measures of the worlds is 1. Thus,

$$\mu_e(\omega) = \begin{cases} \frac{1}{P(e)} \times \mu(\omega) & \text{if } \omega \models e \\ 0 & \text{if } \omega \not\models e \end{cases}$$

The conditional probability of formula h given evidence e is the sum of the measures (using μ_e) of the possible worlds in which h is true. That is,

$$P(h|e) = \sum_{\omega \models h} \mu_e(w)$$

$$= \frac{1}{P(e)} \times \sum_{\omega \models h \wedge e} \mu(w)$$

$$= \frac{P(h \wedge e)}{P(e)}$$

The last form above is usually given as the definition of conditional probability.

For the rest of this chapter, assume that if e is the evidence, $P(e) > 0$. If $P(e) = 0$, then e is impossible (false in all possible worlds) and thus can't be observed.

Proposition
10.3

(Chain rule) Conditional probabilities can be used to decompose conjunctions:

$$\begin{aligned} P(f_1 \wedge f_2 \wedge \ldots \wedge f_n) = \ & P(f_1) \times \\ & P(f_2|f_1) \times \\ & P(f_3|f_1 \wedge f_2) \times \\ & \vdots \\ & P(f_n|f_1 \wedge \cdots \wedge f_{n-1}) \end{aligned}$$

where the left-hand side is zero if any of the elements of the right-hand side are zero (even if some of the conjuncts on the right-hand side are undefined).

| Evidence Versus Background Knowledge |

Two sets of knowledge are important when using probability:

- Background knowledge in terms of conditional probabilities
- Evidence of what is true in the situation under consideration

Within probability, there are two ways to state that a is true:

- The first is to state that the probability of a is 1. You can write $P(a) = 1$.
- The second is to condition on a—to use a on the right-hand side of the conditional bar, as in $P(\cdot|a)$.

The first states that a is true in all possible worlds. The second says that you are only interested in worlds where a happens to be true.

Suppose you had the probabilities

$$P(flies|bird) = 0.8,$$
$$P(bird|emu) = 1.0,$$
$$P(flies|emu) = 0.001.$$

If you decide you are talking about an emu, you can't add the statement $P(emu) = 1$. No probability distribution satisfies these four assertions. The problem is that if emu were true in all possible worlds, it wouldn't be the case that in 0.8 of the possible worlds, the individual flies. You instead need to condition on the fact that the individual is an emu.

To build a probability model, you need to take some knowledge into consideration and build a probability model based on the possibilities given this knowledge. All subsequent knowledge is observations that need to be conditioned on.

You could imagine building a probability model that covered the case before an agent knew anything. If you did this, all of the agent's experiences could be conditioned on. You don't need to do this. You can build a model based on all of the agent's current knowledge and condition only on future observations. This shouldn't affect the derived probabilities. You can either construct a theory that embodies a measure, μ, and then observe and condition on k, or else, after experiencing k, build a probability theory which should embody the measure μ_k. All subsequent probabilities will be identical no matter which of these was done. Building μ_k directly is typically easier because it doesn't need to cover the cases of when k is false.

What is important is that there is a coherent stage where the probability model is reasonable and where every subsequent observation is conditioned on.

Note that any theorem about unconditional probabilities is a theorem about conditional probabilities if you add the same evidence to each probability. This is because the conditional probability is another probability measure.

Bayes' Theorem

When reasoning under uncertainty, you want to update the agent's beliefs in the light of new evidence. As new evidence arrives, the measure of belief needs to take into account this evidence. A new piece of evidence is conjoined to the old evidence to form the complete set of evidence. Bayes' theorem specifies how to update your belief in a proposition based on a new piece of evidence.

Suppose you have a current belief in proposition h based on evidence k, namely $P(h|k)$, and observe e. You are now interested in $P(h|e \wedge k)$. Bayes' theorem or Bayes' rule tells us how to update your belief in h as evidence is accumulated.

Proposition 10.4

Bayes' Theorem: As long as $P(e|k) \neq 0$, we have

$$P(h|e \wedge k) = \frac{P(e|h \wedge k) \times P(h|k)}{P(e|k)}.$$

This is often written with the background knowledge k implicit. In this case, if $P(e) \neq 0$, then

$$P(h|e) = \frac{P(e|h) \times P(h)}{P(e)}.$$

$P(e|h)$ is the **likelihood** of the hypothesis; $P(h)$ is the **prior** of the hypothesis. Bayes' theorem states that the posterior is proportional to the likelihood times the prior.

Proof:

Using the rule for multiplication in two different ways (using the commutativity of conjunction), we have

$$
\begin{aligned}
P(h \wedge e|k) &= P(h|e \wedge k) \times P(e|k) \\
&= P(e|h \wedge k) \times P(h|k).
\end{aligned}
$$

The theorem follows from dividing the right-hand sides by $P(e|k)$.
Q.E.D.

Often Bayes' theorem is used to compare various hypotheses (different h_i's), where it can be noticed that the denominator $P(e|k)$ is a constant normalizing factor that doesn't depend on the hypothesis. If you are comparing only the relative posterior probabilities of hypotheses, you can ignore the denominator. If you actually want the posterior probability, the denominator can be computed by reasoning by cases: If $\{h_i\}$ is the exclusive and covering set of propositions representing all possible hypotheses, then

$$P(e|k) = \sum_{h_i} P(e|h_i \wedge k) \times P(h_i|k).$$

Thus the denominator of Bayes' theorem is obtained by summing the numerators for all the hypotheses.

Generally, one of $P(e|h \wedge k)$ or $P(h|e \wedge k)$ is much easier to estimate than the other. This often occurs when you have a causal theory of a domain, and the predictions of different hypotheses—the $P(e|h_i \wedge k)$ for each hypothesis h_i—can be derived from the domain theory.

Example 10.7

Suppose the diagnostic assistant is interested in the diagnosis of the light switch s_1 of Figure 3.1 (page 71). Suppose the status of s_1 is represented as a variable s_1_st with domain $\{ok, upside\_down, short, intermittent, broken\}$, where $s_1\_st = ok$ means switch s_1 is working normally. $s_1\_st = upside\_down$ means switch s_1 is installed upside down. $s_1\_st = short$ means switch s_1 is shorted and acting as a wire. $s_1\_st = intermittent$ means switch s_1 sometimes lets power flow and sometimes doesn't, independently of the switch's position. $s_1\_st = broken$ means switch s_1 is broken and doesn't allow electricity to flow.

Suppose your background knowledge is that the input wire, w_3, to the switch s_1 is live and that the switch s_1 is down. We will write the formula $(s_1\_pos = down \wedge live(w_3))$ as k. The following probabilities specify how the switch works when wire w_3 is live and switch s_1 is down:

$$P(live(w_2)|k \wedge s_1\_st = ok) = 1$$
$$P(live(w_2)|k \wedge s_1\_st = upside\_down) = 0$$
$$P(live(w_2)|k \wedge s_1\_st = short) = 1$$
$$P(live(w_2)|k \wedge s_1\_st = intermittent) = 0.5$$
$$P(live(w_2)|k \wedge s_1\_st = broken) = 0$$

These probabilities encode the meaning of the switch's states.

Suppose you have the prior probabilities

$$P(s_1\_st = ok) = 0.9$$
$$P(s_1\_st = upside\_down) = 0.02$$
$$P(s_1\_st = short) = 0.02$$
$$P(s_1\_st = intermittent) = 0.02$$
$$P(s_1\_st = broken) = 0.04$$

Assume these probabilities don't depend on the switch's position or on whether w_3 is live. Thus, you also know probabilities such as $P(s_1\_st = ok|k) = 0.9$.

Suppose the switch's output, namely w_2, is observed to be not live. This is represented by conditioning on the proposition $\neg live(w_2)$. Given this and the background knowledge, you can determine the posterior probabilities for the switch's status:

$$P(s_1\_st = ok|k \wedge \neg live(w_2))$$

$$= \frac{P(\neg live(w_2)|k \wedge s_1\_st = ok) \times P(s_1\_st = ok|k)}{P(\neg live(w_2)|k)}$$

$$= \frac{(1 - P(live(w_2)|k \wedge s_1\_st = ok)) \times P(s_1\_st = ok|k)}{P(\neg live(w_2)|k)}$$

$$= \frac{0 \times P(s_1\_st = ok|k)}{P(\neg live(w_2)|k)}$$

$$= 0$$

$$P(s_1\_st = upside\_down|k \wedge \neg live(w_2))$$

$$= \frac{P(\neg live(w_2)|k \wedge s_1\_st = upside\_down) \times P(s_1\_st = upside\_down|k)}{P(\neg live(w_2)|k)}$$

$$= \frac{1 \times 0.02}{P(\neg live(w_2)|k)}$$

$$P(s_1\_st = short|k \wedge \neg live(w_2))$$

$$= \frac{0 \times 0.02}{P(\neg live(w_2)|k)}$$

$$P(s_1\_st = intermittent|k \wedge \neg live(w_2))$$

$$= \frac{0.5 \times 0.02}{P(\neg live(w_2)|k)}$$

$$P(s_1\_st = broken|k \wedge \neg live(w_2))$$

$$= \frac{1 \times 0.04}{P(\neg live(w_2)|k)}$$

The denominator, $P(\neg live(w_2)|k)$, can be computed by adding the numerators

$$P(\neg live(w_2)|k)$$

$$= \sum_{s \in D} P(\neg live(w_2)|k \wedge s_1\_st = s) \times P(s_1\_st = s|k)$$

$$= 0 + 0.02 + 0 + 0.01 + 0.04$$

$$= 0.07.$$

This is a normalizing constant to make sure that the posterior probabilities sum to 1. Finally you have

$$P(s_1\_st = ok|k \wedge \neg live(w_2)) = 0$$

$$P(s_1\_st = upside\_down|k \wedge \neg live(w_2)) = \frac{2}{7}$$

$$P(s_1\_st = short|k \wedge \neg live(w_2)) = 0$$

$$P(s_1\_st = intermittent|k \wedge \neg live(w_2)) = \frac{1}{7}$$

$$P(s_1\_st = broken|k \wedge \neg live(w_2)) = \frac{4}{7}$$

Example
10.8

Suppose you have information about the reliability of fire alarms. You may know how likely it is that an alarm will work if there is a fire. If you would like to know the probability that there is a fire, given that there is an alarm, you can use Bayes' theorem:

$$P(\text{fire}|\text{alarm}) = \frac{P(\text{alarm}|\text{fire}) \times P(\text{fire})}{P(\text{alarm})}.$$

Here $P(\text{alarm}|\text{fire})$ is the probability that the alarm worked assuming that there was a fire. It's a measure of the alarm's reliability. The expression $P(\text{fire})$ is the probability of a fire given no other information. It's a measure of how fire-prone the building is. $P(\text{alarm})$ is the probability of the alarm sounding, given no other information.

Expected Values

You can use the probabilities to give the expected value of any numerical random variable (i.e., one whose domain is a subset of the reals). A variable's expected value is the variable's weighted average value, where its value in each possible world is weighted by the measure of the possible world.

Suppose V is a random variable whose domain is numerical, and suppose ω is a possible world. Define $\rho(V, \omega)$ to be the value x in the domain of V such that $\omega \models V = x$. That is, ρ is a function that returns the value of a variable in a world.

The **expected value** of numerical variable V, written $\mathbf{E}(V)$, is

$$\mathbf{E}(V) \quad = \quad \sum_{\omega \in \Omega} \rho(V, \omega) \times \mu(\omega).$$

Example
10.9

If *number_of_broken_switches* is an integer-valued random variable,

$$\mathbf{E}(\textit{number\_of\_broken\_switches})$$

would give the expected number of broken switches. If the world acted according to the probabilities, this would give the long-run average number of broken switches.

In a manner analogous to the semantic definition of conditional probability (page 354), the **conditional expected value** of variable V conditioned on evidence e, written $\mathbf{E}(V|e)$, is

$$\mathbf{E}(V|e) \quad = \quad \sum_{\omega \in \Omega} \rho(V, \omega) \times \mu_e(\omega)$$

which is equivalent to

$$\mathbf{E}(V|e) \quad = \quad \frac{1}{P(e)} \sum_{\omega \models e} \rho(V, \omega) \times \mu(\omega).$$

Example
10.10

> The expected number of broken switches given that light l_1 isn't lit is given by:
>
> $$\mathbf{E}(\textit{number\_of\_broken\_switches}|\neg\textit{lit}(l_1)).$$
>
> This is obtained by averaging the number of broken switches over all of the worlds in which light l_1 isn't lit.

Information Theory

A **bit** is a binary digit. Because a bit has two possible values, it can be used to distinguish two items. Often the two values are written as 0 and 1, but they can be any two different values.

Two bits can distinguish four items, each associated with either 00, 01, 10, or 11. Similarly, three bits can distinguish eight items. In general, n bits can distinguish 2^n items. Thus you can distinguish n items with $\log_2 n$ bits. It may be surprising, but you can do better than this if you take probabilities into account.

Example
10.11

> Suppose you want to design a code to distinguish the elements of the set $\{a, b, c, d\}$, with $P(a) = \frac{1}{2}$, $P(b) = \frac{1}{4}$, $P(c) = \frac{1}{8}$, and $P(d) = \frac{1}{8}$. Consider the following code:
>
> | | | | |
> |---|-----|---|-----|
> | a | 0 | c | 110 |
> | b | 10 | d | 111 |
>
> This code sometimes uses 1 bit and sometimes uses 3 bits. On average, it uses
>
> $$P(a) \times 1 + P(b) \times 2 + P(c) \times 3 + P(d) \times 3$$
> $$= \frac{1}{2} + \frac{2}{4} + \frac{3}{8} + \frac{3}{8} = 1\frac{3}{4} \text{ bits.}$$
>
> For example, the string $aacabbda$ has code 00110010101110 which uses 14 bits.
>
> With this code, you need $-\log_2 P(a) = 1$ bit to distinguish a from the other symbols. To distinguish b, you need $-\log_2 P(b) = 2$ bits. To distinguish c, you need $-\log_2 P(c) = 3$ bits.

In general, to identify x, you need $-\log_2 P(x)$ bits. If you have a distribution you want to identify, you need at least the expected number of bits it takes to identify a member, which is

$$\sum_x -P(x) \times \log_2 P(x).$$

This is called the **information content** or **entropy** of the distribution.

Analogously to conditioning in probability, you can determine the expected number of bits it takes to describe a distribution given evidence e:

$$I(e) = \sum_x -P(x|e) \times \log_2 P(x|e).$$

If you have a test that can distinguish the cases where α is true from the cases where α is false, the **information gain** from this test is:

$$I(true) - (P(\alpha) \times I(\alpha) + P(\neg\alpha) \times I(\neg\alpha)).$$

Here $I(true)$ is the expected number of bits needed before the test and $P(\alpha) \times I(\alpha) + P(\neg\alpha) \times I(\neg\alpha)$ is the expected number of bits after the test.

Applications of Information Theory include:

- In diagnosis, you can choose a test that provides the most information.

- In decision-tree learning (page 403), information theory provides a useful criteria for choosing which property to split on: Split on the property that provides the most information gain. The elements you need to distinguish are the different values in the target concept, and the probabilities are obtained from the proportion of each value in the training set remaining at each node.

- In Bayesian learning (page 424), information theory provides a basis for deciding which is the best model given some data.

10.3 Independence Assumptions

The axioms of probability are very weak, providing few constraints on allowable conditional probabilities. For example, if there are n binary variables, there are $2^n - 1$ different numbers to be assigned to give a complete probability distribution from which arbitrary conditional probabilities can be derived. To be able to determine any probability, you may have to start with an enormous database of conditional probabilities or of probabilities of possible worlds.

Two main approaches can overcome the need for so many numbers:

Independence The knowledge of the truth of one proposition doesn't affect the belief in another.

Maximum entropy or random worlds Given no other knowledge, we assume that everything is as random as possible. That is, the probabilities are distributed as uniformly as possible consistent with the available information.

We consider here in detail the first of these (but see the box on page 362).

As long as the value of $P(h|e)$ isn't 0 or 1, it doesn't constrain the value of $P(h|f \wedge e)$. This latter could have any value in the range $[0, 1]$: It is 1 when f implies h, and it is 0 if f implies $\neg h$.

Reducing the Numbers

The distinction between allowing representations of independence and using maximum entropy or random worlds highlights an important difference between views of a knowledge representation:

- The first view is that a knowledge representation provides a high-level symbolic modeling language that lets you model any domain in a reasonably natural way. According to this view, it is ok for knowledge representation designers to prescribe how to use the knowledge representation language. It is expected that they provide a user manual on how to describe domains of interest.

- The second view is that a knowledge representation should allow someone to add whatever knowledge they may have about a domain. The knowledge representation should fill in the rest in a common-sense manner. According to this view, it is unreasonable for a knowledge representation designer to specify how particular knowledge should be encoded.

These two different views need to be explicitly noted because judging a knowledge representation by the wrong criteria won't result in an adequate assessment.

Belief networks are a representation for a particular independence amongst variables. They should be judged as a modeling language. The main claim is that many domains can be concisely and naturally represented by exploiting the independencies that belief networks compactly represent. This doesn't mean that you can just throw in lots of facts (or probabilities) and expect a reasonable answer. One needs to think about a domain; and to consider exactly what variables are involved and what are the dependencies amongst the variables. When judged by this criterion, belief networks form a useful representation scheme.

Once the network structure and the domains of the variables for a belief network are defined, exactly which numbers are required (the conditional probabilities) is prescribed. The user can't simply add arbitrary conditional probabilities, but must follow the network's structure. If the numbers that are required of a belief network are provided and are locally consistent—for each assignment of values to the parents of a variable, the probabilities of the values for the variable given the assignment to the parents sum to one—then the whole network will be consistent. Each possible world will have a unique probability.

The maximum entropy or random worlds approaches (Bacchus, Grove, Halpern & Koller, 1996) form a probabilistic knowledge representation of the second type. Given particular probabilistic knowledge, they infer the most random worlds that are consistent with the knowledge.

For the random worlds approach, any numbers that happen to be available can be added and used. However, if you allow someone to add arbitrary probabilities, it's easy for the knowledge to be inconsistent with the axioms of probability.

Example 10.12	As far as probability theory is concerned, there is no reason why the name of the Queen of Canada shouldn't be as significant as a light switch's position in determining whether the light is on. Knowledge of the domain, however, may tell us that it's irrelevant.

In this section we present a representation that allows us to model the structure of the world where relevant propositions are local and where not-directly-relevant variables can be ignored when specifying probabilities. This structure can be exploited for efficient reasoning.

A common kind of qualitative knowledge is of the form $P(h|e) = P(h|f \wedge e)$. This equality says that f is irrelevant to the probability of h given e. For example, the fact that Elizabeth is the Queen of Canada is irrelevant to the probability that w_2 is live given that switch s_1 is down. This idea can apply to random variables as in the following definition:

Random variable x is **independent** of random variable y **given** random variable z if for all values of the random variables (i.e., for all a_i, b_j and c_k),

$$P(x = a_i|y = b_j \wedge z = c_k) = P(x = a_i|z = c_k).$$

That is, knowledge of y's value doesn't affect your belief in the value of x, given a value of z.

The following theorem gives an alternative characterization of independence:

Proposition 10.5	If random variable x is independent of random variable y given random variable z, then for all a_i, b_j, c_k,

$$P(x = a_i \wedge y = b_j|z = c_k) = P(x = a_i|z = c_k) \times P(y = b_j|z = c_k).$$

The proof is left as an exercise. (See Exercise 10.3.)

Belief Networks

The notion of conditional independence can be used to give a concise representation of many domains. The idea is that given a random variable v, there may be a small set of variables that directly affect the variable's value in the sense that every other variable is independent of v given values for the directly affecting variables. This locality is what is exploited in what are called **belief networks**.

A belief network, also called a **Bayesian network**, is a graphical representation of conditional independence. The independence allows us to depict direct effects within the graph and prescribes which probabilities need to be specified. Arbitrary posterior probabilities can be derived from the network.

Recall (page 116) that a directed graph consists of a set N of nodes and a set A of ordered pairs of nodes, called arcs. A directed acyclic graph, or DAG (page 116), is a graph such that there is no directed path from a node back to itself. If there is an arc

$\langle n_i, n_j \rangle \in A$, then n_i is said to be a **parent** of n_j and n_j is a **child** of n_i. Node n_i is an **ancestor** of node n_j if there is a directed path from n_i to n_j. Node n_i is a **descendant** of node n_j if $n_i = n_j$ or there is a directed path from n_j to n_i.

Formally, a **belief network** is a directed acyclic graph with nodes labeled with random variables, together with a domain for each random variable and a set of conditional probability tables for each variable given its parents. These conditional probability tables include prior probabilities for nodes with no parents.

The independence assumption embedded in a belief network is: *Each random variable is independent of its nondescendants given its parents*. That is, if x is a random variable with parents $y_1, \ldots, y_n$, all random variables that are not descendants of x are independent of x given $\langle y_1, \ldots, y_n \rangle$:

$$P(x = a | y_1 = v_1 \wedge \ldots \wedge y_n = v_n \wedge R)$$
$$= P(x = a | y_1 = v_1 \wedge \ldots \wedge y_n = v_n)$$

if R doesn't involve a descendant of x, including x itself. The right-hand side of this equation is the form of the probabilities that are specified as a part of the belief network.

R may involve ancestors of x and other nodes. The independence assumption states that all of the influence of nondescendant variables is captured by knowing the value of x's parents.

Often we refer to just the labeled DAG as a belief network. When this is done, it's important to remember that there is also a domain for each variable and a set of conditional probability tables associated with the network.

The number of probabilities that needs to be specified for each variable is exponential in the number of parents of the variable. The independence assumption is useful insofar as the number of variables that directly affect another variable is small. You want to structure your knowledge base so that nodes have as few parents as possible.

Example 10.13

Consider the wiring example of Figure 3.1 (page 71). Whether light l_1 is lit depends only on whether there is power in wire w_0 and whether light l_1 is ok. Other variables, such as the position of switch s_1, whether light l_2 is lit, or who the Queen of Canada is, are irrelevant. The only way to affect whether light l_1 is lit is to affect whether there is power in wire w_0 or to affect whether light l_1 is ok.

It's easy to see how switch s_1's position affects whether light l_1 is lit: changing the position affects whether there is power in wire w_0. The knowledge of whether l_2 is lit also affects your belief in whether l_1 is lit. Suppose the outside power is very erratic. Knowing that l_2 is lit provides evidence that there is power in wire w_3, which in turn affects your belief as to whether there is power in wire w_0. It's difficult to imagine how the identity of the Queen of Canada could affect whether the light is lit. However, to affect light l_1, any random variable either needs to affect whether there is power in wire w_0 or affect whether light l_1 is ok. This intuition

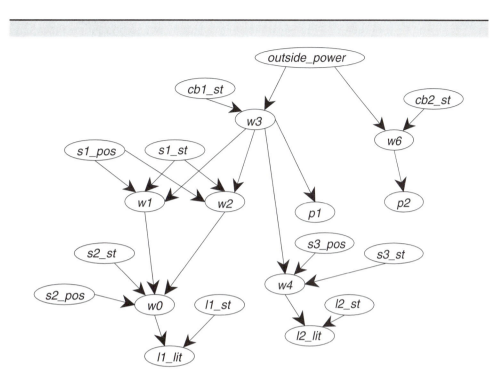

- w_0, a random variable with domain {*live, dead*}, denotes whether there is power in wire w_0. $w_0 = live$ means wire w_0 has power. $w_0 = dead$ means there is no power in wire w_0. Variables $w_1 \ldots w_6$ have the same domain.

- s_1_pos denotes the position of switch s_1. It has domain {*up, down*}. $s_1\_pos = up$ means switch s_1 is up.

- s_1_st is a random variable denoting the state of switch s_1. It has domain {*ok, upside_down, short, intermittent, broken*}. $s_1\_st = ok$ means switch s_1 is working normally. $s_1\_st = upside\_down$ means switch s_1 is installed upside down. $s_1\_st = short$ means switch s_1 is shorted and acting as a wire. $s_1\_st = broken$ means switch s_1 is broken and doesn't allow electricity to flow. The other switches have the same domain.

- cb_1_st denotes the state of circuit breaker cb_1. It has domain {*on, off*}, where $cb_1\_st = on$ means power can flow through cb_1 and $cb_1\_st = off$ means that power can't flow through cb_1. The same is true for cb_2.

- Variable l_1_st with domain {*ok, intermittent, broken*} denotes the state of light l_1. $l_1\_st = ok$ means light l_1 will light if powered, $l_1\_st = intermittent$ means light l_1 intermittently lights if powered, and $l_1\_st = broken$ means light l_1 doesn't work. Light l_2 is modeled analogously.

Figure 10.1: Belief network for the electrical domain of Figure 3.1

that there are few variables that affect the value of another is what a belief network exploits.

Figure 10.1 shows a belief network for the electrical domain of Figure 3.1. In this graph, nodes correspond to random variables. There is a random variable for whether there is power in each wire, for the position of each switch, for whether each light is lit, and for the state of each component (whether it is broken and, if so, how). Arcs correspond to what variables directly affect another variable. For example, there are arcs from variable w_0, representing whether there is power in wire w_0, and from variable l_1_st, representing the state of light l_1, into variable l_1_lit, representing whether l_1 is lit—these are the two variables that directly affect whether light l_1 is lit. Similarly, the only way to affect whether there is power in wire w_0 is to affect switch s_2's position (variable s_2_pos), whether s_2 is ok (variable s_2_st), whether there is power in wire w_1 (variable w_1), or whether there is power in wire w_2 (variable w_2). Thus there are arcs from these four variables into the variable w_0.

Note the restriction "each random variable is independent of its nondescendants given its parents" in the definition of the independence encoded in a belief network (page 364). If R contains a descendant of x, the independence assumption isn't directly applicable.

Example 10.14

In Figure 10.1, variables s_1_pos, s_1_st, and w_3 are the parents of variable w_1. If you know the values of s_1_pos, s_1_st, and w_3, then knowing whether or not l_2 is lit or knowing the value of cb_2_st won't affect your belief in whether there is power in w_1. However, even if you knew the values of s_1_pos, s_1_st, and w_3, learning whether l_1 is lit potentially changes your belief in whether there is power in wire w_1. The independence assumption isn't directly applicable.

The variable s_1_pos has no parents. Thus the independence embedded in the belief network specifies that $P(s_1\_pos = up|A) = P(s_1\_pos = up)$ for any A that doesn't involve a descendant of s_1_pos.

As part of a belief network, you provide a set of conditional probability tables that gives the probability of each value of a variable for each value of the variable's parents. Suppose $y_1, \ldots, y_n$ are the parents of random variable x. The sort of information you provide is of the form

$$P(x = a|y_1 = v_1 \wedge \ldots \wedge y_n = v_n)$$

for each value v_i of y_i and for each value a of x. You only need to give the value for all but one of the values of x because you can determine the probability of the remaining one from the others.

Belief Networks and Causality

Belief networks have often been called **causal networks**, and claimed to be a representation of causality. Although this is controversial, there is a good intuitive motivation for this. Suppose you have in mind a causal model of a domain, where the domain is specified in terms of a set of random variables. For each pair of random variables X_1 and X_2, if there is a direct causal inference from X_1 to X_2 (i.e., if there is some direct causal mechanism to give X_2 a value, and it depends on some value of X_1), add an arc from X_1 to X_2. You would expect that the causal model would obey the independence assumption of the belief network. Thus all of the conclusions of the belief network would be valid.

You would also expect such a graph to be acyclic: You don't want something eventually causing itself. This is reasonable if you consider that the random variables represent particular events rather than event types. For example, consider a causal chain that "being stressed" causes you to "work inefficiently," which in turn causes you to "be stressed." To break the apparent cycle, we can represent "being stressed" at different stages as different propositions that refer to different times. Being stressed in the past causes you to not work well at the moment which causes you to be stressed in the future. The variables have one value and shouldn't be seen as any sort of event types.

The belief network itself has nothing to say about causation, and it can represent noncausal independence, but it seems particularly appropriate when there is causality and locality in a domain. The notion of causality, however, makes it very natural to build a belief network. The idea is to determine what variables are relevant to the domain you want to represent, and add arcs that represent the local causality. The belief network of Figure 10.1 shows how this can be done for a simple domain.

There is one other feature of a network that you would like for a causal network, and this is the notion of an **intervention**. If someone were to come and artificially change one of the values of a variable, you would expect that the variable's descendants—but no other nodes—would be affected. (See the discussion on page 191.) These ideas have been pursued by Pearl & Verma (1991), Spirtes, Glymour & Scheines (1993), Druzdzel & Simon (1993), Heckerman & Shachter (1995), and Pearl (1995).

Finally, you can see how the causality in belief networks relates to causal and evidential reasoning discussed in Section 9.5 (page 335). A causal belief network can be seen as a way of axiomatizing in one direction, abducing to causes, and then predicting from there. A direct mapping exists between the logic-based abductive view discussed in Section 9.5 and belief networks: Belief networks can be modeled as logic programs with probabilities over possible hypotheses (Poole, 1993).

Example 10.15

For the network on Figure 10.1, part of the belief network is the table of probabilities of the form

$$P(w_1 = live|s_1\_pos = up \wedge s_1\_st = ok \wedge w_3 = live)$$
$$P(w_1 = live|s_1\_pos = up \wedge s_1\_st = ok \wedge w_3 = dead)$$
$$P(w_1 = live|s_1\_pos = up \wedge s_1\_st = upside\_down \wedge w_3 = live)$$
$$\vdots$$
$$P(w_1 = live|s_1\_pos = down \wedge s_1\_st = broken \wedge w_3 = dead)$$

There are two values for s_1_pos, five values for s_1_ok and two values for w_3, so there are $2 \times 5 \times 2 = 20$ different cases where you have to assign a value for $w_1 = live$. As far as probability theory is concerned, the probability for $w_1 = live$ for these 20 cases can be assigned arbitrarily. Of course, knowledge of the domain will constrain what values make sense. The values for $w_1 = dead$ can be computed from the values for $w_1 = live$ for each of these cases.

Because the variable s_1_st has no parents, you have to provide the prior distribution for the variable. You need to specify probabilities for all but one of the values; the other value is whatever number is needed to make the probabilities sum to one. Thus you need to specify four of the following five numbers:

$$P(s_1\_st = ok)$$
$$P(s_1\_st = upside\_down)$$
$$P(s_1\_st = short)$$
$$P(s_1\_st = intermittent)$$
$$P(s_1\_st = broken)$$

The other variables are represented analogously.

Belief Networks as a Factorization of Probabilities

Here we give a different intuition for belief nets in terms of a factorization of a joint probability distribution that leads to an alternative definition. This is used to build efficient implementations. (See the section on implementation on page 376.)

Order the variables $x_1, \ldots, x_n$ in a belief network so that the parents of each variable are before the variable. Let π_{x_i} be the tuple of the parents of variable x_i—this forms a complex random variable (page 348). Let π_{v_i} be the corresponding values.

The chain rule (Proposition 10.3 on page 354) shows how to decompose a conjunction into conditional probabilities. For a belief network, you have

$$P(x_1 = v_1 \wedge x_2 = v_2 \wedge \cdots \wedge x_n = v_n)$$
$$= P(x_1 = v_1) \times$$
$$\quad P(x_2 = v_2|x_1 = v_1) \times$$

$$P(x_3 = v_3 | x_1 = v_1 \wedge x_2 = v_2) \times$$

$$\vdots$$

$$P(x_n = v_n | x_1 = v_1 \wedge \cdots \wedge x_{n-1} = v_{n-1}).$$

Each of the $P(x_i = v_i | x_1 = v_1 \wedge \cdots \wedge x_{i-1} = v_{i-1})$ has the property that you are not conditioning on a descendant of variable x_i, given the ordering of the variables in a belief network. Thus you have, for each variable x_i,

$$P(x_i = v_i | x_1 = v_1 \wedge \cdots \wedge x_{i-1} = v_{i-1}) \quad = \quad P(x_i = v_i | \pi_{x_i} = \pi_{v_i})$$

You can now rewrite the chain rule as

$$P(x_1 = v_1 \wedge \cdots \wedge x_n = v_n) \quad = \quad \prod_{i=1}^{n} P(x_i = v_i | \pi_{x_i} = \pi_{v_i}).$$

This is often given as a definition of a belief network.

Reasoning in a Belief Network

This section gives some intuition behind how belief networks work. This description is designed for pedagogical purposes. Although it can be directly translated into code, it isn't the most efficient way to implement belief networks. (How to implement belief networks is discussed on page 376.)

Suppose you have Boolean variable y, which is the only parent of x. If e doesn't contain any descendants of x, then:

$$\begin{aligned}
P(x|e) &= P(x|y \wedge e) \times P(y|e) + P(x|\neg y \wedge e) \times P(\neg y|e) \\
&= P(x|y) \times P(y|e) + P(x|\neg y) \times P(\neg y|e) \\
&= P(x|y) \times P(y|e) + P(x|\neg y) \times (1 - P(y|e)).
\end{aligned}$$

The belief network specifies values for $P(x|y)$ and $P(x|\neg y)$. The only other value needed to compute $P(x|e)$ is $P(y|e)$. This can be computed analogously to how $P(x|e)$ is computed.

The case with multiple parents is similar. Suppose $y_1, \ldots, y_n$ are parents of x, and each y_i has domain D_i. Let $y = \langle y_1, \ldots, y_n \rangle$ and $D = D_1 \times \cdots \times D_n$. If e doesn't involve any descendants of x, then:

$$\begin{aligned}
p(x|e) &= \sum_{d \in D} P(x|y = d \wedge e) \times P(y = d|e) \\
&= \sum_{d \in D} p(x|y = d) \times P(y = d|e).
\end{aligned}$$

Recall that the probabilities $P(x|y = d)$ are specified as part of the belief network.

The definition of the independence in a belief network explicitly excluded conditioning on a descendant of a variable. To derive $P(a|e)$ where e is a conjunction that contains a descendant of variable a, the independence assumption isn't directly

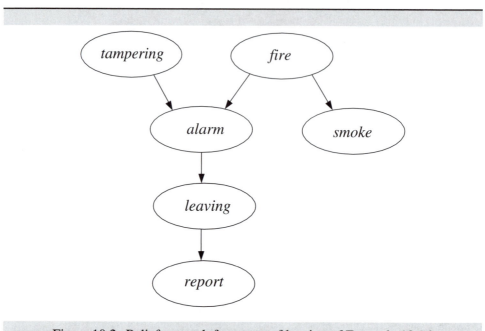

Figure 10.2: Belief network for report of leaving of Example 10.16

applicable, even if all of the parents of a are also part of e. If e specifies a value for a—for example, contains $a = v_1$—then that value is all that is needed to determine the probability of a. Otherwise, suppose e is a conjunction $e_1 \wedge e_2$ where e_1 involves only descendants of a, and e_2 contains no descendants of a. Bayes' rule can be used to compute $P(a|e)$:

$$P(a|e_1 \wedge e_2) = \frac{P(e_1|a \wedge e_2) \times P(a|e_2)}{P(e_1|e_2)}.$$

The right-hand side of this equation has no instance of conditioning on a descendant: Neither a nor e_2 involves a descendant of e_1, and e_2 doesn't involve a descendant of a.

Example 10.16

Suppose you want to use the diagnostic assistant to diagnose whether there is a fire in a building, based on noisy sensor information and possibly conflicting explanations of what could be going on. The agent receives a report about whether everyone is leaving the building. Suppose the report sensor is noisy: It sometimes reports leaving when there is no exodus, a false positive, and sometimes doesn't report when everyone is leaving, a false negative. Suppose the fire alarm going off can cause the leaving. Either tampering or fire could cause the alarm. Fire also causes smoke to rise from the building. The belief network of Figure 10.2 can express such knowledge.

The variable *report* denotes the sensor report that people are leaving. This variable is introduced to allow conditioning on unreliable sensor data. The infobot knows what the sensor reports, but it only has unreliable evidence about people leaving the building. It can condition only on what it knows—in this case, on what the sensor reports.

As well as the graph depicted in Figure 10.2, you also need to specify the domain of each variable and the conditional probabilities of each variable given each assignment of values to its parent. For this example, assume that the variables are all Boolean, with the following probabilities:

$$P(alarm|fire \land tampering) = 0.5$$
$$P(alarm|fire \land \neg tampering) = 0.99$$
$$P(alarm|\neg fire \land tampering) = 0.85$$
$$P(alarm|\neg fire \land \neg tampering) = 0.0001$$
$$P(smoke|fire) = 0.9$$
$$P(smoke|\neg fire) = 0.01$$
$$P(leaving|alarm) = 0.88$$
$$P(leaving|\neg alarm) = 0.001$$
$$P(report|leaving) = 0.75$$
$$P(report|\neg leaving) = 0.01$$
$$P(fire) = 0.01$$
$$P(tampering) = 0.02$$

The probabilities of a variable given nondescendants can be computed using the "reasoning by cases" rule (page 352). The probability of people leaving the building, given there is smoke, can be derived using

$$
\begin{aligned}
&P(leaving|smoke) \\
&= \ P(leaving|alarm \land smoke) \times P(alarm|smoke) \\
&\quad + P(leaving|\neg alarm \land smoke) \times (1 - P(alarm|smoke)) \\
&= \ P(leaving|alarm) \times P(alarm|smoke) \\
&\quad + P(leaving|\neg alarm) \times (1 - P(alarm|smoke)).
\end{aligned}
$$

The probabilities $P(leaving|alarm)$ and $P(leaving|\neg alarm)$ are provided as part of the belief network.

It remains to compute $P(alarm|smoke)$. A case analysis on $P(alarm|smoke)$ gives

$$
\begin{aligned}
&P(alarm|smoke) \\
&= \ P(alarm|fire \land tampering) \times P(fire \land tampering|smoke) \\
&\quad + P(alarm|fire \land \neg tampering) \times P(fire \land \neg tampering|smoke) \\
&\quad + P(alarm|\neg fire \land tampering) \times P(\neg fire \land tampering|smoke)
\end{aligned}
$$

$$+ \ P(alarm|\neg fire \wedge \neg tampering) \times P(\neg fire \wedge \neg tampering|smoke).$$

The left-hand side of each product is given as part of the belief network. The right-hand sides can be derived using the multiplicative rule. For example,

$P(fire \wedge tampering|smoke)$

$= \ P(fire|tampering \wedge smoke) \times P(tampering|smoke).$

Random variable *smoke* isn't a descendant of *tampering*, and tampering has no parents; thus $P(tampering|smoke)$ is the prior probability $P(tampering)$ which is provided as part of the network.

To derive $P(fire|tampering \wedge smoke)$, you can't use the independence assumption because *smoke* is a descendant of *fire*. You can, however, use Bayes' rule:

$P(fire|tampering \wedge smoke)$

$$= \ \frac{P(smoke|fire \wedge tampering) \times P(fire|tampering)}{P(smoke|tampering)}.$$

The resulting formula has no instances of conditioning on a descendant.

The remainder of this example gives some numerical examples of what can be derived from the above belief network.

The prior probabilities (with no evidence) of each variable are (to about three decimal places):

$P(tampering) = 0.02$

$P(fire) = 0.01$

$P(report) = 0.028$

$P(smoke) = 0.0189$

Observing the report gives

$P(tampering|report) = 0.399$

$P(fire|report) = 0.2305$

$P(smoke|report) = 0.215$

As expected, the probability of both *tampering* and *fire* are increased by the report. Because *fire* is increased, so is the probability of *smoke*.

Suppose instead that *smoke* was observed:

$P(tampering|smoke) = 0.02$

$P(fire|smoke) = 0.476$

$P(report|smoke) = 0.320$

Note that the probability of *tampering* isn't affected by observing *smoke*, however, the probabilities of *report* and *fire* are increased.

Suppose that both the *report* and *smoke* were observed:

$$P(tampering|report \wedge smoke) = 0.0284$$

$$P(fire|report \wedge smoke) = 0.964$$

Observing both makes *fire* even more likely. However, in the context of the *report*, the presence of *smoke* makes *tampering* less likely. This is because the *report* is "explained away" by *fire*, which is now more likely.

Suppose instead that you had the *report*, but no *smoke*:

$$P(tampering|report \wedge \neg smoke) = 0.501$$

$$P(fire|report \wedge \neg smoke) = 0.0294$$

In the context of the *report*, *fire* becomes much less likely and so the probability of *tampering* increases to explain the *report*.

This example illustrates how the belief net independence assumption can be used to derive common-sense conclusions, and also demonstrates "explaining away" that is a consequence of the independence assumption of a belief network.

Constructing Belief Networks

To represent a domain in a belief network, you need to consider:

- What are the relevant variables?
- What is the relationship between them? This should be expressed in terms of local influence.
- What values should these variables take? This involves considering the level of detail at which you want to reason.
- How does the value of one variable depend on the variables that locally influence it (its parents)? This is expressed in terms of the conditional probability tables.

Example 10.17

Suppose you want the diagnostic assistant to be able to reason about the possible causes of a patient's aching hands and aching elbows. For this domain, there are the following possible answers to the above questions:

- The relevant variables are whether patients have aching elbows, whether they have aching hands, whether they have arthritis, whether they have tennis elbow, and whether they have dishpan hands.
- Whether patients have aching hands depends only on whether they have arthritis and whether they have dishpan hands. It doesn't directly depend on whether they have an aching elbow or whether they have tennis elbow. Whether the patient has dishpan hands doesn't directly depend on any of the other variables. Whether patients have an aching elbow depends only on whether they have tennis elbow and whether they have arthritis. Whether patients have

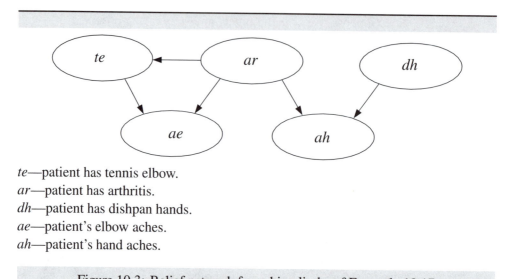

te—patient has tennis elbow.
ar—patient has arthritis.
dh—patient has dishpan hands.
ae—patient's elbow aches.
ah—patient's hand aches.

Figure 10.3: Belief network for aching limbs of Example 10.17

tennis elbow depends on whether they have arthritis. Whether patients have arthritis doesn't directly depend on any of the other variables. The belief network diagram of Figure 10.3 depicts these dependencies.

- Choosing the values for the variables involves considering the level of detail to reason at. You could encode the severity of each of the diseases and symptoms as values for the variables. You could, for example, use the values *severe*, *moderate*, *mild*, or *absent* for the arthritis variable. You could even model the disease at a lower level. For ease of exposition, we will model the domain at a very abstract level, only considering the presence or absence of symptoms and diseases. Each of the variables will be Boolean, representing the presence or absence of the associated disease or symptom.

- Next you need to assess how each variable depends on its parents. You specify this as conditional probabilities, is in

$$P(ah|ar \wedge dh) = 0.1$$
$$P(ah|ar \wedge \neg dh) = 0.99$$
$$P(ah|\neg ar \wedge dh) = 0.99$$
$$P(ah|\neg ar \wedge \neg dh) = 0.00001$$
$$P(ae|ar \wedge te) = 0.1$$
$$P(ae|ar \wedge \neg te) = 0.99$$
$$P(ae|\neg ar \wedge te) = 0.99$$
$$P(ae|\neg ar \wedge \neg te) = 0.00001$$

$$P(te|ar) = 0.0001$$
$$P(te|\neg ar) = 0.01$$
$$P(ar) = 0.001$$
$$P(dh) = 0.01.$$

The DAG of Figure 10.3 allows for arbitrary interaction between how dishpan hands (dh) and arthritis (ar) interact to produce aching hands (ah). The probabilities specify the interaction. In this example, if the patients have one of dishpan hands or arthritis, there is a good chance they will have aching hands. However, a patient with both has a low chance of aching hands. The network also specifies that the diseases are not independent: A patient with arthritis is less likely to have tennis elbow than one who didn't have arthritis.

The process of **diagnosis** is carried out by conditioning on the observed symptoms and deriving posterior probabilities of the faults or diseases.

This example also illustrates another example of "explaining away," and the preference for simpler diagnoses over complex ones.

Before any observations, $P(te) = 0.00999$. Once aching elbow is observed, tennis elbow becomes more likely, $P(te|ae) = 0.908$, arthritis becomes more likely, $P(ar|ae) = 0.0909$, and the probability of dishpan hands doesn't change, $P(dh|ae) = P(dh) = 0.01$.

Suppose both aching elbow and aching hands are observed: $P(te|ae \wedge ah) = 0.0917$, $P(ar|ae \wedge ah) = 0.908$, and $P(dh|ae \wedge ah) = 0.0926$. Once both symptoms are observed, the single cause that can explain both symptoms becomes very likely.

Example 10.18

Consider the belief network depicted in Figure 10.1 (page 365). Note the independence assumption embedded in this model: The DAG specifies that the lights, switches, and circuit breakers break independently. To model dependencies amongst how the switches break, you can add more arcs and perhaps more nodes. For example, if lights don't break independently because they come from the same batch, you can add an extra node that conveys the dependency. You would add a node that represents whether the lights come from a good or bad batch, which is made a parent of l_1_st and l_2_st. The lights can now break dependently. When you have evidence that one light is broken, the probability that the batch is bad may increase and thus make it more likely that the other light is bad. If you're not sure whether the lights are indeed from the same batch, you can add a node representing this, too. The important point is that the belief network provides a specification of independence that lets us model dependencies in a natural and direct manner.

The model implies that there is no possibility that there are shorts in the wires or that the house is wired differently from the diagram. In particular, it implies that w_0 can't be shorted to w_4 so that wire w_0 can get power from wire w_4. You could add extra dependencies that let us model each possible short. An alternative is to add an extra node that indicates that the model is appropriate. Arcs from this node would lead to each variable representing power in a wire and to each light. When the model is appropriate, you can use the probabilities of Example 10.15 (page 368). When the model is inappropriate, you can, for example, specify that each wire and light works at random. When there are weird observations that don't fit in with the original model—they are impossible or extremely unlikely given the model—the probability that the model is inappropriate will increase.

This network can be used in a number of ways, for example:

- By conditioning on the knowledge that the switches and circuit breakers are ok, and on the values of the outside power and the position of the switches, you can use this network to simulate how the lighting should work.

- Given values of the outside power and the position of the switches, you can determine how likely any outcome is—for example, how likely it is that l_1 is lit.

- Given values for the switches and whether the lights are lit, you can determine the posterior probability that each switch or circuit breaker is in any particular state.

- Given some observations, you can use the network to reason backwards to determine the most likely position of switches.

- Given some switch positions and some outputs and some intermediate values, you can determine the probability of any other variable in the network.

Implementing Belief Networks

The problem of determining posterior distributions—the problem of computing conditional probabilities given the evidence—is one that has been widely researched.

The problem of estimating the posterior probability in a belief network within an absolute error (of less than 0.5), or within a constant factor, is NP-hard, so general efficient implementations will not be available.

Three main approaches are taken to implement belief networks:

- Exploiting the structure of the network. This approach is typified by the *clique tree propagation method*, where the network is transformed into a tree with nodes labeled with sets of variables. Evidential reasoning is carried out by passing messages between the nodes in the tree. The values passed between the nodes are the distributions on the variables in common between the nodes. These distributions on a subset of the variables are called **marginal distributions**. One piece of evidence (an observation) can be entered with time complexity linear in the size of the tree. The size of the tree may, however, be exponential in the size of the belief network, but it's small when there are few multiple paths between nodes the belief networks. A related approach that exploits structure is detailed below.

- Search-based approaches. You enumerate some of the possible worlds, and you estimate posterior probabilities from the worlds generated. You can bound the probability mass of the worlds not considered, and you use this bound to estimate the error in the posterior probability. This approach works well when the distributions are extreme (all probabilities are close to zero or close to one).

- Stochastic simulation. In these approaches, random cases are generated according to the probability distributions. By treating these random cases as a set of samples, you can estimate the marginal distribution on any combination of variables.

The following gives a general algorithm for exploiting structure. Many of the efficient methods can be seen as optimizations of this algorithm.

An Algorithm For Evaluating Belief Networks

This section gives an algorithm for finding the posterior distribution for a variable in an arbitrarily structured belief network. The algorithm is based on the notion that a belief network specifies a factorization of the joint probability distribution (page 368). A Prolog implementation of this algorithm is given on page 521.

Before we give the algorithm, we need to define factors and the operations that will be performed on them.

A **factor** is a representation of a function from a tuple of random variables into a number. We will write factor f on variables $x_1, \ldots, x_j$ as $f(x_1, \ldots, x_j)$. The variables $x_1, \ldots, x_j$ are the variables **of** the factor f, and f is a factor **on** $x_1, \ldots, x_j$.

Suppose $f(x_1, \ldots, x_j)$ is a factor and each v_i is an element of the domain of x_i. $f(x_1 = v_1, x_2 = v_2, \ldots, x_j = v_j)$ is a number that is the value of f when each x_i has value v_i. You can assign some of the variables of a factor and make a new factor. For example, $f(x_1 = v_1, x_2, \ldots, x_j)$, sometimes written as $f(x_1, x_2, \ldots, x_j)_{x_1 = v_1}$, where v_1 is an element of the domain of variable x_1, is a factor on $x_2, \ldots, x_j$.

You can multiply factors together. Suppose f_1 and f_2 are factors, where f_1 is a factor that contains variables $x_1, \ldots, x_i$ and $y_1, \ldots, y_j$, and f_2 is a factor with variables $y_1, \ldots, y_j$ and $z_1, \ldots, z_k$, where $y_1, \ldots, y_j$ are the variables in common to f_1 and f_2. The **product** of f_1 and f_2 is a factor on the union of the variables, namely $x_1, \ldots, x_i, y_1, \ldots, y_j, z_1, \ldots, z_k$, defined by:

$$(f_1 \times f_2)(x_1, \ldots, x_i, y_1, \ldots, y_j, z_1, \ldots, z_k)$$
$$= f_1(x_1, \ldots, x_i, y_1, \ldots, y_j) f_2(y_1, \ldots, y_j, z_1, \ldots, z_k).$$

You can sum out a variable in a factor. Given factor $f(x_1, \ldots, x_j)$, summing out a variable, say x_1, results in a factor on $x_2, \ldots, x_j$ defined by:

$$\left(\sum_{x_1} f\right)(x_2, \ldots, x_j) = f(x_1 = v_1, \ldots, x_j) + \cdots + f(x_1 = v_k, \ldots, x_j)$$

where $\{v_1, \ldots, v_k\}$ is the set of possible values of variable x_1.

Given this definition, a conditional probability distribution $P(x|y_1, \ldots, y_j)$ can be seen as a factor f on $x, y_1, \ldots, y_j$, where

$$f(x = u, y_1 = v_1, \ldots, y_j = v_j) = P(x = u|y_1 = v_1 \wedge \cdots \wedge y_j = v_j).$$

Usually we will use the $P(\cdot|\cdot)$ notation rather than use f. The fact that it's a probability means that for all values u and $v_1, \ldots, v_j$,

$$P(x = u|y_1 = v_1, \ldots, y_j = v_j) \geq 0 \text{ and}$$
$$\sum_x P(x|y_1 = v_1, \ldots, y_j = v_j) = 1.$$

The **belief network inference problem** is the problem of computing the posterior distribution of a variable given some evidence.

The problem of computing posterior probabilities can be reduced to the problem of computing the probability of conjunctions. Given evidence $y_1 = v_1, \ldots, y_j = v_j$, and query variable z:

$$
\begin{aligned}
&P(z|y_1 = v_1, \ldots, y_j = v_j) \\
&= \frac{P(z, y_1 = v_1, \ldots, y_j = v_j)}{P(y_1 = v_1, \ldots, y_j = v_j)} \\
&= \frac{P(z, y_1 = v_1, \ldots, y_j = v_j)}{\sum_z P(z, y_1 = v_1, \ldots, y_j = v_j)}.
\end{aligned}
$$

So all you need to do is compute the factor $P(z, y_1 = v_1, \ldots, y_j = v_j)$ and normalize. Note that this is a factor only of z; given a value for z, this returns a number that is the probability of the conjunction of the propositions.

Suppose the variables of the belief network are $x_1, \ldots, x_n$. To compute the factor $P(z, y_1 = v_1, \ldots, y_j = v_j)$, you can sum out the other variables from the joint distribution. Suppose $z_1, \ldots, z_k$ is an enumeration of the other variables in the belief network—that is,

$$\{z_1, \ldots, z_k\} = \{x_1, \ldots, x_n\} - \{z\} - \{y_1, \ldots, y_j\}.$$

You can construct the desired factor by summing out the z_i. The order of the z_i is an **elimination ordering**.

$$P(z, y_1 = v_1, \ldots, y_j = v_j) = \sum_{z_k} \cdots \sum_{z_1} P(x_1, \ldots, x_n)_{y_1 = v_1, \ldots, y_j = v_j}.$$

Note how this is related to the semantics of probability (page 349): There is a possible world for each assignment of a value to each variable. The **joint probability distribution**, $P(x_1, \ldots, x_n)$ gives the probability (or measure) for each possible world. All you are doing is selecting the worlds with the observed values for the y_i's and summing over the possible worlds with the same value for z. This is the definition of conditional probability (page 353).

By the rule for conjunction of probabilities and the definition of a belief network,

$$P(x_1, \ldots, x_n) = P(x_1 | \pi_{x_1}) \times \cdots \times P(x_n | \pi_{x_n}),$$

where π_{x_i} is the set of parents of variable x_i.

We have now reduced the belief network inference problem to a problem of summing out a set of variables from a product of factors. To compute the posterior distribution of a query variable given observations:

1. Construct the joint probability distribution in terms of a product of factors.

2. Set the observed variables to their observed values.

3. Sum out each of the other variables (the $\{z_1, \ldots, z_k\}$).

4. Multiply the remaining factors and normalize.

To sum out a variable z from a product $f_1, \ldots, f_k$ of factors, you first partition the factors into those that don't contain z, say $f_1, \ldots, f_i$, and those that contain $z, f_{i+1}, \ldots, f_k$; then

$$\sum_z f_1 \times \cdots \times f_k = f_1 \times \cdots \times f_i \times \left(\sum_z f_{i+1} \times \cdots \times f_k \right).$$

You explicitly construct a representation (in terms of a multidimensional array, a tree, or a set of rules) of the rightmost factor. The factor's size is exponential in the number of variables of the factor.

Example 10.19	Consider Example 10.16 (page 370). Here two-letter abbreviations are used for the variables. The joint probability distribution is given by

$$P(ta, fi, sm, al, le, re)$$
$$= P(ta) \times P(fi) \times P(sm|fi) \times P(al|ta, fi) \times P(le|al) \times P(re|le).$$

Suppose that each variable has possible values $\{yes, no\}$. Given the query

$$P(ta|sm = yes \wedge re = yes)$$

and the elimination ordering fi, al, le, use the following equation:

$$P(ta \wedge sm = yes \wedge re = yes)$$
$$= \sum_{le} \sum_{al} \sum_{fi} \begin{array}{l} P(ta) \times P(fi) \times P(sm = yes|fi) \times \\ P(al|ta, fi) \times P(le|al) \times P(re = yes|le) \end{array}$$

You first sum out fi:

$$\sum_{fi} P(ta) \times P(fi) \times P(sm=yes|fi) \times$$
$$P(al|ta,fi) \times P(le|al) \times P(re=yes|le)$$
$$= P(ta) \times P(le|al) \times P(re=yes|le) \times$$
$$\sum_{fi} P(fi) \times P(sm=yes|fi) \times P(al|ta,fi)$$
$$= P(ta) \times P(le|al) \times P(re=yes|le) \times f_1(al,ta).$$

f_1 is a newly created factor. f_1 only depends on al and ta. For each combination of values of these variables there is a number obtained by summing the product $P(fi) \times P(sm=yes|fi) \times P(al|ta,fi)$ for each value of fi. Note how the factor's size depends on how many variables are connected to the summed-out variable.

Next sum out al:

$$\sum_{al} P(ta) \times P(le|al) \times P(re=yes|le) \times f_1(al,ta)$$
$$= P(ta) \times P(re=yes|le) \times \sum_{al} P(le|al) \times f_1(al,ta)$$
$$= P(ta) \times P(re=yes|le) \times f_2(le,ta),$$

where f_2 is a newly created factor that depends on le and ta.

You next sum out le:

$$\sum_{le} P(ta) \times P(re=yes|le) \times f_2(le,ta)$$
$$= P(ta) \times \sum_{le} P(re=yes|le) \times f_2(le,ta)$$
$$= P(ta) \times f_3(ta),$$

where f_3 is a newly created factor that depends on ta.

The posterior distribution on ta can be computed by multiplying these factors and normalizing.

Modern exact algorithms use what is essentially this method, and they speed it up by preprocessing as much as possible into a secondary structure before any evidence arrives. This is appropriate when, for example, the same belief network may be used for many different cases. They save intermediate results so that evidence can be incrementally added and so that each variable's probability can be derived after each addition of evidence.

This algorithm can be speeded up by pruning the irrelevant nodes from the network before the query starts. An approximation to what is relevant can be stated as follows: The query node is relevant, ancestors of relevant nodes are relevant, and observed descendants of relevant nodes are relevant. All other nodes are irrelevant. This will not remove relevant nodes but misses some irrelevant nodes. For example, if *le* were observed and the query variable were *re*, all other variables really are irrelevant. A more detailed specification is left as an exercise.

Unfortunately, extensive preprocessing, allowing arbitrary sequences of observations and deriving the posterior on each variable, precludes pruning the network. So for each application you need to choose whether you will save more by pruning irrelevant variables for each query and observation or by preprocessing before you have any observations.

10.4 Making Decisions Under Uncertainty

The main reason you need probabilities is to make decisions under uncertainty. An agent's decision on what to do depends on two things:

- *What the agent believes.* You may be tempted to say "what is true in the world," but when an agent doesn't know what is true in the world, it can act based only on its beliefs. Sensing the world updates the agent's beliefs by conditioning on what is sensed.

- *The agent's goals.* When an agent has to reason under uncertainty, it has to consider not only what will most likely happen but also what may happen. Some possible outcomes may have much worse consequences than others. The notion of a "goal" here needs to be richer than the goals considered in Chapter 8 because you must specify the tradeoffs between different outcomes. For example, if some action results in a good outcome most of the time, but sometimes results in a disastrous outcome, it needs to be compared with doing an alternative action that results in the good outcome less often and the disastrous outcome less often. Decision theory specifies how to trade off the desirability of outcomes with the probabilities.

| Example 10.20 | Consider the problem of the delivery robot when there is uncertainty in the outcome of its actions. In particular, consider the problem of going from position *o*109 in Figure 8.1 (page 285) to the *mail* position, where there is a chance that the robot will slip off course and fall down the stairs. Suppose that you can get pads for the robot that won't change the probability of an accident, but will make it less severe. Unfortunately, the pads add extra weight. The robot could also go the long way around, which would reduce the probability of an accident but make the trip much slower. |

Figure 10.4 shows a **decision tree** that depicts the different choices and outcomes. To read the decision tree, you start from the left. From each node one of

Whose Values?

Any computer program or person who acts or gives advice is using some value system of what is important and what isn't.

> *Alice ... went on "Would you please tell me, please, which way I ought to go from here?"*
>
> *"That depends a good deal on where you want to get to," said the Cat.*
> *"I don't much care where —" said Alice.*
> *"Then it doesn't matter which way you go," said the Cat.*

Lewis Carroll, 1832–1898
Alice's Adventures in Wonderland, 1865

We all, of course, want computers to work on *our* value system, but they can't act according to everyone's value system! When you build programs to work in a laboratory, this isn't usually a problem. The program is acting according to the goals and values of the program's designer, who will also be the program's user. When there are multiple users of a system, you need to be aware of whose value system is incorporated into a program. If a company sells a medical diagnostic program to a doctor, does the advice the program gives reflect the values of society, the company, the doctor, or the patient (all of whom may have very different value systems)? Does it determine the doctor's or the patient's values?

If you want to build an expert system that gives advice to someone, you need to find out what is true as well as what their values are. For example, in a medical diagnostic system, the appropriate procedure depends not only on patients' symptoms, but also on their priorities. Are they prepared to put up with some pain in order to be more aware of their surroundings? Are they willing to put up with a lot of discomfort in order to live a bit longer? What risks are they prepared to take? Always be suspicious of a program or person that tells you what to do if it doesn't ask you what you want to do! As builders of programs that do things or give advice, you need to be aware of whose value systems are incorporated into the actions or advice.

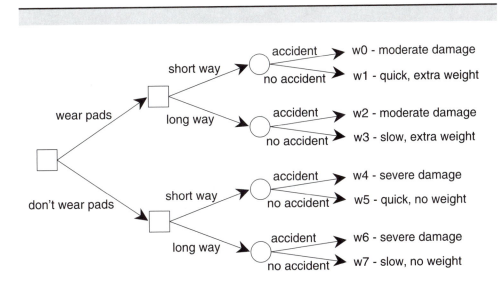

accident → w0 - moderate damage
no accident → w1 - quick, extra weight

accident → w2 - moderate damage
no accident → w3 - slow, extra weight

accident → w4 - severe damage
no accident → w5 - quick, no weight

accident → w6 - severe damage
no accident → w7 - slow, no weight

Square boxes represent decisions that the robot can make. Circles represent random variables that the robot can't observe before making its decision.

Figure 10.4: Decision tree for delivery robot decision of Example 10.20

the branches can be followed. In the decision trees studied in this chapter, for some of the branches (square boxes), the agent gets to choose which branch to take. For other nodes (circles), there is a probability that either branch will be be taken in the current context. The nodes at the right (the w_i) specify the outcome; what will be true if the path to that node is followed.

This example is analogous to the problem (discussed in Example 10.1) of deciding whether to wear a seat belt when driving or whether to stay at home. The agent has to decide whether to wear the pads and which way to go (the long way or the short way). What isn't under its direct control is whether there is an accident, although this probability can be reduced by going the long way around. For each combination of the agent's choices and whether there is an accident, there is an outcome, ranging from severe damage to arriving quickly without the extra weight of the pads. How should an agent decide what to do? This depends on how important it is to arrive quickly, how much the pads' weight matters, how much it's worth to reduce the damage from severe to moderate, and how likely an accident is, given either way the robot could go. All of these may seem incomparable, but when the agent makes a decision it compares them, implicitly or explicitly.

Utility

What an agent decides to do should depend on what it expects the outcomes of the action will be. The **utility** is a real-valued random variable that reflects the desirability of the outcome (the world that results from the decision). The value of the utility U in world ω, namely $\rho(U, \omega)$, reflects the desirability of world ω. (See page 359 for a definition of the function ρ.)

The higher the utility, the more desirable the world. You could alternatively use the concept of **cost**, which is the negative of utility. In other words, the lower the cost, the more desirable the world.

Before we describe a general notion of utility, we assume that we can equate utility with money. Let's initially assume that each agent is trying to maximize its **expected monetary value** (EMV). The best decision is the one given which the agent expects to end up with the most money. This may be sensible if the values are those of a company whose only concern is making money, or if the outcomes are monetary and the amount of money is small compared to the total amount an agent has. Nothing about decision theory forces money to be the measure of utility. A more general notion of utility that can be used as a measure of any sort of outcome is discussed later (page 392).

Decision Variables

As well as having a utility that specifies how desirable a world is, you also have to have a way for agents to affect what the utility will be. In addition to random variables, there are also **decision variables**. A decision variable is like a random variable, with a domain, but it doesn't have an associated probability distribution. Instead, an agent chooses a value for each decision variable. A possible world specifies values for both random and decision variables, and for each combination of values to decision variables, there is a probability distribution over the random variables. That is, for each assignment of one value to each decision variable, the measures of the worlds that satisfy that assignment sum to 1. Conditional probabilities will only be defined when a value for every decision variable is part of what is conditioned on.

Example 10.21

In Example 10.20, the decision variables correspond to the choices of whether to wear pads and which route to take. There are eight possible worlds, corresponding to the eight paths in the decision tree of Figure 10.4. Each of these worlds has a utility for the agent. The agent picks values for its decision variables. For each choice there is a probability over the two possible worlds that the robot could end up in.

In **diagnosis**, decision variables correspond to various treatments and tests. The utility may depend on, for example, the drugs the patient takes, the patient's diseases and allergies, what interventions occurred, and when the treatment occurs. The utilities typically depend on both decision variables and uncertain values. They could depend on, for example, which disease a patient has (all you have is the symptoms) or what they really are allergic to (you only have a record of their previous reactions).

Simple Decisions

In a simple decision, the agent chooses a value for each decision variable. You can treat all the decision variables as a single decision variable that is the tuple of all the decision variables. We call the resulting compound decision variable d.

A **single decision** is an assignment of a value to the decision variable. The **expected utility** of single decision $d = d_i$ is

$$\mathbf{E}(U|d = d_i) = \sum_{\omega \models (d=d_i)} \rho(U, \omega) \times \mu(\omega),$$

where $\mu(\omega)$ is the probability measure of world ω, and $\rho(U, \omega)$ is the value of the utility U in world ω.

An **optimal simple decision** is the decision whose expected utility is maximal. That is, $d = d_{max}$ is an optimal decision if

$$\mathbf{E}(U|d = d_{max}) = \max_{d_i \in dom(d)} \mathbf{E}(U|d = d_i).$$

where $dom(d)$ is the domain of decision variable d. Where d is a complex decision, $dom(d)$ is the cross product of the domains of the individuals decisions that make up d (as for random variables; see page 348).

Example 10.22

The delivery robot problem of Example 10.20 is a simple decision problem where the robot has to decide on the values for the variables *wear_pads* and *which_way*. The single decision is the complex decision variable ⟨*wear_pads*, *which_way*⟩. Each assignment of values to a decision has an expected value. For example, the expected utility of *wear_pads* ∧ *which_way* = *short* is given by

$$\mathbf{E}(U|wear\_pads \wedge which\_way = short)$$
$$= P(accident|wear\_pads \wedge which\_way = short) \times utility(w_0)$$
$$+ (1 - P(accident|wear\_pads \wedge which\_way = short)) \times utility(w_1)$$

where the worlds w_0 and w_1 are as in Figure 10.4.

Sequential Decisions

Generally, agents don't make decisions in the dark, without observing something about the world, nor do they make just a single decision. A more typical scenario is where the agent makes an observation, decides on and carries out an action, observes parts of the world, and then makes another decision conditioned on the observation, and so on. Subsequent actions can depend on what is observed, and what is observed can depend on previous actions. In this scenario, it's often the case that the sole reason for carrying out an action is to provide information for future actions. An example of this is in diagnosis, where one action may be to test a patient (which may even have detrimental effects on the patient), so that future treatment can be contingent on the test's results.

A **sequential decision problem** is a sequence of decision variables $d_1, \ldots, d_n$, where each d_i has an associated set of variables called the **information set** of d_i, written π_{d_i}, which is the tuple of variables whose value will be known at the time decision d_i is made. Acyclic restrictions are also needed because an agent can't observe the outcome of future actions when making a decision. These are formalized in the next section.

Given a sequential decision problem, a **policy** is a set δ of functions, one for each d_i, $\delta_i : dom(\pi_{d_i}) \rightarrow dom(d_i)$. This policy means that when the agent has observed $O \in dom(\pi_{d_i})$, it will do $\delta(O)$.

Decision Networks

A decision network (also called an **influence diagram**) is a graphical representation of a simple decision problem or a finite sequential decision problem. Decision networks are extensions of belief networks to include decision variables and utility.

A **decision network** is a directed acyclic graph (DAG) with three types of nodes:

- **Chance nodes** (drawn as ellipses) are the same type of nodes that are in a belief network. As in a belief network, chance nodes have random variables, domains, and probability tables associated with them. The parents of a chance node represent the variables that directly influence the variable corresponding to the chance node.

- **Decision nodes** (drawn as rectangles) are labeled with decision variables whose values can be set by the decision maker. Arcs coming into these nodes represent the information that will be available when the decision is made. The parents of decision node d will be the nodes corresponding to the elements of the information set π_d. We assume that the decision nodes are totally ordered, and we assume a **no-forgetting** agent where, if d_i is before d_j in the total ordering, then $d_i \in \pi_{d_j}$ and $\pi_{d_i} \subset \pi_{d_j}$. That is, the agent remembers which choices it has made and remembers what values it previously observed.

- A **value node** (drawn as a diamond) represents the utility. There is only one value node. The arcs coming into the value node represent the variables on which the utility depends.

Example 10.23 | Figure 10.5 shows a decision network for the delivery robot decision of Example 10.20. The agent has two decisions to make. The first is whether to wear pads, the second is which way to go. The agent's decision on which way to go can depend on

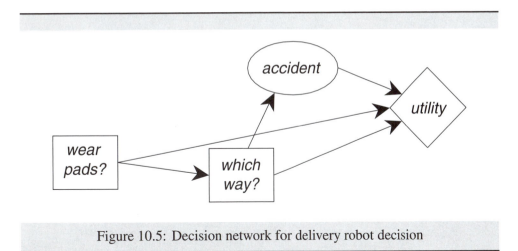

Figure 10.5: Decision network for delivery robot decision

whether it will wear pads or not. There is a random node (that isn't observed when the decisions are made), namely whether there is an accident. Whether there is an accident depends on which way the agent goes and doesn't depend on whether the agent is wearing pads. Compare the decision network to the decision tree representation in Figure 10.4 (page 383). The decision network is more compact, and it makes explicit the random and decision variables that influence other random variables and are observed when a decision is made.

Example 10.24

Figure 10.6 gives a decision network that is an extension of the belief network of Figure 10.2. In this sequential decision problem, there are two decisions to be made. First, the agent must decide whether to check for smoke. The information that will be available when it makes this decision is whether there is a report of people leaving the building. Second, the agent must decide whether or not to call the fire department. When making this decision, the agent will know whether there was a report, whether it checked for smoke, and whether it can see smoke.

The information that is needed, apart from that of the belief network and the domains of the new variables, is:

- how seeing smoke depends on whether the agent looks for smoke and whether there is smoke. We assume here that the agent has a perfect sensor for smoke: It will see smoke if and only if it looks for smoke and there is smoke. (See Exercise 10.6.)
- how the utility depends on whether the agent checks for smoke, whether there is a fire, and whether the fire department is called. Figure 10.7 provides this value information.

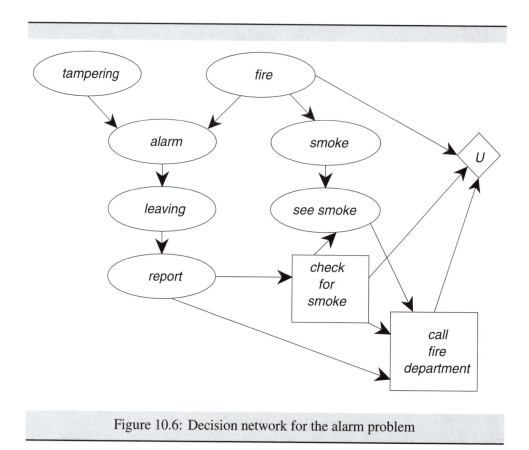

Figure 10.6: Decision network for the alarm problem

check_for_smoke	fire	call_fire_department	value (dollars)
yes	yes	call	−210
yes	yes	don't call	−5010
yes	no	call	−210
yes	no	don't call	−10
no	yes	call	−200
no	yes	don't call	−5000
no	no	call	−200
no	no	don't call	0

Figure 10.7: Value for alarm decision network

Expected Value of a Policy

Recall (page 386) that a policy δ is a tuple of functions, $\delta_i : dom(\pi_{d_i}) \rightarrow dom(d_i)$, one for each decision variable d_i. A policy specifies what the agent will do for each value that it could sense. The general idea is for the agent to choose the policy that maximizes its expected utility. In order to do this you have to define the expected utility of a policy.

Possible world ω **satisfies** policy δ, written $\omega \models \delta$ if

$$\forall \delta_i \in \delta, \omega \models \delta_i(\rho(\pi_{d_i}, \omega)),$$

where $\rho(\pi_{d_i}, \omega)$ is the value of the parents of decision d_i in world ω.

It's important to realize that a possible world here corresponds to a complete history, and specifies the values of all random and decision variables, including all observed variables for a complete sequence of actions. Possible world ω satisfies δ if ω is one possible unfolding of history given that the agent follows policy δ. The satisfiability constraint enforces the intuition that the agent will actually do the action prescribed by δ for each of the possible observations.

The **expected value of policy** δ is

$$\mathbf{E}(\delta) = \sum_{\omega \models \delta} \rho(U, \omega) \times P(\omega),$$

where $P(\omega)$ is the product of the probabilities of the values of the random nodes given their parents' values in ω, and $\rho(U, \omega)$ is the value of the utility U in world ω.

An **optimal policy** is a policy δ^* such that $\mathbf{E}(\delta^*) \geq \mathbf{E}(\delta)$ for all policies $\mathbf{E}(\delta)$.

Suppose a binary decision node has n binary parents. There are 2^n different assignments of values to the parents, and consequently there are 2^{2^n} different possible decision functions for this decision node. Thus an algorithm that enumerates the set of policies looking for the best one will be very inefficient.

Fortunately, you don't have to enumerate all of the policies, but can use **dynamic programming** (page 145) to find the optimal policy. The idea is to first consider the *last* decision and find the optimal decision for each value of its parents. Once you have found the optimal choices for the last decision, you can replace this decision node by a **deterministic node**—a random node that, for each assignment of values to its parents, has probability 1 for the value that the agent will choose if it observes those values and probability 0 for the alternatives the agent won't choose. You then have a new decision network with a new last decision, and you can solve this recursively.

| Example 10.25 | Let's continue Example 10.24. The last decision is *call_fire_department*. You have to determine the optimal decision for each of the values of the parents of *call_fire_department*.

Consider the case where there is a report and where the agent checked for smoke but didn't see smoke. In this case,

$$\mathbf{E}(V|call \wedge check \wedge report \wedge \neg see\_smoke) = -210$$

(because it costs \$210 when it calls, independently of whether there is a fire). You also have to consider the expected cost of not calling:

$$\mathbf{E}(V|\neg call \wedge check \wedge report \wedge \neg see\_smoke)$$
$$= -5010 \times P(fire|check \wedge report \wedge \neg smoke)$$
$$\quad - 10 \times P(\neg fire|check \wedge report \wedge \neg smoke)$$
$$= -5010 \times 0.0294 - 10 \times 0.9706$$
$$= -157.$$

When there is a report and the agent checks and there is no smoke, it expects to lose less when it doesn't call than when it does call. Thus the agent shouldn't call the fire department when there is a report, it checks for smoke, and observes no smoke.

Consider the case where there is a report and it checks and observes smoke. In this case,

$$\mathbf{E}(V|call \wedge check \wedge report \wedge see\_smoke) = -200.$$

The expected cost of not calling can be computed using

$$\mathbf{E}(V|\neg call \wedge check \wedge report \wedge see\_smoke)$$
$$= -5010 \times P(fire|report \wedge smoke) - 10 \times P(\neg fire|report \wedge smoke)$$
$$= -5010 \times 0.964 - 10 \times 0.036$$
$$= -4830.$$

When there is a report and smoke, you expect to lose a lot more when you don't call than when you do call. Thus, the agent should call the fire department when there is a report, it checks, and it observes smoke.

The two cases where there is no report can be similarly considered.

You also need to consider the two cases when the agent doesn't check given there is a report or not. In both cases, the agent doesn't see smoke, independently of whether there is smoke.

The agent can thus determine the optimal last action based on each of the information sets of its parents. It can then use similar reasoning to determine what to do for the first action, given that it knows what it will do for the last action.

The result of evaluating the decision network is an optimal policy. The policy for the alarm decision network can be given as the program that specifies what the agent will do based on what it observes, as in:

$$check\_for\_smoke \leftarrow report$$
$$call\_fire\_department \leftarrow report \wedge check\_for\_smoke \wedge see\_smoke.$$

This specification assumes negation as failure. If the agent doesn't receive a report, it doesn't check for smoke, and it doesn't call the fire department. If the agent doesn't see smoke, it doesn't call the fire department.

The Value of Information

Example 10.26

In the above example, all that the action *check_for_smoke* does for the agent is to provide information about fire. Even though checking for smoke costs $10 and doesn't provide any direct reward, in the optimal policy it's worthwhile to check for smoke when there is a report because the agent can condition its further actions on the information obtained. The information about smoke is valuable to the agent. Even though smoke provides imperfect information as to whether there is fire, that information is still very useful for making decisions.

One of the important lessons from this example is that an information-seeking action, such as the *check_for_smoke* action, doesn't need to be treated differently from other actions, such as *call_fire_department*. As long as the agent can act, affect the world (in the example, checking for smoke affects the value of *see_smoke*), observe the results of the action, and condition future actions on the observation, an optimal policy will often include actions whose only purpose is to seek information. Most actions are not pure information actions but have both an information effect and an effect on the world.

Determining how much it would be worth to the agent to have some information is often useful:

The **value of information** i at decision d is the expected value of the optimal policy that can condition decision d and subsequent decisions on knowledge of i minus the expected value of the optimal policy that can't observe i. Thus in a decision network, it's the value of the optimal policy with i as a parent of d and subsequent decisions minus the value of the optimal policy without i as a parent of d.

This would be a bound on the amount the agent would be willing to pay for the information i at stage d. It is an upper bound on the amount that imperfect information about the value of i at decision d would be worth. Imperfect information is, for example, information available from a noisy sensor of i. It isn't worth paying more for a sensor of i than the value of information i.

The value of information has some interesting properties including:

- The value of information is never negative. The worst that can happen is that the agent can ignore the information.

- If an optimal decision is to do the same thing whether or not i is true, the value of information i is always zero.

Within a decision network, the value of information i at decision d can be evaluated by considering both

- the decision network with arcs from i to d and from i to subsequent decisions, and

- the decision network without such arcs.

The differences in the values of the optimal policies of these two decision networks is the value of information i at d. Note that something more sophisticated needs to be done when adding the arc from i to d causes a cycle.

Example 10.27

In the alarm problem, the agent may be interested in knowing whether it was worthwhile installing a relay for the alarm so that the alarm can be heard directly, instead of relying on the noisy sensor of people leaving. To determine how much a relay could be worth, you consider how much perfect information about the alarm would be worth. If the information is worth less than the cost of the relay, it isn't worthwhile installing the relay.

The value of information about the alarm can be obtained by solving the decision network of Figure 10.6 (page 388) and determining the value of the optimal policy as was done in Example 10.25. You also solve a decision network as in Figure 10.6, but with an arc from *alarm* to *check_for_smoke* and an arc from *alarm* to *call_fire_department*. This new decision network will have an optimal policy whose value is at least as good as the policy of Example 10.25. The differences in the values of the optimal policies for the two decision networks is the value of *alarm* at the decision *check_for_smoke*. (See Exercise 10.7.)

Utility

Example 10.28

Suppose you were given a choice between (a) guaranteed $1,000,000 or (b) a bet on a coin toss where you receive $2,000,000 if a head turns up and $10 if a tail turns up. The bet has a higher expected monetary value, but most people would prefer the guaranteed $1,000,000 rather than the gamble.

In this section we develop a way to specify the value to an agent of a decision outcome. This needs to be more than just a preference ordering of outcomes, as the values need to be combined with probabilities when an agent is uncertain about which outcome will arise. In the example above the agent must be able to compare a known outcome with an uncertain outcome.

The notion of **utility**, as developed in economics, is a good way to summarize values. Utility is a reflection of relative worth to the agent of different outcomes. Note that there is nothing objective about utilities: Each agent can have its own utilities.

Utilities are defined in terms of lotteries:

A **lottery** is a triple $\langle p, \alpha_0, \alpha_1 \rangle$, where $p \in [0, 1]$ and α_0 and α_1 are outcomes, where an outcome is either a possible world or another lottery. The lottery $\langle p, \alpha_0, \alpha_1 \rangle$ represents the outcome of receiving α_0 with probability p or α_1 with probability $(1-p)$.

Any *finite* distribution over possible worlds can be represented as a lottery or, equivalently, as a binary tree with possible worlds at the leaves, and a probability at each branch which gives the relative probabilities of the worlds that are descendants of each subtree. These binary trees are like decision trees (as in Figure 10.4), but with with binary random variables at the internal nodes.

The utility for an agent is defined in terms of a lottery. Let ω_0 be the worst possible world for the agent, and let ω_1 be the best possible world for that agent. For each other world ω, consider the agent having to decide between ω and the lottery $\langle p, \omega_1, \omega_0 \rangle$. Under reasonable assumptions, unless world ω is the worst possible, the agent would prefer ω to the lottery with $p = 0$ (in which case it's guaranteed to get the worst outcome). Similarly, unless world ω is the best possible, the agent would prefer the lottery with $p = 1$ (in which case it's guaranteed to get the best outcome). Under reasonable assumptions, there should be some number, u, such that when $p > u$ the agent prefers the lottery to ω, and when $p < u$ it prefers ω to the lottery. This number, u, is called the agent's **utility** for world ω.

An alternative definition is that an outcome has utility u if the agent is indifferent between that outcome and a lottery where it receives the best possible outcome with probability u and the worst possible outcome with probability $(1 - u)$.

It's now easy to see why utilities combine linearly with probabilities. If any outcome has utility u, the agent is indifferent between this outcome and the lottery $\langle u, \omega_1, \omega_0 \rangle$. Suppose outcome α_0 has utility u_0 and outcome α_1 has utility u_1. The agent is indifferent between α_0 and the lottery $\langle u_0, \omega_1, \omega_0 \rangle$ and between α_1 and the lottery $\langle u_1, \omega_1, \omega_0 \rangle$. Consider the lottery $\langle p, \alpha_0, \alpha_1 \rangle$. Presumably the agent will be indifferent between this lottery and the lottery $\langle p, \langle u_0, \omega_1, \omega_0 \rangle, \langle u_1, \omega_1, \omega_0 \rangle \rangle$. All we have done is replace outcomes that the agent is indifferent about. In this lottery, the agent expects to receive ω_1 with probability $u_0 \times p + u_1 \times (1 - p)$ and otherwise receive ω_0. This lottery is the same as the lottery $\langle u_0 \times p + u_1 \times (1 - p), \omega_1, \omega_0 \rangle$. So we expect the lottery $\langle p, \alpha_0, \alpha_1 \rangle$ to have utility $u_0 \times p + u_1 \times (1 - p)$. This one step can be repeated, reducing any distribution to a lottery between ω_1 and ω_0, where the utility represents the probability of receiving ω_1. Thus, we have argued that utilities defined in this way (unlike money) should be able to be combined with probabilities to produce an expected value that has some meaning. (See Exercise 10.8.)

A linear relationship doesn't usually exist between money and utility, even when the outcomes have a monetary value. People often are **risk-averse** when it comes to money. They would rather have $n in their hand than some randomized setup where they expect to receive $n, but could possibly receive more or less.

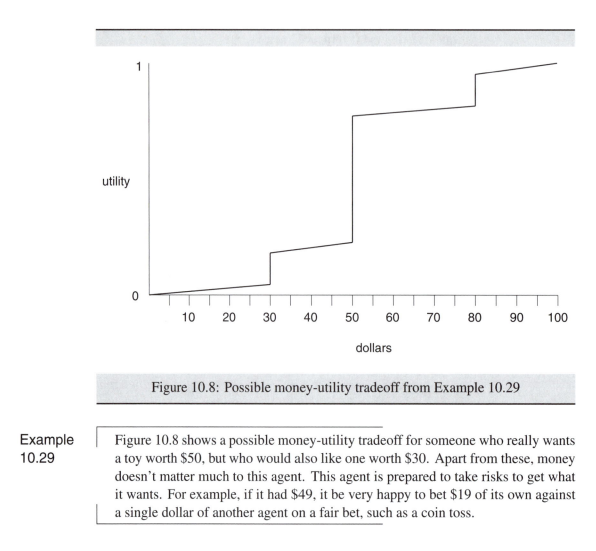

Figure 10.8: Possible money-utility tradeoff from Example 10.29

Example
10.29

Figure 10.8 shows a possible money-utility tradeoff for someone who really wants a toy worth $50, but who would also like one worth $30. Apart from these, money doesn't matter much to this agent. This agent is prepared to take risks to get what it wants. For example, if it had $49, it be very happy to bet $19 of its own against a single dollar of another agent on a fair bet, such as a coin toss.

10.5   References and Further Reading

Shafer & Pearl (1990) contains a collection of papers on reasoning under uncertainty. Jaynes (1985) and Loredo (1990) present introductions to Bayesian methods in science, with historical perspectives, and with comparisons to orthodox frequentist statistics. Halpern (1997) reviews the relationship between logic and probability.

Introductions to probability theory from an AI perspective, and belief (Bayesian) networks are in Pearl (1988b), Jensen (1996) and Castillo, Gutiérrez & Hadi (1996). The algorithm for evaluating belief networks is based on the algorithms of Dechter (1996) and Zhang & Poole (1996).

Decision networks or influence diagrams are due to Howard & Matheson (1981). See the papers in Oliver & Smith (1990). A method using dynamic programming for solving influence diagrams can be found in Shachter & Peot (1992). The value of information is discussed in Matheson (1990). For a comprehensive review of information theory see Cover & Thomas (1991). The notion of utility is due to Von Neumann & Morgenstern (1953), who also introduce the multiagent version of decision theory, *game theory.* See Feldman & Sproull (1977) for an early advocation of decision theory in AI. The annual conferences on *Uncertainty in Artificial Intelligence* provide up-to-date research results.

10.6 Exercises

Exercise 10.1
Prove Proposition 10.1 (page 352), namely that the axioms of probability (Section 10.2) are sound and complete with respect to the semantics of probability. Hint: for soundness, show that each of the axioms is true based on the semantics. For completeness, construct a probability measure from the axioms.

Exercise 10.2
Prove Proposition 10.2 (page 352).
Hint: The formulae $f \vee \neg f$ and $\neg(f \wedge \neg f)$ are tautologies, as are $f \leftrightarrow ((f \wedge g) \vee (f \wedge \neg g))$ and $\neg((f \wedge g) \wedge (f \wedge \neg g))$ and $(f \vee g) \leftrightarrow ((f \wedge \neg g) \vee g)$.

Exercise 10.3
Using only the axioms of probability and the definition of conditional independence, prove the following:

(a) Proposition 10.5 (page 363)

(b) Independence is symmetric; that is, if x is independent of y given z, then z is independent of x given y

Exercise 10.4
Consider the belief network of Figure 10.9 that extends the electrical domain to include the overhead projector. Answer the following questions about how knowledge of the values of some variables would affect the probability of another variable:

(a) Can knowledge of the value of *projector_plugged_in* affect your belief in the value of *alan_reading_book*? Explain.

(b) Can knowledge of *screen_lit_up* affect your belief in *alan_reading_book*? Explain.

(c) Can knowledge of *projector_plugged_in* affect your belief in *alan_reading_book* given that you have observed a value for *screen_lit_up*? Explain.

(d) Which variables could have their probabilities changed if *lamp_works* were observed?

(e) Which variables could have their probabilities changed if *power_in_projector* were observed?

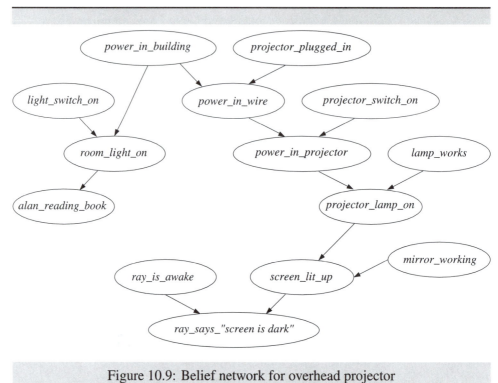

Figure 10.9: Belief network for overhead projector

Exercise 10.5 How sensitive are the answers from the decision network of Example 10.24 (page 387) to the probabilities? Test the program with different conditional probabilities, and see what effect this has on the answers produced. Discuss both the sensitivity to the optimal policy and to the expected value of the optimal policy.

Exercise 10.6 In Example 10.24 (page 387) suppose that the fire sensor was noisy in that it had a 20% false-positive rate:

$$P(see\_smoke|report \land \neg smoke) = 0.2$$

and a 15% false negative-rate:

$$P(see\_smoke|report \land smoke) = 0.85$$

Is it still worthwhile to check for smoke?

Exercise 10.7 What is the value of information of *alarm* at *check_for_smoke* and the subsequent decision in Example 10.27 (page 392)?

Exercise 10.8 List the assumptions that were used in the argument that preferences for outcomes can be based on maximizing expected utility (page 392)—for example, that preferences are totally ordered and that you can substitute equivalent outcomes in lotteries. (See, e.g., Pearl, 1988b, Section 6.1.4.)

Chapter 11

Learning

11.1 Introduction

Learning is an essential part of intelligence. Learning is the ability to improve one's behavior based on experience. This could mean the following:

- The range of behaviors is expanded: the agent can do more.
- The accuracy on tasks is improve: the agent can do things better.
- The speed is improved: the agent can do things faster.

The following components are part of any learning problem:

task The behavior or task that's being improved

data The experiences that are being used to improve performance in the task

measure of improvement How can the improvement be measured—for example, increasing accuracy in prediction, new skills that were not present initially, or improved speed

You can consider a learning algorithm as the box with thick borders depicted in Figure 11.1. This black box can be used to solve new problems, based on the new problem description, the previous experience (or data), and the background knowledge. This box is like the previous description of problem solving, but now you have experience to draw upon in solving a new problem.

If you look inside this black box, you see that you need a representation of experiences in order to use them. This could be the raw experiences themselves, but is more typically some other representation of the data. The problem of inferring an internal representation based on examples is often called **induction**, and can be contrasted with deduction, which is deriving consequences of a knowledge base, and abduction, which

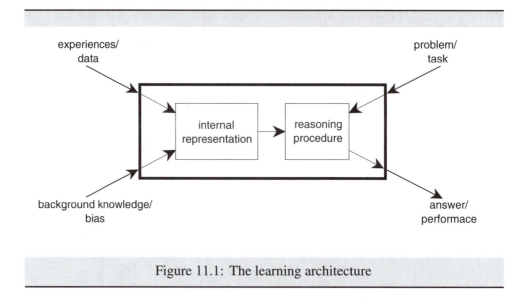

Figure 11.1: The learning architecture

is hypothesizing what may be true about a particular case. There are two principles that are at odds in choosing a representation:

- *The richer the representation, the more useful it is for subsequent problems solving.* If there is some accurate description of the world, which the representation does not allow you to state, then you cannot learn the description. Similarly, if you want to learn a way to solve a problem, then the representation that you learn must be rich enough to express a way to solve the problem. This observation has led computational intelligence researchers to make representations the primary focus of study. We study what knowledge it takes to solve a problem. Only when this is understood do we consider how this knowledge can be acquired: from experts or learned from experience or some mixture. If in some representation it is difficult to express the knowledge required to solve a problem, then we conjecture that it's difficult to learn that knowledge.

- *The richer the representation, the more difficult it is to learn.* Much learning has concentrated on supervised classification tasks (given a set of features, classify a new example into one of a predefined set of classes). While these tasks can become very sophisticated (e.g., learning to recognize hand-written words), they still fall short of much of the reasoning in this book. To learn these more sophisticated tasks is difficult, requiring a mix of expert knowledge and data to make any progress.

These principles are illustrated in the following sections. First we demonstrate learning based on choosing the best representations, where the representations are either decision trees (page 403) or neural networks (page 408). Section 11.3 (page 414) considers inferring properties of a new case directly from stored examples. This is called *case-based reasoning*. Section 11.4 (page 416) considers learning as selecting the subset of the hypothesis space consistent with the examples, rather than choosing a single best representation. Section 11.5 (page 424) considers learning as finding a posterior probability on the hypothesis space. This allows for noisy examples and richer hypotheses such as that some correlations occur by chance. Finally in Section 11.6 (page 433), we consider speed-up learning where the agent learns to compute similar answers more quickly, once it has computed an answer to a query.

Issues

This section outlines some of the issues in machine learning. All of the learning techniques face the same issues:

Task   Virtually any task that can be carried out by a computer can be learned. The most commonly studied learning task is supervised classification: Given a set of pre-classified examples, try to classify a new instance into its appropriate class. Other problems that have been considered include: learning classifications when the examples are not already classified (unsupervised learning), learning what to do based on rewards and punishments (reinforcement learning), learning to reason faster (analytic learning), and learning richer representations such as logic programs (inductive logic programming).

Figure 11.2 shows data typical of a classification task. There are a number of examples, each of which has a value on a number of different attributes. One of these attributes is designated to be the classification. In Figure 11.2 this is "User Action." The aim is to predict the classification of a new example given all attributes but the classification.

Feedback   Learning tasks can be characterized by the feedback given to the learner. In **supervised learning**, what has to be learned is specified for each example. Supervised learning of concepts occurs when a trainer provides the classification for each example. Supervised learning of actions occurs when the agent is given immediate feedback about the value of each action. **Unsupervised learning** occurs when no classifications are given and the learner has to discover categories and regularities in the data. Feedback is often between these extremes, such as in **reinforcement learning**, where the feedback occurs after a sequence of actions. This leads to the **credit-assignment problem** of determining what was responsible for the reward or punishment. For example, you could give rewards to the delivery robot without telling it exactly what it was being rewarded for. It then has to either learn what it's being rewarded for or learn which actions are preferred in which situations. It's possible that it can learn what actions to

Example	User Action	Author	Thread	Length	Where read
e1	skips	known	new	long	home
e2	reads	unknown	new	short	work
e3	skips	unknown	follow_up	long	work
e4	skips	known	follow_up	long	home
e5	reads	known	new	short	home
e6	skips	known	follow_up	long	work
e7	skips	unknown	follow_up	short	work
e8	reads	unknown	new	short	work
e9	skips	known	follow_up	long	home
e10	skips	known	new	long	work
e11	skips	unknown	follow_up	short	home
e12	skips	known	new	long	work
e13	reads	known	follow_up	short	home
e14	reads	known	new	short	work
e15	reads	known	new	short	home
e16	reads	known	follow_up	short	work
e17	reads	known	new	short	home
e18	reads	unknown	new	short	work

This is some example data that the infobot obtained from observing a user deciding whether to read articles depending on whether the author was known or not, whether the article started a new thread or was a follow-up, the length of the article, and whether it was read at home or at work.

Figure 11.2: Some classification data for learning user preferences

perform without actually knowing which of the consequences of the actions are being rewarded.

Representation   If you want an agent to use its experiences, then the experiences must affect the agent's internal representation. Much of machine learning is studied in the context of particular representations (e.g., decision trees, neural networks, or case bases). We present some example representations in order to show the common features behind learning.

Measuring success   Learning is defined in terms of improving performance based on some measure; in order to know you have learned, you need to define a measure of success. The measure is usually not how well the agent performs on the training experiences, but how well the agent performs for new experiences.

In classification tasks, in order to be able to learn a concept based on a set of training examples, being able to correctly classify all training examples isn't really the problem. For example, consider the problem of learning a Boolean concept based on a set of examples. Suppose that there were two agents P and N. P claims that all of the negative examples seen were the only negative examples and that every other instance is positive. N claimed that the positive examples in the training set were the only positive examples and that every other instance is negative. Both of these agents correctly classify every example in the training set, but disagree on every other example! Success in learning should not be judged on correctly classifying the training set but being able to correctly classify unseen examples.

A standard way to measure success is to divide the examples into a training set and a test set. A representation is built using the training set, and then the predictive accuracy is measured on the test set. Of course, this is only an approximation of what you want; the real measure is its performance on some future task.

Bias The tendency to prefer one hypothesis over another is called a **bias**. Consider the agents N and P defined above. Saying that a hypothesis is better than N's or P's hypothesis isn't something that's obtained from the data—both N and P accurately predict all of the data given—but is external to the data. Without a bias, you will not be able to make any predictions on unseen examples, as P and N disagree on all further examples. If you can't choose some hypotheses as better, then you won't be able to resolve this disagreement. To have any inductive process make predictions on unseen data, you need a bias. What constitutes a good bias is an empirical question about which biases work best in practice; we don't imagine that either of P's or N's biases work well in practice.

Learning as search Given a representation and a bias, the problem of learning can be reduced to one of search. Learning is search through the space of possible representations looking for the representation or representations that best fits the data, given the bias. Unfortunately, these search spaces are typically prohibitively large for systematic search, except for the simplest of examples. Nearly all of the search techniques we consider can be seen as a form of hill climbing (page 156) through a space of representations. The definition of the learning algorithm then becomes one of defining the search space, defining the heuristic function, and the search method.

Noise In most real-world situations the data aren't perfect. There is noise in the data (some of the attributes have been assigned the wrong value), there are inadequate attributes (the attributes given do not predict the classification), and often there are examples with missing attributes. One of the important properties of a learning algorithm is the ability to handle noisy data in all of its forms.

Why Should You Believe an Inductive Conclusion?

When learning from data, you typically make predictions beyond what the data give us. From observing the sun rising each morning, you conclude that the sun will rise tomorrow. From observing unsupported objects repeatedly falling you may conclude that unsupported objects always fall (until you come across helium-filled balloons). From observing many swans, all of whom were black, you may conclude that all swans are black. From the data of Figure 11.2 (page 400) you learn a representation that will let you predict the user action based on a case where author is unknown, the thread is new, the length is long, and it was read at work. The data do not tell you what the user does in this case. The question arises as to why you should ever believe any conclusion that isn't a logical consequence of your knowledge.

When you adopt a bias, or choose a hypothesis, you are going beyond the data—even making the same prediction about a new case that's identical to an old case in all measured respects is going beyond the data. So why should you believe one hypothesis over another? By what criteria can you possibly go about choosing a hypothesis?

The most common justification is to choose the simplest hypothesis that fits the data and appeal to *Ockham's razor*. William of Ockham was an English philosopher who was born in about 1285 and died, apparently of the black plague, in 1349. He argued for economy of explanation: "What can be done with fewer [assumptions] is done in vain with more" (Edwards, 1967, p. 8-307).

Why should one believe the simplest hypothesis, especially as which hypothesis is simplest depends on the language used to express the hypothesis?

First, it's reasonable to assume that there is structure in the world that you need to discover. A reasonable way to discover the structure of the world is to search for it. An efficient search strategy is to search from simpler hypotheses to more complicated ones. If there is no structure in the world to be discovered, then nothing will work! The fact that much structure has been found in the world (e.g., all of the structure discovered by physicists) would lead you to believe that this isn't a futile search.

The fact that simplicity is language-dependent should not necessarily make you suspicious. Language has evolved because it's useful — because it can let you express the structure of the world. Thus you would expect that simplicity in everyday language would be a good measure of complexity.

The most important reason for believing inductive hypotheses is that it's useful to believe them. It helps you interact with the world and avoid being killed: an agent that doesn't learn that it should not fling itself from heights will not survive long. The simplest hypothesis heuristic is useful because it works!

11.2   Learning as Choosing the Best Representation

Much of learning has been studied from the perspective of different representations. This chapter is structured along these lines. We first consider two cases where the goal is to find the best representation among a space of all representations. Although what's the best representation given some examples is defined, the algorithms search for the best representation but aren't guaranteed to find it.

Learning Decision Trees

Decision trees are a simple representation for classifying examples. Decision tree learning is one of the most successful techniques for supervised classification learning. For this section, we assume all of the attributes of individuals have finite discrete ranges.

A **decision tree** is a tree where the nonleaf nodes are labeled with attributes. The arcs emanating from a node labeled with attribute A are labeled with each of the possible values of the attribute A. The leaves of the tree are labeled with classifications. To classify an example, filter it down the tree: For each attribute encountered, the arc corresponds to the value of the example for that attribute if followed; when a leaf is reached, the classification corresponding to that leaf is returned.

Example 11.1	Figure 11.3 shows a simple decision tree. This decision tree can be used to classify examples according to the user action. To classify a new example, first determine whether the author is known or unknown. If the author is unknown, then determine whether the thread is new or a follow-up. This decision tree can correctly classify all examples in Figure 11.2.

A decision tree can be simply mapped into a set of rules, with each leaf of the tree corresponding to a rule. The object has the classification at the leaf if all of the conditions on the path from the root to the leaf are true.

Example 11.2	Using the *prop* relation of Definition 5.4 (page 184), the decision tree of Figure 11.3 can be represented as the rules:

$$prop(Obj, user\_action, skips) \leftarrow$$
$$prop(Obj, length, long).$$
$$prop(Obj, user\_action, reads) \leftarrow$$
$$prop(Obj, length, short) \land$$
$$prop(Obj, thread, new).$$

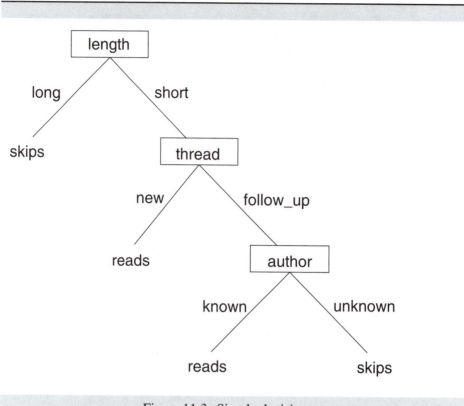

Figure 11.3: Simple decision tree

$prop(Obj, user\_action, reads) \leftarrow$
    $prop(Obj, length, short) \wedge$
    $prop(Obj, thread, follow\_up) \wedge$
    $prop(Obj, author, known).$
$prop(Obj, user\_action, skips) \leftarrow$
    $prop(Obj, length, short) \wedge$
    $prop(Obj, thread, follow\_up) \wedge$
    $prop(Obj, author, unknown).$

Once you have decided to use decision trees as a target representation for learning, there are a number of questions that arise:

- Given some data, what decision tree should be generated? As a decision tree can represent any function of the attributes, the bias that is necessary to learn is incorporated into the preference of one decision tree over another. One proposal is that you want the smallest tree that's consistent with the data; this could mean the tree with the least depth or the tree with the fewest nodes. Which decision trees are best predictors of unseen data is an empirical question.

- How should you go about building a decision tree? One way to do this is to search the space of decision trees for the smallest decision tree that fits the data. Unfortunately the space of decision trees is enormous (see Exercise 11.1). A practical solution is to carry out a hill-climbing search on the space of decision trees. This is the idea behind the algorithm described below.

Searching for a Good Decision Tree

You can incrementally build a decision tree top down by repeatedly splitting the training data. A simple version for binary attributes is shown in Figure 11.4. In this figure the decision regarding how to choose which variable to split on is left undefined.

The general idea is as follows: The input is a target attribute (the *Goal*), a set of examples, and a set of attributes. You stop if all of the examples have the same classification. Otherwise you choose an attribute to split on, and for each of the values of this attribute, you build a subtree for those examples with this attribute value. Thus you build the tree top-down.

Example 11.3

Suppose you apply the algorithm of Figure 11.4 to the classification data of Figure 11.2. To use this program you issue the (appropriately completed) query

?*dtlearn*(*user_action*, [*e*1, *e*2, ..., *e*18],

[*author*, *thread*, *length*, *where_read*], *DT*).

Not all of the examples agree on the user action, so you choose the second clause of *dtlearn*. Suppose you select the attribute *length* to split on first. The variable *Yes* gets unified with the list of those examples with *length* = *long*. Then you call

?*dtlearn*(*user_action*, [*e*1, *e*3, *e*4, *e*6, *e*9, *e*10, *e*12],

[*where_read*, *thread*, *author*], *YT*).

All of these examples agree on the user action, and so *YT* is unified with *skips*. You then call

?*dtlearn*(*user_action*, [*e*2, *e*5, *e*7, *e*8, *e*11, *e*13, *e*14, *e*15, *e*16, *e*17, *e*18],

[*where_read*, *thread*, *author*], *NT*).

Not all of the examples agree on the user action so, again you select an attribute to split on. Suppose you choose *thread*. Eventually this returns the tree corresponding to the case when length is short, such as: *NT* = *if* (*thread* = *new*, *reads*, *if* (*author* = *unknown*, *skips*, *reads*)), and the final result is:

if (*length* = *long*, *skips*,

if (*thread* = *new*, *reads*, *if* (*author* = *unknown*, *skips*, *reads*))),

which is a representation of the tree of Figure 11.3 (page 404).

% *dtlearn(Goal, Examples, Attributes, DT)* is true if given *Examples* and *Attributes*
% induces a decision tree *DT* for *Goal*.

> *dtlearn(Goal, Examples, Atts, Val)* ←
>> *all_examples_agree(Goal, Examples, Val).*
>
> *dtlearn(Goal, Examples, Atts, if(Cond = YesVal, YT, NT))* ←
>> ~*all_examples_agree(Goal, Examples, Val)* ∧
>> *select_split(Goal, Examples, Atts, Cond, Rem_Atts)* ∧
>> *split(Examples, Cond, YesVal, Yes, No)* ∧
>> *dtlearn(Goal, Yes, Rem_Atts, YT)* ∧
>> *dtlearn(Goal, No, Rem_Atts, NT).*

% *all_examples_agree(Goal, Examples, Val)* is true if all examples agree that *Val* is
% the value of attribute *Goal*.

> *all_examples_agree(G, [], V).*
>
> *all_examples_agree(Att, [Obj|Rest], Val)* ←
>> *val(Obj, Att, Val)* ∧
>> *all_examples_agree(Att, Rest, Val).*

% *split(Examples, Att, YesVal, T, F)* is true if *T* is the examples in *Examples* with
% attribute *Att* having value *YesVal* and *F* is the examples with attribute *Att* having the
% other value.

> *split([], As, YesVal, [], []).*
>
> *split([Obj|Rest], Cond, YesVal, [Obj|Yes], No)* ←
>> *val(Obj, Cond, YesVal)* ∧
>> *split(Rest, Cond, Yes, No).*
>
> *split([Obj|Rest], Cond, YesVal, Yes, [Obj|No])* ←
>> *val(Obj, Cond, NoVal)* ∧
>> *NoVal ≠ YesVal* ∧
>> *split(Rest, Cond, YesVal, Yes, No).*

% *choose_split(Goal, Examples, Atts, Cond, Rem_Atts)* is true if *Cond* is an attribute
% chosen from *Atts* to split on, and *Rem_Atts* is the set of remaining attributes.

Figure 11.4: Simple decision tree learning program for binary attributes

The algorithm of Figure 11.4 makes a number of simplifying assumptions, including the following:

- Each attribute has only two values. This restriction can easily be lifted, although the representation of the decision tree becomes more complicated than the simple if-then-else form used for binary attributes. (See Exercise 11.2.)

- The attributes are adequate to represent the concept. This means that it never gets to the stage where it has no attributes left and the examples disagree on the target attribute. The attributes are not adequate if and only if there are two examples, both of which agree on all of the attributes but have different classifications. The algorithm can easily be extended to the case where the attributes are inadequate by returning a probability at the leaves. (See Exercise 11.3.)

- Which attribute to select to split on isn't defined. You really want to choose the attribute that will result in the smallest tree. The best attribute is one that divides the examples into classes, each of which have a unique classification. Failing this, you would like the attribute that makes the most progress towards this. This can be seen as a searching problem, where you are searching for the smallest tree. We present two heuristics. The first is to choose the attribute with the maximum **information gain** (page 360). The second, based on the so-called *GINI index*, is investigated in Exercise 11.4.

Example 11.4

Consider learning the user action from the data of Figure 11.2 (page 400). Let's split on the attribute with the maximum information gain (page 360).

The information content of all examples with respect to the attribute *user_action* is 1.0—there are 9 examples with *user_action = reads*, and 9 examples with *user_action = skips*.

The information gain for the test *author* is zero. A test on *author* will partition the examples into [*e1, e4, e5, e6, e9, e10, e12, e13, e14, e15, e16, e17*] with *author = known* and [*e2, e3, e7, e8, e11, e18*] with *author = unknown*, each of which are evenly split between the different user actions. In this case finding out whether the author is known, by itself, tells you nothing as to what the user action will be.

The test *thread* will partition the examples into [*e1, e2, e5, e8, e10, e12, e14, e15, e17, e18*] and [*e3, e4, e6, e7, e9, e11, e13, e16*]. The first set of examples, those with *thread = new*, contain 3 examples with *user_action = skips* and 7 examples with *user_action = reads*; thus the information content of this set with respect to the user action is

$$-0.3 \times \log_2 0.3 - 0.7 \times \log_2 0.7 = 0.881$$

Similarly, the examples with *thread = old* divide up $6 : 2$ according to the user action and thus have information content 0.811. The expected information gain is thus $1.0 - (10/18) \times 0.881 + (8/18) \times 0.811 = 0.150$.

The test *length* divides the examples into [*e*1, *e*3, *e*4, *e*6, *e*9, *e*10, *e*12] and [*e*2, *e*5, *e*7, *e*8, *e*11, *e*13, *e*14, *e*15, *e*16, *e*17, *e*18]. The former all agree on the value of *user_action*, and so have information content zero. The user action divides the second set 9 : 2, and so the information is 0.684. Thus the expected information gain by the test *length* is $1.0 - 11/18 * 0.684 = 0.582$. This is the highest information gain of any test, and so *length* is chosen to split on.

Note that in choosing which property to split on, the information content before the test is the same for all tests, and so you need only choose the test which results in the minimum expected information after the test.

If there is noise in the data, one major problem of the above algorithm is that of **overfitting** the data. This occurs when you make a distinction in the tree that appears in the data, but that distinction doesn't appear in the unseen examples. This occurs because there may be random correlations in the training set that are not reflected in the data as a whole. There are two ways to overcome this problem:

- You can restrict the splitting, so that you split only when the split is useful.
- You can allow unrestricted splitting and then try to prune the resulting tree where it makes unwarranted distinctions.

The second method seems to work better in practice, but in any case you need to trade off tree size with fit to the examples. A principled way to trade these off is presented when we consider Bayesian learning of decision trees (page 426).

Neural Networks

Neural networks are a popular target representation for learning. These networks are inspired by the neurons in the brain, but do not actually simulate neurons. They typically contain many fewer than the approximately 10^{11} neurons that are in the human brain, and the artificial neurons, called *units*, are much simpler than their biological counterparts.

As in decision tree learning, the problem is: Given a set of labeled training data, find an internal representation that can classify new examples. The main difference is that the target representation is a parameterized network, made up of simple computational units, connected together such that the output of a unit is a function of its inputs and some adjustable parameters. Learning is accomplished by adjusting the parameters to fit the training data.

Artificial neural networks are interesting to study for a number of reasons:

- As part of neuroscience, in order to understand real neural systems, researchers are simulating the neural systems of simple animals such as worms. This promises to lead to an understanding about what aspects of neural systems are necessary to explain the behavior of these animals.

- Some researchers seek to automate not only the functionality of intelligence (which is what the field of computational intelligence is about), but also the mechanism of the brain, suitably abstracted. The hypothesis that you can only get the functionality of the brain via the mechanism of the brain can be tested by attempting to build intelligence using the mechanism of the brain, as well as without using the mechanism of the brain. Experience with building other machines, such as flying machines (see page 2) where you do not need to automate the mechanism that birds use to fly in order to build the functionality of flying, would indicate that this hypothesis is not true, but this doesn't mean that this hypothesis doesn't need to be tested.

- The brain inspires a new way to think about computation that contrasts with currently available computers. Unlike current computers that have one or a few processors and a large but essentially inert memory, the brain consists of a huge number of asynchronous distributed processes, all running concurrently with no master controller. One should not think that the current computers are the only architecture available for computation. The desire for parallel algorithms isn't restricted to neural networks, however; for each of the algorithms in this book you could consider how to build parallel versions.

- As far as learning is concerned, neural networks provide a different measure of simplicity as a learning bias than, for example, decision trees. Multilayer neural networks, like decision trees, can represent any function from a set of Boolean attribute values into another attribute value; however, the functions that correspond to simple neural networks do not necessarily correspond to simple decision trees. Thus neural network learning imposes a different bias than decision tree learning; which is better, in practice, is an empirical question that can be tested on different domains.

There are many different types of neural networks. This book considers only one kind of neural network, the "feed-forward" network. Feed-forward networks are like forward chaining logic programs, with no loops. Associated with the logic programs are a set of parameters that need to be learned.

A generic unit with k inputs is like the logic program

$$prop(Obj, output, V) \leftarrow$$
$$prop(Obj, in_1, I_1) \wedge$$
$$prop(Obj, in_2, I_2) \wedge$$
$$\vdots$$
$$prop(Obj, in_k, I_k) \wedge$$
$$V \text{ is } f(w_0 + w_1 \times I_1 + w_2 \times I_2 + \cdots + w_k \times I_k).$$

where $prop$ is the object-attribute-value relation of (page 184). The I_j are real-valued

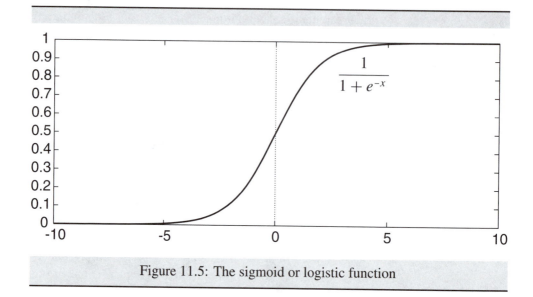

Figure 11.5: The sigmoid or logistic function

inputs, the w_j are the adjustable real parameters, and f is an activation function.   A
typical activation function is the **sigmoid** or **logistic** function:

$$f(x) = \frac{1}{1 + e^{-x}}$$

This function, as depicted in Figure 11.5, squashes the real line into the interval $(0, 1)$.
It also is differentiable, with a simple derivative, namely $f'(x) = f(x)(1 - f(x))$, which
is important for the gradient descent algorithms used to tune the parameters.

   Neural networks of this type can have as inputs any real numbers, and they have a
real number in the range $(0, 1)$ as output. They can be used for approximating arbitrary
functions, including Boolean functions. You can translate Boolean learning problems
into neural network learning problems by, for example, associating 1 with the Boolean
value *true* and 0 with the Boolean value *false*.

Example
11.5

Figure 11.6 shows a simple neural network for the classification data of Figure
11.2. In this example, there are four Boolean inputs corresponding to whether the
author is known, the thread is new, the length is short, and the article was read at
home, and there is a single output corresponding to the user action of Figure 11.2.
You have to convert the discrete variables into a real number, with 1 corresponding
to true, and 0 corresponding to to false. For example, *known* $= 1$ corresponds
to *author* $=$ *known*, and *known* $= 0$ corresponds to *author* $=$ *unknown*. In this
network, there are two hidden units that have no a priori meaning. This neural
network corresponds to the logic program

   predicted_prop(Obj, reads, V) $\leftarrow$

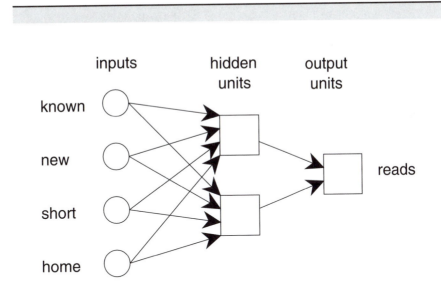

inputs   hidden   output
                units   units

known

new   reads

short

home

Figure 11.6: Simple neural network with one hidden layer

$$prop(Obj, h_1, I_1) \wedge$$
$$prop(Obj, h_2, I_2) \wedge$$
$$V \text{ is } f(w_0 + w_1 \times I_1 + w_2 \times I_2).$$
$$prop(Obj, h_1, V) \leftarrow$$
$$prop(Obj, known, I_1) \wedge$$
$$prop(Obj, new, I_2) \wedge$$
$$prop(Obj, short, I_3) \wedge$$
$$prop(Obj, home, I_4) \wedge$$
$$V \text{ is } f(w_3 + w_4 \times I_1 + w_5 \times I_2 + w_6 \times I_3 + w_7 \times I_4).$$
$$prop(Obj, h_2, V) \leftarrow$$
$$prop(Obj, known, I_1) \wedge$$
$$prop(Obj, new, I_2) \wedge$$
$$prop(Obj, short, I_3) \wedge$$
$$prop(Obj, home, I_4) \wedge$$
$$V \text{ is } f(w_8 + w_9 \times I_1 + w_{10} \times I_2 + w_{11} \times I_3 + w_{12} \times I_4).$$

The values of the attributes are real numbers, and the parameters $w_0, \ldots, w_{12}$ are real numbers. The attributes h_1 and h_2 correspond to the hidden units. The value of h_i for an example is the value of the corresponding hidden unit for that example.

For this example, there are 13 real numbers to be learned. The hypothesis space is thus a 13-dimensional real space. Each point in this 13-dimensional space corresponds to a particular logic program that implies a particular predicted value for *reads* for every example with *known*, *new*, *short*, and *home* given.

Given particular values for the parameters, the neural network predicts a particular value for the output value. Given a set of examples, you can define a measure of the error of the neural network for the examples:

For particular values for the parameters $\overline{w} = w_0, \ldots w_m$ of the neural network, and a set E of examples, the **sum-of-squares error** of the parameter setting with respect to the examples is given by

$$Error_E(\overline{w}) = \sum_{e \in E} (p_e^{\overline{w}} - o_e)^2,$$

where $p_e^{\overline{w}}$ is the predicted output by a neural network with parameter values given by $\overline{w}$ for example e, and o_e is the observed output for example e.

The aim of neural network learning is, given a set of examples, to find parameter settings that minimize the error. Unfortunately, if there are m parameters, finding the parameters settings with minimum error involves searching through an m-dimensional Euclidean space. One method that works well in practice is back-propagation:

Back-propagation learning is gradient descent search (page 157) through the parameter space to minimize the sum-of-squares error.

Intuitively, back-propagation involves, for each example, simulating the network on that example (corresponding to forward chaining on the corresponding logic program) and then passing the error back through the network, updating the parameters in proportion to how much a small change in the parameter value would reduce the error. You start with random values for the parameters and iterate the gradient descent until you have reached convergence or some stopping criteria. At each step you update each parameter in proportion to its derivative. The constant of proportionality—the η of the gradient descent (page 157)—is called the **learning rate**. The learning rate, as well as the network and the data, is given as input to the learning algorithm.

Code is given in Section C.4 (page 511) that implements this algorithm, where the derivatives are estimated by simulating the network with each of the parameters changed slightly. For the restricted case of the logistic function applied to a linear combination of inputs, the derivative can be derived analytically; this will enable the algorithm to run faster.

Gradient descent search (page 157) involves repeated evaluation of the function to be minimized, in this case the error term, and its derivative. An evaluation of the error term involves iterating through all of the examples. Back-propagation learning thus involves repeatedly evaluating the network on all examples. Fortunately, with the

Para-	iteration 0		iteration 40		iteration 80	
meter	Value	Deriv	Value	Deriv	Value	Deriv
w_0	0.2	0.768	-2.23	0.0426	-2.98	0.0159
w_1	0.12	0.373	4.44	-0.203	6.88	-0.0708
w_2	0.112	0.425	-1.11	0.0936	-2.10	0.0228
w_3	0.22	0.0262	-2.69	0.230	-5.25	0.0455
w_4	0.23	0.0179	0.454	-0.207	1.98	-0.0148
w_5	0.26	-0.0111	1.18	-0.0556	1.86	-0.0176
w_6	0.27	-0.0277	3.43	0.0692	4.71	-0.0227
w_7	0.211	0.0123	-0.422	0.0400	-0.389	-0.0153
w_8	0.245	0.0217	-0.0929	-0.0451	0.493	-0.0182
w_9	0.152	0.0148	-0.533	0.0522	-1.03	0.0044
w_{10}	0.102	-0.0132	-0.548	0.0423	-1.06	0.0076
w_{11}	0.105	-0.0303	-0.400	0.00573	-0.749	0.0034
w_{12}	0.205	0.00996	0.0293	-0.0120	0.126	0.00
Error:	4.6121		1.201		0.179	

This figure shows the parameter values from Example 11.6. The random initial values are given as iteration 0; other values are given for iterations 40 and 80. This algorithm used a gradient decent step size $\eta = 0.5$ (see page 157).

Figure 11.7: Simulation of the neural net learning algorithm

logistic function the derivative is easy to determine given the value of the output for each unit.

Example 11.6

Consider the neural net learning problem using the examples of Figure 11.2 on page 400 to train the network given in Figure 11.6. First, select random initial values for the parameters. In Figure 11.7, under "iteration 0" are the initial values for each of the parameters. The total sum-of-squares error is given at the bottom of the column. For each of the parameters, we compute its derivative, which specifies how much changing the parameter's value will change the total error. This is given in the next column. The experiment reported in this figure used a gradient decent step size of 0.5, which means that we add the value and half the value of the derivative to get the next value. After 40 iterations, the total error has been reduced, and after 80 iterations the network has converged—it gives all of the training examples the correct classification, assuming that we predict true when the value is closer to one and false if the value is closer to zero. A finer grained analysis shows that during the training the error doesn't reduce monotonically; sometimes an iteration makes the error larger.

There is one crucial difference between the use of logic and neural networks. According to the semantics of logic (page 31), each symbol can be assessed some meaning. Part of the appeal of neural networks is that, while meaning is attached to the input and output units, you have no idea as to what meaning, if any, will be associated with the hidden units. What the hidden units actually represent is something that's learned. After a neural network has been trained, it's often possible to look inside the network to determine what a particular hidden unit actually represents. Sometimes it's easy to tell what it represents, but often it isn't.

11.3 Case-Based Reasoning

In case-based reasoning the previous experiences are stored and accessed to solve a new problem. This is at one extreme of the learning problem where, unlike decision trees and neural networks, relatively little work needs to be done at acquisition time, and virtually all of the work goes on at query time.

The key defining feature of case-based reasoning is that the previous experiences, the cases, are stored and used when trying to classify a new example. The general idea is to find the case or cases that are like the new example, and use these to predict properties of the new example.

Case-based reasoning takes many forms, depending on the complexity of the cases and on how dense the space of cases is.

- At one extreme, the cases are simple and for each new example the agent has seen many identical instances. In this case the agent can just use exact matches and use the statistics of these matches for the new case. The most straightforward solution works well when the sample size is large. When the sample size is smaller, you need to consider your prior distribution (page 424).

- If the cases are simple but there are few exact matches, such as for the feature learning of Figure 11.2 (page 400), given a new case, you can use the closest case or cases for the prediction. One algorithm that works well is to use the **k-nearest neighbors** of the new case, for some number k. Given the k nearest classified training examples, you predict the classification of a new example according to the mode, average, or some interpolation between what these k training examples predict, perhaps by weighting closer examples more than the more distant of the k closest examples.

  For this to work you need a distance metric that measures how close any two examples are from each other. First define a metric for each dimension, where the values of the attributes are converted to a numerical scale and the difference between the two examples are used for the measure in this dimension. Suppose

x_A is a numerical representation of the value of attribute A for example x. Then $(x_A - y_A)$ is the distance between example x and y on the dimension defined by attribute A. The Euclidean distance, the square root of the sum of the squares of the dimension differences, can be used as the distance between two examples. One issue that needs to be addressed is the relative scales of different dimensions; increasing the scale of one dimension increases the importance of that feature. Let w_A be a nonnegative real-valued parameter that specifies the relative weight of attribute A. The distance between examples x and y is then:

$$d(x, y) = \sqrt{\sum_A w_A(x_A - y_A)^2}$$

You can treat the relative scales of the dimensions as learn-able parameters. You can try to find a parameter setting that minimizes the error that is used to predict the value of each element of the training set, based on every other instance in the training set. This is called the leave-one-out cross-validation error measure.

- If the cases are complicated and the information needed from such cases is more than a classification, more sophisticated techniques are called for. Such reasoning occurs in the legal domain, where lawyers want to find cases that are precedents for a current case. In case-based planning, given a new goal, an agent wants to find previous experiences that solved the goal, and then adapt the plan in previous cases for the new circumstances.

For any of these methods you may want more than the single best match, even if there is an exact match. In general, you may give a probabilistic answer. Predicting from a single case is risky, as the example in the case base may be an anomalous case. Often you want to combine data from other cases even if these are poorer matches.

Example 11.7

Consider using case-based reasoning on the data of Figure 11.2 (page 400). Rather than converting the data to a secondary representation as in decision tree or neural network learning, you can use the examples directly to predict the value for the user action in a new case.

Suppose you want to classify a new case where the author is unknown, the thread is a follow-up, the length is short, and it's read at home. First you try to find similar cases. There is an exact match in example $e11$, so you may want to predict that the user does the same action as for example $e11$ and thus skips the article.

Suppose you see a new case where author is unknown, the thread is new, the length is long, and it was read at work. In this case there are no exact matches. Consider the close matches. Examples $e2$, $e8$, and $e18$ agree on *author*, *thread*, and *where_read*. Examples $e10$ and $e12$ agree on *thread*, *length*, and *where_read*. Example $e3$ agrees on *author*, *length*, and *where_read*. Examples $e2$, $e8$ and $e18$ predict *reads*, but the other examples predict *skips*. So what should be predicted? The decision tree algorithm says that *length* is the best predictor, and so $e2$, $e8$, and $e18$ should be ignored. In the neural network learning algorithm, the final parameter values in Figure 11.7 (page 413) similarly predicts that the reader skips the article. To use the case-based reasoning to predict whether the user will or won't read this article, you need to determine the relative importance of the dimensions.

One of the problems in case-based reasoning is accessing the relevant cases. Imagine a decision tree, but where cases are stored at the leaves rather than probabilities for a particular classification. Such a structure is called a **kd-tree**. The tree doesn't need to be built with respect to any particular classification (as a decision tree is), but just needs to index examples so that training examples that are close to a given example can be found quickly. In building a *kd*-tree from a set of examples, you try to find a test that partitions the examples into two approximately equal sized sets, and then build *kd*-trees for the examples in each partition. This division stops when all of the examples are the same or when there are limited numbers of tests. Given a new example, you can filter the example as in a decision tree search. However, the examples at the leaves of the *kd*-tree could possibly be quite distant from the example to be classified; they agree on the values down the branch of the tree, but could disagree on the values of all other attributes. The same tree can be used to search for those examples that have one attribute different from the ones tested in the tree. (See Exercise 11.9.)

11.4 Learning as Refining the Hypothesis Space

So far we have considered learning as either choosing the best representation—for example, the best decision tree or the best values for parameters in a neural network—or predicting a new case from a database of previous cases. This section considers a different notion of learning. We consider learning as delineating those hypotheses which are consistent with the examples. Rather than choosing a hypothesis, you identify all hypotheses that are consistent. This investigation will shed light on the role of a bias and provide a mechanism for a theoretical analysis of the learning problem.

Again consider supervised learning of a classification from noise-free data.

A **Boolean concept** is a function from individuals into the set {*true, false*}. Boolean concept c can be represented as a formula f with a free variable X. To define c is to give a formula $c(X) \leftrightarrow f(X)$, where f is a formula composed of atoms whose truth value is given for each new instance.

Example 11.8 | The decision tree of Figure 11.3 (page 404) can be seen as a representation of the Boolean concept *prop*$(X, user\_action, reads)$ defined by

$$prop(X, user\_action, reads) \leftrightarrow$$

$$prop(X, length, short) \wedge$$
$$(prop(X, thread, new) \vee prop(X, author, known)).$$

The rest of this section uses a simple relational notation rather than the *prop* relation.

The general idea is to induce a Boolean function from data. Assume that there is a Boolean function f which can be used to classify all of the individuals—you have to try to find this concept from the training examples.

Example 11.9

Consider the infobot trying to infer which article the user reads based on keywords supplied in the article. Suppose you have the following data:

article	crime	academic	local	music	reads
a_1	t	f	f	t	t
a_2	t	f	f	f	t
a_3	f	t	f	f	f
a_4	f	f	t	f	f
a_5	t	t	f	f	t

The target concept is to learn which articles the user reads.

In this example, *reads* is the target Boolean concept, and you are looking for a definition such as

$$reads(X) \leftrightarrow crime(X) \wedge (\neg academic(X) \vee \neg music(X))$$

This definition can be used to classify the training examples as well as future examples.

Hypothesis space learning assumes the following sets:

- I, the **instance space**, is the set of all possible instances (individuals).
- **H**, the **hypothesis space** is a set of Boolean concepts. **H** consists of possible definitions for the concept in terms of Boolean functions on the given attributes.
- $E \subseteq I$ is the set of **training examples**. You are given $c(e)$ for each $e \in E$.

Example 11.10

In Example 11.9, I is the set of the $2^4 = 16$ possible examples, one for each combination of values for the attributes that could be used to determine the value of *reads*.

The hypothesis space **H** could be all Boolean combinations of the attributes or could be more restricted, such as conjunctions or formulae containing less than three attributes. This section assumes that **H** is the set of conjunctions of atoms. This assumes that the concept can be described as a conjunction.

In Example 11.9, the training examples are $E = \{a_1, a_2, a_3, a_4, a_5\}$ and the target concept c is "*reads*." You are given some truth values for the target concept such as $read(a_1) = true$, and $reads(a_3) = false$, but are missing the values for the other 11 elements of I. The aim is to induce the truth value of the concept for these unseen cases. To be able to give truth values for these unseen cases you need a bias (see page 401). In this case the bias is imposed by the hypothesis space.

Note that the hypothesis space here is different from the set of possible hypotheses given in Section 9.2 (page 321). Here the hypothesis space already contains all of the formulae; you are looking for a single element $\mathbf{H}$. The set of possible hypotheses could be seen as the building blocks of a hypothesis space. For abductive reasoning, the hypothesis space corresponds to the set of all conjunctions of elements of H, but in the abductive framework there is no "if and only if" definition.

Hypothesis h is **consistent** with labeled examples E, if $\forall e \in E$, $h(e) \leftrightarrow c(e)$—in other words, when h and c agree on the truth value of all of the examples.

The problem is to find the subset of $\mathbf{H}$, or just an element of $\mathbf{H}$, that's consistent with the training examples.

Example 11.11

In Example 11.9, $\mathbf{H}$ is the set of conjunctions of literals. An example concept in $\mathbf{H}$ that's consistent with $\{a_1\}$ is

$$reads(X) \leftrightarrow \neg academic(X) \wedge music(X),$$

which we will write as $\neg academic \wedge music$. This concept isn't the target concept because it's inconsistent with $\{a_1, a_2\}$.

Version-Space Learning

Rather than enumerating and pruning the hypothesis space, the subset of $\mathbf{H}$ consistent with the examples can be found more efficiently by imposing some structure on the hypothesis space.

Hypothesis h_1 is **more general** than hypothesis h_2 if $\forall i \in I\ h_2(i) \rightarrow h_1(i)$. In this case, h_2 is **more specific** than h_1. Any hypothesis is both more general than itself and more specific than itself.

The "more general than" relation forms a partial ordering over the hypothesis space. The version-space algorithm below exploits this partial ordering to search for hypotheses that are consistent with the training examples.

Given hypothesis space $\mathbf{H}$ and examples E, the **version space** is the subset of $\mathbf{H}$ that's consistent with the examples.

The **general boundary** G of a version space is the set of maximally general members of the version space (i.e., those members of the version space such that there is no element of the version space that's more general). The **specific boundary** S of a version space is the set of maximally specific members of the version space.

These concepts are useful because the general boundary and the specific boundary completely determine the version space:

Let $G = \{true\}$, $S = \{false\}$;

For each example $e \in E$:

- If e is a positive example:
    - The elements of G that classify e as negative are removed from G.
    - Each element s of S that classifies e as negative is removed and replaced by the minimal generalizations of s that classify e as positive and are less general or equal to some member of G.
    - Nonmaximal hypotheses are removed from S.

- If e is a negative example:
    - The elements of S that classify e as positive are removed from S.
    - Each element g of G that classifies e as positive is removed and replaced by the minimal specializations of g that classifies e as negative and are more general or equal to some member of S.
    - Nonminimal hypotheses are removed from G.

Figure 11.8: Candidate elimination algorithm

Proposition 11.1   The version space, given hypothesis space **H** and examples E, can be derived from its general boundary and specific boundary. In particular, the version space is the set of $h \in \mathbf{H}$ such that h is more general than or equal to an element of S and more specific than or equal to an element of G.

Candidate Elimination Algorithm

You can incrementally build the version space given an hypothesis space **H** and a set E of examples. You add the examples one by one; each example possibly shrinks the version space by removing the hypotheses that are inconsistent with the example. The candidate elimination algorithm does this by updating the general and specific boundary for each new example. This is described in Figure 11.8.

Example 11.12   Let's consider how the candidate elimination algorithm handles Example 11.9 (page 417), where **H** is the set of conjunctions of literals:

Before it has seen any examples, $G_0 = \{true\}$—the user reads everything—and $S_0 = \{false\}$—the user reads nothing.

After considering the first example, a_1, $G_1 = \{true\}$ and

$$S_1 = \{crime \wedge \neg academic \wedge \neg local \wedge music\}.$$

Thus the most general hypothesis is that the user reads everything, and the most specific hypothesis is that the user only reads articles exactly like this one.

After considering the first two examples, $G_2 = \{true\}$ and

$$S_2 = \{crime \wedge \neg academic \wedge \neg local\}.$$

Since a_1 and a_2 disagree on music, it has concluded that music cannot be relevant.

After considering the first three examples, the general boundary becomes

$$G_3 = \{crime, \neg academic\}$$

and $S_3 = S_2$. Now there are two most general hypotheses; the first is that the user reads anything about crime, and the second is that the user reads anything nonacademic.

After considering the first four examples,

$$G_4 = \{crime, \neg academic \wedge \neg local\}$$

and $S_4 = S_3$.

After considering all five examples, we have

$$G_5 = \{crime\},$$
$$S_5 = \{crime \wedge \neg local\}.$$

Thus, after five examples, there are only two hypotheses in the version space. They differ only on their prediction on an example that has $crime \wedge local$ true. If the target concept can be represented as a conjunction, only an example with $crime \wedge local$ true will change G or S. This version space can make predictions about all other examples.

The Bias Involved in Version Space Learning

Recall (page 401) that you need a bias for any successful learning to generalize beyond the training data. There must have been a bias in Example 11.12 because after observing only 5 of the 16 possible examples, you were able to make predictions about examples you had not seen.

The bias involved in version space learning is a called a **language bias** or a **restriction bias** as the bias is obtained from a restriction on the allowable hypotheses. The conclusion that's reached about a new example with crime false and music true is that if the target concept is in **H**, then the example will be classified as false (the user will not read the article). No such example has been seen—the restriction on the hypothesis space is enough to predict its value.

This bias should be contrasted with the bias involved in decision-tree learning (page 403). The decision tree can represent any Boolean function. Decision tree learning involves a **preference bias**, in that some Boolean functions are preferred over others (those with smaller decision trees are preferred over those with larger decision trees).

A decision-tree learning algorithm that builds a single decision tree top-down also involves a **search bias** in that which decision tree is returned depends on the search strategy used.

The candidate elimination algorithm is sometimes said to be an unbiased learning algorithm because the learning algorithm doesn't impose any bias beyond the language bias involved in choosing **H**. Note that's it easy for the version space to collapse to the empty set—for example, if the user reads an article with crime false and music true. This means that the target concept isn't in **H**. Version-space learning isn't tolerant to noise; just one misclassified example can throw off the whole system.

You could consider the bias-free hypothesis space, namely where **H** is the set of all Boolean functions. In this case, G always contains one concept: the concept that says that all negative examples have been seen and every other example is positive. Similarly, S contains the single concept that says that all unseen examples are negative. The version space is incapable of concluding anything about examples it hasn't seen. This forms the basis of a technical argument as to why you need a bias for learning.

Probably Approximately Correct Learning

Some relevant questions that you can ask about a theory of computational learning include:

- Is the learner guaranteed to converge to the correct hypothesis as the number of examples increases?
- How many examples are needed to identify a concept?
- How much computation is needed to identify a concept?

In general the answer to the first question is "no," unless it's guaranteed that the examples will eventually rule out all but the correct hypothesis. However, there is something that you can show, namely that randomly chosen examples will get arbitrarily close to the correct hypothesis. This requires a notion of closeness and a specification of what randomly chosen examples mean.

In this section, we assume that an algorithm will choose a hypothesis consistent with the data. We use the notion of a version space to answer the above questions.

The **error** of a hypothesis $h \in \mathbf{H}$ is the proportion of instances it wrongly categorizes:

$$error(h) = P(h(i) \neq c(i) | i \in I)$$

where P denotes probability. Note that $c(i)$ is the actual concept being learned; you typically don't know this and thus don't actually know the error of a particular hypothesis. Intuitively the error is the weighted proportion of elements of i such that $h(i) \neq c(i)$. To define the error properly you need a probability distribution over elements of I, which forms a weighting of elements of I. The following analysis is independent of the distribution; the results will hold no matter what the distribution is.

The error of a hypothesis measures the closeness to the concept being searched for. Given $\epsilon > 0$, hypothesis h is **approximately correct** if $error(h) \leq \epsilon$.

You cannot guarantee that a learning algorithm will converge to an approximately correct concept unless you know something about the examples chosen. If the examples were chosen by someone trying to trick the learning algorithm, they may never show examples that will distinguish close hypotheses from other hypotheses that are not close. Thus, you cannot guarantee that the learner will converge on a correct hypothesis for arbitrary sets of training examples. If, however, the training examples are not chosen by someone out to trick us, then you can show that the learner will converge on the correct hypothesis. You can thus assume the following:

Assumption 11.2   The training examples are chosen independently from the same probability distribution as the population.

It's still possible that the examples do not distinguish hypotheses that are far away from the concept—it's just very unlikely that they do not. Thus you want to show that the learner is **probably approximately correct**. This means that for an arbitrary number $\delta, 0 < \delta \leq 1$, the algorithm can be wrong (not approximately correct) at most δ of the time. That is, the hypothesis generated is approximately correct at least $1 - \delta$ of the time.

You will see that for arbitrary ϵ and δ, you can guarantee that an algorithm that returns a consistent hypothesis will find a hypothesis with error less than ϵ, in at least $1 - \delta$ of the time. Moreover, this doesn't depend on the distribution.

Suppose you are given $\epsilon > 0$ and $\delta > 0$. Partition the hypothesis space $\mathbf{H}$ into:

$$\mathbf{H}_0 = \{h \in \mathbf{H} : error(h) \leq \epsilon\}$$
$$\mathbf{H}_1 = \{h \in \mathbf{H} : error(h) > \epsilon\}$$

You want to guarantee that you do not choose an element of $\mathbf{H}_1$ more than δ of the time.

Suppose $h \in \mathbf{H}_1$, then

$P(h$ is wrong for a single example$) \geq \epsilon$

$P(h$ is correct for a single example$) \leq 1 - \epsilon$

$P(h$ is correct for m examples$) \leq (1 - \epsilon)^m$

Thus you have the following overestimates:

$P(\mathbf{H}_1$ contains an hypothesis that's correct for m examples$)$
$$\leq |\mathbf{H}_1| (1 - \epsilon)^m$$
$$\leq |\mathbf{H}| (1 - \epsilon)^m$$
$$\leq |\mathbf{H}| e^{-\epsilon m}$$

using the inequality that $(1 - \epsilon) \leq e^{-\epsilon}$ if $0 \leq \epsilon \leq 1$.

Thus, if you can ensure that $|\mathbf{H}|\, e^{-\epsilon m} \leq \delta$, then you can guarantee that you do not choose an element of $\mathbf{H}_1$ more than δ of the time.

Solving for m, you obtain

$$m \geq \frac{1}{\epsilon}\left(\ln |\mathbf{H}| + \ln \frac{1}{\delta}\right).$$

Thus you can conclude the following:

Proposition 11.3
If a hypothesis is consistent with at least

$$\frac{1}{\epsilon}\left(\ln |\mathbf{H}| + \ln \frac{1}{\delta}\right)$$

training examples, you can guarantee that it has error at most ϵ, in at least $1 - \delta$ of the time. The number of examples needed to guarantee this error bound is called the **sample complexity**.

The number of examples required according to this model is a function of the size of the hypothesis space, as well as ϵ and δ.

Example 11.13
Suppose the hypothesis space $\mathbf{H}$ is the set of conjunctions of literals on n Boolean variables. In this case $|\mathbf{H}| = 3^n + 1$ as, for each conjunction, each variable in is one of three states: it's unnegated in the conjunction, it's negated or it doesn't appear; the "+1" is needed to represent false, which is the conjunction of any atom and its negation. Thus you need more than $\frac{1}{\epsilon}\left(n \ln 3 + \ln \frac{1}{\delta}\right)$ examples. The sample complexity is polynomial in n, $\frac{1}{\epsilon}$, and $\ln \frac{1}{\delta}$.

If you want to guarantee at most a 5% error 99% of the time and have 30 Boolean variables, then $\epsilon = 1/20$, $\delta = 1/100$, and $n = 30$. The bound says that you can guarantee this performance if you find a hypothesis that's consistent with $20 \times (30 \ln 3 + \ln 100) \approx 752$ examples.

Example 11.14
If the hypothesis space $\mathbf{H}$ is the set of all Boolean functions on n variables, then $|\mathbf{H}| = 2^{2^n}$, thus you need $\frac{1}{\epsilon}\left(2^n \ln 2 + \ln \frac{1}{\delta}\right)$ examples. The sample complexity is exponential in n.

If you want to guarantee at most a 5% error 99% of the time and have 30 Boolean variables, then $\epsilon = 1/20$, $\delta = 1/100$, and $n = 30$. The bound says that you can guarantee this performance if you find a hypothesis that's consistent with $20 \times (2^{30} \ln 2 + \ln 100) \approx 14,885,222,452$ examples.

You can now consider the third question raised at the start of this section, namely how quickly you can find the probably approximately correct hypothesis. First, if the sample complexity is exponential in the size of some parameter (e.g., n above), then so must the computational complexity, because an algorithm must at least consider each example. To show an algorithm with polynomial complexity, you need to find a hypothesis space with polynomial sample complexity and then show the algorithm uses polynomial time for each example.

11.5 Learning Under Uncertainty

So far we have not considered the problem of noise, where the given attributes are inadequate to accurately predict the classification or where some examples in the training set are misclassified. Some of the methods, in particular the version space learning and the PAC learning presented here, do not work with noise. For the other methods, noise is at best a nuisance, but causes problems with overfitting. This section confronts the problem of noise and uncertainty head-on.

We consider the Bayesian view of learning. Rather than choosing a hypothesis or delineating the space of consistent hypotheses, you derive a posterior distribution over the hypotheses. The hypotheses can be noisy in that they predict a probability distribution over classifications. The idea of **Bayesian learning** is simple: Condition on the examples, and derive the posterior probability of the hypotheses using Bayes' theorem.

The probability of a model (or a hypothesis) given some data is obtained by using Bayes' theorem (page 356):

$$P(model|data) \quad = \quad \frac{P(data|model) \times P(model)}{P(data).}$$ (11.1)

The likelihood, $P(data|model)$, is the probability that this model would have produced this data. It's high when the model is a good fit to the data, and it's low when the model would have predicted different data. The prior $P(model)$ encodes the learning bias; this factor is needed in order to bias the theory towards simple models. The denominator $P(data)$ is a normalizing constant to make sure that the probabilities sum to one.

There are two different approaches that can be taken given the above instance of Bayes' rule. One is to choose the model that's most likely given the data (page 427), and the other is to average over all models based on their posterior probabilities (page 428).

Example 11.15

Consider the simplest learning task under uncertainty, where one of two outcomes A and $\neg A$ occurs for each example, but you have no idea how likely it produces outcome A. You would like to learn the probability of A given some data.

To do this, you can treat the probability of A as a real-valued random variable on the interval $[0, 1]$, called *probA*.

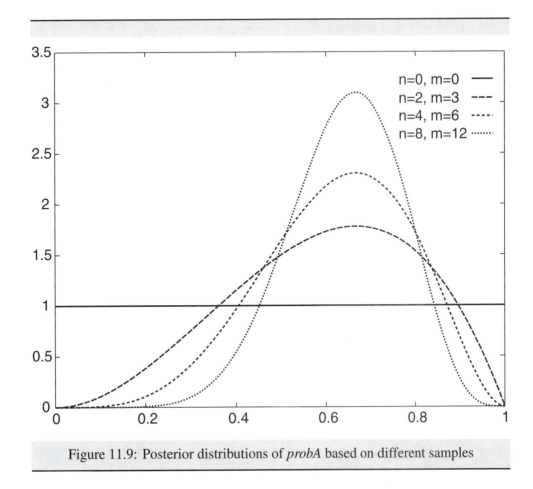

Figure 11.9: Posterior distributions of *probA* based on different samples

Suppose you have no prior information about the probability of Boolean variable *A*, and no knowledge beyond the data. This ignorance can be modeled by having the prior probability distribution as a uniform distribution of the variable *probA* over the interval [0, 1], as in the probability density function labeled $n = 0, m = 0$ in Figure 11.9.

You can update the probability distribution on variable *probA* given some data. Assume that the data are obtained by running a number of independent trials (experiments). Suppose, in *m* trials, you have collected *n* cases where *A* is true, and so $m - n$ cases where *A* is false.

The posterior distribution for *probA* given the experimental data can be derived by Bayes' rule. Let the proposition *data* be the particular sequence of observation that resulted in *n* occurrences of *A* in *m* trials. Bayes' rule gives us

$$P(probA\!=\!p|data) = \frac{P(data|probA\!=\!p) \times P(probA\!=\!p)}{P(data)}$$

Given that the trials are independent,

$$P(data|probA = p) = p^n \times (1 - p)^{m-n}$$

using the multiplicative rule for independent events: If $P(A) = p$ then the probability of observing a particular sequence of n A's in m independent trials is $p^n \times (1-p)^{m-n}$. This is related to the **binomial distribution**. The value of $P(probA = p)$ is assumed to be a uniform distribution on the interval $[0, 1]$. The denominator is a normalizing constant to make sure the area under the curve is one.

Figure 11.9 gives some posterior distributions of the variable *probA* based on different sample sizes. These are the cases: $(n = 2, m = 3)$, $(n = 4, m = 6)$, and $(n = 8, m = 12)$. Each of these peak at the same place, namely at $\frac{2}{3}$, but more data make the curve sharper.

Bayesian learning of decision trees

Consider a Bayesian approach to decision tree learning (page 403). Suppose you have many decision trees that accurately fit the data (as does the tree in Example 11.1, page 403). When *model* denotes one of those decision trees, $P(data|model) = 1$. The preference of one decision tree over the other depends on the prior probabilities of the decision trees; the prior probability encodes the learning bias (page 399). The preference for simpler decision trees over more complicated decision trees is because simpler decision trees have a higher prior probability.

Bayes' theorem gives you a way to trade off simplicity and ability to handle noise. Note that decision trees can handle noisy data by having probabilities at the leaves. While you prefer simpler decision trees, where there is noise, larger decision trees fit the training data better, because the tree can account for random regularities (noise) in the data. In learning with noisy data, you run into the problem of **overfitting** the data. This problem arises when there are correlations in the data that are there by chance and are not reflected in the world. Overfitting occurs when the model predicts correlations in the data that are not present in the world. In decision-tree learning, the likelihood favors bigger decision trees; the more complicated the tree, the better it can fit the data. The prior distribution can favor smaller decision trees. When you have a prior distribution over decision trees, Bayes' theorem tells you how to trade off model complexity and accuracy: The posterior probability of the model given the data is proportional to the product of the likelihood and the prior.

Example 11.16 Consider the data of Figure 11.2 (page 400) where you want to predict the readers actions from observing what articles they read.

One possible decision tree is the one given in Figure 11.3 (page 404). Call this decision tree d_2. The likelihood of the data is $P(data|d_2) = 1$. That is, d_2 accurately fits the data.

Another possible decision tree is one with no internal nodes, and a leaf that says to predict *reads* with probability $\frac{1}{2}$. This is the most likely tree with no internal nodes, given the data. Call this decision tree d_0. The likelihood of the data given this model is:

$$P(data|d_0) = \left(\frac{1}{2}\right)^9 \times \left(\frac{1}{2}\right)^9 \approx 0.00000149.$$

Another possible decision tree is one that just splits on *length*, and with probabilities on the leaves given by $P(reads|length=long) = 0$ and $P(reads|length=short) = \frac{9}{11}$. Call this decision tree d_{1a}. The likelihood of the data given this model is:

$$P(data|d_{1a}) = 1^7 \times \left(\frac{9}{11}\right)^9 \times \left(\frac{2}{11}\right)^2 \approx 0.0543.$$

Another possible decision tree is one that just splits on *thread*, and with probabilities on the leaves given by $P(reads|thread=new) = \frac{7}{10}$ (as 7 out of the 10 examples with *thread=new* have *user_action=reads*), and $P(reads|thread=follow\_up) = \frac{2}{8}$. Call this decision tree d_{1t}. The likelihood of the data given d_{1t} is:

$$P(data|d_{1t}) = \left(\frac{7}{10}\right)^7 \times \left(\frac{3}{10}\right)^3 \times \left(\frac{6}{8}\right)^6 \times \left(\frac{2}{8}\right)^2 \approx 0.0000247.$$

These are just four of the possible decision trees. Which is best depends on the prior on trees. You need to multiply these likelihoods by the priors of the decision trees to determine the ratio of posterior probabilities.

Maximum A Posteriori Probability and Minimum Description Length

It seems reasonable to choose the model that maximizes $P(model|data)$ — this is known as the **maximum a posteriori probability** model, or the MAP model. As the denominator of Equation (11.1) (page 424) is independent of the model, it can be ignored when choosing the model. Thus the MAP model is the model that maximizes

$$P(data|model) \times P(model).$$

You can take $-\log_2$ of this equation and choose the model that minimizes

$$(-\log_2 P(data|model)) + (-\log_2 P(model)).$$

The left-hand side of this expression is the number of bits it takes to describe the data given the model (page 360). The right hand-side is the number of bits it takes to describe the model. Thus the MAP model is the same as the **minimal description length** (MDL) model. The idea of choosing the model with the highest posterior probability is the same as the minimal description length principle: Choose the model that minimizes the number of bits it takes to describe both the model and the data given the model.

One way to think about the MDL principle is that the aim is to communicate the data as succinctly as possible. The only use of the model is to make the communication shorter. In order to communicate the data, you can communicate the model and then communicate the data in terms of the model. The number of bits it takes to communicate the data using a model is the number of bits it takes to communicate the model plus the number of bits it takes to communicate the data in terms of the model. The MDL principle is to choose the model that lets you communicate the data in as few bits as possible.

Example 11.17

In Example 11.15 (page 424) the model that's most likely given n occurrences of A in m trials is $probA = \frac{n}{m}$. This is the point at which the curve peaks. For each of the curves of Figure 11.9 (except the $n = 0, m = 0$ curve), the most likely value of $probA$ is $\frac{2}{3}$.

Example 11.18

In Example 11.16, the definition of the priors on decision trees was left unspecified. The notion of a description length gives us some basis for assigning priors to decision trees—you can consider how many bits it takes to describe a decision tree (see Exercise 11.5). You have to be careful, because you want each code to describe a decision tree, and you want each decision tree to be described by a code. It may seem that the description length is more objective than the prior probability; however, this is illusory—for every prior distribution you can invent a code and for every code you can determine the prior probability; defining the code is just as arbitrary as defining the prior probability, although it often helps to think about the problem in terms of defining codes.

Averaging Over Models

Rather than choosing the most likely model, another approach is to average over all models, weighted by their posterior probabilities given the data.

For many cases this is difficult to do, as the models may be complicated (e.g., if they are decision trees or even belief networks). This section examines one simple case where you can average over the models analytically.

Example 11.19

Consider Example 11.15, which determines the value of $probA$ based on n A's out of m trials. You can derive the posterior distribution as in Figure 11.9. What's interesting about this is that whereas the most likely posterior value of $probA$ is $\frac{n}{m}$, the *expected value* (page 359) of this distribution is $\frac{n+1}{m+2}$.

Thus the expected value of the $n = 2, m = 3$ curve is $\frac{3}{5}$, for the $n = 4, m = 6$ case it's $\frac{5}{8}$, and for $n = 8, m = 12$ case it is $\frac{9}{14}$. As you get more data this value approaches $\frac{n}{m}$.

Why is this a better estimate than $\frac{n}{m}$? First, it tells us what to do if you have no data: Use the uniformed prior of $\frac{1}{2}$. This is the expected value of the $n = 0, m = 0$ case. Second, consider the case where $n = 0$ and $m = 3$. You don't want to use $P(A) = 0$, as this says that A is impossible, and you certainly do not have evidence for this!

As well as using the posterior distribution of *probA* to derive the expected value, you can use it to answer other questions such as: What is the probability that the posterior probability of *probA* is in the range $[a, b]$? In other words, derive $P(probA \geq a \wedge probA \leq b | data)$. This is the exact problem that the Reverend Thomas Bayes solved more than 200 years ago (Bayes, 1763). The solution he gave—although in much more cumbersome notation—was

$$\frac{\int_a^b p^n \times (1-p)^{m-n}}{\int_0^1 p^n \times (1-p)^{m-n}}$$

This kind of knowledge is used in surveys when it may be reported that a survey is correct $\pm 5\%$, 19 times out of 20. It's also the same type of information that's used by PAC learning (page 421), where you guarantee an error at most ϵ at least $1 - \delta$ of the time. If you choose the midpoint of the range $[a, b]$, namely $\frac{a+b}{2}$, as your hypothesis, then you have error less than or equal to $\frac{b-a}{2}$, just when the hypothesis is in $[a, b]$. $1 - \delta$ corresponds to $P(probA \geq a \wedge probA \leq b | data)$. If $\epsilon = \frac{b-a}{2}$ and $\delta = 1 - P(probA \geq a \wedge probA \leq b | data)$, then choosing the midpoint will result in an error at most ϵ in $1 - \delta$ of the time. PAC learning gives worst-case results, whereas the Bayesian view gives the expected number. Typically, the Bayesian estimate is more accurate, but the PAC results give a guarantee of the error. The sample complexity (see Section 11.4) required for Bayesian learning is typically much less than that of PAC learning—much less data are needed to *expect to achieve* the desired accuracy than are needed to *guarantee* the desired accuracy.

Naive Bayesian Classifier

So far, the Bayesian methods discussed have been much more abstract than the other methods. This section discusses a particular probabilistic representation that can be used for classification tasks. This is the **Bayesian classifier**.

The idea behind a Bayesian classifier is to build a probabilistic model of the data and to use that model to predict the classification of a new example. The **naive**

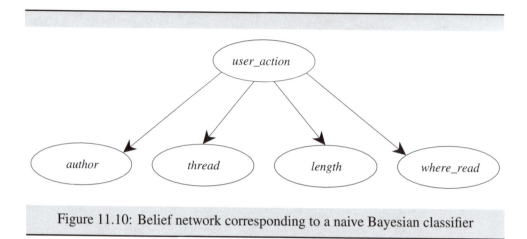

Figure 11.10: Belief network corresponding to a naive Bayesian classifier

Bayesian classifier makes the independence assumption that the values of the attributes are independent of each other given the classification. (See page 363 for a definition of independence.) The training examples are used to determine the probabilities $P(classification)$ and $P(att_i|classification)$ for each attribute att_i. These probabilities can be determined by counting the number of instances.

Suppose you observe a new case with $att_1 = v_1, \ldots, att_k = v_k$. You can use Bayes' theorem (page 356) to determine the posterior probability of the classification of the new case conditioned on the values of its attributes:

$$P(cl|att_1 = v_1, \ldots, att_k = v_k)$$
$$= \frac{P(att_1 = v_1, \ldots, att_k = v_k|cl)P(cl)}{P(att_1 = v_1, \ldots, att_k = v_k)}$$
$$= \frac{P(att_1 = v_1|cl) \cdots P(att_k = v_k|cl)P(cl)}{P(att_1 = v_1, \ldots, att_k = v_k)}$$

where the denominator $P(att_1 = v_1, \ldots, att_k = v_k)$ is a normalizing factor to ensure the probabilities add to one.

The independence of the naive Bayesian classifier is embodied in a particular belief network (page 364), namely one where the classification has no parents, and the classification is the only parent of each attribute. The data are used to derive the prior probability of the classification and the conditional probabilities for the attributes.

Example
11.20

Let's try to predict the user action given the data of Figure 11.2 (page 400). For this example, the user action is the classification. The naive Bayesian classifier for this example corresponds to the belief network of Figure 11.10. The training examples are used to determine the probabilities required for the belief network.

Let's just use the frequencies as the probabilities for this example. The probabilities that can be derived from these data are:

$$P(user\_action = reads) = \frac{9}{18} = 0.5$$

$$P(author = known | user\_action = reads) = \frac{2}{3}$$

$$P(author = known | user\_action = skips) = \frac{2}{3}$$

$$P(thread = new | user\_action = reads) = \frac{7}{9}$$

$$P(thread = new | user\_action = skips) = \frac{1}{3}$$

$$P(length = long | user\_action = reads) = 0$$

$$P(length = long | user\_action = skips) = \frac{7}{9}$$

$$P(where\_read = home | user\_action = reads) = \frac{4}{9}$$

$$P(where\_read = home | user\_action = skips) = \frac{4}{9}$$

Based on these probabilities you can see that the attributes *author* and *where_read* have no predictive power; knowing either will not change the probability that the user will read the article. The rest of this example ignores these attributes.

Suppose you want to classify a new case where the author is unknown, the thread is a follow-up, the length is short, and it's read at home.

$$P(user\_action = reads | thread = follow\_up \wedge length = short)$$
$$= P(follow\_up | reads) \times P(short | reads) \times P(reads) \times c$$
$$= \frac{2}{9} \times 1 \times \frac{1}{2} \times c$$
$$= \frac{1}{9} \times c$$

$$P(user\_action = skips | thread = follow\_up \wedge length = short)$$
$$= P(follow\_up | skips) \times P(short | skips) \times P(skips) \times c$$
$$= \frac{2}{3} \times \frac{2}{9} \times \frac{1}{2} \times c$$
$$= \frac{2}{27} \times c$$

c is a normalizing constant that ensures these add up to one. Thus c must be $\frac{27}{5}$, so

$$P(user\_action = reads | thread = follow\_up \wedge length = short) = 0.6.$$

Note that this prediction is at odds with example $e11$. The naive Bayesian classifier summarizes the data into a few parameters. It predicts the article will be read because being short is a stronger indicator that the article will be read than being a follow-up is an indicator that the article will be skipped.

Suppose you see a new case where author is unknown, the thread is new, the length is long, and it was read at work. In this case $P(length = long|user\_action = reads) = 0$ and the conditional probabilities associated with skipping are not zero, so the posterior probability that the *user_action = reads* is zero.

While there are some cases where the naive Bayesian classifier doesn't produce good results, it's extremely simple, easy to implement, and often works very well. It's a good method to try for a new problem.

In general, the naive Bayesian classifier works well when the independence assumption is appropriate. This is often the case where the classification is good at actually classifying the data, as opposed to being some arbitrary attribute. This may be appropriate for *natural kinds* where the classifications have evolved because they are useful in distinguishing particular objects. Natural kinds are often associated with nouns, such as the class of dogs or the class of chairs. When the classifications correspond to natural kinds, you would expect that the classification would be a good predictor of the attributes.

Probabilities from Experts

Apart from getting probabilities from data, the most normal source is from experts. You can use the above ideas of Bayesian learning as a way to combine probabilities from various experts and with real data.

There are a number of problems with obtaining probabilities from experts. Among these are:

- Experts' reluctance to give an exact probability value that cannot be refined
- Representing the uncertainty of a probability estimate
- Combining the numbers from multiple experts
- Combining expert opinion with actual data

One idea that works quite well is to allow an expert, rather than giving an exact real number such as 0.667 for the probability of A, to give a pair of numbers as $\langle n, m \rangle$ that can be interpreted as though they had observed n A's out of m trials.

These pairs can be combined with actual data. When an A is encountered, both n and m can be incremented. When $\neg A$ is encountered, m can be incremented alone. In this way the probabilities can be continually improved, while still being able to use the expert's knowledge.

These sample estimates from different experts can be easily combined together by adding the two components.

Whereas the ratio reflects the probability, different levels of confidence can be reflected in the absolute values. $\langle 2, 3 \rangle$ reflects extremely low confidence that would quickly be dominated by data or other expert's estimates. The pair $\langle 20, 30 \rangle$ reflects more confidence—a few examples would not change it much, but tens of examples would. Even hundreds of examples would have little effect on an estimate embodied in the pair $\langle 2000, 3000 \rangle$. The completely ignorant ratio would be $\langle 0, 0 \rangle$; this would then give the same probability estimate as when no expert is involved.

11.6 Explanation-Based Learning

Section 11.1 considered the problem of inductive learning, where, given some data, you hypothesize a representation that would account for the data and allow you to predict values for unseen data. This section shows how the simple delaying meta-interpreter (page 210) can be used for a form of **speed-up learning**, or **analytical learning**, where the agent is learning how to solve queries that are similar to the ones it has already solved.

The simplest form of speed-up learning is **caching**, where previously computed answers are stored so that if the same query is asked again, the answer can be looked up in the cache. This is effective if the the cache can be kept small and if the same queries are asked repeatedly.

The problem with caching is that it only works for identical queries. **Explanation-based learning** is a form of speed-up learning that lets the cache be used for similar queries, not just identical queries.

Assume that the clauses are divided into general rules, the background knowledge, and specific facts about a particular example. Each query is about a particular example. The general idea is, given a query that's answered using the domain theory, you determine what aspects of the example were important for the proof to complete. If you find a new example with these aspects, you can use the cached result instead of using the domain theory.

In explanation-based learning, a cached result is a rule whose head contains the predicate being learned (this corresponds to the predicate in the initial query) and whose body contains only predicates that appear in the facts or are built-in. Given a proof for the original query, you want to build the most general rule that captures the essence of that proof.

Figure 11.11 gives a meta-interpreter that returns a representation of the generalized rules needed for explanation-based learning. It assumes that the base-level theory is divided into

- Particular facts about the examples

% $eprove(G, G_1, D_0, D_1)$ is true if G is an instance of G_1, and D_1 is a list of possible
% facts, and D_2 is a tail of D_1 such that, given the base-level theory, G_1 follows from
% the elements of D_1.

$eprove(true, true, D, D)$.

$eprove((A \ \& \ B), (A_1 \ \& \ B_1), D_1, D_3) \leftarrow$
   $\quad eprove(A, A_1, D_1, D_2) \wedge$
   $\quad eprove(B, B_1, D_2, D_3)$.

$eprove(G, G_1, D, [G_1|D]) \leftarrow$
   $\quad fact(G)$.

$eprove(G, G_1, D, [G_1|D]) \leftarrow$
   $\quad built\_in(G) \wedge$
   $\quad call(G)$.

$eprove(H, H_1, D_1, D_2) \leftarrow$
   $\quad (N :: H \Leftarrow B) \wedge$
   $\quad clause(N, H_1, B_1) \wedge$
   $\quad eprove(B, B_1, D_1, D_2)$.

Figure 11.11: A meta-interpreter for explanation-based learning

- Background knowledge that's true for all examples

The base-level fact F about a particular example is represented as the meta-level fact

$\quad fact(F)$.

The background knowledge is represented in terms of labeled clauses of the form

$\quad N :: H \Leftarrow B$

where N is a ground atom representing the name of the base-level clause "H if B."
The clauses are labeled because you need two different instances of the clause in the
same proof: one for the original query and one for the generalized rule that's being
generated.

The general idea of the *eprove* in Figure 11.11 is that the first argument to *eprove*
is the goal to be proved. Which clauses are selected and whether a proof succeeds or
fails is determined completely by this argument. The other three arguments are there
to extract the most general rule from the proof of the first argument. These can be seen
as an instance of the delaying meta-interpreter (page 210) that always chooses exactly

the same rules as does the proof of the example, but collects the facts and the built-in predicates needed for the proof and stores them in a list.

If q is the initial query, then a call to

$?eprove(q, H, [\,], B)$

will succeed if and only if the query q would succeed or fail in the base-level; and if q succeeds, then the learned cache rule has head H and the list B contains the elements of the body of the rule.

Example 11.21

Suppose the infobot needs to access multiple large knowledge sources to determine whether a student can enroll in a course. To answer the first such query, it needs to access all of the knowledge sources. Once it has done this for one student, it can cache the results so that a query about a similar student can be accessed quickly.

Figure 11.12 shows part of a background knowledge base for the university example (page 86), slightly simplified. We intend that this knowledge base is complicated and that much inference is required to determine whether a student has fulfilled their electives. Otherwise there is not much point in using explanation-based learning.

Suppose you have the following information about a student:

$fact(grade(sally, phys101, 89))$.
$fact(grade(sally, hist101, 77))$.
$fact(grade(sally, cs126, 84))$.
$fact(grade(sally, cs202, 88))$.
$fact(grade(sally, cs333, 77))$.
$fact(grade(sally, cs312, 74))$.
$fact(grade(sally, math302, 41))$.
$fact(grade(sally, math302, 87))$.
$fact(grade(sally, cs304, 79))$.
$fact(grade(sally, cs804, 80))$.
$fact(grade(sally, psyc303, 85))$.
$fact(grade(sally, stats324, 91))$.
$fact(grade(sally, cs405, 77))$.

From the facts about *sally* and the background knowledge base, it can be determined that *sally* has indeed fulfilled the electives in the *cs* department. If you were to cache this result, then you could answer the same question quickly about *sally*, but for anyone else, you would need to access the background knowledge base. The EBL meta-interpreter will help you determine what it is about *sally* that made you conclude that she had fulfilled her electives, so that if other students are similar

1 :: *fulfilled_electives*(*St*, *SciDept*) ⇐
    dept(*SciDept*, *science*) &
    has_arts_elective(*St*) & *has_sci_elective*(*St*, *SciDept*).
2 :: *has_arts_elective*(*St*) ⇐
    passed(*St*, *ArtsEl*, 49) &
    course(*ArtsEl*, *ArtsDept*, _) & *dept*(*ArtsDept*, *arts*).
3 :: *has_sci_elective*(*St*, *Major*) ⇐
    passed(*St*, *SciC*, 59) &
    course(*SciC*, *Dept*, *Level*) & *dept*(*Dept*, *science*) &
    diff_dept(*Dept*, *Major*) & *member*(*Level*, [*third*, *fourth*]).
4 :: *passed*(*St*, *C*, *MinPass*) ⇐
    grade(*St*, *C*, *Gr*) & *Gr* > *MinPass*.
5 :: *member*(*X*, [*X*|_]) ⇐ *true*.
6 :: *member*(*X*, [_|*Y*]) ⇐ *member*(*X*, *Y*).
7 :: *dept*(*history*, *arts*) ⇐ *true*.
8 :: *dept*(*english*, *arts*) ⇐ *true*.
9 :: *dept*(*psychology*, *science*) ⇐ *true*.
10 :: *dept*(*statistics*, *science*) ⇐ *true*.
11 :: *dept*(*mathematics*, *science*) ⇐ *true*.
12 :: *dept*(*cs*, *science*) ⇐ *true*.
13 :: *course*(*hist101*, *history*, *first*) ⇐ *true*.
14 :: *course*(*psyc303*, *psychology*, *third*) ⇐ *true*.
15 :: *course*(*stats324*, *statistics*, *third*) ⇐ *true*.
16 :: *course*(*math302*, *mathematics*, *third*) ⇐ *true*.
17 :: *course*(*phys101*, *physics*, *first*) ⇐ *true*.
18 :: *diff_dept*(*psychology*, *cs*) ⇐ *true*.
19 :: *diff_dept*(*statistics*, *cs*) ⇐ *true*.
20 :: *diff_dept*(*mathematics*, *cs*) ⇐ *true*.
built_in((_ > _)).

Figure 11.12: Example background KB for explanation-based learning

enough to *sally*, you will be able to conclude they have have also fulfilled their electives.

The query

      ?ebl(fulfilled_electives(sally, cs), H, [], B).

returns (with variables renamed here):

      $H = fulfilled\_electives(S, cs)$
      $B = [grade(S, hist101, M_1), M_1 > 49, grade(S, math302, M_2), M_2 > 59]$

which can be interpreted as the rule:

      $fulfilled\_electives(S, cs) \Leftarrow$
            $grade(S, hist101, M_1)$ &
            $M_1 > 49$ &
            $grade(S, math302, M_2)$ &
            $M_2 > 59.$

This tells you that anyone who got more than 49 in *history*101 and more than 59 in *math*302 has fulfilled the electives in *cs*. It was these particular facts about *sally* that were used to conclude that she had fulfilled her electives.

11.7 References and Further Reading

For a good overview of machine learning see Mitchell (1997). The collection of papers by Shavlik & Dietterich (1990) contains many classic learning papers. Michie, Spiegelhalter & Taylor (1994) give empirical evaluation of many learning algorithms on many different problems. Briscoe & Caelli (1996) discuss many different machine learning algorithms. Weiss & Kulikowski (1991) overview techniques for classification learning.

Decision tree learning is discussed by Quinlan (1986). For a more recent overview of a mature decision tree learning tool see Quinlan (1993). The GINI index (Exercise 11.4) is based on the splitting criteria in CART (Breiman, Friedman, Olshen & Stone, 1984).

For overviews of neural networks see Bishop (1995), Hertz, Krogh & Palmer (1991), and Jordan & Bishop (1996). Back propagation is introduced in Rumelhart, Hinton & Williams (1986). Minsky & Papert (1988) analyze the limitations of neural networks.

For details on case-based reasoning see Riesbeck & Schank (1989) and Kolodner (1993). For a review of nearest-neighbor algorithms see Duda & Hart (1973) and Dasarathy (1991). The dimension-weighting learning nearest-neighbor algorithm is from Lowe (1995).

Version spaces are due to Mitchell (1977). PAC learning was introduced by Valiant (1984). The analysis here is due to Haussler (1988). Kearns & Vazirani (1994) give a good introduction to computational learning theory and PAC learning. For more details on version spaces and PAC learning see Mitchell (1997).

An accessible introduction to Bayesian learning is in Cheeseman (1990). See also books on Bayesian statistics, such as Pratt, Raiffa & Schlaifer (1995). Bayesian learning of decision trees is described in Buntine (1992). The approach to combining expert knowledge and data is due to Spiegelhalter, Franklin & Bull (1990).

Bayesian classifiers are discussed by Duda & Hart (1973) and Langley, Iba & Thompson (1992). Friedman & Goldszmidt (1996) discuss how the naive Bayesian classifier can be generalized to allow for more appropriate independence assumptions.

For different overviews of explanation-based learning see Mitchell, Keller & Kedar-Cabelli (1986) and DeJong & Mooney (1986).

For research results on machine learning see the journal *Machine Learning*, the annual *International Conference on Machine Learning*, or general AI journals such as *Artificial Intelligence*, *Computational Intelligence*, and *Journal of Artificial Intelligence Research*.

This chapter has concentrated on supervised classification learning. There are other paradigms that you may want to investigate. For overviews of inductive logic programming see Muggleton (1995) and Quinlan & Cameron-Jones (1995). Unsupervised learning is discussed by Fischer (1987) and Cheeseman, Kelly, Self, Stutz, Taylor & Freeman (1988). For an introduction to reinforcement learning, see Kaelbling, Littman & Moore (1996).

11.8 Exercises

Exercise 11.1 The aim of this exercise is to determine the size of the space of decision trees. Suppose there are n binary attributes in a learning problem. How many different decision trees are there? How many different functions are represented by these decision trees? Is it possible that two different decision trees give rise to the same function.

Exercise 11.2 Extend the decision-tree learning algorithm of Figure 11.4 so that multivalued attributes can be represented. Make it so that the rule form of the decision tree is returned.

One problem that must be overcome is when there are no examples that correspond to one particular value of a chosen attribute. You need to make a reasonable prediction for this case.

Exercise 11.3 The decision tree learning algorithm of Figure 11.4 fails if it runs out of attributes and not all examples agree.

Suppose that you are building a decision tree where the attributes are inadequate and you have come to the stage where there are no remaining attributes to split on, and there are examples in the training set, n of which are positive, and m of which are negative. Two strategies have been suggested:

i) Return whichever value has the most examples—return *true* if $n > m$, *false* if $n < m$, and either if $n = m$.

ii) Return a probability $n/(n + m)$, which is the probability that the class is true.

Which of these two strategies has the least error on the training set where

(a) The error is defined as the sum of the absolute differences between the value of the example ($1 = true$ and $0 = false$) and the predicted values in the tree (either $1 = true$ and $0 = false$ or the probability).

(b) The error is defined as the sum of the squares of the differences in values.

Implement whichever of the two strategies you think is most sensible.

Exercise 11.4 In choosing which attribute to split on in decision tree search, an alternative heuristic to the max information split of Section 11.2 is to use the GINI index.

The GINI index of a set of examples (with respect to an attribute) is a measure of the impurity of the examples:

$$gini_{Att}(Examples) = 1 - \sum_{Val} \left(\frac{|\{Obj \in Examples : val(Obj, Att, Val)\}|}{|Examples|} \right)^2$$

where $|\{Obj \in Examples : val(Obj, Att, Val)\}|$ is the number of examples with value *Val* of attribute *Att*, and $|Examples|$ is the total number of examples. The GINI index is always nonnegative and has value zero only if all of the examples have the same value on the attribute. The GINI index reaches it maximum value when the examples are evenly distributed among the values of the attributes.

One heuristic for which property to split on is to choose the split that minimizes the total impurity, $gini_{Goal}(Yes) + gini_{Goal}(No)$, where *Goal*, *Yes*, and *No* are as in Figure 11.4.

(a) Implement a decision-tree searching algorithm that uses the GINI index.

(b) Try both GINI index algorithm and the maximum information split algorithm on some databases, and see which results in better performance.

(c) Find an example database where the GINI index finds a different tree than the maximum information gain heuristic. Which heuristic seems to be better for this example? Consider which heuristic seems more sensible for the data at hand.

(d) Try to find an example database where the maximum information split seems more sensible than the GINI index, and try to find another example for which the GINI index seems better. *Hint* try extreme distributions.

Exercise 11.5 As outlined in Example 11.16 (page 426), define a code for describing decision trees. Make sure that each code corresponds to a decision tree (for every sufficiently long sequence of bits, the initial segment of the sequence will describe a unique decision tree), and each decision tree has a code. How does this code translate into a prior distribution on trees? In particular, how much does the likelihood of introducing a new split have to increase in order to offset the reduction in prior probability of the split (assuming that smaller trees are easier to describe than large trees in your code)?

Exercise 11.6 Consider the neural network learning data of Figure 11.7 (page 413).

(a) Suppose that you decide to use any predicted value from the neural network greater than 0.5 as true, and any value less than 0.5 as false. How many examples are misclassified initially? How many examples are misclassified after 40 iterations? How many examples are misclassified after 80 iterations?

(b) Try the same example and the same initial values, with different step sizes for the gradient descent. Try at least $\eta = 0.1$, $\eta = 1.0$, $\eta = 5.0$. Comment on the relationship between step size and convergence.

(c) Given the final parameter values in Figure 11.7, give a logical formula for what each of the units is computing. You can do this by considering, for each of the units, the truth tables for the input values and determining the output for each combination, then reducing this formula. Is it always possible to find such a formula?

(d) All of the parameters were set to different initial values. What happens if the parameter values were all set to the same (random) value? Test it out for this example, and hypothesize what occurs in general.

(e) For the neural network algorithm, comment on the following stopping criteria:

i) Learn for a limited number of iterations, where the limit is set initially.

ii) Stop when the error is less than 0.25. Explain why 0.25 may be an appropriate number.

iii) Stop when the derivatives all become within some ϵ of zero.

iv) Split the data into training data and test data, and stop when the error on the test data increases.

Which would you expect to handle overfitting better? Which criteria guarantees the hill climbing will stop? Which criteria would guarantee that, if it stops, the network can be used to predict the test data accurately?

Exercise 11.7 In the neural net learning algorithm, each example affects the total error independently, so the parameters can be updated for each example rather than after all examples have been seen. Implement such a learning algorithm, and compare it to the batch learning algorithm that considers all examples before updating the parameters, with respect to both rate of convergence and to speed of the algorithm.

Exercise   Implement a neural network learning algorithm that determines the derivatives analyt-
11.8   ically. *Hint:* If f is the sigmoid function then $f'(x) = f(x) \times (1 - f(x))$.

Exercise   (a) Draw a *kd*-tree for the data of Figure 11.2 (page 400). The topmost attribute to
11.9   split on should be the one that most divides the examples into two equal classes.
                   Assume that you know that *user_action* attribute doesn't appear in subsequent
                   queries, and so should be split on. Show which training examples are at which
                   leaf nodes.

               (b) Show the locations in this tree that contain the closest training examples to a new
                   case where the author is unknown, the thread is new, the length is long, and it
                   was read at work.

               (c) Based on this example, discuss which examples should be returned from a lookup
                   on a *kd*-tree. Why is this different from a lookup on a decision tree?

Exercise   Implement a nearest-neighbor learning system that stores the training examples in a
11.10   *kd*-tree, and uses the neighbors that differ in the fewest number of attributes, weighted
             evenly. How well does this work in practice?

Chapter 12

Building Situated Robots

12.1   Introduction

So far we have discussed agents and how they can reason. We have not emphasized how agents interact with their environment. Chapter 8 considered representations of time and concentrated on reasoning about time. This chapter is about how an intelligent agent can perceive, reason, and act in time in an environment.

Within this framework, agents act in environments and receive information from the environment. The goal is to determine what the agent should do based on its knowledge, its preferences, and what it observes about its environment. These agents can be, for example, physical agents in controlled office or laboratory environments, robots in uncontrolled natural environments, or even software agents in a software environment such as the Internet.

An **agent** is something that acts in the world. An agent can, for example, be a person, a robot, a dog, a worm, the wind, gravity, or a lamp. **Purposive agents** have preferences. They prefer some states of the world to other states, and they act in order to try to achieve worlds they prefer. The nonpurposive agents are grouped together and called *nature*. Whether an agent is purposive or not is a modeling assumption that may or may not be appropriate. For example, for some applications it may be appropriate to model a dog as purposive, and for others it may suffice to model a dog as nonpurposive.

A **robot** is an artificial purposive agent. It's these you get to build. A robot acts with certain goals and tries to achieve some outcome, perhaps trying also to avoid other possible outcomes. The preferences of the robot are usually the preferences of the designer of the robot, but sometimes the robot can be given goals and preferences at run time.

Agents can have sensors and effectors as well as limited memory and limited computational capabilities. Agents reason and act in time.

A robot should react to the world. Its actions depend not only on its goals, but on the information it receives from its sensors. These sensors may or may not reflect what's true in the world. Sensors can be noisy, unreliable, or broken, and even when sensors are reliable there is still ambiguity about the world from sensor readings. An agent must act on the information it has available. Often this information is very weak such as "sensor a appears to be producing value v." Agents can often carry out actions to find more information about the world—for example, opening a cupboard door to find out what's in the cupboard. Actuators can also be noisy, unreliable, slow, or broken. What an agent can control is what message (command) it sends to its actuators.

If an agent doesn't have preferences, by definition it doesn't care what state it ends up in, and so it doesn't matter what it does. The only reason to design an agent is to instill it with preferences—that is, to make it prefer some states and try to achieve these. A robot doesn't have to know its preferences. For example, a thermostat is a robot that senses the world and either turns a heater on or off. There are preferences embedded in the thermostat, such as to keep the occupants of a room at a pleasant temperature, even though the thermostat doesn't know these are its preferences.

12.2 Robotic Systems

Figure 12.1 depicts the general interaction between a robot and its environment. Together the whole system is known as a robotic system:

A **robotic system** is made up of a robot together with an environment. The robot receives **stimuli** from the environment and carries out **actions** in the environment.

A **robot** is made up of a **body** plus a **controller**. The controller receives **percepts** from the body and sends **commands** to the body.

A **body** is made up of **sensors** that convert stimuli into percepts, and **actuators** that convert commands to actions.

Stimuli include such things as light, sound, and electronic commands, as well as physical bumps to the robot.

Common sensors include touch sensors, vision sensors, infrared sensors, sonar, hearing, and text commands. As a prototypical sensor, a vision sensor's sensory data is light coming into a camera, this light gets converted into a 2-dimensional array of brightness values called **pixels**. Sometimes there are multiple pixel arrays for different colors or for multiple cameras. Such pixel arrays could be the percepts for our controller. More often, percepts consist of higher-level features such as lines and edges and depth information such as from stereo, where there are two cameras, or from

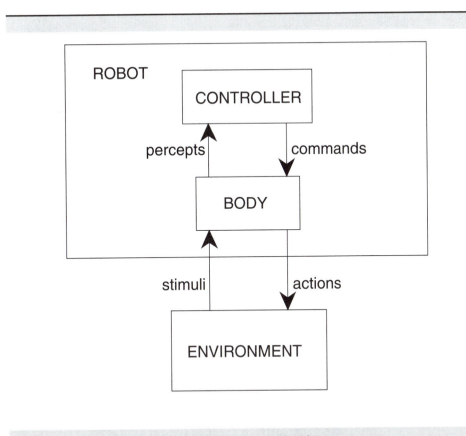

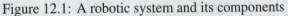

Figure 12.1: A robotic system and its components

motion parallax, where if you move your head, closer objects move more in the field of vision than distant objects. Often the percepts are more specialized—for example, the positions of bright orange dots, or the hand signals given by a human.

Actions include steering, accelerating wheels, moving links of arms and legs, speaking, and sending sonar messages that can be sensed when they bounce off objects. Commands can be low-level commands such as to set the voltage to a motor to some particular value. Commands could be high-level specifications of the desired motion of the robot, such as "stop" or "travel at 1 meter per second due east" or "go to room $o103$." The actuators, like the sensors, are typically noisy. For example, stopping takes time; the robot is governed by the laws of physics and has momentum, and messages take time to travel. The robot may only end up going approximately 1 meter per second, approximately east, and both the speed and direction may fluctuate. Even traveling to a particular room may fail for a number of reasons.

The controller is that part which you get to build.

12.3   The Agent Function

Agents are situated in time, they receive sensory data in time, and do actions in time. To consider controlling robots you first need to consider time (page 281).

Let T be the set of time points. Assume that T is totally ordered and has some metric so that you can measure the temporal distance between any two time points. Basically, you assume that T is isomorphic to some subset of the real line. This subset is often discrete, such as every hundredth of a second or every day. Assume that T has a starting point, which you can arbitrarily call 0.

A **percept trace**, or percept stream, is a function from T into P where P is the set of all possible percepts. A **command trace** is a function from T into C where C is the set of all commands.

A percept trace for an agent is the sequence of all past, present, and future percepts received by the agent. Because you and our agents are situated in time, you can't actually observe percept traces; at any time you can only see the part of the trace up to *now*. You can only observe the value of the trace at time $t \in T$ when you get to time t.

A **transduction** is a function from percept traces into command traces that's **causal** in the sense that the action trace up to time t depends only on percepts up to t. The causality restriction is needed because agents are situated in time; their commands at time t can't depend on percepts after time t.

A **controller** is an implementation of a transduction.

The **history** of an agent at time t is the percept trace and the command trace of the agent up to time t.

Thus a transduction specifies a function from the agent's history at time t and percepts at time t into the command at time t.

If you consider implementing such a transduction, you realize that an agent doesn't have access to its entire history. It only has access to what it has remembered.

The **state** of an agent at time t is a representation in the agent that encodes all of the agent's history that it has access to. The state encapsulates all of the information about its history that it can use for current and future actions.

In terms of probabilistic independence (page 361), the state is the information that makes the agent's future at time t independent of its history at time t given its state at time t.

The state can contain any information. Some instances of state include:

- The simplest state is just a program counter that records the position that the robot is up to in a plan. (Plans were discussed in Chapter 8.)

- The state can contain specific facts that are useful—for example, where the delivery robot left the parcel in order to go and get the key, or where it has already checked for the key. It may be useful for the agent to remember anything that's reasonably stable and that can't be immediately observed.

- The state can encode a model or a partial model of the world. This is often called a **belief state**. An agent could maintain its best guess about the current state of the world, using default reasoning described in Chapter 9, or could have a probability distribution over possible world states, using probability as described in Chapter 10. The belief state is updated by temporal projection—estimating the current state from the beliefs about the previous state—and by observing part of the world. Temporal projection increases uncertainty while observations decrease uncertainty.

- The state can encode **goals**, what the agent is trying to achieve, and **intentions**, the steps it intends to take to achieve those goals. By explicitly storing these it can notice when some intentions can be removed because they are no longer needed, or replaced because they no longer serve a purpose.

Thus a controller must implement a state transition function and a command function. For discrete time, we have the following definitions:

A **state transition function** is a function $\sigma : S \times P \to S$, where S is the set of states and P is the set of possible percepts. $s_{t+1} = \sigma(s_t, p_t)$ means that s_{t+1} is the state following state s_t when p_t is observed.

A **command function** is a function $\chi : S \times P \to C$, where S is the set of states, P is the set of possible percepts, and C is the set of possible commands. $c_t = \chi(s_t, p_t)$ means that the controller issues command c_t when the state is s_t and p_t is observed.

Analogous definitions can be made for continuous time, where the state transition function returns the derivative and the state prediction is obtained by integration.

The rest of this chapter considers how to implement state transition functions and command functions, and thus how to build a controller for a robot. It only considers discrete time.

12.4   Designing Robots

Robots act in an environment in time. In deciding what a robot will do, there are two aspects of computation that need to be distinguished: the computation that goes into the design of the robot and the computation that's done by the robot as it's acting.

On-line computation is the computation that's done by the robot in order to determine what to do. **Off-line computation** is computation that is done toward the design of the robot. It's done before the robot needs to act on the information and needs to be compiled into a usable form.

You need to distinguish between the knowledge in the mind of the designer and in the mind of the robot. Consider the extreme cases:

- At one extreme is a highly specialized robot that works well in the environmental niche for which it was designed. The designer may have done considerable work in building the robot, but the robot may not need to do very much in order to operate well. An example is a thermostat. It may be difficult to design a thermostat so that it turns on and off at exactly the right temperatures, but the thermostat itself doesn't need to do much computation. These very specialized robots don't adapt well to different environments or to changing goals.

- At the other extreme is a very flexible robot that can survive in arbitrary environments and accept new tasks at run time. Such a robot may need to have a lot of knowledge in order to act. The robot will know much more about the particulars of a situation than the designer. Even biology hasn't produced many such agents. Humans may be the only extant example, but even humans need time to adapt to new environments.

Even if the flexible robot is our ultimate dream, perhaps we should try to reach this goal via more mundane goals. Rather than building a universal robot, which can adapt to any environment and solve any task, it may be better to build particular robots for particular world niches. Then the designer can exploit the structure of the particular world niche and the robot need not represent it explicitly at all.

Two different approaches have been pursued in building robots:

- The first is to simplify environments and build complex reasoning systems for these simple environments. Much of the complexity of the problem can be reduced by simplifying the environment. This is also important for building practical systems because many environments can be engineered to make them simpler for robots.

- The second is to build simple robots in natural environments. This is inspired by seeing how insects can survive in complex environments, even though they have very limited reasoning abilities. You can then make the robots have more reasoning abilities as their tasks become more complicated.

One of the advantages of simplifying environments is that it may enable us to prove properties of robots or to optimize robots for particular situations. This requires a model of the robot and the environment. The robot may do a little or a lot of reasoning, but an observer or designer of the robot may be able to reason about the robot and the environment. For example, the designer may be able to prove whether the robot can achieve a goal, whether it can get into situations that may be bad for the robot (safety goals), whether it will get stuck somewhere (liveness), or whether it will eventually get around to each of the things it needs to do (fairness). Of course, the proof is only as good as the model.

The advantage of building robots for complex environments is that these are the types of environments in which you live and want your robots to live. Fortunately, research along both lines is being carried out.

Finally it should be noted that a physical robot has to obey laws of physics. No matter what you do, our robot will not disobey these laws. You want to put the laws of physics to the advantage of the robot. There are other laws that may be desirable of robots. As an example, Isaac Asimov's laws of robotics, paraphrased, are:

(i) Don't allow any human to be harmed as a result of your action or inaction.

(ii) Don't disobey the orders of a human.

(iii) Don't allow yourself to be damaged.

In any case of conflict of the laws, (i) takes precedence over (ii), which in turn, takes precedence over (iii).

12.5 Uses of Agent Models

Before considering how to go about building a robot, let's consider various uses for a robot specification. The program that specifies a robot can be used in a number of different ways:

- In **embedded mode** you run the robot in an actual environment, doing actions in the environment and receiving sense data from the environment.

- In **simulation mode** you use a model of the environment as well as the robot specification. The robot interacts with the model of the environment rather than the environment itself. This entails building a model of the environment as well as a model of the robot. You would like to use the same language to model environments as well as to model robots. How good the simulation is depends on how good the model of the environment is. Models always have to abstract some aspect of the world. Appropriate abstraction is important for simulations to be able to tell us whether the robot will work in a real environment. Whether the robot is run in simulation mode or not is independent of whether the robot itself models the environment internally based on its perception.

- In **verification mode** you have both a model for the robot and a model of the set of possible environments, and you prove theorems about the models that will tell you how the robot will work in such environments. For example, you may want to prove that your robot will always get within a certain distance of the goal, that it will never get stuck in mazes, or that it will never crash. Of course, whether what's proved turns out to be true depends on how accurate the models are.

- In **optimization mode**, you model the robot and the environment such that the robot model leaves some aspects unspecified. The goal is to choose values for these unspecified aspects, from tuning parameters to choosing what commands to issue depending on the percepts, so that the choices optimize some value function. One case of a robot with a fixed sequence of choices was considered when building influence diagrams (page 381).

- In **learning mode** the robot interacts with the real world, and rather than being able to analyze the environment model, as in optimization mode, it must extract the relevant tuning information from the environment by acting in the environment.

- In **design mode** you are not given a controller at all, but must synthesize one from the environment model, the body model, and the robot's goals or preferences.

12.6   Robot Architectures

The definition of a robot doesn't entail that you need to implement a big perception system which feeds the percepts into a reasoning engine implementing the controller that outputs actions to actuators. This turns out to be a bad architecture for intelligent systems. It's too slow, and it's difficult to reconcile the slow high-level goals with the fast reaction that a robot needs to avoid obstacles.

A more sensible architecture is to have a hierarchy of controllers as depicted in Figure 12.2. Each of the controllers of the robot sees the controllers below it as a *virtual body* from which it gets percepts and sends commands. The lower-level controllers can run much faster, react to those aspects of the world that need to be reacted to quickly, and deliver a simpler view of the world to the higher-level controllers. This is an instance of the general levels of representation (page 177), such as is depicted in Figure 5.4 (page 178).

Example
12.1

For the delivery robot, one of the controllers could be to get from one location to an adjacent location and avoid obstacles (e.g., between the neighboring nodes in Figure 8.1 on page 285). This could deliver a body to a higher-level controller for which the commands are like "move from position $o103$ to position $o109$," corresponding to the *move(rob, o103, o109)* action (page 286). It would need percepts about whether its way is blocked and what its location is, and it could deliver percepts about whether it was at the desired location. A lower-level controller could fuse multiple sensor readings to decide whether there is an obstacle in the way of the robot.

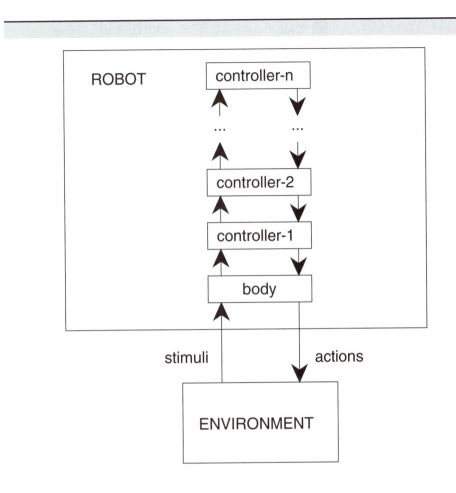

Figure 12.2: A hierarchical robotic system architecture

12.7   Implementing a Controller

This section considers how to implement the controller layers. There are two parts involved in implementing a controller:

- The command function specifies how the commands depend on the current state and the current observations.
- The state transition function specifies how the next state of the robot depends on its current state and its observations.

We'll use logic programs to specify both of these functions.

Maintaining State Using the Event Calculus

The most natural way to represent state is to be able to assign values to fluents that persists until another value is assigned. A **fluent** is a predicate whose value depends on the time. To represent the evolution of fluents, you can use the **event calculus** (page 294). The event of assigning a value to a fluent makes that fluent have that value until there is another event that gives the fluent a different value. The notion of a fluent is analogous to variables in imperative programming languages, where you can assign a variable a value and it has that value until it's assigned another value. Its value depends on the time.

In axiomatizing assignment, you have a choice: Either assigning a value to a variable affects the value in the next state, or the value is changed for the current state. You typically want the latter, as you want the robot to react to changes in goals and remembered values as quickly as possible. To implement this you need to be able to refer to the previous value of a fluent. For example, to increment a value of a fluent, you have to be able to say that a fluent has a value that's one more than its previous value.

Assume the relation $assign(Fl, Val, T)$ is true if fluent Fl was assigned value Val at time T. The predicate was can be used to determine a fluent's previous value. $was(Fl, Val, T_1, T)$ is true if fluent Fl was assigned a value at time T_1, and this was the latest time it was assigned a value before time T:

$$was(Fl, Val, T_1, T) \leftarrow$$
$$T_1 < T \land$$
$$assign(Fl, Val, T_1) \land$$
$$\sim reset\_between(Fl, T_1, T).$$

where $reset\_between(Fl, T_1, T)$ is true if fluent Fl was assigned a value in the interval (T_1, T):

$$reset\_between(Fl, T_1, T) \leftarrow$$
$$T_1 < T_2 \land$$
$$T_2 < T \land$$
$$assign(Fl, V_2, T_2).$$

The predicate val is used to determine the current value of a fluent at any time. $val(Fl, Val, T)$ is true if fluent Fl was assigned value Val at time T or else Val was the value it had immediately before time T.

It is very inefficient to implement this logic program using top-down search, as the program needs to access all of its history to determine the current values of its state variables. Either you can forward chain on these clauses (page 253), or you can actively maintain a state and always access the last value assigned. An implementation of the last idea is presented in Appendix C (page 529).

Implementing a Layered Controller

This section shows how to use the event calculus as a runnable specification of a layered controller.

Example 12.2	Suppose you want to implement the delivery robot (page 284) so that it can use the sorts of plans produced in Chapter 8. You would also want it to sense the environment and, for example, avoid obstacles. Assume the delivery robot is wheeled, like a car, and at each time can turn either straight, right, or left. Turning the wheels is instantaneous, but arriving in a certain direction takes time. Thus the robot can only travel straight ahead or go around in circular arcs. Also assume that the velocity is constant and the only control is the steering angle. Of course, these assumptions are unrealistic, but are made to make exposition clearer.

A layered control for the delivery robot is depicted in Figure 12.3. The robot is given a high-level plan to execute. The robot needs to sense the world and to move in the world in order to complete the plan.

The top layer, which could be called *follow plan*, is modeled in Example 12.4. That layer takes in a plan to execute. The plan is just a sequence of locations to travel to. The top level maintains state of what actions the robot still needs to perform, and it issues commands to the middle layer in terms of the coordinates of the current goal for that layer.

The middle layer, which could be called *go to location and avoid obstacles*, tries to keep traveling towards the current goal position, as provided by the *follow plan* layer, avoiding obstacles. (The middle layer is implemented in Example 12.3.) The current goal position *goal_pos* is a shared variable between the two layers. When the middle level has achieved the goal, it signals to the higher level that it has achieved the goal. This can be seen either as the middle layer issuing an interrupt to the higher level, or as the higher level monitoring the lower level. The higher level then changes the goal position to the next one on the plan.

The middle layer assumes that it can access the robot's current position and direction and can determine whether its single whisker sensor is on or off. You can use a simple strategy of the robot trying to head towards the goal unless it's blocked, in which case it should go left around the obstacle.

The middle layer is built upon a lower layer that provides a simple model of the robot. This lower layer could be called *steer robot and report obstacles and position*. It just takes in steering commands and reports the robot's position, orientation, and whether the sensor is on or not.

Example 12.3	Let's implement the middle *go to location and avoid obstacles* layer for a simple robot that can perceive its position and direction of travel. It has a single whisker sensor that can detect obstacles that touch the whisker. The whisker points 30° to

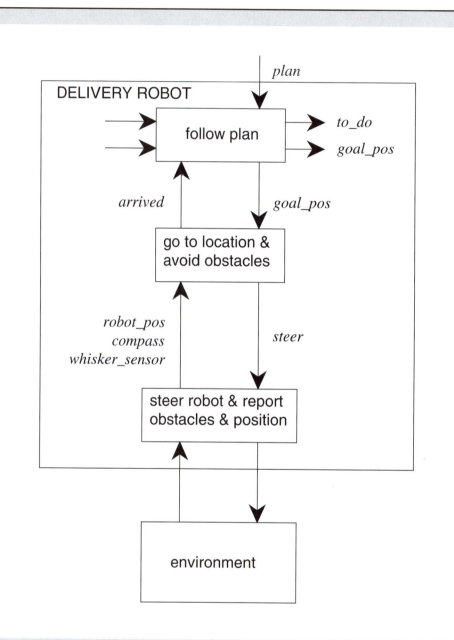

Figure 12.3: A hierarchical decomposition of the delivery robot

the right of the direction the robot is facing. These sensed values are provided by the lower layers. The whisker sensor only returns one bit of information at each time: whether the whisker hit something or not. All the robot can do is to steer left, steer right or go straight. The aim of this layer is to make the robot head towards its current goal position, avoiding obstacles.

As with any controller, you need to implement the command function and the state transition function. This layer of the controller maintains no internal state, so the state transition function is vacuous. To define the command function, you need to axiomatize what the robot's actions are, depending on what it senses and the position of the goal.

$steer(D, T)$ means that the robot will steer in direction D at time T, where $D \in \{left, straight, right\}$. The robot steers towards the goal, except when the sensor is on, in which case it turns left.

$$steer(left, T) \leftarrow whisker\_sensor(on, T).$$

$$steer(D, T) \leftarrow goal\_is(D, T) \wedge \sim whisker\_sensor(on, T).$$

$goal\_is(D, T)$, where $D \in \{left, straight, right\}$ is true if the goal is D-wards from where the robot is pointing. $11°$ is an arbitrary threshold for when the goal is straight ahead enough.

$$goal\_is(left, T) \leftarrow goal\_direction(G, T) \wedge val(compass, C, T) \wedge$$
$$(G - C + 540) \bmod 360 - 180 > 11.$$
$$goal\_is(straight, T) \leftarrow$$
$$goal\_direction(G, T) \wedge val(compass, C, T) \wedge$$
$$|(G - C + 540) \bmod 360 - 180| \leq 11.$$
$$goal\_is(right, T) \leftarrow$$
$$goal\_direction(G, T) \wedge val(compass, C, T) \wedge$$
$$(G - C + 540) \bmod 360 - 180 < -11.$$

$goal\_direction(G, T)$ is true if the goal is in direction G from the robot's position. G is in degrees anticlockwise from due east.

$$goal\_direction(G, T) \leftarrow$$
$$robot\_pos((X_0, Y_0), T) \wedge val(goal\_pos, (X_1, Y_1), T) \wedge$$
$$direction((X_0, Y_0), (X_1, Y_1), G).$$

$direction(P1, P2, Dir)$ is true if Dir is the direction from position $P1$ to position $P2$. The definition of $direction$ is left as an exercise.

This level needs to tell the higher level when it has arrived. $arrived(T)$ is true if the robot has arrived at, or is close enough to, the previous goal position:

$$arrived(T) \leftarrow$$
$$was(goal\_pos, Goal\_Coords, T_0, T) \wedge$$
$$robot\_pos(Robot\_Coords, T) \wedge$$
$$close\_enough(Goal\_Coords, Robot\_Coords).$$
$$close\_enough((X_0, Y_0), (X_1, Y_1)) \leftarrow$$
$$\sqrt{(X_1 - X_0)^2 + (Y_1 - Y_0)^2} < 3.0.$$

Here 3.0 is an arbitrarily chosen threshold that determines when the robot is close enough to its goal position.

Example 12.4

The highest-level *follow plan* controller needs to tell the middle controller what the robot's current goal position is. It assumes that it's given a list of goals to achieve. These are the kinds of goals that would be produced by a planner, such as those developed in Chapter 8.

As usual, you need a state transition function and a command function. This layer maintains internal state. It remembers its short-term goal and what actions it still has to do. The fluent *to_do* has as value a list of all pending actions. The fluent *goal_pos* maintains the goal position for the lower layer.

Once the robot is close enough to its previous goal, the next goal position is the coordinate of the next action to do. Assume for simplicity that the only actions are *goto* actions. The robot is communicated to in terms of named locations, but these need to be translated into coordinates in order for the middle layer, which knows nothing about names of locations, to travel to the goal. The following code shows how the goal position and the *to_do* list are maintained when the robot has arrived at it previous goal position:

$$assign(goal\_pos, Coords, T) \leftarrow$$
$$arrived(T) \wedge$$
$$was(to\_do, [goto(Loc)|R], T_0, T) \wedge$$
$$at(Loc, Coords).$$
$$assign(to\_do, R, T) \leftarrow$$
$$arrived(T) \wedge$$
$$was(to\_do, [C|R], T_0, T).$$

In this axiomatization, if the *to_do* list becomes empty, the robot doesn't change its goal position. It just keeps going around in circles. (See Exercise 12.1.)

You need a database that gives the coordinates of the named locations, such as

$$at(mail, (0, 10)).$$
$$at(o103, (50, 10)).$$

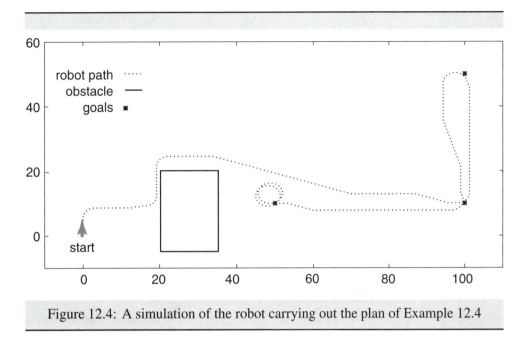

Figure 12.4: A simulation of the robot carrying out the plan of Example 12.4

$at(o109, (100, 10))$.

$at(storage, (100, 50))$.

In order for the robot to carry out a plan, it needs to be assigned the plan to do, and to be interrupted to start carrying out the plan. This can be done by adding, for example,

$assign(to\_do, [goto(o109), goto(storage), goto(o109), goto(o103)], 0)$.

$arrived(1)$.

A simulation of the above plan with one obstacle is given in Figure 12.4. The robot starts at position $(0, 5)$ facing 90° (north), and there is a rectangular obstacle between the positions $(20, 20)$ and $(35, -5)$. This has used a model of the environment that can be given in the same language as the model of the robot. (See Exercise 12.3.)

12.8 Robots Modeling the World

The definition of the robot doesn't prescribe what should be remembered by the robot. Often it's useful for the robot to maintain some model of the world, even if its model is incomplete and inaccurate.

One method is for the robot to maintain its belief about the world and to update these beliefs based on its actions. This is known as **dead reckoning**. For example, a robot could maintain its estimate of its position and update it based on its actions. When the world is dynamic or when there are noisy actuators (e.g., a wheel slips, it isn't of exactly the right diameter, or acceleration isn't instantaneous) the noise accumulates, so that the estimates of position very soon become so inaccurate that they are useless.

An alternative is to use **perception** to build a model of the relevant part of the world. This could, for example, involve using vision to detect features of the world and use these features to determine the position of the robot and obstacles or packages to be picked up. Perception tends to be ambiguous and noisy. It's difficult to build a model of a three-dimensional world based on a single image of the world.

A more promising approach is to combine the robot's prediction of the world with sensing information. This is known as **active perception**. In active perception, the prediction of the state is combined with perception of the state to give an updated state model. This can take a number of forms:

- If both the noise of forward prediction and sensor noise are modeled, the next state can be estimated using Bayes' theorem (page 356). One particular use of this is when the system dynamics, in terms of the state transition function, and the sensors can be modeled as a linear system with Gaussian noise (see the Box on page 351). In this case the prediction of Bayes' theorem can be solved analytically. This is known as a **Kalman filter**.

- An alternative way to do this using more complicated sensors such as vision is to use the forward prediction to guess at where visual features can be found, and then use vision to look for these features around this point. This makes the vision task much simpler, and the vision can greatly reduce the errors in position that arise from the forward prediction alone.

12.9   Reasoning in Situated Robots

This section discusses how the reasoning technologies developed in earlier chapters relate to the development of reactive robots in this chapter.

It may seem that the reactive architectures developed in this chapter are the antithesis of the planning for robots developed in Chapter 8. This is not so. When an intelligent agent (robot or human) faces a situation in which it has a goal to achieve and has to do something, it needs to consider what it will do in the future, as what it intends to do in the future affects what it should do now. The role of planning is to determine what to do now by thinking about goals and what to do in the future. The reactive systems developed in this chapter allow a robot to condition on what it knows about the past and the current environment. The role of planning is to think about the future. An intelligent agent needs to do both. The relationship between reacting and planning becomes less distinct when developing conditional plans, for then the robot must consider what it will perceive and how it will react to these perceptions.

The use of diagnosis, such as consistency-based diagnosis (page 243), abduction (page 332), or Bayesian networks (page 363), is important for reactive robots. The problem of determining what is in the world based on observations about the world, the diagnosis problem, is the problem that a robot faces in trying to act in the world based on imperfect sensors. It's usually too complicated to specify the function from observations into actions, but it's much more convenient and modular to determine a world model from observations and use the world model to determine what to do.

A control problem is called **separable** if the best action can be obtained by first finding the best model of the world and then using that model to determine the best action. Unfortunately, most control problems are not separable. This means that you need to consider multiple models in order to determine what to do, and what information you get from the world depends on what you will do with that information. There is no best model of the world that's independent of what the robot will do with the model.

The science of computational intelligence is rapidly evolving. The traditional Good Old-Fashioned Artificial Intelligence and Robotics model, known as *GOFAIR*, makes certain assumptions about the agent and its environment. The purely *reactive* model made different assumptions. The view presented here is a *situated agent* model that captures the insights and achievements of both the GOFAIR and reactive models in a single formal framework.

The development of tools for designing, modeling, and building embedded, intelligent agents is a rewarding challenge. Much remains to be done: we invite you to contribute.

12.10 References and Further Reading

The model of dynamical systems is based on constraint nets (Zhang & Mackworth, 1995), and also on Rosenschein & Kaelbling (1995). The layered control is based on Albus (1981) and the subsumption architecture of Brooks (1986). The particular axiomatization of constraint nets is based on Poole (1995). The *Turtle Geometry* book (Abelson & DiSessa, 1981) investigates mathematics from the viewpoint of modeling simple reactive agents; it has a detailed analysis of escaping a maze. Luenberger (1979) is a readable introduction to classical dynamic systems theory.

For more detail on many of the issues in this chapter see Dean & Wellman (1991), Latombe (1991), and Agre (1995).

For arguments about methodologies for building intelligent agents, see Haugeland (1985), Brooks (1991b), Kirsh (1991b), Brooks (1991a), and Mackworth (1993). Asimov's laws of robotics are from Asimov (1950).

12.11   Exercises

Exercise 12.1   The definition of the goal position in Example 12.4 (page 456) means that when the plan ends, the robot will just keep the last goal position as its goal position and keeps circling forever. Change the definition so that the robot goes back to its home and stops there.

Exercise 12.2   The obstacle avoidance implemented in Example 12.3 can easily get stuck.

(a) Show an obstacle and goal for which the robot using the controller of Example 12.3 would not be able to get around (it will loop).

(b) Suppose the robot designer has no idea how the robot's sensors will work when the robot is part of the way through a turn. Axiomatize a controller that doesn't check for obstacles part of the way through a turn. *Hint:* You will need to define a new fluent that represents the robot's local goal direction, and while the robot isn't facing its goal direction it just turns without checking its sensors.

(c) Define a controller that can get out of any maze, assuming that the robot has room to steer around corners. *Hint:* Use the *right-hand rule*, where the robot turns left when it hits an obstacle and keeps following a wall, with the wall always on its right. It can keep track of how many degrees it has turned and must unravel in order to keep following its path.

(d) Even without obstacles, the robot may never reach its destination if, for example, it's right next to its goal position. It may keep circling forever, without reaching its goal. Design a controller that can detect this and find its way to the goal.

Exercise 12.3   Axiomatize a model for the environment in Figure 12.4 using the event calculus.

(a) Axiomatize how the steering of the robot affects the direction and the position of the robot.

(b) Axiomatize how the sensors of the robot are affected by obstacles in the environment.

(c) Axiomatize noisy sensors and noisy actuators. This may involve adding random values to the positions and directions to the robot and having sensors that don't necessarily always act exactly as specified. For example, they may fail occasionally or they may detect obstacles at slightly different distances or orientations.

Exercise 12.4   Axiomatize a model for a local environment where you are now.

Appendix A

Glossary

Abduction A reasoning strategy based on making assumptions to explain an observation. Given a set of assumables, a background knowledge base, and an observation, you find a subset of the assumables that, together with the background knowledge imply the observations. See page 322. Contrast with *deduction* and *induction*.

Abductive Diagnosis A method for diagnosis where you hypothesize what may be wrong in order to explain the observations made of a patient or artifact. See page 334. Contrast with *consistency-based diagnosis.*

Action An action is something that is done by an agent. An action can be seen as a function from one state of the world, the world before the action, into a new state of the world, the world after the action. See page 281.

Add List The set of primitive fluents that are made true by an action in the STRIPs representation. See page 288. See also *delete list.*

Answer The response given to a query on a knowledge base. The answer tells you whether some instance of a query follows from the knowledge base, given an appropriate proof procedure. See pages 40, 42, 242, and 260. Also an answer clause with empty body (page 49).

Answer Clause A clause with *yes* in the head. The predicate *yes* can also include arguments that represent the instances of the variables in the initial query. If the body of the answer clause can be derived, there is an associated answer. See page 49.

Arc Consistent Given two variables X and Y, with domains $\mathbf{D}_X$ and $\mathbf{D}_Y$, and a relation $r(X, Y)$ between them, the relation is arc consistent if for each value of X in $\mathbf{D}_X$ there is some value for Y in $\mathbf{D}_Y$ such that $r(X, Y)$ is satisfied. A constraint network is arc consistent if all its arcs are arc consistent. See page 153.

Artificial Intelligence See *Computational Intelligence.*

461

Assumable A formula that can be assumed as part of a proof. This is used in consistency-based diagnosis (page 243), in default reasoning (page 323), and in abduction (page 332).

Atomic Symbol A syntactic item of the form p or $p(t_1, \ldots, t_n)$ where p is a predicate symbol and each t_i is a term. It is often just called an atom. It is atomic in the sense that it is the smallest level of structure which is assigned truth values. See page 30.

Autoepistemic Logic A logic that deals with reasoning about one's own knowledge. Often used for default reasoning (page 323) as it allows the expression of statements such as "if something is a bird and you don't know it doesn't fly, then it flies."

Axiom A statement that is given as true. This should be easily verified as being true in the intended interpretation. See page 38.

Axiom Schema A form that maps into a number of axioms. See page 237 for an example.

Backtracking A way to describe depth-first search (page 125). If nodes n_i and n_j are together on the frontier, with n_i first, in depth-first search all paths from n_i are explored before n_j is taken from the frontier. When all paths from n_i have been explored, depth-first search "backtracks" to node n_j.

Backward Chaining A form of reasoning where, given a query to solve, you select an atom in the query, nondeterministically choose a clause with that atom as the head, and replace the atom with the body of the clause, and repeat. Rather than going forwards from what is known, you go backwards from the goal. See page 49.

Bayes' Theorem A way to find the probability of a hypothesis given evidence. The probability of a hypothesis h given evidence e is equal to the probability of e given h times the prior probability on h divided by the prior probability of e. This is used for probabilistic inference and for learning. See page 356.

Belief Statements that are inside the mind of an agent or can be inferred by the agent. They don't have to be true, and can be believed to varying degrees.

Belief Network A graphical representation of dependence amongst random variables. Often called "Bayesian network." A belief network is a DAG, with each node representing a random variable, together with a domain for each variable and a set of conditional probability tables for each variable given its parents. It is a representation of the independence assumption that a node is independent of its nondescendants given its parents. Belief networks allow you to quantify, using probabilities, any belief about the random variables conditioned on any evidence you may have. See page 363.

Best-First Search A search strategy that uses heuristic knowledge about a problem to guide us to a solution, by always searching from the point that is regarded to be the closest to the solution. See page 133.

Binary Resolution This is a form of inference for general clauses. For the variable-free case, from $c \vee a$ and $\sim a \vee d$ infer $c \vee d$, where c and d can be arbitrary disjunctions of literals. Where there are variables, from $c \vee a_1$ and $\sim a_2 \vee d$ infer $(c \vee d)\theta$, where θ is the most general unifier of atoms a_1 and a_2. See page 260.

Blackboard A mechanism for implementing best-first search (page 133). A number of different agents can use various mechanisms to determine which is the best node to expand next.

Body The body of a rule in datalog and the definite clause language has the form $a_1 \wedge \ldots \wedge a_m$ where each a_i is an atom. See page 30.

   The body of a robot is that part that carries out the commands of the controller and provides percepts to the controller. The body acts in the environment and receives stimuli from the environment. A body and a controller make up a robot. See page 444.

Branching Factor The (forward) branching factor of a node is the number of arcs going out of a node. See page 116.

Breadth First Search A search strategy where you always choose to search from the shortest path on the frontier. That is, all paths of length k are tried before all paths of length $k + 1$. See page 128.

Certainty Factor A number which expresses the degree of belief in a statement, used in the expert system Mycin (Buchanann & Shortliffe, 1984). They were used before we knew how to use probabilistic knowledge effectively (Chapter 10). Now only of historical interest. They were mainly abandoned because they use unrealistic independence assumptions (Heckerman, 1986).

Chance Node A node in a decision network (page 386) that corresponds to a node in a belief network (page 363). A chance node represents a random variable. The parents of a chance node represent those variables on which the probability depends. Associated with a chance node is a probability of the variable given its parents. See page 386.

Church–Turing Thesis The hypothesis that any symbol manipulation can be carried out on a *Turing machine*. See page 4.

Clark Normal Form The Clark normal form of a clause is an equivalent clause with only variables in the head such that each variable appears at most once in the head. See page 250.

Clark's Completion Clark's completion of a knowledge base is a logical formula that incorporates the complete knowledge assumption into the knowledge base.

The completion says that an atom is false if the bodies of all of the clauses defining the atom are false. See page 251.

Clause   Either a fact or a rule. See page 30. This can be extended to having the atom *false* in the head, forming an integrity constraint (page 241), to having negation as failure in the body (page 252), and to having disjunctions in the head and literals in the head and body (page 257).

Closed World Assumption   A fact may be assumed to be false if it cannot be proven true. See page 249.

CNF   See *conjunctive normal form*.

Cognitive Science   The study of cognition in biological and artificial systems. See page 7.

Complete   A property of a proof procedure. A proof procedure is complete with respect to a semantics if there is a proof of each logical consequence of the knowledge base. See pages 25 and 46. Contrast with *sound*.

Complete Knowledge Assumption   The assumption that the agent knows all relevant positive facts about the world. Hence the agent may assume a fact is false if it cannot infer that it is true. See page 249.

Complexity   The (time) complexity of an algorithm is the number of operations required for termination parameterized by the size of the input. The complexity of a problem is the greatest lower bound of the complexity of any algorithm for the problem. The complexity of a logic is the complexity of the most difficult problem that can be expressed in that logic. See page 130.

Computational Intelligence   The study of the design of intelligent agents using computational models.

Computational Linguistics   The study of natural language syntax, semantics and pragmatics using computational models. See page 91.

Computer Science   The study of algorithms to solve problems using digital computers.

Conditional Expected Value   A random variable v's conditional expected value given evidence e is v's weighted average value of all of the worlds in which e holds. See page 359.

Conditional Probability   The measure of belief in a formula h given evidence e, written $P(h|e)$. See page 353.

Conflict   A set of assumables that is incompatible with a theory T. See page 242.

Conjunction   A conjunction is a logical *and*. A formula is a conjunction of subformulas if it has the form $A \wedge B$, read "A and B." See page 30.

Conjunctive Normal Form   A formula expressed as a conjunction of disjunctions of literals. Or equivalently as a conjunction of clauses. See page 273. Compare with *disjunctive normal form*.

leaf have the associated values. Sometimes there are probability distributions in the children of the nodes, sometimes the agent gets to choose the value, and sometimes the values are observed. See pages 381 and 403.

Default Reasoning Reasoning with generalized knowledge that may have exceptions. It involves making assumptions that an individual isn't exceptional unless the individual can be shown to be exceptional. See page 323.

Deduction A process to conclude logical consequences of a knowledge base. Contrast with *abduction* and *induction*.

Definite Clause Either a fact or of the form $h \leftarrow a_1 \wedge \cdots a_k$ where h and the a_i are atoms. It is true in an interpretation if h is true or if one of the a_i is false. If all of the a_i are derived, it can be used to derive h. See page 30.

Definite Clause Resolution A theorem proving technique that determines if a query follows from a set of definite clauses. There are bottom-up (page 47) and top-down (page 49) variants.

Definite Knowledge Assumption The assumption that an agent's knowledge base contains only definite clauses. See page 28.

Delete List The set of primitive fluents that are no longer true after an action, as used in the STRIPs representation. See page 288. See also *add list* and *preconditions*.

Denotation A mapping from ground terms in a knowledge base to individuals in the world. The symbols are said to denote the individuals. See page 32.

Depth-First Search A search strategy that follows a path until the solution or a dead end is found, in which case the search tries the last alternative remaining. It is implemented by treating the frontier on the generic search strategy as a a stack. See page 125.

Derivation A derivation of a query from a set of rules is a sequence of answer clauses such that the first answer clause corresponds to the query, each subsequent answer clause is obtained by resolving the previous answer clause with a rule and the last answer clause is an answer. See page 49.

Derived Knowledge Knowledge defined by the use of rules as opposed to facts. See pages 188 and 283.

Diagnosis The process of hypothesizing internal states of a system to explain its faulty behavior. See pages 243, 334, 375, and 384.

Directed Acyclic Graph (DAG) A graph with directed arcs but no cycles. See page 116.

Disjunction A disjunction is a logical "*or*." A formula is a disjunction of subformulas if it has the form $A \vee B$, read "A or B." Note that this always means inclusive or: $A \vee B$ is true in a model if A is true in the model, B is true in the model, or both A and B are true in the model. See pages 207 and 242.

Consistent A set of beliefs is consistent if *false* (or a contradiction) can't be d from the beliefs.

Consistency-based diagnosis A method of diagnosis where each diagnosis is junction of atoms consistent with the state of the system and the knov base. See page 243. Contrast with *abductive diagnosis*.

Constant A constant is part of the language of definite clause language ai predicate calculus. It is a word starting with a lower case letter or a consisting only of digits (a numeral). Constants are used to denote indivi See page 30.

Constraint Satisfaction A constraint satisfaction problem is a problem of ch(a value for each of a set of variables such that the chosen values satisfy given constraints. See page 147.

Context-Free Grammar A grammar in which each rule rewrites a single non nal into a sequence of terminal and nonterminal symbols. See page 93.

Contrapositive A contrapositive form of a clause is an equivalent clause with e one literal on the left-hand side. Thus it is a rule form of a general claus(page 259.

Cost The cost of an arc in a graph is a positive number representing the c traversing the arc. The cost of a path is the sum of the costs of the a the path. See page 116. In decision making, the cost of a world reflec desirability of the world. See page 384.

DAG See *directed acyclic graph.*

Database A knowledge base of ground facts. See page 75.

Decidable A decision problem is decidable if there exists an algorithm for i always terminates with a correct yes/no answer.

Decision Network A graphical representation of a decision problem. Also call influence diagram. It is a DAG with three types of nodes: chance nodes the same meaning as the nodes in a Bayesian network), decision nodes into these nodes represent information availability) and a values node (v represents the utility of various outcomes). The idea is to choose valu(decision nodes, conditioned on the information available when the decisioi be made, that maximizes expected utility. See page 386.

Decision Node In a decision network, a decision node represents a decision thɛ agent can make. The arcs coming into a decision node represent the inform. that is available when the decision is made. See page 386.

Decision Tree A decision tree is a tree such that the internal nodes are labelled random variables or attributes, the children of an internal node correspond t different values in the domain of the variable. The leaves of the tree corresɛ to outcomes or classifications that arise when the variables along the path t(

Disjunctive and Negative Knowledge Assumption (DNK)   The assumption that the agent's knowledge base may contain clauses with disjunction in the head of the clause and negated atoms. See page 257.

Disjunctive Normal Form   A formula expressed as a disjunction of conjunctions of literals. See page 244. Compare with *conjunctive normal form*.

Domain   Informally, a task domain is some world that you want to reason about. Formally, we use the term domain to be the set of individuals (things) in a world. See page 32.

Domain Expert   In a knowledge-based system a domain expert is the role of the person who has the knowledge of the task domain. They don't necessarily know anything about the reasoning procedures in the computer or about a particular patient or artifact. They are the ones who know what the symbols mean. See page 201.

Epistemological Adequacy   The extent to which a representation and reasoning system can express those concepts and relations needed to solve a problem. See page 193.

Epistemology   The study of knowledge.

Equal   Two terms are equal if they denote the same individual in the interpretation. See page 236.

Euclidean Distance   The straight-line distance between points in an n-dimensional geometric space. See page 140.

Euclidean Relation   A relation r is Euclidean if $r(A, B)$ and $r(A, C)$ implies $r(B, C)$. See page 276.

Evidence   An observation about a particular case that needs to be explained or conditioned on. See pages 337 and 353.

Evidential Reasoning   The process of inferring the internal state of a system from its external behavior. See page 335.

Expected Value   A random variable v's expected value is v's weighted average value in all possible worlds. See page 359.

Expert System   A system which can perform a task at the level of a human expert and which can explain its reasoning. See page 199.

Explanation   Informally, an explanation is a description of how the world could be to account for an observation. Formally, an explanation of an observation g is a consistent set of assumables that, together with the background knowledge, logically implies g. See page 322. An explanation could also be a description of how the system proved a goal. See page 217.

Fair   A fair scheduling policy will ensure that all requests for service will eventually receive service. See page 50.

Finite Domain Assumption   There are only a finite number of individuals of interest in the domain. See page 29.

First-Order Predicate Calculus   A logic with quantification over individuals. See pages 270 and 271. See also *Predicate Calculus*.

Fixed Point   If f is a function from domain D into D, a fixed point of f is a value x such that $f(x) = x$. This is used in defining the set of atoms that follow from some set of clauses (page 48). The f in that case is a function on sets of atoms defined by $f(A) = \{h \mid \text{``}h \leftarrow b_1 \wedge \ldots \wedge b_m\text{''} \in KB, \forall i \, b_i \in A\}$.

Formal Language   The set of sentences generated by a particular *grammar*. See page 24.

Forward Chaining   A form of reasoning from a given set of facts to determine whether a particular set of consequences can be derived. See page 47.

Frame   A data structure used to represent an object based on the object-attribute-value representation. See page 186.

Frame Axiom   In the *situation calculus*, frame axioms specify the relations that are unchanged by an action. See page 293.

Frame Problem   The problem of specifying the effect of an action. The problem exists because we don't want to have to specify the effect of every action on every predicate. See page 315.

Function Symbol   A word starting with a lower case letter. Function symbols let us refer to an infinite number of individuals. See page 58.

Fuzzy Logic   A logic for representing fuzzy concepts, such as "tall" and "pretty", and incorporating fuzzy quantifiers, such as "very" and "slightly." It is mainly used for the specification of heuristic control rules in applications where the rules can be approximate, but should be understandable by humans.

Goal Node   In a search problem, a goal node is a node you are trying to find a path to. See page 116.

Grammar   A set of rules that generates a language. Each rule rewrites a string of non-terminals and terminals as another string of non-terminals and terminals. See page 93.

Graph   A directed graph is a set of nodes connected by arcs. An arc is an ordered pair of nodes. An undirected graph is a set of nodes connected by edges. An edge is an unordered pair of nodes. See page 116.

Ground   A ground term, atom, or clause is one without any variables. See page 31.

HK   See *Horn Knowledge Assumption*.

Head   In the definite clause $a \leftarrow b$, the atom a is the head and b is the body. See page 30.

Herbrand Interpretation   An interpretation where the domain is the set of all ground terms constructible from constants and function symbols in the clauses (inventing a new constant if there are no constant symbols in the clauses) and where each ground term denotes itself. See page 55.

Heuristic Adequacy   The extent to which a representation and reasoning system can use a problem representation to solve the problem within reasonable computational resource limits. See page 193.

Heuristics   Extra knowledge about a domain that you can exploit to solve a particular problem. See page 114.

Hill Climbing   A heuristic search strategy that maintains a single node. At each stage, it chooses the most promising neighbor of the node. It is used for optimization and satisfiability problems. See page 156.

Horn Clause   A clause of the from $h \leftarrow b_1 \wedge \ldots \wedge b_n$, where h is either an atom or is *false* and each b_i is an atomic symbol. If $n = 0$, the clause is written as h. In other words, a Horn clause is either a definite clause or an integrity constraint. See page 241.

Horn Knowledge Assumption   An agent's knowledge of the world can be described in terms of Horn clauses. See page 241.

Hypothetical Reasoning   A process of positing consistent hypotheses to explain observations about the world, or to make predictions. See page 321.

Implicate   An implicate of a theory is a clause that logically follows from the theory. See page 260.

Inconsistent   A set of clauses is inconsistent if *false* can be derived from the clauses. See page 243. See also *unsatisfiable*.

Independent   Two propositions are independent if knowledge of the truth of one proposition does not affect the belief in the other. See page 361.

Individuals   The objects posited to exist in a domain. See pages 11 and 32.

Individuals and Relations Assumption   An agent's knowledge can be usefully described in terms of individuals and relations amongst individuals. See page 28.

Induction   A process of inferring from individual cases to general rules. Also called *learning*. Contrast with *deduction* and *abduction*. See page 397.

Influence Diagram   See *decision network*.

Instance   An instance of a *clause* is obtained by uniformly substituting *terms* for *variables* in the clause. See page 42.

Integrity Constraint   A clause with *false* in the head where *false* is a special atom that is false in all interpretations. Note that this cannot be defined in datalog. See page 241.

Intelligent Agent   A system that represents, uses and acquires knowledge in order to perceive, reason, plan, act, or use language. See page 1.

Intentional Logic   A logic which distinguishes the "extension" of a statement (those objects for which the expression is true), and the "intention" of a statement (what is meant by the statement).

Interpretation   An assignment of a world to a set of symbols. An assignment of individuals to constants, mappings to function symbols, and relations to predicate symbols. See page 31 for an informal notion and page 32 for the formal definition.

IR   See *Individuals and Relations Assumption*.

Knowledge   We use the term knowledge to mean any information an agent uses to solve a problem. It doesn't need to be true, justified, or even believed. Knowledge can be of the form of facts, can be hypotheses, can be probabilistic, or can be heuristic.

Knowledge Base   The place where the knowledge of an agent is stored. The set of all knowledge of an agent. See page 24.

Knowledge Engineer   A person who designs, builds and debugs a knowledge base in consultation with domain experts. They know about the computer system, but not necessarily anything about the domain of expertise. See page 201.

Knowledge Representation   The form of the knowledge allowable in a *knowledge base*. A knowledge representation specifies the syntax, the semantics and the allowed operations on the knowledge.

Language Parser   A program that recognizes legal sentences in a formal language and assigns a description in terms of constituent structures to each legal sentence. See page 25.

Learning   The ability of an agent to improve its task behavior based on experience. See page 397.

Lisp   A programming language, developed by John McCarthy in the late 1950's, is a list programming language loosely based on the lambda calculus and using a very simple syntax.

Literal   An atomic formula or the negation of an atomic formula. See page 257.

Logic   A language (syntax), together with a way to ascribe meaning to statements in the language (semantics), and a way to derive other statements (proof procedure). See pages 9, 27, and 175.

Logic Programming   A methodology of programming, whereby statements in the language have both a denotational semantics as well as a way to be interpreted procedurally.

Logical Consequence A sentence S_2 is a logical consequence of a sentence S_1, written as $S_1 \models S_2$, if S_2 is true in all models of S_1. See page 36. This should be contrasted with S_2 can be proven from S_1, written $S_1 \vdash S_2$.

Maximum Entropy The assumption that the world is as random as possible given your knowledge. See page 361.

Minimal An element of a partial order is minimal if there is no element that is strictly less than that element. There can be many minimal elements. We use the term *minimum* when there is only one minimal element, for example when the ordering is a total order. Minimal is often used with the subset partial ordering where a set is minimal with respect to a property if no subset also has the property.

Minimal Conflict A *conflict* of a theory such that no strict subset of the conflict is a conflict. See page 242.

Minimal Explanation A minimal explanation of formula g is an explanation of g such that no strict subset is also an explanation of g. See page 322.

Minimal Model A model of a set of axioms such that there is no other model that is lower in some partial ordering. The partial ordering is typically with respect to what propositions are true in the models. See pages 49 and 332.

Modal Logic A logic with special-purpose operators over formulae. These operators represent concepts such as necessary, possible, known, or believed. See page 275.

Model A model of a knowledge base is an interpretation in which every element of the knowledge base is true. See page 36.

Model-Theoretic Semantics A semantics defined in terms of truth functions and interpretations. The truth of an expression in an interpretation can be derived from the truth or denotation of its components in that interpretation. Also called a *Tarskian semantics*. The semantics given on page 31 is an example of a model-theoretic semantics.

Most General Unifier A most general unifier of two terms or atoms is a substitution that unifies the terms or atoms such that there is no other unifier that makes fewer commitments to the values of variables. See page 53. See page 230 for a unification algorithm.

Mycin An early expert system for bacterial blood infections developed in the 1970's at Stanford University. It was influential in being one of the first expert systems, and for introducing the idea that an expert system should be able to justify its answer in terms of how and why questions (page 217). See Buchanann & Shortliffe (1984) for a retrospective.

Negation The negation of a formula f is true in an interpretation if and only if f is false in the interpretation. See page 242.

Negation as Failure   Under the *Complete Knowledge Assumption*, if the proof for a formula fails, the negation of the formula can be concluded. See page 252.

Negation Normal Form   A formula with arbitrary conjunctions and disjunctions of literals but no other operators. See page 272.

Negative Ancestor Rule   The proof of a subgoal succeeds if the subgoal unifies with the negation of an ancestor literal. See page 263.

Neighbor   A node n_2 in a graph is a neighbor of node n_1 if there is an arc from n_1 to n_2. See page 116.

Neural network   An artificial neural network is a model of computation based on a simplified model of neurons. See page 408.

NNF   See *Negation Normal Form.*

Nogood   A *conflict* that has been detected and stored in an assumption-based reasoning system. See page 340.

NP-Complete   A decision problem in the complexity class *NP*, the class of those problems whose solution can be verified in polynomial time, that is as hard as any problem in *NP*. See page 50.

Node   A node is the primitive component of a directed graph. Sometimes called a vertex in an undirected graph. See page 116.

Normal Form   The normal form of a *clause* is a disjunction of literals. See page 259.

Ontology   The study of what exists. See page 11.

Optimal   The best according to some measure. For example, the shortest path in a search problem (page 132) or the lowest-cost solution in a constraint satisfaction problem (page 147). See page 171.

Paramodulation   A way to handle equality that allows a term to be replaced by a term equal to it. You thus rewrite the term to a canonical representation for the individual denoted by the term. See page 237.

Path   A sequence of nodes in a graph connected by arcs; equivalently, the sequence of arc. See page 116.

Path Consistency   In a *constraint satisfaction* problem, a path in the constraint graph is path consistent if for each pair of values allowed by the binary constraint on the pair of nodes at the beginning and end of the path there is a sequence of values at the nodes along the path consistent with that pair of values. If every path in the graph is path consistent the graph is path consistent. See page 155.

Planning   The activity of finding a sequence of actions to achieve some goal. See page 298.

Policy   A specification of what an agent will do based on what it observes. See page 386.

Polynomial Time An algorithm is a polynomial time algorithm if there exists a function $f(n)$ such that the run time is less than $f(n)$, where f is a polynomial function of n, and n is the size of the input. A polynomial function is one of the form $a_0 + a_1 n + a_2 n^2 + \cdots + a_k n^k$. See page 130.

Positive Facts Assumption The agent's database consists only of positive, function-free ground facts (atomic formulas). See page 149.

Possible World A possible world specifies the truth of propositions, and can have some other structure, for example an accessibility relation with other possible worlds (page 275) or a measure (page 350). Thus, it is like an interpretation but possibly with more structure.

Pragmatics That part of understanding natural language that goes beyond just understanding the words. It incorporates understanding of the beliefs, goals, and intentions of the speaker. See page 92.

Preconditions In the STRIPS representation of an action the preconditions are the atomic formulas that must be true for the action to occur. See page 288.

Predicate Calculus A logic with conjunction, disjunction, negation and with universal and existential quantification over individuals, together with a two-valued model-theoretic semantics. See page 270.

Predicate Symbol A syntactic category in the definite clause language. A world starting with a lower case letter. Predicate symbols denote relations. See page 30.

Predict You predict something is true if you would expect it to be true based on the evidence presented. In default reasoning, you predict a proposition if there are good arguments why it should be true. See page 330 for a formal definition and page 331 for some alternate suggestions for what should be predicted.

Prime Implicate A prime implicate of a theory is a minimal implicate that is not a tautology of the form $a \vee \sim a$ for some atom a. See page 260.

Primitive Relation In a planning system the relations that can change over time are modeled with primitive relations whose truth values can be determined by considering previous times and derived relations which depend on primitive relations. See page 283.

Prior Probability The probability of a hypothesis given no evidence. See page 353.

Probability A measure of an agent's belief in a proposition. See page 346.

Prolog A programming language, based on the definite clause language and top-down depth-first search, and with extra-logical predicates for input/output, database management, meta-level facilities, and some control over the search space. See page 477.

Proof   A sequence of propositions, each of which is an axiom (given), or follows from previous statements via some rule of inference. The last result of the sequence is a theorem. See page 46.

Proposition   A proposition is something that is either true or false. Ground atoms denote propositions.

Propositional Calculus   Variable-free *predicate calculus*. See page 271.

Query   A way to ask whether something is a consequence of a knowledge base. A query has the form ?*B* where *b* is a *body*. See page 30.

Random Variable   A ground term with a domain whose values are exclusive and exhaustive. See page 348.

Reasoning Procedure   An implementation of a reasoning theory with a search strategy. See page 25.

Reasoning Theory   A possibly nondeterministic specification of how an answer can be derived from a knowledge base. See page 25.

Reflexive   A binary relation r is reflexive if $r(X, X)$ holds for all X. See page 276.

Regression   A technique used in planning. The planner regresses a goal through an action to determine the weakest precondition that holds before the action. See page 305.

Relation   A k-ary relation is a set of k-tuples of individuals. These elements are said to be true of the relation. See page 33.

Resolution   An inference procedure. Given the two clauses $R \vee L$ and $S \vee \sim L$ their binary resolution is $R \vee S$. See page 260. This appear in various guises, for example, resolution is used to infer $a \leftarrow c \wedge d$ from $a \leftarrow b \wedge d$ and $b \leftarrow c$, as in SLD resolution (page 49).

Rule of Inference   A method of deriving new sentences in a logic. See page 47.

Scenario   A scenario, in assumption-based reasoning, is a set of ground hypotheses consistent with the facts. See page 322.

SE   See *Static Environment Assumption*.

Semantic Network   A graphical notation for taxonomic knowledge. See page 185.

Semantics   The study of meaning. A description of what some expression or sentence means. See page 31.

Semidecidability   A decision problem is semidecidable if it has a decision procedure that always terminates correctly on positive instances but need not terminate on negative instances.

Situation Calculus   A state-action oriented representation for planning and reasoning about change in which the states and actions are reified as terms in the language. See page 290.

Skolemization A way of removing existentially quantified variables by naming the objects said to exist. This introduces Skolem constants for existentially quantified variables that aren't in the scope of any universally quantified variable, otherwise a Skolem function is introduced. See page 269.

SLD Resolution A top-down proof procedure for definite clauses. SLD stands for Linear resolution with a Selector function on definite clauses. See page 49.

Software Engineer Someone who designs, builds, tests, and debugs software using formal specification techniques. See page 201.

Sound A property of a proof procedure. A proof procedure is sound with respect to a semantics if every statement that can be proved is a logical consequence. That is, it can only produce correct answers. See page 46.

Start Node In a search problem, the start nodes represent those nodes that the search starts from. The aim is to find a path from a start node to a goal node. In a state-space graph the start notes represent the possible initial states of the world. See page 116.

Starvation The situation that arises when a request for services is indefinitely ignored by the scheduler. See page 50.

Static Environment Assumption The assumption that the environment does not change or that you are reasoning about the environment at one particular time. See page 28. This assumption is relaxed in Chapter 8.

State-space Graph A directed graph in which each node represents a state of the world and each arc represents an action changing the world from one state to another. See page 115.

Statistical Independence See *Independence*.

STRIPS Stanford Research Institute Problem Solver; a planning system built in the early 1970s at SRI to guide the robot called Shakey. This has inspired the STRIPS representation (page 288) and the STRIPS planning algorithm (page 302).

STRIPS Assumption The assumption, used in STRIPS, that any primitive relations not specifically affected by an action remain unchanged. See page 288.

Substitution A set of bindings of values to variables. See page 52.

Symmetric A relation r is symmetric if $r(A, B)$ implies $r(B, A)$. See page 276.

Syntax A specification of what a legal sentence in a language is. The syntax is often specified by a grammar. See pages 29 and 92.

Tarskian Semantics See *Model-Theoretic Semantics*.

Term In predicate calculus, a term is a constant, a variable or of the form $f(t_1, \ldots, t_n)$ where f is a function symbol and each t_i is a term. Ground terms denote individuals. See page 30.

Theorem   A statement which can be proven by some proof procedure from a set of axioms. See page 46.

Theorem Prover   An algorithm or a program that uses a *reasoning procedure* to decide if a statement is a theorem.

Transitive   A relation r is transitive if $r(A, B)$ and $r(B, C)$ implies $r(A, C)$. See page 276.

Tree   A tree is a directed acyclic graph in which no node has more than one incoming arc.

Undecidable   A logic is undecidable if there are true statements which can never be proved by a sound proof procedure. In other words, if there can't exist a sound and complete proof procedure.

Unsatisfiable   A set of clauses is unsatisfiable if it has no model. See page 243. See also *inconsistent*.

Unifier   A unifier of to expressions is substitution which, when applied to either expression gives exactly the same answer. See page 53.

Universally Quantified   In the formula $\forall X\ w$, where w is a formula in which X appears free, X is universally quantified. See page 268.

User   A user is someone who uses a knowledge-based system for a particular task. The role of a user should be contrasted with the role of domain expert and a knowledge engineer. See page 201.

User Model   A representation of a user's beliefs and preferences that can be used in a computer system to help the user. See page 334.

Utility   A measure of desirability of an outcome or possible world. See page 384. The higher the utility the more valued it is. This measure can be calibrated in terms of lotteries. See page 392.

Value Node   In a decision network, a value node represents the utility of the outcome. The parents of the value node represent the variables on which the value depends. See page 386.

Variable   A logical variable is a word consisting of a sequence of letters, digits or "_", starting with an upper case letter or "_". Logical variables denote individuals. See page 29. A random variable is a ground term that can have a value. Typically you don't know the value. See page 348. Decision variables are like random variables, but the agent gets to choose the value for the variable. See page 384.

Appendix B

The Prolog Programming Language

B.1   Introduction

The programming language **Prolog** is an implementation of our definite clause representation and reasoning system presented in Chapter 2 and used in Chapter 3. It incorporates a particular search strategy and includes a number of pragmatic choices and nonlogical features intended to make it a practical programming language.

The main features of Prolog as an implementation of our RRS are:

- It uses a top-down search strategy (page 49). It starts looking for a proof only when presented with a query.

- It makes a particular commitment to what subgoal of a conjunction is chosen at any time. It selects the leftmost subgoal (page 49).

- It does a depth-first search, with chronological backtracking. It may go into an infinite loop even when there is a solution.

- It has unsound unification. It does not do the occurs check (page 230). It does this for pragmatic complexity reasons. Without the occurs check, the *append* program can be computed in time linear in the size of the lists to be appended. With the occurs check, each unification takes linear time and so append uses $o(n^2)$ time, where n is the length of the lists.

- It has a number of predefined predicate symbols, for example for arithmetic and list processing.

- It has some debugging and tracing facilities.

- It has a number of extralogical features:
  - database facilities, which allow dynamic assertion and retraction of clauses
  - input/output facilities to read and write files, to read and write from a terminal, and to communicate with other processes
  - meta-level facilities, such as to allow the access of internals of terms and to determine whether some variable is currently bound
  - control predicates that allow for extralogical control of execution, such as succeeding if some other goal fails, collecting all answers, or cutting off part of the search space

Some of these features make Prolog unsound and incomplete as a logic, but are there to make it a practical programming language. It can be used as a logic programming language, where you can interpret the clauses both logically and procedurally. A clause $p \leftarrow q \wedge r$ can either mean

- p is true if q is true and r is true
- to prove p, prove q and prove r, and, if the conjunction fails, then use the next clause or fail if there is no next clause

Prolog can also be used as a nonlogical programming language that allows for pattern-matching and nondeterministic executions of programs.

Note that the code in the chapters of this book is *not* Prolog code. There is nothing in any of the clauses that relies on Prolog's search strategy or any other search strategy. This does not mean that they cannot be typed in as Prolog programs and run. For many of the programs, you do not want to use Prolog's search strategy. You can build much more efficient strategies than the one Prolog uses, such as for the situation calculus rules (Chapter 8) or the robot control rules (Chapter 12). Some of the programs assume a richer language than Prolog's—for example, the use of integrity constraints and disjunction in the head of rules (Chapter 7).

Most of the programs in the book can be translated directly into Prolog; for those that can't, you can write meta-interpreters (page 202) in Prolog to evaluate them reasonably efficiently.

In this appendix we present standard Prolog.

B.2   Interacting with Prolog

The normal way to use Prolog is for the user to ask the system queries and for the system to give back answers. Prolog's prompt depends on the system, but a common one that we'll use here is "| ?-".

Operator	This book	Prolog	Page reference
if	$\leftarrow$	: -	30
and	$\wedge$	,	30
or	$\vee$	;	207
negation as failure	$\sim$	\+	252

Figure B.1: Syntactic translation between this book and standard Prolog

If there are no variables in the query, the system returns "*yes*" if the query succeeds and "*no*" if the query fails. If there are variables in the query, it returns either "*no*" if there are no answers, or a substitution corresponding to an instance which succeeds. When the system has given a satisfying substitution, the user either types

- ";" meaning "retry" to get another answer or
- a return to indicate that no more answers are needed.

After a sequence of retries, a system response of "*no*" means that there are no more answers. Note that the answers correspond to the different proofs; there is one answer for each proof. If multiple proofs give the same answer, that answer will be returned multiple times.

Programs are typically written in files. To **load** a file named "filename" into Prolog, issue the query

```
| ?- ['filename'].
```

To add a definition from the terminal use the query "[user]." Then type in the clauses and end with an end-of-file character (^D).

Files loaded contain definitions of clauses, not queries. To give a query in a file, it should be preceded by ": -". This may be useful, for example, to load another file as a command in a file being loaded.

B.3 Syntax

The syntax of Prolog is very much like the syntax given in Section 2.4 (page 29). In this and the following appendix, Prolog code will be written in fixed-width typewriter font. Figure B.1 summarizes the main syntactic differences between Prolog and the syntax presented in this book for the cases that they both cover.

Comments in Prolog are written between "%" and the end of the line, or between "/*" and "*/". Comments are ignored by the computer, but are not ignored by human readers. Use them to give the intended interpretation of the symbols.

Prolog allows for constants to be written between single quotes. This lets you have special characters in constants, or allows you to start a constant with an upper-case letter. For example, 'X', 'is a constant', and '%^@(!&@^#+-' are all constants.

There is an anonymous variable "_" that represents a variable whose eventual binding you do not care about. Each instance of an anonymous variable can have a different value. Any variable that is only used once in a clause should probably be an anonymous variable. In this way, a singleton variable can be noticed, and it is usually indicative of a typing error. Anonymous variables can also be used in queries when you don't care about the value of a variable.

Prolog allows for declarations in files. These begin with ":-". These are typically compiler directives—for example, to define an infix operator or to declare a predicate to be dynamic.

Infix Operators

Prolog allows for infix, prefix, and postfix operators. If binary predicate symbol or function symbol op is an infix operator, then op(x,y) can be written as x op y. If unary function symbol f is a prefix operator, then f(x) can be written as f x; and if unary function symbol f is a postfix operator, then f(x) can be written as x f.

Example 2.1

The term +(X,Y) can be written as X+Y. The term

```
*(*(+(X,*(a,Y)),Y),+(*(a,W),X))
```

can be written as

```
(X+a*Y)*Y*(a*W+X).
```

Note that in this example we are assuming that + has a higher precedence than *.

You can use either the standard prefix or the infix notation. They have exactly the same internal representation, and they mean exactly the same thing.

You can declare your own infix operators.

Example 2.2

To allow "or" as an infix operator, you can use the declaration

```
:- op( 1100, xfy, [or]).
```

This defines the syntax of or so that, for example, "a or b" is the same as "or(a,b)". It does not define a meaning for or. The "1100" is the precedence. The lower the number, the more tightly the operator is bound. The "xfy" means that or is right associative: The "x" means there can only be one item of the same precedence on the left and the "y" means there can be more than one item of the same precedence on the right. With this declaration, the term "a or b or c" means the same as "a or (b or c)" which means the same as "or(a,or(b,c))".

See your local Prolog manual for details.

Lists

Lists are very common in Prolog. Lists are written using the square bracket notation (page 83), where the list with first element H, and rest of the list T, is written as [X|T]. The list syntax of this book is compatible with Prolog's list syntax.

B.4   Arithmetic

There is one main arithmetic relation, namely is(N,E), which is true if E is an expression that evaluates to number N. An expression can contain numbers and many arithmetic operations including +, *, /, -, // (integer quotient), mod (integer remainder), abs (absolute value). These have their standard meaning.

The predicate is can also be used in infix notation, in which case you write

    N is E.

The arithmetic operations can also be used using infix notation. For example, the atoms is(X,+(3,*(5,7))) and X is 3+5*7 can be used interchangeably and mean the same thing.

There is nothing illogical about "is". It can be defined like any other relation. (See Exercise 3.10 on page 108.) However, in order for it to be computed quickly using hardware arithmetic, without needing arbitrary constraint solving, Prolog restricts the use of is so that, when called, E must not contain any variables. That is, ground terms must have been substituted for the variables in E.

There are arithmetic comparison algorithms, including X=:=Y that is true if X and Y evaluate to the same expression, X<Y that is true if X evaluates to a number less than the number that Y evaluates to, and X=\=Y that is true if X and Y evaluate to different numbers. Note that each of these relations requires both arguments to be ground (variable free) when called. There are also relations >, =< (meaning $\leq$) and

>= (meaning $\geq$). To remember the last two, these are the combinations that don't look like arrows.

Example 2.3

Using these comparisons you can define the minimum relation, min(A,B,C) which is true if C is the minimum of A and B:

```
min(A,B,A)  :- A =< B.
min(A,B,B)  :- A > B.
```

The following is a legal query:

```
| ?- min(2*22,7*3,Y).
```

with answer Y=7*3. Note that the answer is *not* Y=21, as 7*3 does not unify with 21.

Example 2.4

odd(N) is true if N evaluates to an odd integer

```
odd(N)  :-
    1 is abs(N) mod 2.
```

The following is a legal query:

```
| ?- odd(77*3).
```

However, the following query will give an error,

```
| ?- odd(2*N).
```

because, in Prolog, the right-hand side of the is predicate must be ground.
    even(N) is true if N evaluates to an even integer

```
even(N)  :-
    0 is abs(N) mod 2.
```

Example 2.5

power(X,N,P) is true if N is a positive integer, and P is X multiplied by itself N times—that is, if P = X^N, and N $\geq$ 1.

```
power(X,1,X).
power(X,N,V)  :-
    even(N),
    N>1,
    X2 is X*X,
    N2 is N // 2,
    power(X2,N2,V).
power(X,N,V)  :-
    odd(N),
    N>1,
```

```
N1 is N-1,
power(X,N1,V1),
V is V1 * X.
```

Power requires X and N to be bound.

The restriction of the use of variables in arithmetic expressions means that some predicates are not as flexible as possible.

B.5   Database Relations

Prolog includes relations to manipulate the database of rules and facts. Unlike the relations you have seen so far, these are not logical relations; they have side effects that can change the meaning of programs. The most common are:

`asserta(C)` adds clause C to the knowledge base before any other clauses for the same predicate.

Example 2.6

To assert "`foo(X,Y) :- bar(X),baz(Y)`", you issue the query

```
| ?- asserta((foo(X,Y) :- bar(X),baz(Y))).
```

Note that the extra set of parentheses is needed because of the way precedence of operators works.

To add the fact "`foo(c,d)`", you issue the query

```
| ?- asserta(foo(c,d)).
```

`assertz(C)` is like `asserta(C)`, except the clause C is added as the last clause.

`retract(C)` removes the first clause in the knowledge base that matches C. `retract` will retract more clauses when it is retried, and it fails when there are no more clauses that match C.

Note that `asserta` does not undo what was asserted on backtracking, nor is retraction undone on backtracking. (See Example B.13.)

`clause(H,B)` finds a clause in the database that matches H `:-` B. A fact in the database corresponds to a clause with B having value "`true`". Conjunctive bodies are represented using the infix conjunction operator "`,`".

In order to be able to assert and retract a predicate, it needs to be declared as `dynamic`. This declaration allows the compiler to add code to allow for dynamic redefinition of the predicate, and it prevents optimizations that rely on the predicate not changing. Binary predicate `foo` can be made dynamic by adding a declaration:

```
:- dynamic(foo/2).
```

where the 2 is the arity, the number of arguments, for the relation `foo`.

| Example | Given the clauses asserted for `foo` in Example B.6, the query |
| 2.7 | |

```
| ?- clause(foo(X,Y),B)
```

has two answers: `X=c,Y=d,B=true` and `B=(bar(X),baz(Y))`.

Meta-programming

To do meta-level programming (page 202) and access object-level rules, we do not recommend using `clause`. This is for two reasons. First, because what there are clauses for varies amongst implementations. For example, in some implementations there are clauses that match `clause((A,B),C)`, which will mess up many meta-interpreters. Secondly, the database is optimized differently for looking up a clause and executing a clause (the first emphasizing fast pattern matching, and the second by translating the clause into machine language). In many Prolog systems, you cannot use `clause` on predicates that have been optimized for performance, but you must declare them to be dynamic. This means that the clause will not execute as fast as it otherwise would.

We recommend using another notation for object level clauses, as is done in the following code where the infix operator "`<-`" is used for object level "if" (the symbol "$\Leftarrow$" in Section 6.3), and "`&`" is used for object-level conjunction. First make these symbols into infix operators:

```
:- op(1150, xfx, <- ).
:- op(950,xfy, &).
```

Here is an object-level definition of `subset(L1,L2)` that is true if every element of list `L1` is a member of list `L2`.

```
subset([],_) <- true.
subset([A|B],L) <-
    member(A,L) &
    subset(B,L).
```

Our vanilla meta-interpreter (page 205) defining `prove(G)` that is true if G can be proven from a knowledge base of object-level clauses, is:

```
prove(true).
prove(A & B) :- prove(A), prove(B).
prove(G)  :- (G <- B), prove(B).
```

Note that we need `G <- B` in parentheses, otherwise the body of the clause would be parsed as `G <- (B, prove(B))`.

Global Variables

One thing that the database can be used for is recording global variables or global flags.

Example 2.8	`newindex` returns a different integer each time it is called. This is useful, for example, when you want to make sure that some particular clauses are uniquely indexed so that you know you are accessing the right clause.

```
:- dynamic(previndex/1).
previndex(0).
newindex(I1) :-
    retract(previndex(I)),
    I1 is I+1,
    asserta(previndex(I1)).
```

Example 2.9	You can use a similar idea to define a random number generator. Such a random number generator is used for randomized search algorithms (page 158). The following is a random number generator based on the linear congruential method (Knuth, 1969) for generating random numbers.

```
:- dynamic seed/1.
seed(447563523).
```

`rand(R)` generates random real number R in the range [0, 1)

```
rand(R) :-
    retract(seed(S)),
    N is (S * 314159262 + 453816693) mod 2147483647,
    assert(seed(N)),
    R is N / 2147483647.0 .
```

`random(R,M)` generates random integer R in the range 0..M-1

```
random(R,M) :-
    rand(RR),
    R is integer(M * RR).
```

B.6   Returning All Answers

Assert and retract can be used to write a procedure to find all solutions to a query. Here is a very general procedure that accumulates answers to a query, by using the database to store intermediate results as the goal is being retried. When the goal finally fails, the collected answer is retrieved. We use an index to make sure that different accumulations do not interfere with each other.

accumulate(Goal,Init,Prev,CN,Next,Result) accumulates information for each success of Goal. Init is the zero accumulation, the value that should be returned if there were no clauses defining the goal. CN is a relation that shows how to compute Next accumulation from Prev accumulation. CN should succeed exactly once. Result is the final accumulation.

```
accumulate(G,I,P,CN,N,Res) :-
    newindex(Ind),
    iaccumulate(Ind,G,I,P,CN,N,Res).
```

iaccumulate is like accumulate but maintains an index Ind so that separate accumulations do not interfere with each other.

```
:- dynamic(result/2).
iaccumulate(Ind,G,I,P,CN,N,_) :-
    asserta(result(Ind,I)),
    G,
    retract(result(Ind,P)),
    CN,
    asserta(result(Ind,N)),
    fail.
iaccumulate(Ind,_,_,_,_,_,Res) :-
    retract(result(Ind,Res)).
```

where result(Ind,Res) in the database means that Res is the current accumulation for the call with index Ind. You first store the initial accumulation which is Init. You then call G. If G succeeds, you retract the previous accumulation, and run CN, which computes the next accumulation N which is stored. When the fail is executed, the asserta, CN, and the retract have no backtrack points and thus fail immediately. Then G is retried. If G finds another answer, the result is accumulated by CN and stored. When G eventually fails, the second rule to iaccumulate is called, which extracts the accumulated answer and returns it. Note that G and CN have to be bound when they are called; otherwise an error will occur.

Example 2.10

Here are some examples of the use of accumulate. Suppose p is defined as follows:

```
p(3).
p(5).
p(7).
p(1).
```

The following shows the values of Res that is returned by the associated queries.

```
| ?- accumulate(p(X),0,P,N is P+X,N,Res).
Res = 16
| ?- accumulate(p(X),1,P,N is P*X,N,Res).
Res = 105
| ?- accumulate(p(X),[],P,N=[X|P],N,Res).
Res = [1,7,5,3]
| ?- accumulate(p(X),L-L,P1-P2,P2=[X|P],P1-P,Res-[]).
Res = [3,5,7,1]
```

accumulate can be used for a procedure that collects in a list all of the instances of
a predicate that succeeds:

findall(X,G,Res) is true if Res is the list of all instances of X such that G
is true, in the order they are generated. findall is sometimes called allof.

```
findall(X,G,Res) :-
    accumulate(G,L-L,P1-[X|P],true,P1-P,Res-[]).
```

The following shows examples of the use of findall:

```
| ?- findall(X,p(X),L).
L = [3,5,7,1]
| ?- findall(next(X,Y),append(_,[X,Y|_],[a,b,c,d]),L).
L = [next(a,b),next(b,c),next(c,d)]
```

Example
2.11

findall can be used to define a version of negation as failure. This is a nonlogical
version that is different from that defined in the text (page 248), as it does not delay
calls with free variables. It is only correct when its argument contains no free
variables. Whatever variables happen to be bound when it is called are used. Thus
the success of this negation, not1, depends on which order conjunctions are called:

```
not1(G) :- findall(true,G,[]).
```

B.7 Input and Output

There are many built-in predicates to read and write from files and the terminal.
The only ones we will use are:

- read(T) reads a term from the terminal, and unifies it with T. Terms are ended
  by a period ".".

- write(T) writes term T to the terminal.

- `nl` writes a new-line to the terminal.

The code in the next appendix uses two simple but useful relations. `writel(L)` writes the elements of the list L.

```
writel([]).
writel([H|T]) :-
    write(H),
    writel(T).
```

`writeln(L)` writes the elements of the list L followed by a new line.

```
writeln(L) :-
    writel(L),
    nl.
```

B.8   Controlling Search

This section considers two different control primitives: an if-then-else construct and a way to cut out part of the search space.

The if-then-else construct (C -> Y ; N) evaluates C. If C succeeds, it evaluates Y. If C fails, it evaluates N. Only one answer is ever found for C. Thus if C succeeds, the number of answers for the if-then-else is the number of answers for Y, and if C fails the number of answers is the number for N.

Negation as failure can also be defined in terms of if-then-else:

```
not2(G) :- (G -> fail ; true).
```

This again does not delay goals with free variables.

Prolog's *cut* (written "!") is used to prune part of the search-space and to restrict backtracking. The rules for cut are simple: When called, it succeeds immediately; when it is retried (backtracked to), it fails the procedure in which it appears.

Cut can be interpreted as *commit*. It means that you are committing to the choices you have already made within the current predicate. Once the cut is evaluated, you don't make any more choices to the left of the cut, nor do you try any other clauses below the clause where the cut was evaluated. You only make this commitment when the cut is evaluated. Thus you can try other clauses, and can backtrack to the left of a cut, but commit when the cut is evaluated.

Note that

```
P :- (C -> Y ; N).
```

is exactly the same as

```
P :- C , !, Y.
P :- N.
```

In more complicated situations, each has its own advantage. Many cuts can appear in a definition, and the cut is global to the whole definition. The if-then-else is local and can be nested.

Here is another implementation of negation as failure:

```
not3(G)  :- G, !, fail.
not3(_).
```

If G succeeds, cut is executed. `fail` has no clauses defined for it, and retries cut. Cut then fails, and so `not3(G)` fails. If G fails, then the next clause is evaluated and so `not3(G)` succeeds. `not3` is built-in in Prolog and written as the prefix operator "\+". Again this only approximates the correct operation of negation as failure (page 248), because its result depends on what variables are bound and thus depends on the ordering of conjuncts. Combining "\+" with delaying can be used to implement sound negation as failure.

Cut often seems seductive, but the simplistic use of cut can often lead to erroneous conclusions:

| Example 2.12 | It is tempting to write the following definition of the minimum predicate (as opposed to the definition in Example B.3, page 482):

```
min(A,B,A)  :- A =< B, !.
min(A,B,B).
```

on the grounds that the second clause does not need the A>B condition because it is only used when the first clause fails. This definition of min gets the correct answer for the queries

```
| ?- min(5,17,X).
| ?- min(17,5,Y).
```

but gets the wrong answer to the query

```
| ?- min(5,17,17).
```

A general rule is that you should not tell the system lies, as in the second clause in the definition of `min`, otherwise you run the risk of producing incorrect answers. If you follow this maxim, then cut will only make the system incomplete and not unsound. This rule only holds without negation as failure, because unsoundness in the definition of *g* results in incompleteness of $\sim g$, and incompleteness in the definition of *g* results in unsoundness of $\sim g$.

If you *know* that no other clauses will be applicable at any stage, then cut can be used to reduce the space used by Prolog. When a cut is executed, all of the local backtrack points can be removed, thus saving a lot of space. It is also useful when implementing a nondeterministic *select*, as opposed to a *choose* (page 50). If you know that no other choices will lead to a solution if the current one doesn't, then the other alternatives can be cut out. All of the example code in Appendix C uses cut in one of these ways, unless explicitly mentioned.

Sometimes, cut is needed when combined with other nonlogical features of Prolog:

Example 2.13

The relation `bassert(C)`, backtrack-able assert, asserts clause C to the rule base, and undoes the assertion on backtracking. This may rearrange the order of the clauses, and it is not guaranteed to work when there are other clauses that match C that are added between the `asserta` and the `retract`.

```
bassert(C)  :-
    asserta(C).
bassert(C)  :-
    retract(C),!, fail.
```

where `fail` is a predicate that always fails. It is defined by not writing any clauses for it. Note that the second clause for `bassert` is only called if the first clause fails, and it just retracts the first clause that matches C and then fails. Without the cut, it would remove all clauses that match C.

`bretract(C)` retracts clause C from the rule base, and undoes the retraction on backtracking.

```
bretract(C)  :-
    retract(C)  ,
    assert_on_failure(C).
assert_on_failure(_).
assert_on_failure(C)  :-
    asserta(C),
    fail.
```

Appendix C

Some More Implemented Systems

This appendix gives Prolog code for some larger examples. We use the notation conventions of Appendix B. All definitions not given in the particular section can be found in Appendix B, or in Section 3.5 (page 81) for the list predicates. Except where noted, these programs should run on any Prolog system that uses the standard Prolog syntax or the traditional Edinburgh syntax.

C.1 Bottom-Up Interpreters

In this section we implement three forward chaining interpreters that are described in the text. The aim is to show how *select* (page 50) can be implemented and to show how a simple interpreter can be augmented to provide more features.

Bottom-Up Definite Clause Interpreter

The following code implements the bottom-up proof procedure for computing consequences of *KB* presented in Figure 2.4 (page 47).

It assumes that the base-level is implemented using "<-" for "if" and "&" for "and."

```
% "<-" is the object-level "if"
:- op(1150, xfx, <- ).
% "&" is the object level conjunction.
:- op(950,xfy, &).
```

```
% fc(C,R) is true if you can forward chain from
%    atoms in C resulting in R.
fc(C,R) :-
   ( A <- B ),        % there is a rule in the KB
   derived(B,C),      % whose body is derived and
   notin(A,C),        % whose head isn't derived.
   !,                 % commit to this selection.
   fc([A|C],R).       % forward chain with head derived.
fc(R,R).              % no selection is possible.
```

```
% derived(B,C) is true if body B can be directly
% derived from atoms in C.
derived(true,_) :-
   !.    % the other rules aren't applicable for true body
derived((A&B),C) :-
   !,    % the other rule isn't applicable for conjunctive body
   derived(A,C),
   derived(B,C).
derived(A,C) :-
   member(A,C).
```

Note the use of cut to implement nondeterministic selection. Evaluating cut commits you to all of the choices up to the cut. The second clause for `fc` is only called if the first clause fails before the cut is evaluated. This means that no selections are possible.

To execute the interpreter, issue the query

```
| ?- fc([],Ans).
```

The following code is the object level code of Example 2.14 (page 47):

```
a <- b & c.
b <- d & e.
b <- g & e.
c <- e.
d <- true.
e <- true.
f <- a & g.
```

Exercise C.1 Write a more efficient version of this program that uses better indexing of rules. Assume that the knowledge base contains only ground (variable-free) clauses. There are a number of properties that can be exploited for efficiency. Each clause need only be used once. For a clause to be used, the leftmost atom in the body must be used. Thus you need only consider the rule when forward chaining on the leftmost atom in the body. Forward chaining on a rule may result in a new clause with a new leftmost atom in the body, or in an element of *C*.

Exercise
C.2

Write a version that uses better indexing of clauses but also allows variables in the clauses. This is complicated because multiple instances of the same clause can be used.

Bottom-Up Negation as Failure Implementation

The bottom-up procedure for implementing negation as failure, shown in Figure 7.4 (page 254), is a modification of the bottom-up procedure for definite clauses. The difference is that you can add literals of the form $\sim p$ to the set C of derived consequences. $\sim p$ means p fails. $\sim p$ is added to C when you can determine that p must fail. p fails when every body with p as the head fails. A body fails if one of the literals in the body fails. An atom b_i in the body finitely fails if $\sim b_i$ has been derived (is in C). A negation $\sim b_i$ finitely fails if b_i has been derived.

The following Prolog code assumes that the knowledge base consists of ground clauses as well as declarations of the atomic symbols using the atomic_symbol predicate. The meta-level declaration is needed as you want to conclude any atom with no clauses, and you don't want to start concluding all atoms because there are too many.

```
% "<-" is the object-level "if"
:- op(1150, xfx, <- ).
% "&" is the object level conjunction.
:- op(950,xfy, &).
% "~" is the object level negation.
:- op(900,fy, ~).

% fc_nf(C,R) is true if you can forward chain from
% atoms in C resulting in R.
fc_nf(C,R) :-
   ( A <- B ),
   derived(B,C),
   notin(A,C),
   !,
   fc_nf([A|C],R).
fc_nf(C,R) :-
   atomic_symbol(A),
   notin((~ A),C),
   allof(B,(A <- B),Bds),
   allfail(Bds,C),
   !,
   fc_nf([(~ A)|C],R).
fc_nf(C,C).

% derived(B,C) is true if body B can be directly
```

```
% derived from atoms in C
derived(true,_).
derived((A & B),C) :-
    derived(A,C),
    derived(B,C).
derived(A,C) :-
    member(A,C).

% allfail(Bs,C) is true if all of the bodies in
% list Bs are false given the truths in list C.
allfail([],_).
allfail([H|T],C) :-
    fails(H,C),
    allfail(T,C).

% fails(B,C) is true if body B is false given C
fails((A&_),C) :- fails(A,C), !.
fails((_&B),C) :- !,fails(B,C).
fails((~ A),C) :- !,member(A,C).
fails(A,C) :- member((~ A),C),!.
```

To run this program issue the query

```
| ?- fc_nf([],Ans).
```

The following is the representation of Example 7.22 (page 253).

```
p <- q &  ~ r.
p <- s.
q <-   ~ s.
r <-   ~ t.
t <- true.
s <- w.

atomic_symbol(p).
atomic_symbol(q).
atomic_symbol(r).
atomic_symbol(s).
atomic_symbol(t).
atomic_symbol(u).
atomic_symbol(v).
atomic_symbol(w).
```

Exercise C.3 Write a more efficient version of this program that uses better indexing of rules. You can assume that the clauses in the knowledge base are ground.

Exercise C.4 Write a version of this program that indexes rules and allows for incremental addition of clauses the the knowledge base. As a clause is added to the knowledge base, the

$Q := \{\langle h, \{h\}\rangle : h \in H\}$;
repeat
    select "$a \leftarrow b_1 \wedge \ldots \wedge b_n$" in KB such that
        $\langle b_i, D_i \rangle \in Q$ for $i = 1 \ldots n$ and
        $\neg \exists B \; (B \subseteq \bigcup_i D_i) \wedge (\langle a, B \rangle \in Q \vee \langle false, B \rangle \in Q)$;
    if $a = false$
        then $\forall b \; \forall D$ such that $\bigcup_i D_i \subseteq D$
            remove $\langle b, D \rangle$ from Q
        else $\forall D$ such that $\bigcup_i D_i \subset D$
            remove $\langle a, D \rangle$ from Q;
    $Q := Q \cup \{\langle a, \bigcup_i D_i \rangle\}$;
until no more choices

Figure C.1: Forward-chaining assumption-based reasoner

elements of C should be updated. You can assume that the knowledge base is given as a data structure.

Bottom-Up Assumable Interpreter

The bottom-up assumable interpreter that includes assumables and integrity constraints (page 241) is an augmented version of the bottom up algorithm for definite clauses. Figure C.1 gives pseudocode for the algorithm. This is a refinement of the program of Figure 7.2 (page 246).

The conclusions set consists of pairs $\langle a, A \rangle$, where a is an atom and A is a set of assumables that imply a. When you generate the pair $\langle false, A \rangle$, you know that the assumptions in A form a conflict; they can't coexist. Thus if $A = \{a_1, \ldots, a_k\}$, you know that

$$T \models \neg a_1 \vee \ldots \vee \neg a_k.$$

The refinement to the program of Figure 7.2 is to prune redundant supersets of assumptions. If you have $\langle a, A_1 \rangle$ and $\langle a, A_2 \rangle$ in C, where $A_1 \subset A_2$, you can remove $\langle a, A_2 \rangle$ from C, or not add it if you were about to add it. There is no need to make the extra assumptions to imply a. Similarly, if you have $\langle false, A_1 \rangle$ and $\langle a, A_2 \rangle$ in C, where $A_1 \subseteq A_2$, you can remove $\langle a, A_2 \rangle$ from C. You can do this because you already know that A_1 and any superset, including A_2, is inconsistent with the knowledge base, and so nothing more can be learned from considering such sets of assumables.

The following Prolog code implements this for ground knowledge bases:

```
% '<-' is the object-level 'if'
:- op(1150, xfx, <- ).
% '&' is the object level conjunction.
:- op(950,xfy, &).


% An environment is a term of the form env(G,E) where
% E is a list of assumables such that G<-E logically
% follows from the clauses.  Environment E1 subsumes
% environment E2 if the implication represented by E1
% implies the implication represented by E2.


% An explanation of G is a set E of assumables that,
% together with the given rules implies G and is
% consistent (doesn't imply false).  An explanation
% is represented by the corresponding environment
% env(G,E).  A minimal explanation is an explanation
% such that no subset is also an explanation.


% explanations(Ans) is true if Ans is a list of
% minimal explanations.
explanations(Ans) :-
    allof( env(A,[A]),assumable(A),C),
    writeln(['Initial assumables: ',C]),
    fc_ab(C, Ans).


% fc_ab(C,R) means you can forward chain from C
% getting R.  This chooses one rule at a time to
% forward chain on, producing one candidate
% environment. This environment must not already be
% subsumed by an already derived environment.
fc_ab(C,R) :-
    ( A <- B ),
    derived(B,[],H,C),
    \+ subsumed(env(A,H),C),
    !,                     % commit to this selection
    writeln(['Derived: ',env(A,H)]),
    add_n_prune(env(A,H),C,C1),
    fc_ab(C1,R).
fc_ab(C,C).


% derived(B,H0,H,C) is true if body B can be
% directly explained starting from the assumptions
% H0, resulting in the assumptions H1 given list
% of environments C
derived(true,H,H,_) :-!.
derived((A & B),H0,H2,C) :-
```

```
      !,
      derived(A,H0,H1,C),
      derived(B,H1,H2,C).
derived(A,H0,H1,C) :-
    member(env(A,H),C),
    insert_each(H,H0,H1).

insert_each([],L,L).
insert_each([A|R],L0,L1) :-
    member(A,L0),
    !,
    insert_each(R,L0,L1).
insert_each([A|R],L0,L1) :-
    insert_each(R,[A|L0],L1).

% subsumed(E,C) is true if environment E is subsumed
% by an element of C.
subsumed(E,C) :-
    member(E1,C),
    subsumes(E1,E).

% subsumes(E1,E2) is true if environment E1 subsumes
% environment E2
subsumes(env(A,H1),env(A,H2)) :-
    subset(H1,H2),!.
subsumes(env(false,H1),env(_,H2)) :-
    subset(H1,H2),!.

% add_n_prune(E,L,R) is true if R is the resulting
% list of minimal environments obtained by adding
% environment E to L.  We need to remove all
% environments in L that are subsumed by E.
add_n_prune(E,[],[E]).
add_n_prune(E,[E1|R],C) :-
    subsumes(E,E1),
    add_n_prune(E,R,C).
add_n_prune(E,[E1|R],[E1|C]) :-
    \+ subsumes(E,E1),
    add_n_prune(E,R,C).
```

To run this, issue the query

```
| ?- explanations(Ans).
```

The Ans returned is the set of all minimal explanations for all atoms.

The following is a simple example to demonstrate the base-level representation:

```
% test code for the forward chaining abduction
```

```
% interpreter
a <- b & c.
a <- e & t.
b <- d.
b <- e.
c <- w.
c <- e.
f <- e.
h <- i.
false <- f.

assumable(e).
assumable(d).
assumable(g).
assumable(w).
assumable(t).
```

Exercise C.5 Write a more efficient assumable interpreter that uses better indexing of rules. When a new environment is created, you should be able to directly access the rules that could use this environment, instead of testing every rule at each iteration.

Exercise C.6 Write an assumable interpreter that allows for incremental addition of rules and assumables. You need to think about what information needs to be maintained in order to construct new environments, so that you don't recompute everything. *Hint:* It is useful to have a common representation of both the rules in the knowledge base and the environments. One such representation is to maintain a list of triples of the form imp(H,B,A) where B is a body and A is a list of assumables that imply A. When A is the empty list it corresponds to a rule. When B is true, it corresponds to an environment.

C.2   Top-Down Interpreters

In this section we give the details of some useful top-down interpreters that need some extralogical mechanisms, such as input-output, or the ability to find all answers. These are all based on the vanilla meta-interpreter of Figure 6.3 (page 205).

Meta-Interpreter for Traversing Proof Trees

The first meta-interpreter allows the user to pose "HOW" queries (page 217). This is useful for understanding and debugging programs that succeed.

The version here is based on the meta-interpreter that builds a proof tree of Figure 6.9 (page 218). It builds the same proof tree and it provides facilities for the user to traverse the proof tree.

show(*G*) means prove goal *G* and then show the top-level rule that was used to prove *G*. The user can in turn look at the rules that were used to prove the elements of the body of that rule and thus allow the user to traverse the proof tree. When traversing the tree, the program always maintains a current rule from which the user can go up and down.

When a rule used in the proof tree is displayed, the user can issue the commands:

n where *n* is an integer, to show how the *n*th conjunct in the body was proved.

up to show how the parent of the current rule was proved. This is the rule whose body contains the head of the current rule.

retry to find and display another proof.

The following Prolog code implements this:

```
% <- is the object-level 'if'
:- op(1150, xfx, <- ).
% '&' is the object level conjunction.
:- op(950,xfy, &).

% show(G) means prove goal G and show how it was proved
show(G)  :-
    solve(G,T),
    traverse(T).

% solve(Goal,Tree) is true if Tree is a proof tree for Goal
solve(true,true).
solve((A&B),(AT&BT))  :-
    solve(A,AT),
    solve(B,BT).
solve(H,if(H,builtin))  :-
    builtin(H),
    call(H).
solve(H,if(H,BT))  :-
    ( H <- B ),
    solve(B,BT).

% builtin(G) is true if goal G is defined in Prolog
% (as opposed to being defined by rules in the
%  object-level knowledge base.)
builtin((_ is _)).
builtin((_ =< _)).
```

```
builtin((_ >= _)).
builtin((_ = _)).
builtin((_ < _)).
builtin((_ > _)).

% traverse(T) true if T is a tree being traversed
traverse(if(H,true)) :-
   !,                    % no other rules are applicable
   writeln([H,' is a fact']).
traverse(if(H,builtin)) :-
   !,                    % no other rules are applicable
   writeln([H,' is built-in.']).
traverse(if(H,B)) :-
   B \== true,
   B \== builtin,
   writeln([H,' :-']),
   print_body(B,1,Max),
   read(Comm),
   interpret_command(Comm,B,Max,if(H,B)).

% print_body(B,N1,N2) is true if B is a body to be
% printed and N1 is the count of atoms before B
% N2 is the count after
print_body(true,N,N) :-
   !.                    % no other rules are applicable
print_body((A&B),N1,N3) :-
   !,                    % no other rules are applicable
   print_body(A,N1,N2),
   print_body(B,N2,N3).
print_body(if(H,_),N,N1) :-
   writeln(['   ',N,': ',H]),
   N1 is N+1.

% interpret_command(Comm,B,Max,G) interprets the
% command Comm on body B where Max is the number of
% conjunctions in the body and G is the goal
interpret_command(N,B,Max,G) :-
   integer(N),
   N > 0,
   N =< Max,
   !,                    % no other rules are applicable
   nths(B,N,E),
   traverse(E),
   traverse(G).
interpret_command(N,_,Max,G) :-
   integer(N),
```

```
      !,                         % no other rules are applicable
      (N < 1 ; N > Max),
      writeln(['Number out of range: ',N]),
      traverse(G).
interpret_command(up,_,_,_) :-
      !.                         % no other rules are applicable
interpret_command(C,_,_,G) :-
      C \== retry,
      writeln(['Give either a number, up or retry.']),
      traverse(G).
% note that interpret_command(retry,_,_,_) fails.

% nths(S,N,E) is true if E is the N-th element of
% conjunction S. This assumes that '&' is left associated.
nths(A,1,A) :-
      \+ (A = (_&_)).
nths((A&_),1,A).
nths((_&B),N,E) :-
      N>1,
      N1 is N-1,
      nths(B,N1,E).
```

The first clause for interpret_command needs explanation. Notice how it first traverses E then it traverses G. traverse is written such that traverse(E) only succeeds if an up command is issued when displaying E. Traverse is deterministic so that if it fails, solve is retried.

Given the knowledge base of Figure 6.4 (page 206), the following shows the program being used. Note that "| :" is the prompt for reading from the user. User commands follow this prompt.

```
| ?- show(lit(L)).
lit(l2) :-
    1: light(l2)
    2: ok(l2)
    3: live(l2)
|: 3.
live(l2) :-
    1: connected_to(l2,w4)
    2: live(w4)
|: 2.
live(w4) :-
    1: connected_to(w4,w3)
    2: live(w3)
|: 1.
connected_to(w4,w3) :-
    1: up(s3)
|: 1.
```

```
up(s3) is a fact
connected_to(w4,w3) :-
   1: up(s3)
|: up.
live(w4) :-
   1: connected_to(w4,w3)
   2: live(w3)
|: 2.
live(w3) :-
   1: connected_to(w3,w5)
   2: live(w5)
|: 1.
connected_to(w3,w5) :-
   1: ok(cb1)
|: retry.
no
```

The final "no" means that there are no alternative proofs for the query.

Iterative Deepening Definite Clause Interpreter

The following interpreter uses a depth-bounded search (page 209) to implement an iterative deepening (page 140) definite clause interpreter.

The depth is increased by one each time. It reports an answer when it reaches the depth bound because these are the answers that would have failed in the previous depth-bounded search.

One of the tricky things is to detect when to stop increasing the depth bound and fail the procedure. The proof should fail when none of the proofs reached the current depth bounds. In this way you say that the depth-bounded prover failed naturally. It failed unnaturally if it failed because the depth bound was reached.

The following Prolog code implements this:

```
% '<-' is the object-level 'if'
:- op(1150, xfx, <- ).

% '&' is the object level conjunction.
:- op(950,xfy, &).

% iprove(G) is true if we can prove G using
% iterative deepening search
iprove(G) :-
   iprove_from(G,1).

% bprove(G,Bound,CL1,CL2) is true if we can prove G
% with a proof tree of depth less than Bound where
```

```
% CL1 is the closest that we came to the bound before
% proving G, and CL2 is the closest we came to the
% bound after proving G.
bprove(true,Bnd,In,Out) :- !,
   min(Bnd,In,Out).

bprove((A&B),Bnd,C1,C3) :- !,
   bprove(A,Bnd,C1,C2),
   bprove(B,Bnd,C2,C3).

bprove(G,Bnd,C1,C2) :-
   Bnd > 1, !,
   B1 is Bnd-1,
   (G <- Body),
   bprove(Body,B1,C1,C2).

bprove(_,1,_,_) :-
   retract(failed(naturally)),
   assert(failed(unnaturally)),
   fail.

% iprove_from(G,Bnd) is true if we can prove G using
% an iterative deepening search, starting from depth
% Bnd. We succeed if there is a number, such that
% iprove previously failed  unnaturally, such that
% bprove can prove G, getting within one of the depth
% of the previous search (so it must have failed for
% the previous iteration).

iprove_from(G,Bnd) :-
   retractall(failed(_)),
   assert(failed(naturally)),
   bprove(G,Bnd,Bnd,1).

iprove_from(G,B) :-
   failed(unnaturally),
   B1 is B+1,
   iprove_from(G,B1).

% cleanup needs to be done between proofs so the
% proofs do not interfere with each other.
cleanup :-
   retractall(failed(_)).
```

Consider the following object-level program:

```
odd(s(s(N))) <- odd(N).
```

```
odd(s(0)) <- true.
even(s(s(N))) <- even(N).
even(0) <- true.
num(N) <- even(N).
num(N) <- odd(N).
```

Using the standard Prolog search strategy, the query ?num(N) would loop forever,
however the query ?iprove(num(N)) will return all answers.

Querying the User

Section 6.4 (page 212) presented pseudocode for querying the user. In this section
we show how yes/no questions of the user can be implemented in Prolog. This uses
"asserta" to maintain the knowledge base as queries are asked.

This procedure assumes that the asked queries are ground. The user can reply
either "yes," "no," or "unknown." The rules will only be used for an askable query
when the user replies unknown. In the code below, an answer other than yes or no
is interpreted as unknown.

The following Prolog code implements this:

```
% ASK THE USER INTERPRETER - assumes ground questions.
% It only uses rules for an askable goal if the user
% doesn't know the answer.

% '<-' is the object-level 'if'
:- op(1150, xfx, <- ).
% '&' is the object level conjunction.
:- op(950,xfy, &).
% answered(G,Ans) is dynamically added to the database
% when the user answers Ans to question 'Is G true?'.
:- dynamic(answered/2).

% aprove(G) is true if G can be proven perhaps by
% asking the user.
aprove(true).
aprove((A & B)) :-
    aprove(A),
    aprove(B).
aprove(G) :-
    askable(G),
    answered(G,yes).
aprove(G) :-
    askable(G),
    unanswered(G),
    ask(G,Ans),
```

```
        asserta(answered(G,Ans)),
        Ans=yes.
aprove(H) :-
        \+ answered(H,yes),
        \+ answered(H,no),
        (H <- B),
        aprove(B).

% unanswered(G) is true if G has not been answered
unanswered(G) :- \+ been_answered(G).
been_answered(G) :- answered(G,_).

% ask(G,Ans) asks the user G, and the user
% replies with Ans
ask(G,Ans) :-
        writel(['Is ',G,' true? [yes, no, unknown]: ']),
        read(Ans).
```

The following is a trace of Example 6.8 (page 214), with user input after the colon:

```
| ?- aprove(lit(L)).
Is up(s2) true? [yes, no, unknown]: yes.
Is up(s1) true? [yes, no, unknown]: no.
Is down(s2) true? [yes, no, unknown]: no.
Is up(s3) true? [yes, no, unknown]: yes.

L = 12 ?

yes
```

Exercise C.7 Implement an ask-the-user facility that allows for variables in the query. It should ask yes/no questions if there are no variables in the askable, otherwise it should ask for which instances is the goal true. This may require more sophisticated Prolog programming than presented here.

Meta-Interpreter with Search

In this section we implement the meta-interpreter with search (page 226), where unification is handled by the underlying Prolog system.

Let $goal(C, A)$ represent the generalized answer clause (page 56) "$yes(A) \leftarrow C$," where A is a form of an answer, and C is a list of atoms representing a conjunction that needs to be proved in order to infer answer A.

To define the meta-interpreter, you axiomatize $proveall(Qs, As)$ that is true if Qs is a list of generalized answer clauses, and As is the list of all answers. The elements of Qs are implicitly disjoined — they represent the different proofs that are being pursued. You choose a generalized query from *Qs*. If its body is empty, you add the corresponding answer to the set of all answers, and continue with the rest of the queries. If its body isn't empty, you select an atom to resolve against (see page 56) and form new answer clauses for each clause that the atom resolves with.

```
% META INTERPRETER WITH SEARCH

% '<-' is the object-level 'if'
:- op(1150, xfx, <- ).
% '&' is the object level conjunction.
:- op(950,xfy, &).

% proveall(Qs,As) is true if Qs is a list of
% generalized answer clauses, and As is the list of
% all answers to the queries in Qs.
proveall([],[]).
proveall(Qs,[A|Ans]) :-
   select_query(goal(A,[]),Qs,RQs),
   proveall(RQs,Ans).
proveall(Qs,Ans) :-
   select_query(goal(A,C),Qs,RQs),
   select_atom(G,C,C1),
   allof(goal(A,NC),((G <- B), add_body(B,C1,NC)),S),
   add_to_frontier(S,RQs,ND),
   proveall(ND,Ans).

% select_query(Q,Qs,RQs) is true if generalized
% answer clause Q is selected from list Qs, and RQs are
% the remaining elements of Qs.
select_query(Q,[Q|Qs],Qs).

% select_atom(A,As,RAs) is true if selecting atom
% A from list As leaves RAs
select_atom(A,[A|Gs],Gs).

% add_to_frontier(NN,F0,F1) is true if adding
% elements NN to frontier F0 produces frontier F1
add_to_frontier(NN,F0,F1) :-
   append(NN,F0,F1).

% add_body(B,C1,C2) adds the elements of body B to
% the conjuncts in C1 producing the conjuncts in C2
add_body(true,Con,Con) :- !.
add_body((A & B),Con1,Con3) :- !,
   add_body(B,Con1,Con2),
   add_body(A,Con2,Con3).
add_body(A,Con,[A|Con]).
```

Note that *proveall* uses the relation *allof* (X, C, L) that's true if L is the list of instances of X such that C is true (page 485). *allof* has the property that variables in different elements are renamed. This is exactly what's required to make sure different proofs do not interfere with each other.

The object-level program corresponding to Example 6.16 (page 228) is:

```
% app(A,B,C) means appending A to B results in C
app([],L,L) <- true.
app([A|X],Y,[A|Z]) <- app(X,Y,Z).

% sublist(S,C) is true if list S is in the middle of list C
sublist(S,L) <- app(_,B,L) & app(S,_,B).

% Example query:
% ?proveall([goal(yes(X,Y),[sublist([X,Y],[a,b,c,d,e])])],A).
```

Exercise C.8 Change the meta-interpreter so that it does breadth-first search. Note that this is a very simple change to the code.

Exercise C.9 Change the meta-interpreter so that it cycles through the goals in the conjunct. That is, the selection rule should select the first conjunct added. This is again a simple modification. Why might this be a good idea?

Exercise C.10 Write a meta-interpreter that uses the search-space representation of Chapter 4. The nodes of the graph will be labeled with *goal*(C, S). You need to axiomatize the relations *neighbors* and *is_goal*. With such a representation, what is an admissible heuristic for A^* search? How well does it work?

C.3 Constraint Satisfaction Problem Solver

This section presents a CSP solver that uses arc consistency (page 147) and domain splitting. It only allows binary relations.

A binary FCSP query is given by *csp(Domains, Relations)*, where *Domains* is a list of terms of the form *dom(Var, Values)*, where *Var* is a variable and *Values* is a list of possible values for the variable (the domain of the variable). *Relations* is a list of terms of the form *rel([Var1, Var2], Rel)*, where *Var1* and *Var2* are the variables that appear in relations *Rel*. All of the relations are binary (have two variables) and can be evaluated by querying the system. These relations should either succeed once or fail for each ground query.

The following Prolog code solves such CSPs using arc consistency and domain splitting:

```
% csp(Domains, Relations) means that each variable has
% an instantiation to one of the values in its Domain
% such that all the Relations are satisfied.
% Domains represented as list of
% [dom(V,[c1,...,cn]),...]
% Relations represented as [rel([X,Y],r(X,Y)),...]
%  for some r
csp(Doms,Relns) :-
   ac(Doms,Relns).

% ac(Dom,Relns) is true if the domain constraints
% specified in Dom and the binary relations
% constraints specified in Relns are satisfied.
ac(Doms,Relns) :-
   make_arcs(Relns,A),
   consistent(Doms,[],A,A).

% make_arcs(Relns, Arcs) makes arcs Arcs corresponding to
% relations Relns. There are arcs for each ordering of
% variables in a relations.
make_arcs([],[]).
make_arcs([rel([X,Y],R)|O],
          [rel([X,Y],R),rel([Y,X],R)|OA]) :-
   make_arcs(O,OA).

% consistent(Doms,CA,TDA,A) is true if
% Doms is a set of domains
% CA is a set of consistent arcs,
% TDA is a list of arcs to do
% A is a list of all arcs
consistent(Doms,CA,TDA,A) :-
   consider(Doms,RedDoms,CA,TDA),
   solutions(RedDoms,A).

% consider(D0,D1,CA,TDA)
% D0 is the set of initial domains
% D1 is the set of reduced domains
% CA = consistent arcs,
% TDA = to do arcs
consider(D,D,_,[]).
consider(D0,D3,CA,[rel([X,Y],R)|TDA]) :-
   choose(dom(XV,DX),D0,D1),X==XV,
   choose(dom(YV,DY),D1,_),Y==YV, !,
   prune(X,DX,Y,DY,R,NDX),
   ( NDX = DX
   ->
```

```
        consider(D0,D3,[rel([X,Y],R)|CA],TDA)
    ; acc_todo(X,Y,CA,CA1,TDA,TDA1),
        consider([dom(X,NDX)|D1],D3,
                [rel([X,Y],R)|CA1],TDA1)).

% prune(X,DX,Y,DY,R,NDX)
% variable X has domain DX
% variable Y has domain DY
% R is a relation on X and Y
% NDX = {X in DX | exists Y such that R(X,Y) is true}
prune(_,[],_,_,_,[]).
prune(X,[V|XD],Y,YD,R,XD1):-
    \+ (X=V, member(Y,YD), call(R)),!,
    prune(X,XD,Y,YD,R,XD1).
prune(X,[V|XD],Y,YD,R,[V|XD1]):-
    prune(X,XD,Y,YD,R,XD1).

% acc_todo(X,Y,CA,CA1,TDA,TDA1)
% given variables X and Y,
% updates consistent arcs from CA to CA1 and
% to do arcs from TDA to TDA1
acc_todo(_,_,[],[],TDA,TDA).
acc_todo(X,Y,[rel([U,V],R)|CA0],
        [rel([U,V],R)|CA1],TDA0,TDA1) :-
    ( X \== V
    ; X == V,
      Y == U),
    acc_todo(X,Y,CA0,CA1,TDA0,TDA1).
acc_todo(X,Y,[rel([U,V],R)|CA0],
        CA1,TDA0,[rel([U,V],R)|TDA1]) :-
    X == V,
    Y \== U,
    acc_todo(X,Y,CA0,CA1,TDA0,TDA1).

% solutions(Doms,Arcs) given a reduced set of
% domains, Doms, and arcs Arcs, solves the CSP.
solutions(Doms,_) :-
    solve_singletons(Doms).
solutions(Doms,A) :-
    select(dom(X,[XV1,XV2|XVs]),Doms,ODoms),
    split([XV1,XV2|XVs],DX1,DX2),
    acc_todo(X,_,A,CA,[],TDA),
    ( consistent([dom(X,DX1)|ODoms],CA,TDA,A)
    ; consistent([dom(X,DX2)|ODoms],CA,TDA,A)).

% solve_singletons(Doms) is true if Doms is a
```

```
% set of singleton domains, with the variables
% assigned to the unique values in the domain
solve_singletons([]).
solve_singletons([dom(X,[X])|Doms]) :-
    solve_singletons(Doms).

% select(E,L,L1) selects the first element of
% L that matches E, with L1 being the remaining
% elements.
select(D,Doms,ODoms) :-
    remove(D,Doms,ODoms), !.

% choose(E,L,L1) chooses an element of
% L that matches E, with L1 being the remaining
% elements.
choose(D,Doms,ODoms) :-
    remove(D,Doms,ODoms).

% split(L,L1,L2) splits list L into two lists L1 and L2
% with the about same number of elements in each list.
split([],[],[]).
split([A],[A],[]).
split([A,B|R],[A|R1],[B|R2]) :-
    split(R,R1,R2).
```

There is one tricky part of this code. The third clause for prune is used when R succeeds for appropriate values of X and Y. The use of the negation as failure and cut means that it does this without binding the values of X and Y.

The following is an example CSP with 4 answers.

```
% Here is a simple example using the CSP solver
test_csp(X,Y,Z) :-
    csp([dom(X,[1,2,3,4]),
         dom(Y,[1,2,3,4]),
         dom(Z,[1,2,3,4])],
        [rel([X,Y],X<Y),rel([Y,Z],Y<Z)]).

% | ?- test_csp(X,Y,Z).
```

The following is the scheduling problem from Example 4.22 (page 150).

```
% Here's the scheduling problem from Chapter 4:
schedule(A,B,C,D,E) :-
    csp([dom(A,[1,2,3,4]),
         dom(B,[1,2,4]),
         dom(C,[1,3,4]),
         dom(D,[1,2,3,4]),
         dom(E,[1,2,3,4])],
```

```
[rel([A,B], A =\= B),
 rel([B,C], B =\= C),
 rel([C,D], C < D),
 rel([B,D], B =\= D),
 rel([A,D], A =:= D),
 rel([A,E], A > E),
 rel([B,E], B > E),
 rel([C,E], C > E),
 rel([D,E], D > E)]).
```

```
% | ?- schedule(A,B,C,D,E).
```

Exercise C.11   The use of `select` in `solutions` means that an arbitrary domain is split. The aim of this exercise is to experiment with different selection criteria for choosing which domain to split. Is there any advantage to choosing the domain based on its size? Try splitting the smallest/largest domain and measure the performance on some CSPs. Can you think of another heuristic? Test it.

Exercise C.12   Extend and revise the *csp* implementation to handle unary predicates and domain consistency. Test your extension on this problem: Given a set of cubes, all the same size and each a uniform color, build a tower of a given height so that each cube is a different color from the cube above it and the cube below it. The color of any particular cube may also be specified. So, for example, given two red cubes and three green cubes can you build a tower of four cubes with the top red cube that satisfies the adjacent color constraint?

Exercise C.13   Extend and revise the *csp* implementation to allow for *k*-ary constraints. As a sample problem, consider building a magic square: Find a way of placing the integers $1, 2, 3, 4, 5, 6, 7, 8$, and 9 in the 3×3 empty crossword shown in Figure 4.13 (page 167) so that the sum of the numbers in each row, column, and main diagonal is 15.

Exercise C.14   Investigate the ideas behind *path consistency* and, more generally *k-consistency*. Implement path consistency. Is it more efficient than arc consistency and domain splitting? Explain why.

C.4   Neural Network Learner

In this section we present a neural-network learning algorithm (page 408) that can learn parameters for any parameterized definite clause program based on a set of examples. All it requires is a set of numerical parameters and a measure of error for each example. It uses gradient descent search (page 157) on the parameter space to try to find the parameter settings that minimize the sum of the errors for each example.

Gradient descent needs the partial derivative of the error for each parameter. One advantage of the additive sigmoid neural networks (page 410) is that the derivative can be solved analytically. In the code in this section, rather than finding the derivatives analytically, they are estimated numerically. The partial derivative of parameter p_i can be approximated by

$$\frac{error(p_i + dx) - error(p_i)}{dx}$$

where $error(p)$ is the error given parameter setting p_i, for sufficiently small number dx. In the code below, the number dx is an input to the procedure. Another input is the learning rate, the η of the gradient descent (page 157), which is the number the derivatives need to be multiplied by to obtain the next parameter setting.

The axiomatization has two parts. The first is a meta-interpreter that evaluates a logic program with respect to a set of numerical parameters. We assume that the parameters can only be used in expressions as part of an "is." This isn't a big restriction as the object-level program can always include a goal "V is p," where V is a variable and p is a parameter; then the variable X can be used anywhere in the program and has the value of the parameter.

The second part uses the meta-interpreter to derive the total error of a program for a set of examples and then uses gradient descent to minimize the error. Note that the only stopping criteria provided is to carry out the descent a certain number of times. More sophisticated stopping criteria can be used based on this code.

```
% Neural-net style learning for parameterized logic
% programs.  Given a set of examples, a parameterized
% logic program and a measure of error for each
% example, this program used back-propagation to tune
% the parameters. Derivatives are estimated numerically.

% nnlearn(N,DX,LC,Exs,P0,P1)
% N is the number of iterations to do
% DX is the increment to evaluate derivatives
% LC is the learning constant for gradient descent
% Exs is a list of all of the examples
% P0 is the list of parameter settings before the learning
% P1 is the list of parameter settings after the learning
nnlearn(0,_,_,Exs,P0,P0) :-
    total_error(Exs,P0,0,Err0),
    writeln(['Error = ', Err0]).
nnlearn(N,DX,LC,Exs,P0,P2) :-
    update_parms(DX,LC,Exs,P0,P1),
    N1 is N-1,
    nnlearn(N1,DX,LC,Exs,P1,P2).

% update_parms(DX,LC,Exs,P0,P1).
```

```prolog
% updates all of parameter in P0 to P1
update_parms(DX,LC,Exs,P0,P1) :-
   total_error(Exs,P0,0,Err0),
   writeln(['Error = ', Err0]),
   update_each(P0,P0,Err0,Exs,DX,LC,P1).

% update_each(PR,P0,Err0,Exs,DX,LC,P1)
% updates each parameter in PR.
% P0 is the initial parameter setting with error Err0
% Exs is a list of all of the examples
% DX is the increment to evaluate derivatives
% LC is the learning constant for gradient descent
% P1 is the updated parameter settings
update_each([],_,_,_,_,_,[]).
update_each([val(P,V)|RPs],P0,Err0,Exs,DX,LC,
                                  [val(P,NV)|NPs]) :-
   V1 is V+DX,
   total_error(Exs,[val(P,V1)|P0],0,Nerr),
   NV is V+LC*(Err0-Nerr)/DX,
   update_each(RPs,P0,Err0,Exs,DX,LC,NPs).

% total_error(Exs,P,Err0,Err1)
% P is a list of parameter settings
% Exs is a list of examples
% Err1 is Err0 plus the sum of the errors on examples
total_error([],_,Err,Err).
total_error([Ex|Exs],P,PErr,TErr) :-
   pprove(error(Ex,Err),P),!,
   NErr is PErr+Err,
   total_error(Exs,P,NErr,TErr).


% '<-' is the object-level 'if'
:- op(1150, xfx, <- ).
% '&' is the object level conjunction.
:- op(950,xfy, &).

% pprove(G,P) means prove G with parameter settings P.
% P is a list of the value assignments of the
% form val(P,V)

pprove(true,_) :- !.
pprove((A & B),P) :-
   !,
   pprove(A,P),
   pprove(B,P).
```

```
pprove((A is E),P) :-
   !,
   eval(E,A,P).
pprove(H,P) :-
   (H <- B),
   pprove(B,P).

% eval(E,V,P) true if expression E has value V
% given parameter settings P
eval((A+B),S,P) :-
   !,
   eval(A,VA,P),
   eval(B,VB,P),
   S is VA+VB.
eval((A*B),S,P) :-
   !,
   eval(A,VA,P),
   eval(B,VB,P),
   S is VA*VB.
eval((A-B),S,P) :-
   !,
   eval(A,VA,P),
   eval(B,VB,P),
   S is VA-VB.
eval((A/B),S,P) :-
   !,
   eval(A,VA,P),
   eval(B,VB,P),
   S is VA/VB.
eval(sigmoid(E),V,P) :-
   !,
   eval(E,EV,P),
   V is 1/(1+exp(-EV)).
eval(N,N,_) :-
   number(N),!.
eval(N,V,P) :-
   member(val(N,V1),P),!,V=V1.
       % it only sees the first value for the parameter
eval(N,V,_) :-
   writeln(['Arithmetic Error: ',V,' is ',N]),fail.
```

Using the network of Example 11.5 (page 410), you can define the error for each example as

```
error(Obj,E) <-
   predicted_prop(Obj,reads,VC) &
```

```
        prop(Obj,reads,V) &
        E is (VC-V)*(VC-V).
```

Here `prop` gives the actual value for the example, and `predicted_prop` gives the predicted value for the example.

An example query is:

```
| ?- nnlearn(50,0.01,1.0,
        [e1,e2,e3,e4,e5,e6,e7,e8,e9,e10,e11,e12,e13,e14],
        [val(w_0,0.2), val(w_1,0.12), val(w_2,0.112),
         val(w_3,0.22), val(w_4,0.23), val(w_5,0.26),
         val(w_6,0.27), val(w_7,0.211), val(w_8,0.245),
         val(w_9,0.152), val(w_10,0.102)],
     P1).
```

The `P1` returned is the tuned parameter settings.

C.5   Partial-Order Planner

In this section we axiomatize a partial-order planner (page 309) and apply it to a representation of the delivery robot domain (page 284).

The following gives an axiomatization of the delivery robot domain using what is essentially the STRIPS representation.

```
% DEFINITION OF DELIVERY ROBOT WORLD IN STRIPS NOTATION

% ACTIONS
% move(Ag,Pos,Pos_1) is the action: Ag moves from Pos to Pos_1

poss(move(Ag,Pos,Pos_1)) <-
    autonomous(Ag) &
    adjacent(Pos,Pos_1) &
    sitting_at(Ag,Pos).
achieves(move(Ag,_,Pos_1),sitting_at(Ag,Pos_1)).
deletes(move(Ag,Pos,_),sitting_at(Ag,Pos)).

% pickup(Ag,Obj,Pos) is the action: agent Ag picks up Obj
% when they are both at position Pos.
poss(pickup(Ag,Obj,Pos)) <-
    autonomous(Ag) &
    Ag \= Obj &
    sitting_at(Obj,Pos) &
    at(Ag,Pos).
```

```
achieves(pickup(Ag,Obj,Pos), carrying(Ag,Obj)).
deletes(pickup(Ag,Obj,Pos), sitting_at(Obj,Pos)).

% putdown(Ag,Obj,Pos) is the action: Ag puts down Obj at Pos
poss(putdown(Ag,Obj,Pos)) <-
    autonomous(Ag) &
    Ag \= Obj &
    at(Ag,Pos) &
    carrying(Ag,Obj).
achieves(putdown(Ag,Obj,Pos),sitting_at(Obj,Pos)).
deletes(putdown(Ag,Obj,Pos),carrying(Ag,Obj)).

% unlock(Ag,Door) is the action: agent Ag unlocks Door
poss(unlock(Ag,Door)) <-
    autonomous(Ag) &
    blocks(Door,P_1,_) &
    opens(Key,Door) &
    carrying(Ag,Key) &
    at(Ag,P_1).
achieves(unlock(Ag,Door),unlocked(Door)).

% PRIMITIVE RELATIONS
primitive(carrying(_,_)).
primitive(sitting_at(_,_)).
primitive(unlocked(_)).

% DERIVED RELATIONS

at(Obj,Pos) <-
    sitting_at(Obj,Pos).
at(Obj,Pos) <-
    autonomous(Ag) &
    Ag \= Obj &
    carrying(Ag,Obj) &
    at(Ag,Pos).

adjacent(o109,o103) <- true.
adjacent(o103,o109) <- true.
adjacent(o109,storage) <- true.
adjacent(storage,o109) <- true.
adjacent(o109,o111) <- true.
adjacent(o111,o109) <- true.
adjacent(o103,mail) <- true.
adjacent(mail,o103) <- true.
adjacent(lab2,o109) <- true.
adjacent(P_1,P_2) <-
```

```
      blocks(Door,P_1,P_2) &
      unlocked(Door).
blocks(door1,o103,lab2) <- true.
opens(k1,door1) <- true.
autonomous(rob) <- true.

% INITIAL SITUATION
holds(sitting_at(rob,o109),init).
holds(sitting_at(parcel,storage),init).
holds(sitting_at(k1,mail),init).

achieves(init,X) :-
   holds(X,init).
```

The following planner implements a partial-order planner that uses the above representation.

```
% PARTIAL ORDER PLANNER using STRIPS REPRESENTATION
% This assumes the underlying Prolog contains delaying
%     (in particular the "when" construct).

% Definitions:
% An agenda is a list of goals.
% A goal is of the form goal(P,NA) where P is a proposition
%    that is a precondition of action instance index NA.
% A plan is of the form plan(As,Os,Ls) where
%    As is a list of actions instances,
%       An action instance is of the form act(N,A)
%          where N is an integer and A is an action
%          (this is needed in case there are multiple
%          instances of the same action).
%          N is the action instance index.
%    Os is a list of ordering constraints (A1<A2)
%       where A1 & A2 are action instance indexes.
%    Ls is a list of causal links.
% A causal link is of the form cl(A0,P,A1)
%    where A0 and A1 are action instance indexes,
%    and P is a proposition that is a precondition of
%    action A1 --- means that A0 is making P true for A1.

% '<-' is the object-level 'if'
:- op(1150, xfx, <- ).
% '&' is the object level conjunction.
:- op(950,xfy, &).
% '\=' is the object level not equal.
:- op(700,xfx, \=).
```

```
% pop(CPlan, Agenda, FPlan,DB) is true if
%  CPlan is the current plan
%  Agenda is the current agenda
%  FPlan is the final plan (with all subgoals supported)
%  and there are DB or fewer actions in the plan
%          -- DB is the depth-bound for the plan size.

pop(Plan,[],Plan,_).
pop(CPlan,Agenda,FPlan,DB) :-
   select(Goal,Agenda,Agenda1),
   solve_goal(Goal,CPlan,NPlan,Agenda1,NAgenda,DB,NDB),
   pop(NPlan,NAgenda,FPlan,NDB).

% select(Goal,Agenda,NewAgenda) is true if
%    Goal is selected from Agenda with
%    NewAgenda the remaining elements.
select(goal(G,GG),[goal(G,GG)|A],A) :-
   primitive(G),!.
select(goal((X \= Y),GG),[goal((X\=Y),GG)|A],A).
select(P,[goal(G,GG)|R],A) :-
   (G <- B),
   add_to_agenda(B,GG,R,NR),   % agenda is stack
   select(P,NR,A).

add_to_agenda(true,_,R,R) :- !.
add_to_agenda((A&B),GG,R0,R2) :- !,
   add_to_agenda(B,GG,R0,R1),
   add_to_agenda(A,GG,R1,R2).
add_to_agenda(H,GG,R,[goal(H,GG)|R]).

% solve_goal(Goal,CPl,NPl,CAg,NAg,DB,DB1)
%    chooses an action to solve Goal,
%    updating plan CPl to NPl and agenda CAg to NAg,
%    and updating depth bound DB to depth bound DB1

% CASE 0: inequality goal
solve_goal(goal((X \= Y),_),Pl,Pl,Ag,Ag,DB,DB)   :-
   !,
   when(?=(X,Y), X\==Y).

% CASE 1: use existing action
solve_goal(goal(P,A1),
           plan(As,Os,Ls),
           plan(As,NOs,[cl(N0,P,A1)|Ls]),
           Ag,Ag,DB,DB) :-
```

```
        member(act(N0,Act0),As),
        achieves(Act0,P),
        add_constraint(N0<A1,Os,Os1) ,
        incorporate_causal_link(cl(N0,P,A1),As,Os1,NOs).

% CASE 2: add new action.
%    Note that DB acts as the index of the new action instance.
solve_goal(goal(P,A1),
           plan(As,Os,Ls),
           plan([act(DB,Act0)|As],NOs,[cl(DB,P,A1)|Ls]),
           Ag,Ag1,
           DB,NDB) :-
    DB>0,
    achieves(Act0,P),
    writeln(['*** new action ',act(DB,Act0),
             ' to achieve ',P,' for ',A1]),
    add_constraint(DB<A1,Os,Os1),
    add_constraint(start<DB,Os1,Os2),
    incorporate_action(act(DB,Act0),Ls,Os2,Os3),
    incorporate_causal_link(cl(DB,P,A1),As,Os3,NOs),
    NDB is DB-1,
    append(Ag,[goal(poss(Act0),DB)],Ag1).


% add_constraint(A0<A1,Os,Os1) adds ordering constraint
%    A0<A1 to partial ordering Os producing ordering Os1.
% Fails if A0<A1 is inconsistent with Os.
% We represent partial orderings as their transitive closure.
add_constraint(C,L,L) :-
    member(C,L),
    !.                       % don't use other rules
add_constraint(A0<A1,L1,L2) :-
    A0 \== A1,
%    \+ member(A0<A1,L1),
                 % omitted because of the cut.
    \+ member(A1<A0,L1),
    add_constraint1(A0<A1,L1,L1,L2).

% add_constraint1(A0<A1,Os,AOs,Os1) adds constraint A0<A1
%    Os is the list of orderings to be checked
%    AOs is the list of all orderings
%    Os1 is the final set of orderings
add_constraint1(A0<A1,[],AOs,NOs) :-
    insert(A0<A1,AOs,NOs).
add_constraint1(A0<A1,[A1<A2|R],AOs,NR) :-
    A0 \== A2,
```

```
        insert(A0<A2,AOs,AOs1),
        add_constraint1(A0<A1,R,AOs1,NR).
add_constraint1(A0<A1,[A2<A0|R],AOs,NR) :-
     A1 \== A2,
        insert(A2<A1,AOs,AOs1),
        add_constraint1(A0<A1,R,AOs1,NR).
add_constraint1(A0<A1,[A2<A3|R],AOs,NR) :-
     A0 \== A3,
     A1 \== A2,
        add_constraint1(A0<A1,R,AOs,NR).


% incorporate_causal_link(CL, As, Os, NOs)
%   incorporates causal link CL to links Ls
%   producing new links NLs, and updating
%   partial ordering Os to NOs.
incorporate_causal_link(_,[],Os,Os).
incorporate_causal_link(CL,[A|RAs],Os,NOs) :-
     protect(CL,A,Os,Os1),
        incorporate_causal_link(CL,RAs,Os1,NOs).


% incorporate_action(A,Ls,Os,NOs)
% incorporates action A into list Ls is causal lists
% updates partial ordering Os to NOs
% fails if the action cannot be reordered
incorporate_action(_,[],Os,Os).
incorporate_action(A,[CL|Ls],Os,NOs) :-
     protect(CL,A,Os,Os1),
        incorporate_action(A,Ls,Os1,NOs).


% protect(Cl,Action,Os0,Os1) protects
%   causal link CL from Action if necessary
protect(cl(A0,_,_),act(A0,_),Os,Os) :- !.
protect(cl(A0,_,_),act(NA,_),Os,Os) :-
     member(NA<A0,Os),!.
protect(cl(_,_,A1),act(A1,_),Os,Os) :- !.
protect(cl(_,_,A1),act(NA,_),Os,Os) :-
     member(A1<NA,Os), !.
protect(cl(_,P,_),act(_,A),Os,Os) :-
%  NA\==A0, \+ member(NA<A0,Os), NA\==A1, \+ member(A1<NA,Os),
        % deleted because of cuts
     when(ground((A,P)), \+ deletes(A,P)).
protect(cl(A0,P,A1),act(NA,A),Os,Os1) :-
%  NA\==A0, \+ member(NA<A0,Os), NA\==A1, \+ member(A1<NA,Os),
        % deleted because of cuts
     deletes(A,P),
     enforce_order(A0,NA,A1,Os,Os1).
```

```
%enforce_order(A0,A,A1,Os,Os1) extends Os to
% ensure that A<A0 or A>A1 in Os1
enforce_order(_,A,A1,Os,Os1) :-
    add_constraint(A1<A,Os,Os1),
    writeln(['   ... adding constraint ',A,' after ',A1]).
enforce_order(A0,A,_,Os,Os1) :-
    add_constraint(A<A0,Os,Os1),
    writeln(['   ... adding constraint ',A,' before ',A0]).

% solve(Goals,Plan,DB) is true if Plan is a partial-order plan
% to solve Goals with fewer then DB steps
solve(Goals,Plan,DB) :-
    add_to_agenda(Goals,finish,[],Ag),
    pop(plan([act(finish,end),act(start,init)],
              [start<finish],[]),
        Ag,Plan,DB).

% seq(Plan,Seq) extracts a legal sequence Seq from Plan
seq(plan([],_,_),[]).
seq(plan(As,Os,_),[A|P]) :-
    remove(act(N,A),As,As1),
    \+ (member(act(N1,_),As1), member(N1<N,Os)),
    seq(plan(As1,Os,_),P).

% TRY THE FOLLOWING QUERIES with pop_t.pl:
% ? solve(carrying(rob,k1),P,3), seq(P,S).
% ? solve(sitting_at(k1,lab2),P,7), seq(P,S).
% ? solve((carrying(rob,parcel) & sitting_at(rob,lab2)),P,9),
%       seq(P,S).
% ? solve((sitting_at(rob,lab2) & carrying(rob,parcel)),P,9),
%       seq(P,S).
```

C.6   Implementing Belief Networks

Belief networks (page 363) are a representation of independence for reasoning under uncertainty.

The general problem that we consider is to compute

$$P(x|e_1 \wedge \cdots \wedge e_k),$$

where each e_i is evidence in the form *variable* = *value*, and x is a variable. Thus we find a posterior distribution of variable x based on conjunctive evidence.

The following code is based the structure-exploiting algorithm for belief network inference (page 376):

```
% BELIEF NETWORK INTERPRETER

% A belief network is represented with the relations
% variables(Xs) Xs is the list of random variables.
%    Xs is ordered: parents of node are before the node.
% parents(X,Ps) Ps list of parents of variable X.
%    Ps is ordered consistently with Xs
% values(X,Vs) Vs is the list of values of X
% pr(X,As,D) X is a variable, As is a list of Pi=Vi where
%    Pi is a parent of X, and Vi is a value for variable Pi
%    The elements of As are ordered consistently with Ps.


% p(Var,Obs,Dist) is true if Dist represents the
% probability distribution of P(Var|Obs)
% where Obs is a list of Vari=Vali. Var is not observed.
p(Var,Obs,VDist) :-
    relevant(Var,Obs,RelVars),
    to_sum_out(RelVars,Var,Obs,SO),
    joint(RelVars,Obs,Joint),
    sum_out_each(SO,Joint,Dist),
    collect(Dist,DT0),
    normalize(DT0,0,_,VDist).

% relevant(Var,Obs,RelVars) Relvars is the relevant
% variables given query Var and observations Obs.
% This is the most conservative.
relevant(_,_,Vs) :-
    variables(Vs).          % all variables are relevant

% to_sum_out(Vs,Var,Obs,SO),
%    Given all variables Vs, query variable Var
% and observations Obs, SO specifies the elimination
% ordering. Here, naively, the elimination ordering
% is the same as variable ordering
to_sum_out(Vs,Var,Obs,SO) :-
    remove(Var,Vs,RVs),
    remove_each_obs(Obs,RVs,SO).

% remove_each_obs(Obs,RVs,SO) removes each of the
% observation variables from RVs resulting in SO.
remove_each_obs([],SO,SO).
remove_each_obs([X=_|Os],Vs0,SO) :-
    remove_if_present(X,Vs0,Vs1),
```

```
      remove_each_obs(Os,Vs1,SO).

/* A joint probability distribution is represented
as a list of distribution trees, of the form
          dtree(Vars,DTree)
where Vars is a list of Variables (ordered
consistently with the ordering of variables), and
DTree is tree representation for the function from
values of variables into numbers such that if
Vars=[] then DTree is a number. Otherwise
Vars=[Var|RVars], and DTree is a list with one
element for each value of Var, and each element
is a tree representation for RVars. The ordering
of the elements in DTree is given by the ordering
of Vals given by values(Var,Vals). */

% joint(Vs,Obs,Joint) Vs is a list of variables,
% Obs is an observation list returns a list of
% dtrees that takes the observations into account.
% There is a dtree for each non-observed variable.
joint([],_,[]).
joint([X|Xs],Obs,[dtree(DVars,DTree)|JXs]) :-
   parents(X,PX),
   make_dvars(PX,X,Obs,DVars),
   DVars \== [], !,
   make_dtree(PX,X,Obs,[],DTree),
   joint(Xs,Obs,JXs).
joint([_|Xs],Obs,JXs) :-
          % we remove any dtree with no variables
   joint(Xs,Obs,JXs).

% make_dvars(PX,X,Obs,DVars)
% where X is a variable and PX are the parents of
% X and Obs is observation list returns
% DVars = {X} U PX - observed variables
% This relies on PX ordered before X
make_dvars([],X,Obs,[]) :-
   member(X=_,Obs),!.
make_dvars([],X,_,[X]).
make_dvars([V|R],X,Obs,DVs) :-
   member(V=_,Obs),!,
   make_dvars(R,X,Obs,DVs).
make_dvars([V|R],X,Obs,[V|DVs]) :-
   % \+member(V=_,Obs),
   make_dvars(R,X,Obs,DVs).
```

```
% make_dtree(RP,X,Obs,Con,Dtree) constructs a factor
% corresponding to p(X|PX). RP is list of remaining
% parents of X, Obs is the observations, Con is a
% context of assignments to previous (in the
% variable ordering) parents of X - in reverse order
% to the variable assignment, returns DTree as the
% dtree corresponding to values of RP.
make_dtree([],X,Obs,Con,DX) :-
   member(X=OVal,Obs),!,
   reverse(Con,RCon),
   pr(X,RCon,DXPr),
   values(X,Vals),
   select_corresp_elt(Vals,OVal,DXPr,DX).
make_dtree([],X,_,Con,DX) :-
   reverse(Con,RCon),
   pr(X,RCon,DX).
make_dtree([P|RP],X,Obs,Con,DX) :-
   member(P=Val,Obs),!,
   make_dtree(RP,X,Obs,[P=Val|Con],DX).
make_dtree([P|RP],X,Obs,Con,DX) :-
   values(P,Vals),
   make_dtree_for_vals(Vals,P,RP,X,Obs,Con,DX).


% make_dtree_for_vals(Vals,P,RP,X,Obs,Con,DX).
%  makes a DTree for each value in Vals, and
% collected them into DX.  Other variables are as
% for make_dtree.
make_dtree_for_vals([],_,_,_,_,_,[]).
make_dtree_for_vals([Val|Vals],P,RP,X,Obs,Con,[ST|DX]):-
   make_dtree(RP,X,Obs,[P=Val|Con],ST),
   make_dtree_for_vals(Vals,P,RP,X,Obs,Con,DX).


% select_corresp_elt(Vals,Val,List,Elt) is true
% if Elt is at the same position in List as Val is
% in list Vals. Assumes Vals, Val, List are bound.
select_corresp_elt([Val|_],Val,[Elt|_],Elt) :-
   !.
select_corresp_elt([_|Vals],Val,[_|Rest],Elt) :-
   select_corresp_elt(Vals,Val,Rest,Elt).


% sum_out_each(SO,Joint0,Joint1) is true if
% Joint1 is a distribution Joint0 with each
% variable in SO summed out
sum_out_each([],J,J).
sum_out_each([X|Xs],J0,J2) :-
   sum_out(X,J0,J1),
```

```
            sum_out_each(Xs,J1,J2).

    % sum_out_each(V,J0,J1) is true if
    % Joint1 is a distribution Joint0 with
    % variable V summed out.
    sum_out(X,J0,[dtree(CVars1,CTree)|NoX]) :-
       partition(J0,X,NoX,SomeX),
       variables(AllVars),
       find_tree_vars(SomeX,AllVars,CVars),
       remove(X,CVars,CVars1),
       CVars1 \== [], !,
       create_tree(CVars1,CVars1,SomeX,X,[],CTree).
    sum_out(X,J0,NoX) :-
          % remove any dtrees that have no variables
       partition(J0,X,NoX,_).

    % partition(J0,X,NoX,SomeX) partitions J0 into
    % those dtrees that contain variable X (SomeX) and
    % those that do not contain X (NoX)
    partition([],_,[],[]).
    partition([dtree(Vs,Di)|R],X,NoX,[dtree(Vs,Di)|SomeX]) :-
       member(X,Vs),
       !,
       partition(R,X,NoX,SomeX).
    partition([dtree(Vs,Di)|R],X,[dtree(Vs,Di)|NoX],SomeX) :-
       partition(R,X,NoX,SomeX).

    % find_tree_vars(SomeX,AllVars,CVars) is true
    % if CVars is the set of variables that appear in
    % some dtree in SomeX, ordered according to AllVars
    find_tree_vars([],_,[]).
    find_tree_vars([dtree(Vs,_)|RDs],All,Res) :-
        find_tree_vars(RDs,All,Cvars0),
        ordered_union(Vs,Cvars0,Res,All).

    % create_tree(CVars,Vars,SomeX,X,Context,CTree)
    % CTree is the tree corresponding to variables CVars.
    % The values of the leaves of the tree are obtained
    % by multiplying the corresponding values in SomeX.
    create_tree([],Vars,SomeX,X,Context,Num) :-
       reverse(Context,CVals),
       values(X,Vals),
       sum_vals(Vals,X,Vars,CVals,SomeX,0,Num).
    create_tree([Var|CVars],Vars,SomeX,X,Context,CTree) :-
       values(Var,Vals),
       create_tree_vals(Vals,CVars,Vars,SomeX,X,Context,CTree).
```

```
% create_tree_vals(Vals,CVars,Vars,SomeX,X,Context,CTree).
% creates a tree for each value in Vals.
create_tree_vals([],_,_,_,_,_,[]).
create_tree_vals([Val|Vals],CVars,Vars,
                  SomeX,X,Context,[SubTr|CTree]) :-
   create_tree(CVars,Vars,SomeX,X,[Val|Context],SubTr),
   create_tree_vals(Vals,CVars,Vars,SomeX,X,Context,CTree).


% sum_vals(Vals,X,Vars,CVals,SomeX,Acc,Sum).
% sums out X in the context Vars=CVals
% Vals is the remaining set of values to be added
% SomeX is the factors that need to be multiplied
sum_vals([],_,_,_,_,S,S).
sum_vals([Val|Vals],X,Vars,CVals,SomeX,S0,Sum) :-
   mult_vals(SomeX,Val,X,Vars,CVals,1,Prod),
   S1 is S0+Prod,
   sum_vals(Vals,X,Vars,CVals,SomeX,S1,Sum).


% mult_vals(SomeX,Val,X,Vars,CVals,Acc,Prod),
% computes product of SomeX factors given X=Val, Vars=CVals
mult_vals([],_,_,_,_,P,P).
mult_vals([Tree|SomeX],Val,X,Vars,CVals,P0,Prod) :-
   lookup(X,Val,Vars,CVals,Tree,ContextVal),
   P1 is P0*ContextVal,
   mult_vals(SomeX,Val,X,Vars,CVals,P1,Prod).



% lookup(Var0,Val0,Vars,Vals,dtree(DVars,DTree),Prob)
% DVars is a subset of Vars U {Var}. Returns
% the value Prob by looking up "Var0=Val0 & Vars=Vals"
% in DTree.  It assumes that the elements of Vars
% and TreeVars are ordered consistently.

lookup(_,_,[],[],dtree([],P),P).
lookup(Var0,Val0,[Var|RVars],[Val|RVals],
           dtree([Var|TVars],DTree),Prob) :-
   !,
   values(Var,Vals),
   select_corresp_elt(Vals,Val,DTree,Subtree),
   lookup(Var0,Val0,RVars,RVals,dtree(TVars,Subtree),Prob).
lookup(Var0,Val0,RVars,RVals,dtree([Var0|TVars],DTree),Prob):-
   !,
   values(Var0,Vals),
   select_corresp_elt(Vals,Val0,DTree,Subtree),
   lookup(Var0,Val0,RVars,RVals,dtree(TVars,Subtree),Prob).
```

```
lookup(Var0,Val0,[_|RVars],[_|RVals],DT,Prob) :-
   lookup(Var0,Val0,RVars,RVals,DT,Prob).

% collect(Dist,DT) multiplies all of the factors together
% forming a DTRee. This assumes that all of the factors
% contain just the query variable
collect([dtree(_,DT)],DT) :- !.
collect([dtree(_,DT0)|R],DT2) :-
   collect(R ,DT1),
   multiply_corresp_elts(DT0,DT1,DT2).

% multiply_corresp_elts(DT0,DT1,DT2) DT2 is the dot
% product of DT0 and DT1
multiply_corresp_elts([],[],[]).
multiply_corresp_elts([E0|L0],[E1|L1],[E2|L2]) :-
   E2 is E0*E1,
   multiply_corresp_elts(L0,L1,L2).

% normalize(List,CumVal,Sum,NList) makes NList
% the same a list, but where elements sum to 1.
% Sum is the sum of all of the list, and CumVal
% is the accumulated sum to this point.
normalize([],S,S,[]).
normalize([A|L],CV,Sum,[AN|LN]) :-
   CV1 is CV + A,
   normalize(L,CV1,Sum,LN),
   AN is A/Sum.

%  ordered_union(L0,L1,R,RL) is true if R = L0 U L1, where RL
%  is a reference list that provides the ordering of elements.
%  L0, L1, RL must all be bound.
ordered_union([],L,L,_) :- !.
ordered_union(L,[],L,_) :- !.
ordered_union([E|L0],[E|L1],[E|R],[E|RL]) :-
   !,
   ordered_union(L0,L1,R,RL).
ordered_union([E|L0],L1,[E|R],[E|RL]) :-
   !,
   ordered_union(L0,L1,R,RL).
ordered_union(L0,[E|L1],[E|R],[E|RL]) :-
   !,
   ordered_union(L0,L1,R,RL).
ordered_union(L0,L1,R,[_|RL]) :-
   !,
   ordered_union(L0,L1,R,RL).
```

```
% STANDARD DEFINITIONS
% reverse(L,R) is true if R contains same elements
% as list L, in reverse order
reverse(L,R) :-
   rev(L,[],R).
rev([],R,R).
rev([H|T],Acc,R) :-
   rev(T,[H|Acc],R).

% remove(E,L,R) true if R is the list L with
% one occurrence of E removed
remove(E,[E|L],L).
remove(E,[A|L],[A|R]) :-
   remove(E,L,R).

% remove_if_present(E,L,R) true if R is the list
% L with one occurrence of E removed
remove_if_present(_,[],[]).
remove_if_present(E,[E|L],L) :- !.
remove_if_present(E,[A|L],[A|R]) :-
   remove_if_present(E,L,R).

% member(E,L) is true if E is a member of list L
member(A,[A|_]).
member(A,[_|L]) :-
   member(A,L).
```

The following is a representation of Example 10.16 (page 370).

```
% list of all variables, ordered so the parents of a node
% are before the node.
variables([tampering, fire, smoke, alarm, leaving, report]).

% Structure of the graph
parents(report,[leaving]).
parents(leaving,[alarm]).
parents(alarm,[tampering,fire]).
parents(smoke,[fire]).
parents(tampering,[]).
parents(fire,[]).

% values for variables
values(report,[yes,no]).
values(leaving,[yes,no]).
values(alarm,[yes,no]).
values(tampering,[yes,no]).
values(fire,[yes,no]).
```

```
values(smoke,[yes,no]).

% conditional probabilities
pr(report,[leaving=yes],[0.75,0.25]).
pr(report,[leaving=no],[0.01,0.99]).

pr(leaving,[alarm=yes],[0.88,0.12]).
pr(leaving,[alarm=no],[0.001,0.999]).

pr(alarm,[tampering=yes,fire=yes],[0.5,0.5]).
pr(alarm,[tampering=yes,fire=no], [0.85,0.15]).
pr(alarm,[tampering=no, fire=yes],[0.99,0.01]).
pr(alarm,[tampering=no, fire=no], [0.0001,0.9999]).

pr(smoke,[fire=yes],[0.9,0.1]).
pr(smoke,[fire=no], [0.01,0.99]).

pr(tampering,[],[0.02,0.98]).

pr(fire,[],[0.01,0.99]).

% EXAMPLE QUERIES:
% ? p(fire,[report=yes],P).
% ? p(fire,[report=yes,smoke=yes],P).
% ? p(fire,[report=yes,smoke=no],P).
% ? p(report,[],P).
% ? p(report,[smoke=yes],P).
% ? p(report,[smoke=yes,tampering=no],P).
```

C.7 Robot Controller

In this section we present an efficient implementation of the robot controller based
on the event calculus (page 452). In particular we show how to exploit the fact that
time is moving forward and at any time all of the previous assigns have been asserted
into the database in order. Thus the last value assigned to a value is the first value
asserted in the database. This isn't correct if you wanted to make arbitrary queries
about the past, but is appropriate for the robot controller. The top assigned values
represent the *state* of the robot that's being maintained.

```
:- op(1110,xfx,<-).  % object-level "IF"
:- op(1000,xfy,&).   % object-level "AND"
:- op(950,fy,~).     % object-level "NOT"
```

```prolog
:- op(700,xfx,~=).    % object-level "NOT EQUALS"

% assigned(Fl,Val,T) is true if fluent FL was
% assigned value Val at time T. We exploit the fact
% that the simulation runs forward and make sure that
% the first fact in the database always represents
% the latest time that the fluent was assigned a
% value. This means that we can always look up the
% last value assigned to a fluent quickly.
:- dynamic(assigned/3).

% tried(Fl) is true if fluent Fl has been tried to determine
% if it should be assigned a value for the current time.
:- dynamic(tried/1).

% sim(T0,T2,DT) means simulate the system for all
% times in range [T0,T2] in increments of DT.
sim(T0,T2,DT) :-
    T0 =< T2,
    !,
    prove_all_assigns(T0),
    view_all(T0),
    retractall(tried(_)),
    T1 is T0+DT,
    sim(T1,T2,DT).
sim(_,_,_).

% clear clears the database for another simulation
clear :-
    retractall(assigned(_,_,_)).

% prove(G,T) is true if G can be proved at time T, but where
% special care is taken to remember state (assigned
% values) rather than recomputing.
prove(true,_) :- !.
prove((A & B),T) :- !,
    prove(A,T),
    prove(B,T).
prove((A ; B),T) :- !,
    (prove(A,T);
    prove(B,T)).
prove(val(Fl,Val,T),T) :-
    \+ tried(Fl),
    asserta(tried(Fl)),
    prove(assign(Fl,V1,T),T),
    asserta(assigned(Fl,V1,T)),
```

```
      !,
      Val=V1.
prove(val(Fl,Val,T),_) :-
         % either tried or can't currently be assigned
         % look up latest value
      assigned(Fl,V1,T1),
      T1 =< T,
      !,
      Val=V1 .
prove(was(Fl,Val,T1,T),_) :-
      assigned(Fl,V1,T1),
      T1 < T, !,
      Val=V1.
prove((A ~= B),_) :-
      \+ (A = B).
prove((~ G),T) :-!,
      \+ prove(G,T).
prove(G,_) :-
      builtin(G),
      !,
      call(G).
prove(H,T) :-
      (H <- B),
      prove(B,T).

% builtin(G) is true if G is built-in
builtin((_ =< _)).
builtin((_ >= _)).
builtin((_ = _)).
builtin((_ < _)).
builtin((_ > _)).
builtin((_ is _)).

% prove_all_assigns(T) is true if all assignments
% of values to variables are proved and remembered
% for time T
prove_all_assigns(T) :-
      fluent(Fl),
      \+ tried(Fl),
      asserta(tried(Fl)),
      prove(assign(Fl,Val,T),T),
      asserta(assigned(Fl,Val,T)),
      fail.
prove_all_assigns(_).

fluent(Fl) :-
```

```
   (assign(Fl,_,_) <- _).

% view_all(T) lets us print out all of the "view"
% variables for time T. This lets us monitor the
% simulation.  view(G,T,P) is true if the elements
% of list P should be printed when G is proved at
% time T
view_all(T) :-
   view(G,T,P),
   prove(G,T),
   writeln(P),
   fail.
view_all(_).
```

Bibliography

Abelson, H. & DiSessa, A. (1981). *Turtle Geometry: The Computer as a Medium for Exploring Mathematics*, MIT Press, Cambridge, MA.

Abramson, H. & Dahl, V. (1989). *Logic Grammars*, Symbolic Computation, Springer-Verlag, New York, NY.

Abramson, H. & Rogers, M. H. (eds) (1989). *Meta-Programming in Logic Programming*, MIT Press, Cambridge, MA.

Agre, P. E. (1995). Computational research on interaction and agency, *Artificial Intelligence* **72**: 1–52.

Albus, J. S. (1981). *Brains, Behavior and Robotics*, BYTE Publications, Peterborough, NH.

Allen, J. F. (1994). *Natural Language Understanding*, second edition, Benjamin Cummings, Redwood City, CA.

Allen, J., Hendler, J. & Tate, A. (eds) (1990). *Readings in Planning*, Morgan Kaufmann, San Mateo, CA.

Apt, K. & Bol, R. (1994). Logic programming and negation: A survey, *Journal of Logic Programming* **19,20**: 9–71.

Asimov, I. (1950). *I, Robot*, Doubleday, Garden City, NY.

Bacchus, F., Grove, A. J., Halpern, J. Y. & Koller, D. (1996). From statistical knowledge bases to degrees of belief, *Artificial Intelligence* **87**(1-2): 75–143.

Bacchus, F. & Kabanza, F. (1996). Using temporal logic to control search in a forward chaining planner, *in* M. Ghallab & A. Milani (eds), *New Directions in AI Planning*, ISO Press, Amsterdam, pp. 141–153.

Bayes, T. (1763). An essay towards solving a problem in the doctrine of chances, *Philosophical Transactions of the Royal Society of London* **53**: 370–418. Reprinted in *Biometrika* 45, 298–315, 1958. Reprinted in S. J. Press, *Bayesian Statistics*, 189–217, Wiley, New York, 1989.

Bell, J. L. & Machover, M. (1977). *A Course in Mathematical Logic*, North-Holland, Amsterdam.

Bishop, C. M. (1995). *Neural Networks for Pattern Recognition*, Oxford University Press, Oxford, England.

Blum, A. & Furst, M. L. (1995). Fast planning through planning graph analysis, *Proc. 14th International Joint Conf. on Artificial Intelligence (IJCAI-95)*, Montréal, Québec, pp. 1636–1642.

Bobrow, D. G. (1993). Artificial intelligence in perspective: a retrospective on fifty volumes of Artificial Intelligence, *Artificial Intelligence* **59**: 5–20.

Bobrow, D. G. & Winograd, T. (1977). An overview of KRL, a knowledge representation language, *Cognitive Science* **1**(1): 3–46. Reprinted in Brachman & Levesque (1985).

Boddy, M. & Dean, T. L. (1994). Deliberation scheduling for problem solving in time-constrained environments, *Artificial Intelligence* **67**(2): 245–285.

Bondarenko, A., Dung, P., Kowalski, R. & Toni, F. (1997). An abstract, argument-theoretic approach to default reasoning, *Artificial Intelligence* **93**(1–2): 63–101.

Boutilier, C. (1994). Conditional logics of normality: A modal approach, *Artificial Intelligence* **68**: 87–154.

Bowen, K. A. (1985). Meta-level programming and knowledge representation, *New Generation Computing* **3**(4): 359–383.

Brachman, R. J. (1979). On the epistemological status of semantic networks, *in* N. V. Findler (ed.), *Associative Networks: Representation and Use of Knowledge by Computers*, Academic Press, New York, pp. 3–50. Reprinted in Brachman & Levesque (1985).

Brachman, R. J. & Levesque, H. J. (eds) (1985). *Readings in Knowledge Representation*, Morgan Kaufmann, San Mateo, CA.

Bratko, I. (1990). *Prolog Programming for Artificial Intelligence*, second edition, Addison-Wesley, Wokingham, England.

Breiman, L., Friedman, J. H., Olshen, R. A. & Stone, C. J. (1984). *Classification and Regression Trees*, Wadsworth and Brooks, Monterey, CA.

Briscoe, G. & Caelli, T. (1996). *A Compendium of Machine Learning, Volume 1: Symbolic Machine Learning*, Ablex, Norwood, NJ.

Brooks, R. A. (1986). A robust layered control system for a mobile robot, *IEEE Journal of Robotics and Automation* **2**(1): 14–23. Reprinted in Shafer & Pearl (1990).

Brooks, R. A. (1991a). Intelligence without reason, *Proc. 12th International Joint Conf. on Artificial Intelligence (IJCAI-91)*, Sydney, Australia, pp. 569–595.

Brooks, R. A. (1991b). Intelligence without representation, *Artificial Intelligence* **47**: 139–159.

Buchanann, B. & Shortliffe, E. (eds) (1984). *Rule-Based Expert Systems: The MYCIN Experiments of the Stanford Heuristic Programming Project*, Addison-Wesley, Reading, MA.

Buntine, W. (1992). Learning classification trees, *Statistics and Computing* **2**: 63–73.

Castillo, E., Gutiérrez, J. M. & Hadi, A. S. (1996). *Expert Systems and Probabilistic Network Models*, Springer Verlag, New York.

Chang, C. L. & Lee, R. C. T. (1973). *Symbolic Logical and Mechanical Theorem Proving*, Academic Press, New York.

Chapman, D. (1987). Planning for conjunctive goals, *Artificial Intelligence* **32**(3): 333–377.

Charniak, E. & McDermott, D. (1985). *Introduction to Artificial Intelligence*, Addison-Wesley, Reading, MA.

Cheeseman, P. (1990). On finding the most probable model, *in* J. Shranger & P. Langley (eds), *Computational Models of Scientific Discovery and Theory Formation*, Morgan Kaufmann, San Mateo, CA, chapter 3, pp. 73–95.

Cheeseman, P., Kelly, J., Self, M., Stutz, J., Taylor, W. & Freeman, D. (1988). Autoclass: A Bayesian classification system, *Proc. Fifth International Conference on Machine Learning*, Ann Arbor, MI, pp. 54–64. Reprinted in Shavlik & Dietterich (1990).

Clark, K. L. (1978). Negation as failure, *in* H. Gallaire & J. Minker (eds), *Logic and Databases*, Plenum Press, New York, pp. 293–322.

Colmerauer, A., Kanoui, H., Roussel, P. & Pasero, R. (1973). Un systeme de communication homme-machine en français, *Technical report*, Groupe de Researche en Intelligence Artificielle, Université d'Aix-Marseille.

Console, L., Theseider Dupré, D. & Torasso, P. (1991). On the relationship between abduction and deduction, *Journal of Logic and Computation* **1**(5): 661–690.

Copi, I. M. (1982). *Introduction to Logic*, sixth edition, Macmillan, New York.

Cover, T. M. & Thomas, J. A. (1991). *Elements of information theory*, Wiley, New York.

Dahl, V. (1994). Natural language processing and logic programming, *Journal of Logic Programming* **19,20**: 681–714.

Dasarathy, B. V. (1991). NN concepts and techniques, *in* B. V. Dasarathy (ed.), *Nearest Neighbour (NN) Norms: NN Pattern Classification Techniques*, IEEE Computer Society Press, New York, pp. 1–30.

Davis, E. (1990). *Representations of Commonsense Knowledge*, Morgan Kaufmann, San Mateo, CA.

Davis, R. & Hamscher, W. (1988). Model-based reasoning: Troubleshooting, *in* H. E. Shrobe (ed.), *Exploring Artificial Intelligence*, Morgan Kaufmann, San Mateo, CA, chapter 8, pp. 297–346.

de Kleer, J. (1986). An assumption-based TMS, *Artificial Intelligence* **28**(2): 127–162.

de Kleer, J. & Williams, B. C. (1987). Diagnosing multiple faults, *Artificial Intelligence* **32**(1): 97–130.

Dean, T., Allen, J. & Aloimonos, Y. (1995). *Artificial Intelligence: Theory and Practice*, Benjamin/Cummings, Redwood City, CA.

Dean, T. L. & Wellman, M. P. (1991). *Planning and Control*, Morgan Kaufmann, San Mateo, CA.

Dechter, R. (1996). Bucket elimination: A unifying framework for probabilistic inference, *in* E. Horvitz & F. Jensen (eds), *Proc. Twelfth Conf. on Uncertainty in Artificial Intelligence (UAI-96)*, Portland, OR, pp. 211–219.

DeJong, D. & Mooney, R. (1986). Explanation-based learning: An alternative view, *Machine Learning* **1**: 145–176. Reprinted in Shavlik & Dietterich (1990).

Delgrande, J. (1987). A first-order conditional logic for prototypical properties, *Artificial Intelligence* **33**(1): 105–130.

Dijkstra, E. W. (1976). *A discipline of programming*, Prentice-Hall, Englewood Cliffs, NJ.

Doyle, J. (1979). A truth maintenance system, *AI Memo 521*, MIT Artificial Intelligence Laboratory.

Druzdzel, M. & Simon, H. (1993). Causality in Bayesian belief networks, *Proc. Ninth Conf. on Uncertainty in Artificial Intelligence (UAI-93)*, Washington, DC, pp. 3–11.

Duda, R. O. & Hart, P. E. (1973). *Pattern Classification and Scene Analysis*, Wiley, New York.

Dung, P. (1995). On the acceptability of arguments and its fundamental role in nonmonotonic reasoning, logic programming and n-person games, *Artificial Intelligence* **77**(2): 321–357.

Edwards, P. (ed.) (1967). *The Encyclopedia of Philosophy*, Macmillan, New York.

Enderton, H. B. (1972). *A Mathematical Introduction to Logic*, Academic Press, Orlando, FL.

Fagin, R. Y., Halpern, J., Moses, Y. & Vardi, M. Y. (1994). *Reasoning about Knowledge*, MIT Press, Cambridge, MA.

Feldman, J. A. & Sproull, R. F. (1977). Decision theory and artificial intelligence II: The hungry monkey, *Cognitive Science* **1**: 158–192. Reprinted in Allen et al. (1990).

Fikes, R. E. & Nilsson, N. J. (1971). STRIPS: A new approach to the application of theorem proving to problem solving, *Artificial Intelligence* **2**(3-4): 189–208.

Fischer, D. (1987). Knowledge acquisition via incremental conceptual clustering, *Machine Learning* **2**: 139–172. Reprinted in Shavlik & Dietterich (1990).

Forbus, K. D. (1988). Qualitative physics: Past, present and future, *in* H. E. Shrobe (ed.), *Exploring Artificial Intelligence*, Morgan Kaufmann, chapter 7, pp. 239–296.

Forbus, K. & de Kleer, J. (1993). *Building Problem Solvers*, Bradford Books, Cambridge, MA.

Friedman, N. & Goldszmidt, M. (1996). Building classifiers using Bayesian networks, *Proc. 13th National Conference on Artificial Intelligence*, Portland, OR, pp. 1277–1284.

Gabbay, D. M., Hogger, C. J. & Robinson, J. A. (eds) (1993). *Handbook of Logic in Artificial Intelligence and Logic Programming*, Clarendon Press, Oxford, England. 5 volumes.

Gardner, H. (1985). *The Mind's New Science*, Basic Books, New York.

Garey, M. R. & Johnson, D. S. (1979). *Computers and Intractability: A Guide to the Theory of NP-Completeness*, W. H. Freeman, New York.

Geffner, H. & Pearl, J. (1992). Conditional entailment: Bridging two approaches to default reasoning, *Artificial Intelligence* **53**(2–3): 209–244.

Genesereth, M. R. (1984). The use of design descriptions in automated diagnosis, *Artificial Intelligence* **24**(1-3): 411–436.

Genesereth, M. R. & Nilsson, N. J. (1987). *Logical Foundations of Artificial Intelligence*, Morgan Kaufmann, Los Altos, CA.

Gent, I. P. & Walsh, T. (1993). An empirical analysis of search in GSAT, *Journal of Artificial Intelligence Research* **1**: 47–59.

Ginsberg, M. L. (1989). A circumscriptive theorem prover, *Artificial Intelligence* **39**(2): 209–230.

Ginsberg, M. L. (1993). *Essentials of Artificial Intelligence*, Morgan Kaufmann, San Mateo, CA.

Ginsberg, M. L. (ed.) (1987). *Readings in Nonmonotonic Reasoning*, Morgan Kaufmann, Los Altos, CA.

Goldberg, D. E. (1989). *Genetic Algorithms in Search, Optimization and Machine Learning*, Addison-Wesley, Reading, MA.

Green, C. (1969). Application of theorem proving to problem solving, *Proc. 1st International Joint Conf. on Artificial Intelligence*, Washington, DC, pp. 219–237. Reprinted in Webber & Nilsson (1981).

Grosz, B., Jones, K. S. & Webber, B. L. (eds) (1986). *Readings in Natural Language Processing*, Morgan Kaufmann, Los Altos, CA.

Halpern, J. (1997). A logical approach to reasoning about uncertainty: A tutorial, *in* X. Arrazola, K. Kortha & F. Pelletier (eds), *Discourse, Interaction and Communication*, Kluwer. to appear.

Halpern, J. Y. & Moses, Y. (1985). A guide to the modal logics of knowledge and belief: Preliminary draft, *Proc. 9th International Joint Conf. on Artificial Intelligence (IJCAI-85)*, pp. 480–490.

Hamscher, W., Console, L. & de Kleer, J. (eds) (1992). *Readings in Model-Based Diagnosis*, Morgan Kaufmann Publishers, San Mateo, CA.

Hanks, S. & McDermott, D. V. (1987). Nonmonotonic logic and temporal projection, *Artificial Intelligence* **33**: 379–412.

Haugeland, J. (1985). *Artificial Intelligence: The Very Idea*, MIT Press, Cambridge, MA.

Haugeland, J. (ed.) (1997). *Mind Design II: Philosohpy, Psycholgy, Artificial Intelligence*, revised and enlarged edition, MIT Press, Cambridge, MA.

Haussler, D. (1988). Quantifying inductive bias: AI learning algorithms and Valiant's learning framework, *Artificial Intelligence* **36**(2): 177–221. Reprinted in Shavlik & Dietterich (1990).

Hayes, P. J. (1973). Computation and deduction, *Proc. 2nd Symposium on Mathematical Foundations of Computer Science*, Czechoslovak Academy of Sciences, pp. 105–118.

Hayes, P. J. (1979). The logic of frames, *in* D. Metzing (ed.), *Frame Conceptions and Text Understanding*, de Gruyter, Berlin, pp. 46–61. Reprinted in Webber & Nilsson (1981) and Brachman & Levesque (1985).

Heckerman, D. (1986). Probabilistic interpretation of MYCIN's certainty factors, *in* L. Kanal & J. Lemmer (eds), *Uncertainty in Artificial Intelligence*, North-Holland, New York, pp. 167–196.

Heckerman, D. & Shachter, R. (1995). Decision-theoretic foundations for causal reasoning, *Journal of Artificial Intelligence Research* **3**: 405–430.

Hertz, J., Krogh, A. & Palmer, R. G. (1991). *Introduction to the Theory of Neural Computation*, Lecture Notes, Volume I, Santa Fe Institute Studies in the Sciences of Complexity, Addison-Wesley, Reading, MA.

Hewitt, C. (1969). Planner: A language for proving theorems in robots, *Proc. 1st International Joint Conf. on Artificial Intelligence*, Washington, DC, pp. 295–301.

Hobbs, J. R., Stickel, M. E., Appelt, D. E. & Martin, P. (1993). Interpretation as abduction, *Artificial Intelligence* **63**(1–2): 69–142.

Holland, J. H. (1975). *Adaption in Natural and Artificial Systems: an introductory analysis with applications to biology, control, and artificial intelligence*, University of Michigan Press, Ann Arbor, MI.

Horvitz, E. J. (1989). Reasoning about beliefs and actions under computational resource constraints, *in* L. Kanal, T. Levitt & J. Lemmer (eds), *Uncertainty in Artificial Intelligence 3*, Elsevier, New York, pp. 301–324.

Howard, R. A. & Matheson, J. E. (1981). Influence diagrams, *in* R. A. Howard & J. Matheson (eds), *The Principles and Applications of Decision Analysis*, Strategic Decisions Group, CA, pp. 720–762.

Hughes, G. E. & Cresswell, M. J. (1968). *An Introduction to Modal Logic*, Methuen, London and New York.

Jaffar, J. & Maher, M. (1994). Constraint logic programming: A survey, *Journal of Logic Programming* **19,20**: 503–581.

Jaynes, E. (1985). Bayesian methods: General background, *in* J.H. Justice (ed.), *Maximum Entropy and Bayesian Methods in Applied Statistics*, Cambridge University Press, Cambridge, England, pp. 1–25.

Jensen, F. V. (1996). *An Introduction to Bayesian Networks*, Springer Verlag, New York.

Jordan, M. & Bishop, C. (1996). Neural networks, *Memo 1562*, MIT Artificial Intelligence Lab, Cambridge, MA.

Kaelbling, L. P., Littman, M. L. & Moore, A. W. (1996). Reinforcement learning: A survey, *Journal of Artificial Intelligence Research* **4**: 237–285.

Kakas, A. C., Kowalski, R. A. & Toni, F. (1993). Abductive logic programming, *Journal of Logic and Computation* **2**(6): 719–770.

Kambhampati, S., Knoblock, C. A. & Yang, Q. (1995). Planning as refinement search: a unified framework for evaluating design tradeoffs in partial order planning, *Artificial Intelligence* **76**: 167–238. Special issue on Planning and Scheduling.

Katsuno, H. & Mendelzon, A. O. (1991). On the difference between updating a knowledge base and revising it, *Proc. Second International Conf. on Principles of Knowledge Representation and Reasoning*, Cambridge, MA, pp. 387–394.

Kautz, H. (1991). A formal theory of plan recognition and its implementation, *in* J. Allen, H. Kautz, R. Peravin & J. Tenenberg (eds), *Reasoning About Plans*, Morgan Kaufmann, San Meteo, CA, chapter 2.

Kautz, H. & Selman, B. (1996). Pushing the envelope: Planning, propositional logic and stochastic search, *Proc. 13th National Conference on Artificial Intelligence*, Portland, OR, pp. 1194–1201.

Kearns, M. & Vazirani, U. (1994). *An Introduction to Computational Learning Theory*, MIT Press, Cambridge, MA.

Kirkpatrick, S., Gelatt, C. D. & Vecchi, M. P. (1983). Optimization by simulated annealing, *Science* **220**: 671–680.

Kirsh, D. (1991a). Foundations of AI: the big issues, *Artificial Intelligence* **47**: 3–30.

Kirsh, D. (1991b). Today the earwig, tomorrow man?, *Artificial Intelligence* **47**: 161–184.

Knuth, D. E. (1969). *The Art of Computer Programming, Volume 2, Seminumerical Algorithms*, Addison-Wesley, Reading, MA.

Kolodner, J. (1993). *Case-Based Reasoning*, Morgan Kaufmann, San Mateo, CA.

Konolige, K. (1992). Abduction versus closure in causal theories, *Artificial Intelligence* **53**(2-3): 255–272.

Korf, K. E. (1985). Depth-first iterative deepening: An optimal admissible tree search, *Artificial Intelligence* **27**(1): 97–109.

Kowalski, R. (1979a). Algorithm = logic + control, *Communications of the ACM* **22**: 424–431.

Kowalski, R. (1979b). *Logic for Problem Solving*, Artificial Intelligence Series, North-Holland, New York.

Kowalski, R. A. (1974). Predicate logic as a programming language, *Information Processing 74*, North-Holland, Stockholm, pp. 569–574.

Kowalski, R. & Sergot, M. (1986). A logic-based calculus of events, *New Generation Computing* **4**(1): 67–95.

Koza, J. R. (1992). *Genetic Programming: On the Programming of Computers by Means of Natural Selection*, MIT Press, Cambridge, MA.

Kraus, S., Lehmann, D. & Magidor, M. (1990). Nonmonotonic reasoning, preferential models and cumulative logics, *Artificial Intelligence* **44**(1–2): 167–207.

Kuipers, B. (1994). *Qualitative Reasoning: Modeling and Simulation with Incomplete Knowledge*, MIT Press, Cambridge, MA.

Lakatos, I. (1976). *Proofs and Refutations: The Logic of Mathematical Discovery*, Cambridge University Press, Cambridge, England.

Langley, P., Iba, W. & Thompson, K. (1992). An analysis of Bayesian classifiers, *Proc. 10th National Conference on Artificial Intelligence*, San Jose, CA, pp. 223–228.

Laplace, P. (1812). *Théorie Analytique de Probabilités*, Courcier, Paris.

Latombe, J.-C. (1991). *Robot Motion Planning*, Kluwer Academic Publishers, Boston.

Lehmann, D. (1989). What does a conditional knowledge base entail?, *Proc. First International Conf. on Principles of Knowledge Representation and Reasoning*, Toronto, pp. 212–222.

Lenat, D. B. & Feigenbaum, E. A. (1991). On the thresholds of knowledge, *Artificial Intelligence* **47**: 185–250.

Lin, F. & Shoham, Y. (1989). Argument systems: A uniform basis for nonmonotonic reasoning, *Proc. First International Conf. on Principles of Knowledge Representation and Reasoning*, Toronto, pp. 245–255.

Lloyd, J. W. (1987). *Foundations of Logic Programming*, Symbolic Computation Series, second edition, Springer-Verlag, Berlin.

Loredo, T. (1990). From Laplace to supernova SN 1987A: Bayesian inference in astrophysics, *in* P.F. Fougère (ed.), *Maximum Entropy and Bayesian Methods*, Kluwer Academic Press, Dordrecht, The Netherlands, pp. 81–142.

Loui, R. P. (1987). Defeat among arguments: a system for defeasible inference, *Computational Intelligence* **3**(2): 100–106.

Loveland, D. W. (1978). *Automated Theorem Proving: A Logical Basis*, Fundamental Studies in Computer Science, North-Holland, New York.

Lowe, D. G. (1995). Similarity metric learning for a variable-kernel classifier, *Neural Computation* **7**: 72–85.

Luenberger, D. G. (1979). *Introduction to Dynamic Systems: Theory, Models and Applications*, Wiley, New York.

Mackworth, A. K. (1992). Constraint satisfaction, *in* S. C. Shapiro (ed.), *Encyclopedia of Artificial Intelligence*, second edition, Wiley, New York, pp. 285–293.

Mackworth, A. K. (1993). On seeing robots, *in* A. Basu & X. Li (eds), *Computer Vision: Systems, Theory, and Applications*, World Scientific Press, Singapore, pp. 1–13.

Matheson, J. E. (1990). Using influence diagrams to value information and control, *in* R. M. Oliver & J. Q. Smith (eds), *Influence Diagrams, Belief Nets and Decision Analysis*, Wiley, chapter 1, pp. 25–48.

McAllester, D. & Rosenblitt, D. (1991). Systematic nonlinear planning, *Proc. 9th National Conference on Artificial Intelligence*, pp. 634–639.

McCarthy, J. (1986). Applications of circumscription to formalizing common-sense knowledge, *Artificial Intelligence* **28**(1): 89–116.

McCarthy, J. & Hayes, P. J. (1969). Some philosophical problems from the standpoint of artificial intelligence, *in* M. Meltzer & D. Michie (eds), *Machine Intelligence 4*, Edinburgh University Press, pp. 463–502.

McDermott, D. & Hendler, J. (1995). Planning: What it is, what it could be, an introduction to the special issue on planning and scheduling, *Artificial Intelligence* **76**: 1–16.

Mendelson, E. (1987). *Introduction to Mathematical Logic*, third edition, Wadsworth and Brooks, Monterey, CA.

Michie, D., Spiegelhalter, D. J. & Taylor, C. C. (eds) (1994). *Machine Learning, Neural and Statistical Classification*, Series in Artificial Intelligence, Ellis Horwood, Hemel Hempstead, Hertfordshire, England.

Minsky, M. (1986). *The Society of Mind*, Simon and Schuster, New York.

Minsky, M. L. (1975). A framework for representing knowledge, *in* P. Winston (ed.), *The Psychology of Computer Vision*, McGraw-Hill, New York, pp. 211–277. Alternative version is in Haugeland (1997), and reprinted in Brachman & Levesque (1985).

Minsky, M. & Papert, S. (1988). *Perceptrons: An Introduction to Computational Geometry*, expanded edition, MIT Press, Cambridge, MA.

Mitchell, T. (1997). *Machine Learning*, McGraw-Hill, New York.

Mitchell, T. M. (1977). Version spaces: A candidate elimination approach to rule learning, *Proc. 5th International Joint Conf. on Artificial Intelligence*, Cambridge, MA, pp. 305–310.

Mitchell, T. M., Keller, R. M. & Kedar-Cabelli, S. T. (1986). Explanation-based generalization: A unifying view, *Machine Learning* **1**: 47–80. Reprinted in Shavlik & Dietterich (1990).

Muggleton, S. (1995). Inverse entailment and Progol, *New Generation Computing* **13**(3,4): 245–286.

Newell, A. & Simon, H. A. (1976). Computer science as empirical enquiry: Symbols and search, *Communications of the ACM* **19**: 113–126. Reprinted in Haugeland (1997).

Nilsson, N. J. (1971). *Problem-Solving Methods in Artificial Intelligence*, McGraw-Hill, New York.

Nilsson, N. J. (1980). *Principles of Artificial Intelligence*, Morgan Kaufmann, San Mateo, CA.

O'Keefe, R. A. (1990). *The Craft of Prolog*, MIT Press, Cambridge, MA.

Oliver, R. M. & Smith, J. Q. (eds) (1990). *Influence Diagrams, Belief Nets and Decision Analysis*, Series in probability and mathematical statistics, Wiley, Chichester, England.

Pearl, J. (1984). *Heuristics*, Addison-Wesley, Reading, MA.

Pearl, J. (1988a). Embracing causation in default reasoning, *Artificial Intelligence* **35**(2): 259–271.

Pearl, J. (1988b). *Probabilistic Reasoning in Intelligent Systems: Networks of Plausible Inference*, Morgan Kaufmann, San Mateo, CA.

Pearl, J. (1989). Probabilistic semantics for nonmonotonic reasoning: A survey, *in* R. J. Brachman, H. J. Levesque & R. Reiter (eds), *Proc. First International Conf. on Principles of Knowledge Representation and Reasoning*, Toronto, pp. 505–516.

Pearl, J. (1995). Causal diagrams for empirical research, *Biometrika* **82**(4): 669–710.

Pearl, J. & Verma, T. S. (1991). A theory of inferred causation, *Proc. Second International Conf. on Principles of Knowledge Representation and Reasoning*, Cambridge, MA, pp. 441–452.

Peng, Y. & Reggia, J. A. (1990). *Abductive Inference Models for Diagnostic Problem-Solving*, Symbolic Computation – AI Series, Springer-Verlag, New York.

Pereira, F. C. N. & Shieber, S. M. (1987). *Prolog and Natural-Language Analysis*, Center for the Study of Language and Information.

Perrault, C. R. & Grosz, B. J. (1986). Natural language interfaces, *Annual Review of Computer Science* **1**: 47–82.

Pollock, J. (1994). Justification and defeat, *Artificial Intelligence* **67**: 377–407.

Poole, D. (1989). Explanation and prediction: An architecture for default and abductive reasoning, *Computational Intelligence* **5**(2): 97–110.

Poole, D. (1990). A methodology for using a default and abductive reasoning system, *International Journal of Intelligent Systems* **5**(5): 521–548.

Poole, D. (1993). Probabilistic Horn abduction and Bayesian networks, *Artificial Intelligence* **64**(1): 81–129.

Poole, D. (1994). Representing diagnosis knowledge, *Annals of Mathematics and Artificial Intelligence* **11**: 33–50.

Poole, D. (1995). Logic programming for robot control, *Proc. 14th International Joint Conf. on Artificial Intelligence (IJCAI-95)*, Montréal, Québec, pp. 150–157.

Poole, D., Goebel, R. & Aleliunas, R. (1987). Theorist: A logical reasoning system for defaults and diagnosis, *in* N. Cercone & G. McCalla (eds), *The Knowledge Frontier: Essays in the Representation of Knowledge*, Springer-Verlag, New York, NY, pp. 331–352.

Posner, M. I. (ed.) (1989). *Foundations of Cognitive Science*, MIT Press, Cambridge, MA.

Pratt, J., Raiffa, H. & Schlaifer, R. (1995). *Introduction to Statistical Decision Theory*, MIT Press, Cambridge, MA.

Przymusinski, T. C. (1989). An algorithm to compute circumscription, *Artificial Intelligence* **38**(1): 49–73.

Quinlan, J. R. (1986). Induction of decision trees, *Machine Learning* **1**: 81–106. Reprinted in Shavlik & Dietterich (1990).

Quinlan, J. R. (1993). *C4.5 Programs for Machine Learning*, Morgan Kaufmann, San Mateo, CA.

Quinlan, J. R. & Cameron-Jones, R. M. (1995). Induction of logic programs: FOIL and related systems, *New Generation Computing* **13**(3,4): 287–312.

Reiter, R. (1980). A logic for default reasoning, *Artificial Intelligence* **13**(1,2): 81–132.

Reiter, R. (1987). A theory of diagnosis from first principles, *Artificial Intelligence* **32**(1): 57–95.

Reiter, R. (1991). The frame problem in the situation calculus: A simple solution (sometimes) and a completeness result for goal regression, *in* V. Lifschitz (ed.), *Artificial Intelligence and Mathematical Theory of Computation: Papers in Honor of John McCarthy*, Academic Press, San Diego, CA, pp. 359–380.

Riesbeck, C. & Schank, R. (1989). *Inside Case-Based Reasoning*, Lawrence Erlbaum, Hillsdale, NJ.

Robinson, J. A. (1965). A machine-oriented logic based on the resolution principle, *Journal ACM* **12**(1): 23–41.

Ronan, C. A. (1986). *The Shorter Science and Civilization in China*, Vol. 1, Cambridge University Press, Cambridge, England. An Abridgement of Joseph Needham's Original Text.

Rosenschein, S. J. & Kaelbling, L. P. (1995). A situated view of representation and control, *Artificial Intelligence* **73**: 149–173.

Rumelhart, D. E., Hinton, G. E. & Williams, R. J. (1986). Learning internal representations by error propagation, *in* D. E. Rumelhart & J. L. McClelland (eds), *Parallel Distributed Processing*, MIT Press, Cambridge, MA, chapter 8, pp. 318–362. Reprinted in Shavlik & Dietterich (1990).

Russell, S. (1997). Rationality and intelligence, *Artificial Intelligence* **94**: 57–77.

Russell, S. & Norvig, P. (1995). *Artificial Intelligence: A Modern Approach*, Series in Artificial Intelligence, Prentice-Hall, Englewood Cliffs, NJ.

Sacerdoti, E. D. (1975). The nonlinear nature of plans, *Proc. 4th International Joint Conf. on Artificial Intelligence*, Tbilisi, Georgia, USSR, pp. 206–214.

Schank, R. C. (1990). What is AI, anyway?, *in* D. Partridge & Y. Wilks (eds), *The Foundations of Artificial Intelligence*, Cambridge University Press, Cambridge, England, pp. 3–13.

Schank, R. C. & Abelson, R. P. (1977). *Scripts, Plans, Goals and Understanding: An Inquiry into Human Knowledge Structures*, Lawrence Erlbaum, Hillsdale, NJ.

Schubert, L. K. (1990). Monotonic solutions to the frame problem in the situation calculus: An efficient method for worlds with fully specified actions, *in* H. E. Kyburg, R. P. Loui & G. N. Carlson (eds), *Knowledge Representation and Defeasible Reasoning*, Kluwer Academic Press, Boston, MA, pp. 23–67.

Selman, B., Kautz, H. A. & Cohen, B. (1994). Noise strategies for improving local search, *Proc. 12th National Conference on Artificial Intelligence*, Seattle, WA., pp. 337–343.

Selman, B., Levesque, H. J. & Mitchell, D. G. (1992). GSAT: A new method for solving hard satisifability problems, *Proc. 10th National Conference on Artificial Intelligence*, San Jose, CA, pp. 440–446.

Sergot, M. (1983). A query-the-user facility for logic programs, *in* P. Degano & E. Sandewall (eds), *Integrated Interactive Computer Systems*, North-Holland, Amsterdam, pp. 27–44.

Shachter, R. & Peot, M. A. (1992). Decision making using probabilistic inference methods, *Proc. Eighth Conf. on Uncertainty in Artificial Intelligence (UAI-92)*, Stanford, CA, pp. 276–283.

Shafer, G. & Pearl, J. (eds) (1990). *Readings in Uncertain Reasoning*, Morgan Kaufmann, San Mateo, CA.

Shanahan, M. (1989). Prediction is deduction, but explanation is abduction, *Proc. 11th International Joint Conf. on Artificial Intelligence (IJCAI-89)*, Detroit, MI, pp. 1055–1060.

Shanahan, M. (1997). *Solving the Frame Problem: A Mathematical Investigation of the Common Sense Law of Inertia*, MIT Press, Cambridge, MA.

Shapiro, E. Y. (1983). *Algorithmic Program Debugging*, MIT Press, Cambridge, MA.

Shapiro, S. C. (ed.) (1992). *Encyclopedia of Artificial Intelligence*, second edition, Wiley, New York.

Shavlik, J. W. & Dietterich, T. G. (eds) (1990). *Readings in Machine Learning*, Morgan Kaufmann, San Mateo, CA.

Simari, G. & Loui, R. (1992). A mathematical treatment of defeasible reasoning and its implementation, *Artificial Intelligence* **53**: 125–157.

Simon, H. (1996). *The Sciences of the Artificial*, third edition, MIT Press, Cambridge, MA.

Simon, H. A. (1995). Artificial intelligence: an empirical science, *Artificial Intelligence* **77**(1): 95–127.

Smith, B. C. (1991). The owl and the electric encyclopedia, *Artificial Intelligence* **47**: 251–288.

Spiegelhalter, D. J., Franklin, R. C. G. & Bull, K. (1990). Assessment, criticism and improvement of imprecise subjective probabilities for a medical expert system, *in* M. Henrion, R. D. Shachter, L. Kanal & J. Lemmer (eds), *Uncertainty in Artificial Intelligence 5*, North-Holland, Amsterdam, The Netherlands, pp. 285–294.

Spirtes, P., Glymour, C. & Scheines, R. (1993). *Causation, Prediction and Search*, Springer-Verlag, New York.

Sterling, L. & Shapiro, E. (1994). *The Art of Prolog: Advanced Programming Techniques*, second edition, MIT Press, Cambridge, MA.

Stillings, N. A., Feinstein, M. H., Garfield, J. L., Rissland, E. L., Rosenbaum, D. A., Weisler, S. E. & Baker-Ward, L. (1987). *Cognitive Science: An Introduction*, MIT Press, Cambridge, MA.

Tarski, A. (1956). *Logic, Semantics, Metamathematics*, Clarendon Press, Oxford, England. Papers from 1923 to 1938 collected and translated by J. H. Woodger.

Tate, A. (1977). Generating project networks, *Proc. 5th International Joint Conf. on Artificial Intelligence*, Cambridge, MA, pp. 888–893.

Turing, A. (1950). Computing machinery and intelligence, *Mind* **59**: 433–460. Reprinted in Haugeland (1997).

Valiant, L. G. (1984). A theory of the learnable, *Communications of the ACM* **27**: 1134–1142. Reprinted in Shavlik & Dietterich (1990).

van Emden, M. H. & Kowalski, R. A. (1976). The semantics of predicate logic as a programming language, *Journal ACM* **23**(4): 733–742.

Van Hentenryck, P. (1989). *Constraint Satisfaction in Logic Programming*, Logic Programming Series, MIT Press, Cambridge, MA.

Von Neumann, J. & Morgenstern, O. (1953). *Theory of Games and Economic Behavior*, third edition, Princeton University Press, Princeton, NJ.

Waldinger, R. (1977). Achieving several goals simultaneously, *in* E. Elcock & D. Michie (eds), *Machine Intelligence 8: Machine Representations of Knowledge*, Ellis Horwood, Chichester, England, pp. 94–136.

Webber, B. L. & Nilsson, N. J. (eds) (1981). *Readings in Artificial Intelligence*, Morgan Kaufmann, San Mateo, CA.

Wegner, P. (1997). Why interaction is more powerful than algorithms, *Communications of the ACM* **40**(5): 80–91.

Weiss, S. & Kulikowski, C. (1991). *Computer Systems that Learn: Classification and Prediction Methods from Statistics, Neural Nets, Machine Learning, and Expert Systems*, Morgan Kaufmann, San Mateo, CA.

Weld, D. & de Kleer, J. (eds) (1990). *Readings in Qualitative Reasoning about Physical Systems*, Morgan Kaufmann, San Mateo, CA.

Weld, D. S. (1992). Qualitative physics: Albatross or eagle?, *Computational Intelligence* **8**(2): 175–186. Introduction to special issue on the future of qualitative physics.

Weld, D. S. (1994). An introduction to least commitment planning, *AI Magazine* **15**(4): 27–61.

Wilkins, D. E. (1988). *Practical Planning: Extending the Classical AI Planning Paradigm*, Morgan Kaufmann, San Mateo, CA.

Winograd, T. (1983). *Language as a Cognitive Process – Volume 1: Syntax*, Addison-Wesley, Reading, MA.

Winograd, T. (1990). Thinking machines: Can there be? Are we?, *in* D. Partridge & Y. Wilks (eds), *The Foundations of Artificial Intelligence: A Sourcebook*, Cambridge University Press, Cambridge, England, pp. 167–189.

Woods, W. A. (1975). What's in a link: Foundations for semantic networks, *in* D. G. Bobrow & A. M. Collins (eds), *Representation and Understanding: Studies in Cognitive Science*, Academic Press, New York, pp. 35–82. Reprinted in Brachman & Levesque (1985).

Yang, Q. (1997). *Intelligent Planning: A Decomposition and Abstraction-Based Approach*, Springer–Verlag, New York.

Zhang, N. & Poole, D. (1996). Exploiting causal independence in Bayesian network inference, *Journal of Artificial Intelligence Research* **5**: 301–328.

Zhang, Y. & Mackworth, A. K. (1995). Constraint nets: A semantic model for hybrid dynamic systems, *Theoretical Computer Science* **138**: 211–239.

Zilberstein, S. (1996). Using anytime algorithms in intelligent systems, *AI Magazine* **17**(3): 73–83.

Index